Methods in Enzymology

Volume 180
RNA PROCESSING
Part A
General Methods

METHODS IN ENZYMOLOGY

EDITORS-IN-CHIEF

John N. Abelson   Melvin I. Simon

DIVISION OF BIOLOGY
CALIFORNIA INSTITUTE OF TECHNOLOGY
PASADENA, CALIFORNIA

FOUNDING EDITORS

Sidney P. Colowick and Nathan O. Kaplan

Methods in Enzymology

Volume 180

RNA Processing

Part A
General Methods

EDITED BY

James E. Dahlberg

DEPARTMENT OF PHYSIOLOGICAL CHEMISTRY
UNIVERSITY OF WISCONSIN, MADISON
MADISON, WISCONSIN

John N. Abelson

DIVISION OF BIOLOGY
CALIFORNIA INSTITUTE OF TECHNOLOGY
PASADENA, CALIFORNIA

ACADEMIC PRESS, INC.
Harcourt Brace Jovanovich, Publishers
San Diego   New York   Berkeley   Boston
London   Sydney   Tokyo   Toronto

ACADEMIC PRESS, INC.
San Diego, California 92101

United Kingdom Edition published by
ACADEMIC PRESS LIMITED
24-28 Oval Road, London NW1 7DX

LIBRARY OF CONGRESS CATALOG CARD NUMBER: 54-9110

ISBN   0-12-182081-5   (alk. paper)

PRINTED IN THE UNITED STATES OF AMERICA
89 90 91 92   9 8 7 6 5 4 3 2 1

Table of Contents

CONTRIBUTORS TO VOLUME 180   ix

PREFACE   xiii

VOLUMES IN SERIES   xv

Section I. Preparation of Substrates

A. Purification of RNA Molecules

1. Isolation of RNA   SHELBY L. BERGER AND
                                                       JOHN M. CHIRGWIN   3

2. Purification of RNA Molecules by Gel Techniques   TOSHIMICHI IKEMURA   14

3. Purifying RNA by Column Chromatography   N. KYLE TANNER   25

B. *In Vitro* Synthesis of RNA

4. Synthesis of Long, Capped Transcripts *in Vitro* by JOEL K. YISRAELI
   SP6 and T7 RNA Polymerases   AND DOUG A. MELTON   42

5. Synthesis of Small RNAs Using T7 RNA Poly-   JOHN F. MILLIGAN AND
   merase   OLKE C. UHLENBECK   51

6. *In Vitro* Synthesis of End-Mature, Intron-Contain- VICENTE M. REYES AND
   ing Transfer RNAs   JOHN N. ABELSON   63

C. *In Vivo* Preparation of Substrates

7. Preparation of Precursors to mRNA from Mam-   SELINA CHEN-KIANG AND
   malian Cell Nuclei   DANIEL J. LAVERY   69

8. Pulse Labeling of Heterogeneous Nuclear RNA in   SELINA CHEN-KIANG AND
   Isolated Nuclei   DANIEL J. LAVERY   82

9. Isolation and Characterization of Yeast Ribosomal   JACOBUS KLOOTWIJK AND
   RNA Precursors and Preribosomes   RUDI J. PLANTA   96

10. Preparation of Yeast Transfer RNA Precursors *in* GAYLE KNAPP   110
    Vivo

v

Section II. Characterization of RNAs

A. Primary Structure

11. Dideoxy Sequencing of RNA Using Reverse Transcriptase — CHANG S. HAHN, ELLEN G. STRAUSS, AND JAMES H. STRAUSS   121

12. RNA Fingerprinting — ANDREA D. BRANCH, BONNIE J. BENENFELD, AND HUGH D. ROBERTSON   130

13. Enzymatic RNA Sequencing — Y. KUCHINO AND S. NISHIMURA   154

14. Characterization of Cap Structures — YASUHIRO FURUICHI AND AARON J. SHATKIN   164

15. Isolation and Characterization of Branched Oligonucleotides from RNA — J. DAVID REILLY, JOHN C. WALLACE, RANDA F. MELHEM, DAVID W. KOPP, AND MARY EDMONDS   177

B. Secondary Structure

16. Enzymatic Approaches to Probing of RNA Secondary and Tertiary Structure — GAYLE KNAPP   192

17. A Guide for Probing Native Small Nuclear RNA and Ribonucleoprotein Structures — ALAIN KROL AND PHILIPPE CARBON   212

18. Phylogenetic Comparative Analysis of RNA Secondary Structure — BRYAN D. JAMES, GARY J. OLSEN, AND NORMAN R. PACE   227

19. Mapping the Genetic Organization of RNA by Electron Microscopy — LOUISE T. CHOW AND THOMAS R. BROKER   239

20. Computer Prediction of RNA Structure — MICHAEL ZUKER   262

21. RNA Pseudoknots: Structure, Detection, and Prediction — C. W. A. PLEIJ AND L. BOSCH   289

22. Absorbance Melting Curves of RNA — JOSEPH D. PUGLISI AND IGNACIO TINOCO, JR.   304

C. RNA Functions

23. Nuclease Digestion: A Method for Mapping Introns — SHELBY L. BERGER   325

24. Characterization of RNA Molecules by S1 Nuclease Analysis — ARNOLD J. BERK   334

25. Primer Extension Analysis of RNA — WILLIAM R. BOORSTEIN AND ELIZABETH A. CRAIG   347

26. Modification Interference Analysis of Reactions LAURA CONWAY AND
   Using RNA Substrates MARVIN WICKENS 369

Section III. RNA Interactions

A. Cross-Linking

27. Photoaffinity Cross-Linking Methods for Studying MICHELLE M. HANNA 383
   RNA–Protein Interactions

28. Ultraviolet-Induced Cross-Linking of RNA to SERAFÍN PIÑOL-ROMA,
   Proteins *in Vivo* STEPHEN A. ADAM,
       YANG DO CHOI, AND
       GIDEON DREYFUSS 410

29. Analysis of Ultraviolet-Induced RNA–RNA ANDREA D. BRANCH,
   Cross-Links: A Means for Probing RNA Struc- BONNIE J. BENENFELD,
   ture–Function Relationships CYNTHIA P. PAUL, AND
       HUGH D. ROBERTSON 418

B. Other Methods

30. Analysis of Splicing Complexes and Small Nu- MARIA M. KONARSKA 442
   clear Ribonucleoprotein Particles by Native Gel
   Electrophoresis

31. Determination of RNA–Protein and RNA–Ribo- KATHERINE A. PARKER
   nucleoprotein Interactions by Nuclease Probing AND JOAN A. STEITZ 454

32. Immunoprecipitation of Ribonucleoproteins Using JOAN A. STEITZ 468
   Autoantibodies

33. Electron Microscopy of Ribonucleoprotein Com- YVONNE N. OSHEIM AND
   plexes on Nascent RNA Using Miller Chroma- ANN L. BEYER 481
   tin Spreading Method

34. Genetic Methods for Identification and Character- ROY PARKER 510
   ization of RNA–RNA and RNA–Protein Inter-
   actions

Section IV. Appendix

35. Compilation of Small Nuclear RNA Sequences RAM REDDY 521

36. Sequences and Classification of Group I and JOHN M. BURKE 533
   Group II Introns

37. Compilation of Self-Cleaving Sequences from GEORGE BRUENING 546
   Plant Virus Satellite RNAs and Other Sources

AUTHOR INDEX 559

SUBJECT INDEX 579

Contributors to Volume 180

Article numbers are in parentheses following the names of contributors.
Affiliations listed are current.

JOHN N. ABELSON (6), *Division of Biology, California Institute of Technology, Pasadena, California 91125*

STEPHEN A. ADAM (28), *Department of Molecular Biology, Research Institute of Scripps Clinic, La Jolla, California 92037*

BONNIE J. BENENFELD (12, 29), *Cornell University Medical College, New York, New York 10021*

SHELBY L. BERGER (1, 23), *Section on Genes and Gene Products, National Cancer Institute, Bethesda, Maryland 20892*

ARNOLD J. BERK (24), *Department of Microbiology and Molecular Biology Institute, University of California, Los Angeles, California 90024*

ANN L. BEYER (33), *Department of Microbiology, University of Virginia School of Medicine, Charlottesville, Virginia 22908*

WILLIAM R. BOORSTEIN (25), *Molecular and Cellular Biology Program, University of Wisconsin, Madison, Madison, Wisconsin 53706*

L. BOSCH (21), *Department of Biochemistry, Gorlaeus Laboratories, Leiden University, 2333 CC Leiden, The Netherlands*

ANDREA D. BRANCH (12, 29), *The Rockefeller University, New York, New York 10021*

THOMAS R. BROKER (19), *Department of Biochemistry and Cancer Center, University of Rochester School of Medicine, Rochester, New York 14642*

GEORGE BRUENING (37), *Department of Plant Pathology, College of Agricultural and Environmental Sciences, University of California, Davis, California 95616*

JOHN M. BURKE (36), *Department of Microbiology and Molecular Genetics, University of Vermont, Burlington, Vermont 05405*

PHILIPPE CARBON (17), *Institut de Biologie Moleculaire et Cellulaire du CNRS, 67084 Strasbourg Cedex, France*

SELINA CHEN-KIANG (7, 8), *Brookdale Center for Molecular Biology, Mount Sinai School of Medicine, New York, New York 10029*

JOHN M. CHIRGWIN (1), *Division of Endocrinology, Department of Medicine, The University of Texas Health Science Center at San Antonio, San Antonio, Texas 78284*

YANG DO CHOI (28), *Department of Agricultural Chemistry, Seoul National University, Suwon, Korea 170*

LOUISE T. CHOW (19), *Department of Biochemistry and Cancer Center, University of Rochester School of Medicine, Rochester, New York 14642*

LAURA CONWAY (26), *Department of Biochemistry, College of Agriculture and Life Sciences, University of Wisconsin, Madison, Madison, Wisconsin 53706*

ELIZABETH A. CRAIG (25), *Department of Physiological Chemistry, University of Wisconsin, Madison, Madison, Wisconsin 53706*

GIDEON DREYFUSS (28), *Department of Biochemistry, Molecular Biology, and Cell Biology, Northwestern University, Evanston, Illinois 60208*

MARY EDMONDS (15), *Department of Biological Sciences, University of Pittsburgh, Pittsburgh, Pennsylvania 15260*

YASUHIRO FURUICHI (14), *Subdivision of Biotechnology, Nippon Roche Research Center, Kamakura, Kanagawa 247, Japan*

CHANG S. HAHN (11), *Department of Microbiology and Immunology, Washington University School of Medicine, St. Louis, Missouri 63110*

MICHELLE M. HANNA (27), *Department of Biological Chemistry, California College of Medicine, University of California, Irvine, California 92717*

TOSHIMICHI IKEMURA (2), *DNA Research Center, National Institute of Genetics, Mishima, Shizuoka-ken 411, Japan*

BRYAN D. JAMES (18), *Department of Molecular Genetics, MD Anderson Cancer Center, Houston, Texas 77030*

JACOBUS KLOOTWIJK (9), *Biochemisch Laboratorium, Vrije Universiteit de Boelelaan, 1081 HV Amsterdam, The Netherlands*

GAYLE KNAPP (10, 16), *Department of Chemistry and Biochemistry, Utah State University, Logan, Utah 84322*

MARIA M. KONARSKA (30), *The Rockefeller University, New York, New York 10021*

DAVID W. KOPP (15), *Department of Biological Sciences, University of Pittsburgh, Pittsburgh, Pennsylvania 15260*

ALAIN KROL (17), *Institut de Biologie Moleculaire et Cellulaire du CNRS, 67084 Strasbourg Cedex, France*

Y. KUCHINO (13), *Biophysics Division, National Cancer Center Research Institute, 5-1-1, Tsukiji, Chuo-ku, Tokyo 104, Japan*

DANIEL J. LAVERY (7, 8), *Department of Microbiology, Mount Sinai School of Medicine, New York, New York 10029*

RANDA F. MELHEM (15), *Department of Pediatrics, University of Michigan Medical Center, Ann Arbor, Michigan 48109*

DOUG A. MELTON (4), *Department of Biochemistry and Molecular Biology, Harvard University, Cambridge, Massachusetts 02138*

JOHN F. MILLIGAN (5), *Department of Biochemistry and Biophysics, University of California, San Francisco, California 94143*

S. NISHIMURA (13), *Biology Division, National Cancer Center Research Institute, 5-1-1, Tsukiji, Chuo-ku, Tokyo 104, Japan*

GARY J. OLSEN (18), *Department of Microbiology, University of Illinois, Urbana, Illinois 61801*

YVONNE N. OSHEIM (33), *Department of Microbiology, University of Virginia School of Medicine, Charlottesville, Virginia 22908*

NORMAN R. PACE (18), *Department of Biology, Indiana University, Bloomington, Indiana 47405*

KATHERINE A. PARKER (31), *Department of Cell Biology, Baylor College of Medicine, Houston, Texas 77030*

ROY PARKER (34), *Department of Molecular and Cellular Biology, University of Arizona, Tucson, Arizona 85721*

CYNTHIA P. PAUL (29), *Cornell University Medical College, New York, New York 10021*

SERAFÍN PIÑOL-ROMA (28), *Department of Biochemistry, Molecular Biology, and Cell Biology, Northwestern University, Evanston, Illinois 60208*

RUDI J. PLANTA (9), *Biochemisch Laboratorium, Vrije Universiteit de Boelelaan, 1081 HV Amsterdam, The Netherlands*

C. W. A. PLEIJ (21), *Department of Biochemistry, Gorlaeus Laboratories, Leiden University, 2333 CC Leiden, The Netherlands*

JOSEPH D. PUGLISI (22), *Department of Chemistry and Laboratory of Chemical Biodynamics, University of California, Berkeley, California 94720*

RAM REDDY (35), *Department of Pharmacology, Baylor College of Medicine, Houston, Texas 77030*

J. DAVID REILLY (15), *United States Department of Agriculture, East Lansing, Michigan 48823*

VICENTE M. REYES (6), *Laboratory of Tumor Cell Biology, National Cancer Institute, National Institutes of Health, Bethesda, Maryland 20892*

HUGH D. ROBERTSON (12, 29), *Cornell University Medical College, New York, New York 10021*

AARON J. SHATKIN (14), *Center for Advanced Biotechnology and Medicine, Piscataway, New Jersey 08854*

JOAN A. STEITZ (31, 32), *Department of Molecular Biophysics and Biochemistry, Yale University School of Medicine, New Haven, Connecticut 06510*

ELLEN G. STRAUSS (11), *Division of Biology, California Institute of Technology, Pasadena, California 91125*

JAMES H. STRAUSS (11), *Division of Biology, California Institute of Technology, Pasadena, California 91125*

N. KYLE TANNER (3), *Division of Biology, California Institute of Technology, Pasadena, California 91125*

IGNACIO TINOCO, JR. (22), *Department of Chemistry and Laboratory of Chemical Biodynamics, University of California, Berkeley, California 94720*

OLKE C. UHLENBECK (5), *Department of Chemistry and Biochemistry, University of Colorado, Boulder, Colorado 80309*

JOHN C. WALLACE (15), *Department of Botany and Pathology, Purdue University, West Lafayette, Indiana 47907*

MARVIN WICKENS (26), *Department of Biochemistry, College of Agriculture and Life Sciences, University of Wisconsin, Madison, Madison, Wisconsin 53706*

JOEL K. YISRAELI (4), *Department of Anatomy and Embryology, Hadassah Medical School, Hebrew University, Jerusalem, Israel 91010*

MICHAEL ZUKER (20), *Division of Biological Sciences, National Research Council of Canada, Ottawa, Canada K1A 0R6*

Preface

That RNA processing is one of the most active and rapidly developing areas of biology can be seen from the great increase in papers published on the subject and the high attendance at meetings and workshops in this field.

Because RNAs rarely, if ever, function as primary transcription products, they all must undergo various forms of processing. This can be as simple as the removal of one or a few nucleotides at one end or as complex as splicing (*cis* or *trans*), modification of nucleotide bases, and 3' cleavage and polyadenylation. Hence, RNA processing plays a key role in the expression of genetic information.

Numerous methods have been developed or adapted for the study of RNA processing which pertain to preparation of substrates, to preparation and purification of processing enzymes or factors, and to analysis of the resulting products. Methods have also been developed that allow investigators to study the structures of the precursors, the cofactors, and the enzymes or complexes that catalyze function at individual steps in processing. In many instances, approaches or procedures have very wide applicability to other areas of research, such as in cell structure and physiology, in transcription and translation, and in enzymology. In other instances, the methods are defined by the scope of the problem at hand, so the methods are primarily applicable to the processing of RNA. As in most fields it is impossible to predict the breadth of applications for which a procedure or method will be used. Nevertheless, we have tried to assemble the contributions according to whether they are more general (Volume 180) or more specific (Volume 181).

In an effort to keep the size of these *Methods in Enzymology* volumes manageable, we had to make many difficult and arbitrary decisions in the selection of topics to be included, and are gratified by the number of excellent chapters received. The care and dedication of the authors are evident in the quality of their papers and their willingness to share with others the methods used in their laboratories. We thank them for their contributions to these volumes.

As with any publication in a rapidly evolving field, many of the methods described will undoubtedly become supplemented by new, more powerful, and more informative ones. But most, perhaps all, of these methods will remain very useful for several years and will serve as fundamental procedures from which new techniques can be developed in this very exciting field.

JAMES E. DAHLBERG
JOHN N. ABELSON

METHODS IN ENZYMOLOGY

VOLUME I. Preparation and Assay of Enzymes
Edited by SIDNEY P. COLOWICK AND NATHAN O. KAPLAN

VOLUME II. Preparation and Assay of Enzymes
Edited by SIDNEY P. COLOWICK AND NATHAN O. KAPLAN

VOLUME III. Preparation and Assay of Substrates
Edited by SIDNEY P. COLOWICK AND NATHAN O. KAPLAN

VOLUME IV. Special Techniques for the Enzymologist
Edited by SIDNEY P. COLOWICK AND NATHAN O. KAPLAN

VOLUME V. Preparation and Assay of Enzymes
Edited by SIDNEY P. COLOWICK AND NATHAN O. KAPLAN

VOLUME VI. Preparation and Assay of Enzymes (*Continued*)
Preparation and Assay of Substrates
Special Techniques
Edited by SIDNEY P. COLOWICK AND NATHAN O. KAPLAN

VOLUME VII. Cumulative Subject Index
Edited by SIDNEY P. COLOWICK AND NATHAN O. KAPLAN

VOLUME VIII. Complex Carbohydrates
Edited by ELIZABETH F. NEUFELD AND VICTOR GINSBURG

VOLUME IX. Carbohydrate Metabolism
Edited by WILLIS A. WOOD

VOLUME X. Oxidation and Phosphorylation
Edited by RONALD W. ESTABROOK AND MAYNARD E. PULLMAN

VOLUME XI. Enzyme Structure
Edited by C. H. W. HIRS

VOLUME XII. Nucleic Acids (Parts A and B)
Edited by LAWRENCE GROSSMAN AND KIVIE MOLDAVE

VOLUME XIII. Citric Acid Cycle
Edited by J. M. LOWENSTEIN

VOLUME XIV. Lipids
Edited by J. M. LOWENSTEIN

VOLUME XV. Steroids and Terpenoids
Edited by RAYMOND B. CLAYTON

VOLUME XVI. Fast Reactions
Edited by KENNETH KUSTIN

VOLUME XVII. Metabolism of Amino Acids and Amines (Parts A and B)
Edited by HERBERT TABOR AND CELIA WHITE TABOR

VOLUME XVIII. Vitamins and Coenzymes (Parts A, B, and C)
Edited by DONALD B. MCCORMICK AND LEMUEL D. WRIGHT

VOLUME XIX. Proteolytic Enzymes
Edited by GERTRUDE E. PERLMANN AND LASZLO LORAND

VOLUME XX. Nucleic Acids and Protein Synthesis (Part C)
Edited by KIVIE MOLDAVE AND LAWRENCE GROSSMAN

VOLUME XXI. Nucleic Acids (Part D)
Edited by LAWRENCE GROSSMAN AND KIVIE MOLDAVE

VOLUME XXII. Enzyme Purification and Related Techniques
Edited by WILLIAM B. JAKOBY

VOLUME XXIII. Photosynthesis (Part A)
Edited by ANTHONY SAN PIETRO

VOLUME XXIV. Photosynthesis and Nitrogen Fixation (Part B)
Edited by ANTHONY SAN PIETRO

VOLUME XXV. Enzyme Structure (Part B)
Edited by C. H. W. HIRS AND SERGE N. TIMASHEFF

VOLUME XXVI. Enzyme Structure (Part C)
Edited by C. H. W. HIRS AND SERGE N. TIMASHEFF

VOLUME XXVII. Enzyme Structure (Part D)
Edited by C. H. W. HIRS AND SERGE N. TIMASHEFF

VOLUME XXVIII. Complex Carbohydrates (Part B)
Edited by VICTOR GINSBURG

VOLUME XXIX. Nucleic Acids and Protein Synthesis (Part E)
Edited by LAWRENCE GROSSMAN AND KIVIE MOLDAVE

VOLUME XXX. Nucleic Acids and Protein Synthesis (Part F)
Edited by KIVIE MOLDAVE AND LAWRENCE GROSSMAN

VOLUME XXXI. Biomembranes (Part A)
Edited by SIDNEY FLEISCHER AND LESTER PACKER

VOLUME XXXII. Biomembranes (Part B)
Edited by SIDNEY FLEISCHER AND LESTER PACKER

VOLUME XXXIII. Cumulative Subject Index Volumes I–XXX
Edited by MARTHA G. DENNIS AND EDWARD A. DENNIS

VOLUME XXXIV. Affinity Techniques (Enzyme Purification: Part B)
Edited by WILLIAM B. JAKOBY AND MEIR WILCHEK

VOLUME XXXV. Lipids (Part B)
Edited by JOHN M. LOWENSTEIN

VOLUME XXXVI. Hormone Action (Part A: Steroid Hormones)
Edited by BERT W. O'MALLEY AND JOEL G. HARDMAN

VOLUME XXXVII. Hormone Action (Part B: Peptide Hormones)
Edited by BERT W. O'MALLEY AND JOEL G. HARDMAN

VOLUME XXXVIII. Hormone Action (Part C: Cyclic Nucleotides)
Edited by JOEL G. HARDMAN AND BERT W. O'MALLEY

VOLUME XXXIX. Hormone Action (Part D: Isolated Cells, Tissues, and Organ Systems)
Edited by JOEL G. HARDMAN AND BERT W. O'MALLEY

VOLUME XL. Hormone Action (Part E: Nuclear Structure and Function)
Edited by BERT W. O'MALLEY AND JOEL G. HARDMAN

VOLUME XLI. Carbohydrate Metabolism (Part B)
Edited by W. A. WOOD

VOLUME XLII. Carbohydrate Metabolism (Part C)
Edited by W. A. WOOD

VOLUME XLIII. Antibiotics
Edited by JOHN H. HASH

VOLUME XLIV. Immobilized Enzymes
Edited by KLAUS MOSBACH

VOLUME XLV. Proteolytic Enzymes (Part B)
Edited by LASZLO LORAND

VOLUME XLVI. Affinity Labeling
Edited by WILLIAM B. JAKOBY AND MEIR WILCHEK

VOLUME XLVII. Enzyme Structure (Part E)
Edited by C. H. W. HIRS AND SERGE N. TIMASHEFF

VOLUME XLVIII. Enzyme Structure (Part F)
Edited by C. H. W. HIRS AND SERGE N. TIMASHEFF

VOLUME XLIX. Enzyme Structure (Part G)
Edited by C. H. W. HIRS AND SERGE N. TIMASHEFF

VOLUME L. Complex Carbohydrates (Part C)
Edited by VICTOR GINSBURG

VOLUME LI. Purine and Pyrimidine Nucleotide Metabolism
Edited by PATRICIA A. HOFFEE AND MARY ELLEN JONES

VOLUME LII. Biomembranes (Part C: Biological Oxidations)
Edited by SIDNEY FLEISCHER AND LESTER PACKER

VOLUME LIII. Biomembranes (Part D: Biological Oxidations)
Edited by SIDNEY FLEISCHER AND LESTER PACKER

VOLUME LIV. Biomembranes (Part E: Biological Oxidations)
Edited by SIDNEY FLEISCHER AND LESTER PACKER

VOLUME LV. Biomembranes (Part F: Bioenergetics)
Edited by SIDNEY FLEISCHER AND LESTER PACKER

VOLUME LVI. Biomembranes (Part G: Bioenergetics)
Edited by SIDNEY FLEISCHER AND LESTER PACKER

VOLUME LVII. Bioluminescence and Chemiluminescence
Edited by MARLENE A. DELUCA

VOLUME LVIII. Cell Culture
Edited by WILLIAM B. JAKOBY AND IRA PASTAN

VOLUME LIX. Nucleic Acids and Protein Synthesis (Part G)
Edited by KIVIE MOLDAVE AND LAWRENCE GROSSMAN

VOLUME LX. Nucleic Acids and Protein Synthesis (Part H)
Edited by KIVIE MOLDAVE AND LAWRENCE GROSSMAN

VOLUME 61. Enzyme Structure (Part H)
Edited by C. H. W. HIRS AND SERGE N. TIMASHEFF

VOLUME 62. Vitamins and Coenzymes (Part D)
Edited by DONALD B. MCCORMICK AND LEMUEL D. WRIGHT

VOLUME 63. Enzyme Kinetics and Mechanism (Part A: Initial Rate and Inhibitor Methods)
Edited by DANIEL L. PURICH

VOLUME 64. Enzyme Kinetics and Mechanism (Part B: Isotopic Probes and Complex Enzyme Systems)
Edited by DANIEL L. PURICH

VOLUME 65. Nucleic Acids (Part I)
Edited by LAWRENCE GROSSMAN AND KIVIE MOLDAVE

VOLUME 66. Vitamins and Coenzymes (Part E)
Edited by DONALD B. MCCORMICK AND LEMUEL D. WRIGHT

VOLUME 67. Vitamins and Coenzymes (Part F)
Edited by DONALD B. MCCORMICK AND LEMUEL D. WRIGHT

VOLUME 68. Recombinant DNA
Edited by RAY WU

VOLUME 69. Photosynthesis and Nitrogen Fixation (Part C)
Edited by ANTHONY SAN PIETRO

VOLUME 70. Immunochemical Techniques (Part A)
Edited by HELEN VAN VUNAKIS AND JOHN J. LANGONE

VOLUME 71. Lipids (Part C)
Edited by JOHN M. LOWENSTEIN

VOLUME 72. Lipids (Part D)
Edited by JOHN M. LOWENSTEIN

VOLUME 73. Immunochemical Techniques (Part B)
Edited by JOHN J. LANGONE AND HELEN VAN VUNAKIS

VOLUME 74. Immunochemical Techniques (Part C)
Edited by JOHN J. LANGONE AND HELEN VAN VUNAKIS

VOLUME 75. Cumulative Subject Index Volumes XXXI, XXXII, XXXIV–LX
Edited by EDWARD A. DENNIS AND MARTHA G. DENNIS

VOLUME 76. Hemoglobins
Edited by ERALDO ANTONINI, LUIGI ROSSI-BERNARDI, AND EMILIA CHIANCONE

VOLUME 77. Detoxication and Drug Metabolism
Edited by WILLIAM B. JAKOBY

VOLUME 78. Interferons (Part A)
Edited by SIDNEY PESTKA

VOLUME 79. Interferons (Part B)
Edited by SIDNEY PESTKA

VOLUME 80. Proteolytic Enzymes (Part C)
Edited by LASZLO LORAND

VOLUME 81. Biomembranes (Part H: Visual Pigments and Purple Membranes, I)
Edited by LESTER PACKER

VOLUME 82. Structural and Contractile Proteins (Part A: Extracellular Matrix)
Edited by LEON W. CUNNINGHAM AND DIXIE W. FREDERIKSEN

VOLUME 83. Complex Carbohydrates (Part D)
Edited by VICTOR GINSBURG

VOLUME 84. Immunochemical Techniques (Part D: Selected Immunoassays)
Edited by JOHN J. LANGONE AND HELEN VAN VUNAKIS

VOLUME 85. Structural and Contractile Proteins (Part B: The Contractile Apparatus and the Cytoskeleton)
Edited by DIXIE W. FREDERIKSEN AND LEON W. CUNNINGHAM

VOLUME 86. Prostaglandins and Arachidonate Metabolites
Edited by WILLIAM E. M. LANDS AND WILLIAM L. SMITH

VOLUME 87. Enzyme Kinetics and Mechanism (Part C: Intermediates, Stereochemistry, and Rate Studies)
Edited by DANIEL L. PURICH

VOLUME 88. Biomembranes (Part I: Visual Pigments and Purple Membranes, II)
Edited by LESTER PACKER

VOLUME 89. Carbohydrate Metabolism (Part D)
Edited by WILLIS A. WOOD

VOLUME 90. Carbohydrate Metabolism (Part E)
Edited by WILLIS A. WOOD

VOLUME 91. Enzyme Structure (Part I)
Edited by C. H. W. HIRS AND SERGE N. TIMASHEFF

VOLUME 92. Immunochemical Techniques (Part E: Monoclonal Antibodies and General Immunoassay Methods)
Edited by JOHN J. LANGONE AND HELEN VAN VUNAKIS

VOLUME 93. Immunochemical Techniques (Part F: Conventional Antibodies, Fc Receptors, and Cytotoxicity)
Edited by JOHN J. LANGONE AND HELEN VAN VUNAKIS

VOLUME 94. Polyamines
Edited by HERBERT TABOR AND CELIA WHITE TABOR

VOLUME 95. Cumulative Subject Index Volumes 61–74, 76–80
Edited by EDWARD A. DENNIS AND MARTHA G. DENNIS

VOLUME 96. Biomembranes [Part J: Membrane Biogenesis: Assembly and Targeting (General Methods; Eukaryotes)]
Edited by SIDNEY FLEISCHER AND BECCA FLEISCHER

VOLUME 97. Biomembranes [Part K: Membrane Biogenesis: Assembly and Targeting (Prokaryotes, Mitochondria, and Chloroplasts)]
Edited by SIDNEY FLEISCHER AND BECCA FLEISCHER

VOLUME 98. Biomembranes (Part L: Membrane Biogenesis: Processing and Recycling)
Edited by SIDNEY FLEISCHER AND BECCA FLEISCHER

VOLUME 99. Hormone Action (Part F: Protein Kinases)
Edited by JACKIE D. CORBIN AND JOEL G. HARDMAN

VOLUME 100. Recombinant DNA (Part B)
Edited by RAY WU, LAWRENCE GROSSMAN, AND KIVIE MOLDAVE

VOLUME 101. Recombinant DNA (Part C)
Edited by RAY WU, LAWRENCE GROSSMAN, AND KIVIE MOLDAVE

VOLUME 102. Hormone Action (Part G: Calmodulin and Calcium-Binding Proteins)
Edited by ANTHONY R. MEANS AND BERT W. O'MALLEY

VOLUME 103. Hormone Action (Part H: Neuroendocrine Peptides)
Edited by P. MICHAEL CONN

VOLUME 104. Enzyme Purification and Related Techniques (Part C)
Edited by WILLIAM B. JAKOBY

VOLUME 105. Oxygen Radicals in Biological Systems
Edited by LESTER PACKER

VOLUME 106. Posttranslational Modifications (Part A)
Edited by FINN WOLD AND KIVIE MOLDAVE

VOLUME 107. Posttranslational Modifications (Part B)
Edited by FINN WOLD AND KIVIE MOLDAVE

VOLUME 108. Immunochemical Techniques (Part G: Separation and Characterization of Lymphoid Cells)
Edited by GIOVANNI DI SABATO, JOHN J. LANGONE, AND HELEN VAN VUNAKIS

VOLUME 109. Hormone Action (Part I: Peptide Hormones)
Edited by LUTZ BIRNBAUMER AND BERT W. O'MALLEY

VOLUME 110. Steroids and Isoprenoids (Part A)
Edited by JOHN H. LAW AND HANS C. RILLING

VOLUME 111. Steroids and Isoprenoids (Part B)
Edited by JOHN H. LAW AND HANS C. RILLING

VOLUME 112. Drug and Enzyme Targeting (Part A)
Edited by KENNETH J. WIDDER AND RALPH GREEN

VOLUME 113. Glutamate, Glutamine, Glutathione, and Related Compounds
Edited by ALTON MEISTER

VOLUME 114. Diffraction Methods for Biological Macromolecules (Part A)
Edited by HAROLD W. WYCKOFF, C. H. W. HIRS, AND SERGE N. TIMASHEFF

VOLUME 115. Diffraction Methods for Biological Macromolecules (Part B)
Edited by HAROLD W. WYCKOFF, C. H. W. HIRS, AND SERGE N. TIMASHEFF

VOLUME 116. Immunochemical Techniques (Part H: Effectors and Mediators of Lymphoid Cell Functions)
Edited by GIOVANNI DI SABATO, JOHN J. LANGONE, AND HELEN VAN VUNAKIS

VOLUME 117. Enzyme Structure (Part J)
Edited by C. H. W. HIRS AND SERGE N. TIMASHEFF

VOLUME 118. Plant Molecular Biology
Edited by ARTHUR WEISSBACH AND HERBERT WEISSBACH

VOLUME 119. Interferons (Part C)
Edited by SIDNEY PESTKA

VOLUME 120. Cumulative Subject Index Volumes 81–94, 96–101

VOLUME 121. Immunochemical Techniques (Part I: Hybridoma Technology and Monoclonal Antibodies)
Edited by JOHN J. LANGONE AND HELEN VAN VUNAKIS

VOLUME 122. Vitamins and Coenzymes (Part G)
Edited by FRANK CHYTIL AND DONALD B. MCCORMICK

VOLUME 123. Vitamins and Coenzymes (Part H)
Edited by FRANK CHYTIL AND DONALD B. MCCORMICK

VOLUME 124. Hormone Action (Part J: Neuroendocrine Peptides)
Edited by P. MICHAEL CONN

VOLUME 125. Biomembranes (Part M: Transport in Bacteria, Mitochondria, and Chloroplasts: General Approaches and Transport Systems)
Edited by SIDNEY FLEISCHER AND BECCA FLEISCHER

VOLUME 126. Biomembranes (Part N: Transport in Bacteria, Mitochondria, and Chloroplasts: Protonmotive Force)
Edited by SIDNEY FLEISCHER AND BECCA FLEISCHER

VOLUME 127. Biomembranes (Part O: Protons and Water: Structure and Translocation)
Edited by LESTER PACKER

VOLUME 128. Plasma Lipoproteins (Part A: Preparation, Structure, and Molecular Biology)
Edited by JERE P. SEGREST AND JOHN J. ALBERS

VOLUME 129. Plasma Lipoproteins (Part B: Characterization, Cell Biology, and Metabolism)
Edited by JOHN J. ALBERS AND JERE P. SEGREST

VOLUME 130. Enzyme Structure (Part K)
Edited by C. H. W. HIRS AND SERGE N. TIMASHEFF

VOLUME 131. Enzyme Structure (Part L)
Edited by C. H. W. HIRS AND SERGE N. TIMASHEFF

VOLUME 132. Immunochemical Techniques (Part J: Phagocytosis and Cell-Mediated Cytotoxicity)
Edited by GIOVANNI DI SABATO AND JOHANNES EVERSE

VOLUME 133. Bioluminescence and Chemiluminescence (Part B)
Edited by MARLENE DELUCA AND WILLIAM D. MCELROY

VOLUME 134. Structural and Contractile Proteins (Part C: The Contractile Apparatus and the Cytoskeleton)
Edited by RICHARD B. VALLEE

VOLUME 135. Immobilized Enzymes and Cells (Part B)
Edited by KLAUS MOSBACH

VOLUME 136. Immobilized Enzymes and Cells (Part C)
Edited by KLAUS MOSBACH

VOLUME 137. Immobilized Enzymes and Cells (Part D)
Edited by KLAUS MOSBACH

VOLUME 138. Complex Carbohydrates (Part E)
Edited by VICTOR GINSBURG

VOLUME 139. Cellular Regulators (Part A: Calcium- and Calmodulin-Binding Proteins)
Edited by ANTHONY R. MEANS AND P. MICHAEL CONN

VOLUME 140. Cumulative Subject Index Volumes 102–119, 121–134

VOLUME 141. Cellular Regulators (Part B: Calcium and Lipids)
Edited by P. MICHAEL CONN AND ANTHONY R. MEANS

VOLUME 142. Metabolism of Aromatic Amino Acids and Amines
Edited by SEYMOUR KAUFMAN

VOLUME 143. Sulfur and Sulfur Amino Acids
Edited by WILLIAM B. JAKOBY AND OWEN GRIFFITH

VOLUME 144. Structural and Contractile Proteins (Part D: Extracellular Matrix)
Edited by LEON W. CUNNINGHAM

VOLUME 145. Structural and Contractile Proteins (Part E: Extracellular Matrix)
Edited by LEON W. CUNNINGHAM

VOLUME 146. Peptide Growth Factors (Part A)
Edited by DAVID BARNES AND DAVID A. SIRBASKU

VOLUME 147. Peptide Growth Factors (Part B)
Edited by DAVID BARNES AND DAVID A. SIRBASKU

VOLUME 148. Plant Cell Membranes
Edited by LESTER PACKER AND ROLAND DOUCE

VOLUME 149. Drug and Enzyme Targeting (Part B)
Edited by RALPH GREEN AND KENNETH J. WIDDER

VOLUME 150. Immunochemical Techniques (Part K: *In Vitro* Models of B and T Cell Functions and Lymphoid Cell Receptors)
Edited by GIOVANNI DI SABATO

VOLUME 151. Molecular Genetics of Mammalian Cells
Edited by MICHAEL M. GOTTESMAN

VOLUME 152. Guide to Molecular Cloning Techniques
Edited by SHELBY L. BERGER AND ALAN R. KIMMEL

VOLUME 153. Recombinant DNA (Part D)
Edited by RAY WU AND LAWRENCE GROSSMAN

VOLUME 154. Recombinant DNA (Part E)
Edited by RAY WU AND LAWRENCE GROSSMAN

VOLUME 155. Recombinant DNA (Part F)
Edited by RAY WU

VOLUME 156. Biomembranes (Part P: ATP-Driven Pumps and Related Transport: The Na,K-Pump)
Edited by SIDNEY FLEISCHER AND BECCA FLEISCHER

VOLUME 157. Biomembranes (Part Q: ATP-Driven Pumps and Related Transport: Calcium, Proton, and Potassium Pumps)
Edited by SIDNEY FLEISCHER AND BECCA FLEISCHER

VOLUME 158. Metalloproteins (Part A)
Edited by JAMES F. RIORDAN AND BERT L. VALLEE

VOLUME 159. Initiation and Termination of Cyclic Nucleotide Action
Edited by JACKIE D. CORBIN AND ROGER A. JOHNSON

VOLUME 160. Biomass (Part A: Cellulose and Hemicellulose)
Edited by WILLIS A. WOOD AND SCOTT T. KELLOGG

VOLUME 161. Biomass (Part B: Lignin, Pectin, and Chitin)
Edited by WILLIS A. WOOD AND SCOTT T. KELLOGG

VOLUME 162. Immunochemical Techniques (Part L: Chemotaxis and Inflammation)
Edited by GIOVANNI DI SABATO

VOLUME 163. Immunochemical Techniques (Part M: Chemotaxis and Inflammation)
Edited by GIOVANNI DI SABATO

VOLUME 164. Ribosomes
Edited by HARRY F. NOLLER, JR., AND KIVIE MOLDAVE

VOLUME 165. Microbial Toxins: Tools for Enzymology
Edited by SIDNEY HARSHMAN

VOLUME 166. Branched-Chain Amino Acids
Edited by ROBERT HARRIS AND JOHN R. SOKATCH

VOLUME 167. Cyanobacteria
Edited by LESTER PACKER AND ALEXANDER N. GLAZER

VOLUME 168. Hormone Action (Part K: Neuroendocrine Peptides)
Edited by P. MICHAEL CONN

VOLUME 169. Platelets: Receptors, Adhesion, Secretion (Part A)
Edited by JACEK HAWIGER

VOLUME 170. Nucleosomes
Edited by PAUL M. WASSARMAN AND ROGER D. KORNBERG

VOLUME 171. Biomembranes (Part R: Transport Theory: Cells and Model Membranes)
Edited by SIDNEY FLEISCHER AND BECCA FLEISCHER

VOLUME 172. Biomembranes (Part S: Transport: Membrane Isolation and Characterization)
Edited by SIDNEY FLEISCHER AND BECCA FLEISCHER

VOLUME 173. Biomembranes [Part T: Cellular and Subcellular Transport: Eukaryotic (Nonepithelial) Cells]
Edited by SIDNEY FLEISCHER AND BECCA FLEISCHER

VOLUME 174. Biomembranes [Part U: Cellular and Subcellular Transport: Eukaryotic (Nonepithelial) Cells]
Edited by SIDNEY FLEISCHER AND BECCA FLEISCHER

VOLUME 175. Cumulative Subject Index Volumes 135–139, 141–167 (in preparation)

VOLUME 176. Nuclear Magnetic Resonance (Part A: Spectral Techniques and Dynamics)
Edited by NORMAN J. OPPENHEIMER AND THOMAS L. JAMES

VOLUME 177. Nuclear Magnetic Resonance (Part B: Structure and Mechanism)
Edited by NORMAN J. OPPENHEIMER AND THOMAS L. JAMES

VOLUME 178. Antibodies, Antigens, and Molecular Mimicry
Edited by JOHN J. LANGONE

VOLUME 179. Complex Carbohydrates (Part F)
Edited by VICTOR GINSBURG

VOLUME 180. RNA Processing (Part A: General Methods)
Edited by JAMES E. DAHLBERG AND JOHN N. ABELSON

VOLUME 181. RNA Processing (Part B: Specific Methods) (in preparation)
Edited by JAMES E. DAHLBERG AND JOHN N. ABELSON

VOLUME 182. Guide to Protein Purification (in preparation)
Edited by MURRAY P. DEUTSCHER

VOLUME 183. Molecular Evolution: Computer Analysis of Protein and Nucleic Acid Sequences (in preparation)
Edited by RUSSELL F. DOOLITTLE

VOLUME 184. Avidin-Biotin Technology (in preparation)
Edited by MEIR WILCHEK AND EDWARD A. BAYER

VOLUME 185. Gene Expression Technology (in preparation)
Edited by DAVID V. GOEDDEL

Section I

Preparation of Substrates

A. Purification of RNA Molecules
Articles 1 through 3

B. *In Vitro* Synthesis of RNA
Articles 4 through 6

C. *In Vivo* Preparation of Substrates
Articles 7 through 10

[1] Isolation of RNA

By SHELBY L. BERGER and JOHN M. CHIRGWIN

There are two basic approaches to the isolation of intact RNA: those that permit subcellular fractionation and those that dissociate macromolecular complexes in denaturing solutions. Techniques that permit subcellular fractionation are entirely dependent on inhibitors of ribonuclease. Although many commercially available substances diminish or abolish the activity of cellular ribonucleases, none is ideal. For each tissue, the particular inhibitor or mixture of inhibitors that is most efficacious must be determined. In contrast, methods that employ chaotropic agents such as guanidinium salts are inevitably successful, once rapid homogenization of the tissue in solutions containing these agents has been achieved. In this chapter, an example of a widely accepted method of each type will be presented and more recently described variations will be briefly discussed. More detailed information has been compiled.[1]

Use of Chaotropic Agents to Prepare RNA

If purification of large initial transcripts of RNA or rare RNA intermediates undergoing processing is the goal, protocols that utilize guanidium salts to solubilize tissue and denature proteins, including ribonucleases, are the methods of choice.[2-4] For tissues or cells with high ribonuclease levels, homogenization in the presence of guanidinium thiocyanate has been found reproducibly to yield the highest quality RNA.[5-7]

Solutions and Equipment

1. Prepare a solution of 4 M guanidinium thiocyanate by mixing 50 g of guanidinium thiocyanate (Fluka, Ronkonkoma, NY, 50990) with 10 ml

[1] S. L. Berger and A. R. Kimmel, eds., this series, Vol. 152, p. 213.

[2] R. A. Cox, this series, Vol. 12, Part B, p. 120.

[3] J. M. Chirgwin, A. E. Przybyla, R. J. MacDonald, and W. J. Rutter, *Biochemistry* **18,** 5294 (1979).

[4] R. J. MacDonald, G. H. Swift, A. E. Przybyla, and J. M. Chirgwin, this series, Vol. 152, p. 219.

[5] L. S. Huang, S. C. Bock, S. I. Feinstein, and J. L. Breslow, *Proc. Natl. Acad. Sci. U.S.A.* **82,** 6825 (1985).

[6] P. M. Bricknell, D. S. Latchman, D. Murphy, K. Willison, and P. W. J. Rigby, *Nature (London)* **316,** 162 (1985).

[7] T. H. Turpen and O. M. Griffith, *BioTechniques* **4,** 11 (1986).

1 M Tris–HCl at pH 7.5 and water to make 100 ml. Once dissolved, filter the solution through Whatman #1 filter paper and store at room temperature. Just before use, make the solution 20 mM in dithiothreitol or 1% (v/v) in 2-mercaptoethanol.

2. Prepare the following solutions.

10% Sodium lauryl sarcosinate (Sigma L-5125, St. Louis, MO)
2 M Potassium acetate at pH 5.5
2 M Sodium acetate at pH 7.0
1 M Acetic acid
20 mM EDTA
SET buffer: 10 mM Tris–HCl at pH 7.4, 5 mM ethylenediaminetetraacetic acid (EDTA), and 0.1% sodium dodecyl sulfate (SDS)
5.7 M CsCl (BRL, Gaithersburg, MD, 5507UB) in a solution containing 4 mM EDTA[8]

3. Prepare diethyl pyrocarbonate (DEPC)-treated water. Water is made 0.1% (v/v) in DEPC, stirred vigorously for 10–20 min, and either boiled for 1 hr in a fume hood, autoclaved, or heated at 60° overnight to destroy unreacted DEPC.

Sterile glassware (heated for several hours at 180°) and disposable sterile plasticware are recommended for RNA solutions, once strong denaturants have been removed. Buffers should be treated with DEPC and heated as described above for water. However, since Tris and mercaptans react with this compound,[9] solutions containing them are usually prepared in DEPC-treated water and filtered twice through a nitrocellulose membrane (Millipore) to remove ribonuclease (RNase) and other trace proteins. It is not necessary to treat solutions of chaotropic agents with DEPC.

Wear gloves to prevent "finger nucleases" from contaminating purified RNA.

A Tissumizer (Tekmar Ind., Cincinnati, OH) or Polytron (Brinkmann, Westbury, NY) is required for rapid homogenization. An ultracentrifuge accomplishes physical separation of crude RNA from other cellular components.

Guanidinium Thiocyanate Method for Purifying Total RNA

The following method was developed for rat pancreas (0.7 g), a tissue in which ribonuclease abounds.[10] The technique is also suitable for an

---

[8] Impure CsCl must be treated with DEPC to inactivate nucleases. An EDTA concentration as high as 0.1 M may be necessary to chelate contaminating divalent cations.

[9] S. L. Berger, *Anal. Biochem.* **67,** 428 (1975).

[10] J. H. Han, C. Stratowa, and W. J. Rutter [*Biochemistry* **26,** 1617 (1987)] describe refinements applicable to the isolation of RNA from pancreas.

equivalent amount (up to 2 g) of frozen tissue, pulverized in liquid nitrogen, or for washed tissue culture cells.

1. Place a freshly excised rat pancreas on the lip of an appropriate vessel containing 10 ml of the buffered guanidinium thiocyanate solution described in step 1 of the previous section. Force the tissue quickly into solution with the tip of the homogenizer probe and homogenize at full speed for 60 sec at room temperature. Thorough homogenization shears DNA and prevents a mat of DNA from forming during subsequent ultracentrifugation.

2. Add 0.5 ml of 10% sodium lauryl sarcosinate and mix.

3. Centrifuge at room temperature for 5 min at about 5000 g to remove debris and reserve the supernatant fluid.

At this point, the RNA can be recovered by first adding 0.5 ml of 2 M potassium acetate at pH 5.5 and 0.8 ml of 1 M acetic acid. After mixing, a 7.5-ml ($\frac{3}{4}$) volume of ethanol is slowly introduced while vortexing. Precipitation requires chilling to $-20°$ for at least 2 hr. However, for rare RNAs or solutions in which the RNA concentration is low, the use of ultracentrifugation through a layer of CsCl[11] is recommended as an alternative to the ethanol precipitation described here.

4. Layer about 5.8 ml of the cleared homogenate[12] from step 3 onto 2 ml of CsCl solution.

5. Centrifuge at 20° in a SW30.1 rotor for at least 15 hr at 29,000 rpm to pellet RNA; or for 9.7 ml of homogenate and 3.3 ml of CsCl solution, use a SW41 rotor at 30,000 rpm for 23 hr. Conditions for other rotors have been tabulated.[4]

6. Aspirate the supernatant fluid carefuly without disturbing the glasslike pellet and dry the walls of the tube with a sterile cotton-tipped applicator.

7. Dissolve the pellet. Because of the high concentration of CsCl, the process of dissolving the RNA is sometimes difficult. For small pellets of less than 1 mg of RNA, vortex extensively in SET buffer to drive the RNA into solution. To dissolve large pellets, one or more of the following may be required: one cycle of freezing and thawing in SET buffer, homogenization with the small probe of the Tissumizer or Polytron in SET buffer, or prior extraction of the pellet with a mixture of 2-propanol : 0.2 M sodium acetate (7 : 3) to remove CsCl. In each case, proceed to step 8.

[11] V. Glisin, R. Crkvenjakov, and C. Byus, *Biochemistry* **13**, 2633 (1974).
[12] High concentrations of mercaptoethanol are incompatible with some recent batches of ultracentrifuge tubes produced by some manufacturers. Sarstedt tubes are recommended.

Alternatively, the pellet can be redissolved in neutralized 7.5 M guanidinium hydrochloride and reprecipitated at $-20°$ with 0.05 volume of potassium acetate and 0.5 volume of ethanol. The RNA recovered by centrifugation at 10,000 g for 30 min at 4° can then be dissolved in SET buffer and reprecipitated as described in step 9. (Step 8 is rendered unnecessary using this approach.)

8. Remove residual contaminants in the RNA solution in SET buffer by extracting with 2 volumes of chloroform : butanol (4 : 1). Centrifuge briefly to separate the phases. Reserve the upper aqueous phase in a fresh sterile tube in ice and extract the interface and organic layer with 1-ml aliquots of 20 mM EDTA until the interface no longer shrinks; two to three extractions should be sufficient.

9. Combine the aqueous phases and precipitate RNA with 0.1 volume of sodium acetate and 2.5 volumes of ethanol at $-20°$ overnight.

10. Recover the pellet by centrifugation at 10,000 g for 30 min at 4° and dissolve in DEPC-treated water. The RNA concentration can be ascertained by determining the absorbance at 260 nm; an absorbance of 1 reflects a RNA concentration of 40 μg/ml. RNA solutions can be stored for years at $-70°$.

The handling of tissue in which DNA, glycogen, or intractable nucleases persist has been described.[4]

Isolation of RNA from Subcellular Compartments

The preparation of RNA from subcellular fractions requires procedures for gentle disruption of cells and for inhibiting nucleases during subsequent purification of the desired molecules. In this section, we will present a method for obtaining cytoplasm and its included RNA. Subsequently, we will focus on methods for maintaining the integrity of nuclei and nuclear RNAs. The methods as outlined must be considered as points of departure. Since cells and their organelles differ in their sensitivity to salts, detergents, and RNase inhibitors, the investigator must determine how best to accomplish a specific task.

The study of RNA in subcellular fractions relies on nuclease inhibitors which can prevent degradation of RNA until thorough deproteinization has taken place. The most widely used inhibitors for preparative work are ribonucleoside–vanadyl complexes[13] at 10 mM. These materials are particularly attractive because they are transition state analogs of pancreatic

[13] S. L. Berger and C. S. Birkenmeier, *Biochemistry* **18**, 5143 (1979).

ribonuclease A[14] and bind very tightly to RNase A as well as to a broad range of cellular ribonucleases. The sulfated polysaccharide, heparin, at 0.2–2 mg/ml[13] is also effective in many situations. Because of its cost, the commercially available ribonuclease inhibitor protein from placenta at 1000 U/ml is usually reserved for small-scale work. Another potent RNase inhibitor, diethyl pyrocarbonate at 0.1–0.5% (v/v), is used to inactivate enzymatic activities in solutions that are required in the preparation of RNA. Diethyl pyrocarbonate is not usually used directly in cell lysates because it rapidly alkylates proteins and reacts more slowly with exposed adenine residues in RNA or DNA.[15] Classical, nonspecific inhibitors, such as bentonite (USP, City Chemical Corp., New York, NY) at 0.5–1% and macaloid (NL Chemicals, Inc., Hightstown, NJ), are often ignored today, but may be valuable adsorbents of RNase under conditions in which preservation of active RNA–enzyme complexes is crucial. The preparation of these clays can be found elsewhere.[16–18]

Solutions for Cytoplasmic RNAs

1. Prepare the following buffers.
   Low-salt Tris: 20 mM Tris–HCl at pH 7.4, 10 mM NaCl, and 3 mM magnesium acetate
   Lysing buffer: low-salt Tris containing 5% (w/w) sucrose and either 1.2% (w/w) Triton N-101 (Rohm and Haas, Philadelphia, PA) or 1.2% (w/w) Nonidet P-40 (NP-40) (BRL).
   Ace buffer: 10 mM sodium acetate, 50 mM NaCl, and 3 mM EDTA at pH 5.1
   2 M Sodium acetate
   10% SDS
   Ribonucleoside–vanadyl complexes at 200 mM may easily be synthesized in the laboratory[13] or purchased from BRL or New England BioLabs (Beverly, MA).
2. Saturate redistilled phenol with 0.3 volume of Ace buffer and mix well. Phenol should be stored at −20° in the presence of 0.1% 8-hydroxyquinoline. Immediately before use, phenol should be thawed at 60–65°

[14] G. E. Lienhard, I. I. Secemski, K. A. Koehler, and R. N. Lindquist, *Cold Spring Harbor Symp. Quant. Biol.* **36,** 45 (1971).
[15] R. E. L. Henderson, L. H. Kirkegaard, and N. J. Leonard, *Biochim. Biophys. Acta* **294,** 356 (1973).
[16] D. D. Blumberg, this series, Vol. 152, p. 20.
[17] H. Frankel-Conrat, B. Singer, and A. Tsugita, *Virology* **14,** 54 (1961).
[18] R. Poulson, *in* "The Ribonucleic Acids" (P. R. Stewart and D. S. Letham, eds.), p. 333. Springer-Verlag, Berlin and New York, 1977.

and mixed with buffer. Unused, thawed phenol is discarded after a few hours.

3. Mix chloroform with isoamyl alcohol [94 : 6 (v/v)].

Treat buffers and salt solutions with DEPC as described in Solutions and Equipment. Organic reagents and detergents do not require treatment.

Preparation of Cytoplasm Using Ribonucleoside–Vanadyl Complexes

The procedure delineated in this section was designed for the isolation of mRNA from resting and mitogen-stimulated normal lymphocytes and lymphoblastoid cells. It is particularly well suited for use with cells grown in culture or with tissues in which single cell suspensions can be obtained without serious damage to the plasma membrane. Using this technique, intact RNA has been obtained from cells with high ribonuclease levels.[13,19] It is unlikely, however, that intact cytoplasmic RNA can be isolated from nuclease-rich *tissue* because homogenization methods that spare nuclei cannot accomplish rapid mixing of cellular constituents in lumps and particles with nuclease inhibitors in the buffer. The method as written is appropriate for 1 liter of rapidly growing lymphoblastoid cells containing about 2×10^6 cells/ml.

1. Centrifuge the cell suspension at 1600 *g* at room temperature for 10 min. Retain the cells by decanting the supernatant or removing culture fluids by aspiration. Centrifugation conditions will obviously vary with the cell type.

2. Resuspend the cells in 250 ml of ice-cold Eagle's minimal essential medium, Earle's balanced salt solution, or the equivalent and centrifuge again at 4° to recover the cells.

3. Resuspend the washed cells in 12 ml of ice-cold low-salt Tris, 0.8 ml of vanadyl complexes, and after thorough mixing, add 4 ml of ice-cold lysing buffer. Vortex. Once the cells are broken, it is essential to work very quickly; RNA is at risk until mixed with phenol in step 2 of the following section (Isolation of RNA).

4. Sediment the nuclei by centrifugation for 5 min at 1200 *g* and 4° and recover the supernatant fluids.

The cytoplasmic components contained in the supernatant fluids are suitable for analysis of polysomes, enrichment of specific RNAs by immunoisolation of polysomes, isolation of RNA–protein complexes, and puri-

---

[19] S. L. Berger, M. J. M. Hitchcock, K. C. Zoon, C. S. Birkenmeier, R. M. Friedman, and E. H. Chang, *J. Biol. Chem.* **255**, 2955 (1980).

fication of cytoplasmic RNAs. The pelleted nuclei are not usually used for the isolation of RNA. Methods designed specifically for nuclear components are described in Isolation of Nuclear RNA.

Isolation of RNA

This procedure is one of many variations of the now classic phenol extraction of nucleic acids.[20,21] When the isolation of RNA is the object, buffers at pH values near 5 are chosen in order to drive DNA into the interphase. The presence of SDS ensures quantitative recovery of polyadenylated RNAs in the aqueous phase regardless of the pH value. In this case, the volume of phenol is deliberately large to prevent trapping RNA in the interphase.

1. Discharge the cytoplasmic lysate consisting of 16 ml of supernatant fluid (step 4 of Preparation of Cytoplasm Using Ribonucleoside–Vanadyl Complexes) rapidly into 60 ml of Ace buffer mixed with 4 ml of 10% SDS in a 250-ml glass bottle fitted with a ground-glass stopper. The Ace buffer is maintained at 4° and mixed with SDS at room temperature immediately before use.

2. Add 80 ml of Ace-saturated phenol at 50° as quickly as possible and shake vigorously at 25–50° for 5 min. The phenol phase, initially bright yellow owing to dissolved 8-hydroxyquinoline, becomes gray black on contact with vanadyl complexes.

3. Recover the upper, aqueous phase by centrifugation at 4000 *g* for 10 min at 4° in glass or phenol-resistant plastic tubes.

4. Add the cloudy, aqueous phase to a mixture of 40 ml of Ace-saturated phenol and 40 ml of chloroform–isoamyl alcohol in a fresh 250-ml glass-stoppered bottle and shake vigorously at room temperature for 5 min. Transfer the suspension to glass centrifuge tubes and recover the upper, aqueous phase by centrifugation at 10,000 *g* for 10 min at 4°.

The color of phenol in the organic phase indicates the presence (grayish to black) or absence (bright yellow) of vanadyl complexes. Multiple reextractions (step 4) may be required for quantitative removal of these materials.

If the aqueous solution remains cloudy and if there is still a prominent interphase, the aqueous layer may be extracted again as described in step 4, but with 80 ml of chloroform–isoamyl alcohol serving as the organic phase.

5. Precipitate cytoplasmic RNAs in the final aqueous phase by adding

---

[20] K. S. Kirby, this series, Vol. 12, Part B, p. 87.
[21] K. Scherrer and J. E. Darnell, *Biochem. Biophys. Res. Commun.* **7**, 486 (1962).

0.05 volume of sodium acetate and 2.5 volumes of cold ethanol. For best recovery of RNA, chill at −20° overnight.

6. Sediment RNA by centrifugation at 10,000 g and 4° for 30 min. Carefully remove the supernatant fluid, wash the pellet in a small volume of 70% (v/v) ethanol if desired, and dissolve the RNA in DEPC-treated water. Quantify the RNA as described in step 10 of Guanidinium Thiocyanate Method for Purifying Total RNA.

The RNA is suitable for purification of polyadenylated molecules and for electrophoretic evaluation. Residual vanadyl complexes will not interfere with either procedure. Indeed, chromatography in columns of oligo(dT) cellulose or poly(U) Sepharose will remove the complexes from poly(A)⁺ RNA. For cell-free protein synthesis, it is important to exclude both sodium ion and vanadyl complexes. Several ethanol precipitations of RNA in the presence of 0.1 M potassium acetate as the salt are recommended before attempting translation.

Isolation of Nuclear RNA[22]

The isolation of nuclei or nuclear components is usually carried out in order to purify nuclear RNAs or RNA processing complexes or for *in vitro* transcription. Briefly, cultured cells are lysed either by hypotonic swelling and Dounce homogenization or with detergents.[23] Nuclei are recovered by centrifugation and are either used directly or washed and fractionated further. To isolate RNA, nuclei are ruptured with high-salt solutions, treated with DNAse to reduce viscosity, and extracted with hot phenol.[24] RNA can also be extracted directly from intact nuclei.

Solutions and Equipment for Nuclear RNAs

Prepare the following solutions.

Phosphate-buffered saline (PBS): 2.16 g/liter $Na_2HPO_4 \cdot 7H_2O$, 0.2 g/liter KH_2PO_4, 8 g/liter NaCl, and 0.2 g/liter KCl

Modified RSB: 10 mM Tris–HCl at pH 7.9, 10 mM NaCl, and 5 mM $MgCl_2$

NP-40 lysis buffer: 10 mM Tris–HCl at pH 7.9, 140 mM KCl, 5 mM $MgCl_2$, and 1 mM dithiothreitol

High-salt buffer (HSB): 10 mM Tris–HCl at pH 7.4, 0.5 M NaCl, 50 mM $MgCl_2$, and 2 mM $CaCl_2$.

[22] J. R. Nevins, this series, Vol. 152, p. 234.
[23] D. F. Clayton and J. E. Darnell, *Mol. Cell. Biol.* **3,** 1552 (1983).
[24] R. Soeiro and J. E. Darnell, *J. Mol. Biol.* **44,** 551 (1969).

NETS: 10 mM Tris–HCl at pH 7.4, 100 mM NaCl, 10 mM EDTA, and 0.2% SDS.

Extraction buffer: 10 mM Tris–HCl at pH 7.4, 20 mM EDTA, and 1% SDS.

Transcription storage buffer: 20 mM Tris–HCl or HEPES (potassium salt) at pH 7.9, 140 mM KCl, 10 mM MgCl$_2$, 1 mM dithiothreitol, and 20% (v/v) glycerol

2 M KCl

Redistilled phenol saturated with 0.3 volume of NETS buffer and warmed to 65° shortly before use

A Dounce homogenizer with a tight-fitting pestle is needed for detergent-free preparation of nuclei, and RNase-free deoxyribonuclease (DNase) (Worthington, Freehold, NJ) is required for degradation of DNA.

Preparation of Nuclei by Hypotonic Lysis

1. Collect cells by centrifugation at 4° and wash them several times with PBS.

2. Resuspend the cells in modified RSB in a Dounce homogenizer at a concentration of 3–5 × 10^7 cells/ml and allow them to swell in ice for 5–10 min.

3. Break the cells with 10–15 strokes of the pestle and add KCl to make 100 mM. Keep the solution cold.

4. Recover nuclei by centrifugation at 1000 g for 3 min at 4°.

Cells differ in their ability to be lysed by this procedure. For example, for Friend erythroleukemia cells, subject the cells to 1 stroke of the Dounce homogenizer, add KCl to 50 mM, and continue "Douncing" for 9–12 strokes. Add KCl to 100 mM final concentration. For resting lymphocytes, the tightest pestles are essential. Since the allowed tolerance of pestles labeled "tight" from the same manufacturer is quite broad, only the closest fitting pestles must be selected from among many bearing the identical catalog number. Detergents are often used for primary mouse hepatocytes, mouse myelomas, and some mouse lymphomas.[23]

Preparation of Nuclei with Detergents

1A. Resuspend washed cells (see step 1 of Preparation of Nuclei by Hypotonic Lysis) in modified RSB at 4° at a concentration of 3–5 × 10^7 cells/ml and add Triton X-100 to a final concentration of 0.1% (v/v).

2A. Cool in an ice bucket for 5–10 min to allow the cells to swell.

3A. Break the cells with 15–20 strokes of a Dounce homogenizer.
4A. Collect the nuclei at 1000 g for 3 min at 4°. Discard the supernatant fluid.

OR

1B. Resuspend washed cells at the same concentration as in step 1A above in NP-40 lysis buffer at 4° supplemented with 0.2–0.5% (v/v) NP-40.
2B. Mix the cells gently and keep in ice for 5 min.
3B. Collect the nuclei as in step 4A above.

The efficiency of cell lysis and the quality of the nuclei can be ascertained by examination of the cell suspension before and after cell lysis in a phase-contrast microscope. Adjust the number of strokes of the Dounce homogenizer and the amount of detergent to suit the cell type.

Nuclei prepared as described in this section are usually suitable for *in vitro* transcription. They can be frozen in liquid nitrogen at $\geq 10^8$ nuclei/ml in transcription storage buffer and stored at $-70°$.[22]

Lysing of Nuclei and Extraction of RNA

Starting materials for nuclear lysis are the pelleted nuclei described in Preparation of Nuclei by Hypotonic Lysis (step 4) or in Preparation of Nuclei with Detergents (step 4A or 3B).

1. Wash nuclei in modified RSB containing 100 mM KCl.
2. Resuspend them at 5×10^7 nuclei/ml in HSB containing DNase at ≥ 50 U/ml and warm the solution to room temperature by placing the vessel in a beaker of water. Nuclei will lyse when exposed to HSB. To resuspend the now viscous mixture, pipet up and down vigorously at room temperature with a Pasteur pipet to shear DNA. Adjust the DNase concentration such that the incremental increase in viscosity disappears within 30 sec; longer incubations are associated with degradation of RNA.
3. Add an equal volume of extraction buffer.
4. Add phenol saturated with NETS buffer and chloroform, each equal in volume to the total volume in step 3, and shake in a water bath at 65° for 5 min.
5. Cool to 4° and centrifuge at 1000 g for 5 min to separate the phases.
6. Remove and discard the organic, bottom layer and extract the aqueous layer and interphase again with phenol and chloroform as in steps 4 and 5.
7. Remove the organic layer and extract the aqueous phase with 2–3 volumes of chloroform.

8. Centrifuge as described in step 5 and remove the aqueous phase to a fresh tube.

9. Recover the RNA by precipitation with 2.5 volumes of ethanol at $-20°$ overnight. Collect the precipitate, wash it, dissolve the RNA, and characterize the resultant solution using the directions in step 6 of Isolation of RNA.

An alternative that is attractive for nuclease-rich cells makes use of intact nuclei as the starting material for the isolation of RNA. This can be accomplished with guanidinium salts by homogenizing nuclei in buffered guanidinium thiocyanate as described in step 1 of Guanidinium Thiocyanate Method for Purifying Total RNA and continuing with all subsequent steps in that section. Toward the same end, one can substitute nuclei for the cytoplasmic lysate described in step 1 of Isolation of RNA and continue to the end of that section to obtain purified nuclear RNA. However, with nuclei, the phenol extraction should be performed at elevated temperatures (50–60°) with vigorous shaking to shear DNA, and the organic phase and interphase should be reextracted with Ace buffer to recover trapped RNA. The use of either vanadyl complexes or heparin to inhibit nucleases during cell lysis is also recommended.

Other Methods for Isolating RNA

Techniques for extracting RNA have been evolving for over 30 years. Today, they fall into two broad categories: (1) those that make use of guanidinium salts and (2) those that make use of phenol. A third subcategory uses SDS to disrupt cells and organelles and proteinase K to digest partly and fully denatured proteins.[25] There are also multitudinous variations on these themes. In one example, guanidinium–HCl is mixed with phenol in what amounts to a combination of the basic approaches to RNA isolation.[26] In another case, poly(A)+ RNA is purified directly from cell lysates without first isolating RNA.[27] Many methods are streamlined versions of the aforementioned techniques, occasionally with innovative touches.[28] Although these can be highly successful with some systems, none enjoys the time-tested status of the two major methods which we have presented in detail.

[25] H. Hilz, U. Wiegers, and P. Adamietz, *Eur. J. Biochem.* **56,** 103 (1975).
[26] P. Chomczynski and N. Sacchi, *Anal. Biochem.* **162,** 156 (1987).
[27] J. E. Badley, G. A. Bishop, T. St. John, and J. A. Frelinger, *BioTechniques* **6,** 114 (1988).
[28] D. E. Graham, *DNA Protein Eng. Tech.* **1,** 1 (1988).

[2] Purification of RNA Molecules by Gel Techniques

By TOSHIMICHI IKEMURA

There are many articles and textbooks that describe procedures for the purification of RNAs by the use of gel electrophoresis.[1] In a recent volume of this series, Ogden and Adams[2] extensively described gel electrophoresis of nucleic acids in agarose and polyacrylamide, which is routinely applied in gene technology. The scope of this chapter is therefore confined to a rather specialized gel technique for the fractionation of complex RNA mixtures to a level of purity that can be analyzed by a direct sequencing method; thus, two-dimensional gel electrophoresis is described, but not the single-dimensional run that has been described by Ogden and Adams.[2] By using a two-dimensional gel system, wide varieties of tRNA precursor molecules that had accumulated in an *Escherichia coli* mutant defective in tRNA biosynthesis have been purified (Fig. 1).[3]

Several different two-dimensional gel systems for RNA purification have been described. In the acrylamide concentration shift method,[4] the first- and second-dimensional gels have different concentrations of acrylamide (usually in the ratio of 1 : 2); this method has been used mainly for fractionation of RNAs in the size range of 50–500 nucleotides. In the urea shift method,[5] the first- and second-dimensional gels have different concentrations of urea (usually one contains and the other lacks urea); this method has been used mainly for the separation of small RNA molecules such as tRNAs and, in a few cases, for the separation of large RNA molecules.[6] The combination of acrylamide concentration and urea concentration shifts gives good resolution for a wide variety of RNA species, as will be described in this chapter. The combination of pH shift with urea and acrylamide concentration shifts gives good resolution for RNAs of much smaller sizes, such as oligonucleotides and RNA fragments produced by RNase digestion, and therefore, this is used most frequently for the fingerprinting of RNase digests of large RNA molecules.[7]

[1] D. Rickwood and B. D. Hames (eds.), "Gel Electrophoresis of Nucleic Acids: A Practical Approach." IRL Press, Washington, D.C., 1982.
[2] R. C. Ogden and D. A. Adams, this series, Vol. 152, p. 61.
[3] T. Ikemura, Y. Shimura, H. Sakano, and H. Ozeki, *J. Mol. Biol.* **96**, 69 (1975).
[4] T. Ikemura and J. E. Dahlberg, *J. Biol. Chem.* **248**, 5024 (1973).
[5] M. Stein and F. Varricchio, *Anal. Biochem.* **61**, 112 (1974).
[6] J. Burckhardt and M. L. Birnstiel, *J. Mol. Biol.* **118**, 61 (1978).
[7] R. De Wachter and W. Fiers, *Anal. Biochem.* **49**, 184 (1972).

METHODS IN ENZYMOLOGY, VOL. 180

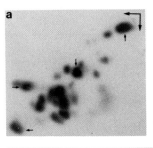

FIG. 1. Two-dimensional gel separation of [^{32}P]tRNAs and their precursors. An *E. coli* temperature-sensitive mutant (TS709) defective in RNase P at high temperature was labeled by $^{32}PO_4^{3-}$ at a permissive (a) or a restrictive (b) temperature. Electrophoresis in the first dimension on 10% (w/v) acrylamide gel was from top to bottom and that in the second dimension on 20% (w/v) acrylamide was from right to left. (a) Labeled at 30°. All tRNAs in a were shown to be mature tRNA molecules. (b) Labeled at 43°. Most RNA molecules in b were shown to be tRNA precursors. For reference, in either autoradiogram, the arrows were added to the positions of mature tRNA^{Glu_2}($\leftarrow$), tRNA^{Asn}($\rightarrow$), tRNA^{Leu_1}($\downarrow$), and 5 S RNA ($\uparrow$). (For details, see Ref. 3.)

De Wachter and Fiers[8] have described extensively the procedures of these two-dimensional methods of RNA fractionation. In this chapter, I will introduce mainly a method used in our laboratory that is based on the combination of acrylamide and urea concentration shifts.[9,10] Table I shows the five types of two-dimensional systems that are routinely used. RNA molecules ranging in size from ~50 to 100 nucleotides are usually separated by system A (Table I). Most tRNA molecules of single organisms (e.g., *E. coli, Salmonella typhimurium, Bacillus subtilis, Saccharomyces cerevisiae,* mouse, and chicken) have been purified by this system.[9,10] Detailed protocols of the two-dimensional electrophoreses are

[8] R. De Wachter and W. Fiers, *in* "Gel Electrophoresis of Nucleic Acids: A Practical Approach" (D. Rickwood and B. D. Hames, eds.), p. 77. IRL Press, Washington, D.C., 1982.

[9] T. Ikemura and M. Nomura, *Cell (Cambridge, Mass.)* **11,** 779 (1977).

[10] T. Ikemura, *J. Mol. Biol.* **158,** 573 (1982).

TABLE I
CONDITIONS FOR TWO-DIMENSIONAL GEL ELECTROPHORESIS

System	Dimension	Concentration (%, w/v)		
		Acrylamide	Urea	Agarose
A	First	14		
	Second	22	41.7	
B	First	10		
	Second	20		
C	First	5		
	Second	10		
D	First	3		1
	Second	10	41.7	
E	First	2.8		1
	Second	5.5	33.3	

presented in this chapter for system A; those for other systems are explained by noting differences from system A.

Equipment and Reagents

To obtain the most advanced and reproducible separation, we use an E–C Vertical Slab Gel Cell (E–C Apparatus Corporation, St. Petersburg, FL), an EC480 cell (12-cm-wide × 44-cm-long gel area) for the first dimension, and an EC490 cell (24-cm-wide × 17-cm-long gel area) for the second dimension. A gel slab is placed between two cooling plates for temperature control. Precise temperature control obtained by the rapid circulation of coolant helps to give reproducible results and also makes it possible to apply a high voltage to the gel. To ensure little change in the pH of the electrophoresis buffer, the buffer is recirculated between the lower and upper buffer tanks. This circulation affords the possibility to use a fairly dilute buffer and thus to apply a high voltage to the gel. When RNA mixtures that are not too complex are to be separated, an EC470 cell (12-cm-wide × 17-cm-long gel area) is used for either dimension because of its ease of handling.

Urea and 3-dimethylaminopropionitrile (DMAPN) are obtained from Schwarz-Mann (Cambridge, MA) and Merck (Darmstadt, FRG), respectively
Concentrated buffer solution (10× TBE)
Tris base (Sigma, St. Louis, MO)   108 g

Boric acid	55 g
Na$_2$EDTA · 2H$_2$O	9.3 g
Distilled water to	1000 ml

The pH of this solution should be about 8.3
40% (w/v) Acrylamide stock solution (acrylamide/bisacrylamide in a
  19 : 1 ratio)

Acrylamide (Bio-Rad, Richmond, CA)	380 g
N,N'-Methylenebisacrylamide (Bio-Rad)	20 g
Distilled water to	1000 ml

  This solution is filtered and stocked in a brown bottle at cold
  temperature

Fractionation of Small RNAs

First Dimension for System A

105 ml	40% (w/v) Acrylamide stock solution
19 ml	6.4% (v/v) DMAPN solution
9 ml	10× TBE buffer solution
157 ml	H$_2$O

The solution (a total of 290 ml) is degassed, mixed with 10 ml of freshly prepared 1% (w/v) ammonium persulfate solution, and poured between the cooling plates of an EC480 cell placed horizontal. A slot former (16- or 12-place slot former) is inserted between the cooling plates. After polymerization (about 30 min), the gel cell is placed in the vertical position and filled with 5 liters of electrophoresis buffer (0.3× TBE). The polymerization time can be adjusted, if necessary, by a change in the concentration of the ammonium persulfate. The excess gel block above the slot former is cut and removed with a bent spatula, and the slot former is also removed. The electrophoresis buffer is recirculated, and the gel cell is cooled by the rapid circulation of water at 15°. We usually perform the overnight prerun at a low voltage (about 100 V/44 cm) and thus can start the first-dimensional electrophoresis the next morning.

    A RNA sample (usually several micrograms) is dissolved in a 5-μl solution of 50% (w/v) urea and 5 mM ethylenediaminetetraacetic acid (EDTA) (pH 7) and then is mixed with a 2-μl solution of 20% (w/v) sucrose, 5 mM EDTA (pH 7), 0.05% (w/v) bromphenol blue, and 0.05% (w/v) xylene cyanol FF. The sample is put into a small slot (3–5 mm wide × 1 mm thick) formed in the top of the gel. Electrophoresis in the first dimension is done at 1400 V (per 44 cm) for 9.5 hr, when tRNA molecules

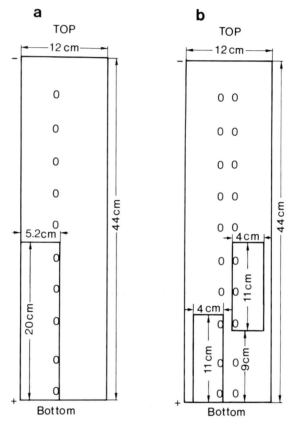

Fig. 2. Schematic diagram to show the gel portion to be cut for the second-dimensional run. Ellipses represent dye spots. (a) A gel strip (5.2 cm wide × 20 cm long) is cut by a sharp, long knife with reference to the trail of dye spots. (b) Two gel strips (4 cm wide × 11 cm long) are cut for the second-dimensional runs with different electrophoresis times (see Fig. 4a and b).

are to be analyzed. A small amount of the sucrose and dye solution mentioned above is applied on the sample slot for each 30-min interval during the run. After this first-dimensional run, a portion of the bottom part of the gel (20 cm long × 5.2 cm wide) is cut out (Fig. 2a). The trail of dye spots enables us to locate the gel portion where RNA molecules are. The temperature of the gel and the pH of the buffer can be kept fairly constant during electrophoresis in this system, so the electrophoresis pattern is reproducible. Thus, once the electrophoresis conditions are decided on after a preliminary shorter run, termination can be dependent solely on the running time, estimated from the preliminary run.

Second Dimension for System A

A gel strip (20 × 5.2 cm) is placed, at right angles to the first run, at the top part of the EC490 cell that is horizontal. The part of the gel where the dye spots (and thus RNAs) are should be completely inserted between the cooling plates. To prevent the gel strip from floating out of position when the gel solution containing urea for the second dimension is poured around it, a plastic weight (e.g., 2 × 3 × 19 cm) is put on the part of the gel not covered by the cooling plate. A 22% (w/v) acrylamide solution containing urea for the second dimension is as follows.

132 ml	40% Acrylamide stock solution
100 g	Urea
7.5 ml	6.4% DMAPN solution
7.2 ml	10× TBE buffer solution
7 ml	H_2O

This solution (a total of 230 ml) is degassed, mixed with 10 ml of 1% (w/v) ammonium persulfate solution, and poured into a EC490 cell; to prevent air bubbles from being trapped around the first-dimensional gel, we tilt the cell by putting a small glass rod (about 1 cm in diameter) under an edge of the top part of the cell. After pouring the gel solution, the cell is restored to the horizontal position and is cooled by rapid circulation of water at 15°. After gelation (about 30 min), the cell is placed in the vertical position and filled with 4.5 liters of the buffer (0.3× TBE). The portion of gel not covered by the cooling plates (and thus above the trail of dye spots) is carefully cut off and removed with the flat end of a bent spatula. Electrophoresis is done at 600 V (per 24 cm) for the first hour and then at 800 V for 15 hr, when tRNA molecules are to be separated.

When a gel with a high percentage of acrylamide such as 22% is used in the E–C gel system, it is difficult to disassemble the cell. To overcome this difficulty, after the one toggle clamp at the sides of the apparatus is opened, air is injected between the gel and the cell surface from a 50-ml syringe with a 20-gauge (70-mm) needle, and the other clamp is opened. Autoradiograms of the separation of ^{32}P-labeled yeast and mouse tRNAs are shown in Fig. 3a and b. If certain tRNAs of interest are not purified, the acrylamide concentration of the first or the second dimension (or both) should be changed[10] (e.g., 12% (w/v) acrylamide can be used for the first dimension and 20% (w/v) acrylamide for the second dimension). The applied voltage and temperature also affect the electrophoretic pattern.

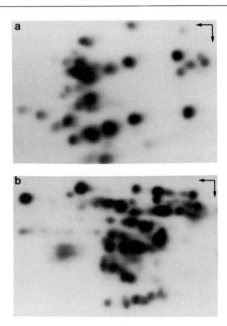

FIG. 3. Two-dimensional gel separation of ^{32}P-labeled yeast and mouse tRNAs. Electrophoresis in the first dimension was from right to left and that in the second dimension was from top to bottom. (a) Yeast tRNAs. (b) Mouse liver tRNAs. (For details, see Ref. 10.)

Fractionation of RNAs of Intermediate Size

RNAs of sizes other than that described above are separated by selection of an appropriate gel concentration, which has to be done by trial and error. RNA molecules ranging in size from ~100 to 500 nucleotides can be separated by system B or C (Table I). For example, tRNA precursors composed of multiple tRNA molecules have been purified by system B,[3] although the electrophoresis had to be long. Recovery of RNAs of intermediate size from a gel with a high percentage of acrylamide by conventional diffusion methods is poor; the electroblotting method described later gives better recovery. To sidestep the inconvenience of prolonged electrophoresis, as well as of the RNA recovery, the acrylamide concentration in either dimension may be reduced (e.g., for system C,[11] 5% for the first dimension and 10% for the second dimension) if the RNA mixture is not too complex. System D, in which a composite gel of acrylamide plus agarose is used in the first dimension, has been found to give good

[11] T. Ikemura and H. Ozeki, *Eur. J. Biochem.* **51,** 117 (1975).

resolution of RNAs of intermediate size, and details of the procedure for making the composite gel are explained in the following section describing system E, which fractionates large RNAs.

Fractionation of Large RNAs

The development of a two-dimensional system for the separation of large RNAs is still in the preliminary stages compared with that for the separation of small RNAs. Large RNAs such as mRNAs are usually fractionated by a single-dimensional gel and are analyzed rather indirectly by Northern hybridization. However, a few mRNA species have been purified with a two-dimensional gel system,[6] based on the urea-shift method. We introduce here an alternative method based on a combination of acrylamide and urea concentration shifts. Although the resolving power is far from satisfactory when compared with that for small RNAs, the separation is reproducible and is useful for purification of RNAs with a relatively high content.[12]

First Dimension for System E

The first dimension is on a composite gel of polyacrylamide (2.8–3%) and agarose (1%). The 2.8% (w/v) polyacrylamide plus 1% (w/v) agarose gel is prepared as follows.

Solution A
21 ml   40% Acrylamide stock solution
16 ml   6.4% DMAPN
9 ml   10× TBE
29 ml   H$_2$O
  Solution A is kept at 45°
Solution B
  Three grams of agarose (Seakem LE, FMC Corp., Rockland, ME)
  is completely dissolved in 215 ml of H$_2$O by being heated, and
  the melted agarose solution is kept at 50°

Solutions A and B are combined and mixed with 10 ml of 3% (w/v) ammonium persulfate solution. The gel solution is poured on an EC480 cell and is cooled by circulation of coolant at 20°. After the solidified gel is kept in this condition for several hours, a prerun is usually done overnight in a 0.3× TBE buffer at a low voltage (e.g., 100 V/44 cm). When RNAs longer than 1000 nucleotides are to be fractionated, electrophoresis is performed at 1000 V (per 44 cm) for about 10 hr. After electrophoresis, a

---

[12] T. Ikemura, *Nucleic Acids Symp. Ser.* **19**, 57 (1988).

portion of the bottom of the gel (11 cm long × 4 cm wide) is cut out, and the gel strip is placed at right angles to the first run at the bottom of the EC480 cell. If necessary, another gel strip (11 × 4 cm) where the longer RNAs are located is also cut out and placed at the bottom of another EC480 cell. Alternatively, to prevent the loss of RNAs present around the cutting line of the gel, two RNA samples are applied on separate slots of the first-dimensional gel, and two gel strips that overlap partially are cut out (Fig. 2b).

Second Dimension for System E

41.25 ml	40% Acrylamide stock solution
9.3 ml	6.4% DMAPN
9 ml	10× TBE
100 g	Urea
157 ml	H_2O

The solution (290 ml) is degassed, mixed with 10 ml of 3% ammonium persulfate, and poured onto an EC480 cell. Electrophoresis is usually done at a constant current of 45 mA (about 1000 V/44 cm at equilibrium) for 35 hr, when RNAs longer than 1000 nucleotides are to be fractionated; the appropriate conditions should be decided after a preliminary short run. The electrophoresis buffer (5.6 liter of 0.3× TBE) is circulated between the buffer tanks and is cooled by rapid circulation of coolant at 15°.

The gel strip of the first dimension is placed at the bottom of the apparatus, and thus the second electrophoresis is from bottom to top. This is because the stiff composite gel of the first dimension can act as a support to prevent the fragile 5.5% acrylamide gel of the second dimension from sliding down in the vertical cell. The long-path EC480 cell (12 cm wide × 44 cm long) is used, as opposed to the EC490 cell (24 cm wide × 17 cm long) for small RNA separation, for this second-dimensional run. This improves resolution, but only a narrow range of the first-dimensional gel (and thus a narrow size range of RNAs) can be analyzed when one apparatus is used. We often use two EC480 cells for the second-dimensional run with different electrophoresis times. Figure 4 shows autoradiograms of such electrophoresis to analyze poly(A)[+] fraction of ^{32}P-labeled *S. cerevisiae* RNAs.[12] The size of RNAs in Fig. 4a is estimated roughly to be from 1600 to 3500 nucleotides and those in Fig. 4b to be from 1200 to 1600 nucleotides. About 45 discrete spots in the former size range and about 15 spots in the latter were detected; stronger exposure gave about 80 spots in total (data not shown). Although the poly(A)[+] fraction was analyzed, a trace amount of 18 S rRNA (the spot located far from other RNAs and marked by an arrow in Fig. 4a) was detected; 25 S rRNA,

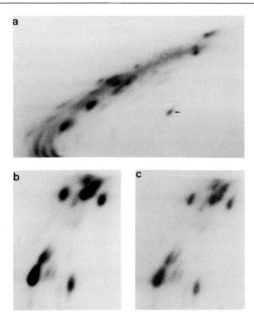

FIG. 4. Two-dimensional gel separation of ^{32}P-labeled yeast poly(A)$^+$ RNAs. Electrophoresis in the first dimension was from top to bottom and that in the second dimension from right to left. Autoradiograms of only the gel portion where RNAs were located are presented. The first-dimensional run was at 1000 V (per 44 cm) for 10 hr. (a) Time of the second run was 40 hr. The arrow is added to 18 S rRNA. (b) Time of the second run was 35 hr. (c) The gel portion presented in b was subjected to electroblotting to diethylaminoethyl (DEAE) paper for recovery of the separated RNAs.

longer than the RNAs analyzed and thus not seen in the autoradiograms of Fig. 4, also had unusual mobility (data not shown). Unusual mobility of rRNAs may reflect their highly folded and stable structures. The cellular concentration of yeast mRNAs for individual genes is clearly variable, and the RNAs made visible here are thought to be fairly abundant.

Recovery of RNAs from Gels

There are many ways to elute RNA and DNA from gels, as Ogden and Adams[2] have described. We introduce in this section a special method that is useful for the extraction of a large number of RNA species (e.g., all tRNA molecules of one organism) by a relatively simple procedure. After the second-dimensional run, RNAs are transferred to diethylaminoethyl (DEAE) paper (Whatman DE-81) by electroblotting (Fig. 4c) as follows. The gel is soaked in the 0.3× TBE buffer for 30 min and is placed on a wet

Whatman 3MM paper placed on a chemical sponge pad; when a Bio-Rad Trans-Blot cell is used, the gel is cut in advance to 15 × 20 cm. A wet DEAE paper and a wet Whatman 3MM paper are placed in this order on the gel, and air bubbles between the gel and the papers are removed by a glass rod being rolled on the Whatman 3MM paper. Electroblotting is done at 60 V (about 7.5 V/cm) for 5–7 hr in the 0.3× TBE buffer. The buffer is cooled by rapid circulation of coolant at 15° and is stirred by a magnetic stirring bar during this period. After the blotting, the DEAE paper is stripped from the gel together with the Whatman 3MM paper, and the Whatman 3MM paper is stripped from the DEAE paper. The DEAE paper should not be lifted up by itself, because it is very fragile. The wet DEAE paper is covered by thin wrapping and is autoradiographed at −80° with an intensifying screen, if necessary. Small RNAs are eluted from the DEAE paper with a small amount (200–300 μl) of a high-salt elution buffer [2 M NaCl, 0.02% (w/v) sodium dodecyl sulfate (SDS), 5 mM EDTA, pH 7.0, and 20 μg/ml of carrier RNA] as follows. The portion of DEAE paper corresponding to the radioactive spot is cut out with reference to the autoradiogram and is incubated in the high-salt elution buffer at 65° for 10 min. During this incubation, it is vortexed a few times and is then centrifuged at 15,000 rpm for 15 min at 5°. The pelleted DEAE paper is reextracted as above with the elution buffer and is centrifuged. The two supernatants are combined and centrifuged again to remove residual DEAE fibers. RNA is precipitated from the supernatant by the addition of ethanol at a volume of 2.5 times, according to the usual ethanol precipitation procedure. RNA is precipitated twice more using a 300 mM NaCl and 5 mM EDTA, pH 7.0, solution by this procedure. When large RNAs are to be eluted, the DEAE paper is incubated in a formamide solution with a high salt concentration [55% (v/v) formamide, 1.8 M sodium acetate, pH 7.0, 2 mM EDTA, 0.1% SDS, and 20 μg/ml of carrier RNA], instead of the high-salt solution.

Concluding Remarks

A rather elaborate apparatus, the E–C Vertical Slab Gel Cell, is used in the present system, mainly because of the reproducibility of the electrophoretic pattern obtained. However, much simpler apparatuses have also been used for two-dimensional gel systems for the separation of small RNAs; tRNA molecules have been purified by a slab gel apparatus of the type used routinely for a single-dimensional run.[13] When a gel cell without a cooling system is used, the voltage applied on the gel has to be lower

[13] M. C. Zuniga and J. A. Steiz, *Nucleic Acids Res.* **4**, 4175 (1977).

than that specified here. Purification techniques, as well as cloning and sequencing techniques, for RNA molecules are far behind such techniques for DNA molecules. Therefore, RNA molecules are converted frequently to DNA molecules *in vitro*, and the resultant DNA molecules are cloned and sequenced (cDNA cloning and sequencing). RNA molecules are also characterized indirectly by hybridization techniques that make use of cloned DNAs as probes (Northern hybridization). This strategy is undoubtedly useful and suitable for most analyses of RNA structures. However, several specific but important characteristics of RNA structures are inevitably omitted from the analysis. For example, modified nucleotides that are present in a wide range of RNA species, as well as nonstandard types of phosphodiester bonds such as those found for lariat molecules produced during RNA splicing, cannot be studied. Purification of the RNA molecule itself is a prerequisite for such studies. The scope of the present chapter is largely confined to this aspect. A two-dimensional gel system based on a combination of acrylamide and urea concentration shifts was described in this chapter, because this system seems to have the highest resolution and can be used for RNAs with a wide size range. The resolution for large RNA species, however, is still far from satisfactory. We hope this chapter proves helpful not only in the progress of studies on RNA processing, but also in the further development of gel techniques for the fractionation of large RNA molecules.

[3] Purifying RNA by Column Chromatography

By N. KYLE TANNER

The recent explosion of work on biologically interesting and, in some cases, catalytically active RNAs has underscored the importance of being able to isolate highly purified RNA. While gel electrophoresis (see Ikemura, this volume [2]) is generally still the preferred method of purifying RNAs, it becomes limiting when large quantities of material are needed for assays and for biophysical studies. Column chromatography, on the other hand, can resolve milligram quantities of material. In addition, it can separate RNAs, which often cannot be separated by gel electrophoresis, on the basis of differences in their primary sequences or in their secondary and tertiary structures. The mild fractionation conditions that are possible can be used to isolate RNAs in a nondenatured conformation or to isolate RNA–protein complexes. Furthermore, recent improvements in sorbents and techniques have eliminated some of the earlier problems associated with column chromatography. In particular, high-performance liquid chromatography (HPLC) and fast-protein liquid chromatography (FPLC) have done much to improve its speed and resolution.

In this chapter, I have limited my discussion to chromatographic techniques that are used widely and that use readily available—generally commercially prepared—sorbents. Thus, many interesting and useful chromatographic techniques have been necessarily omitted. Many examples of these can be found in a recent book by Osterman.[1] I have also not gone into extensive descriptions of chromatographic theory, column construction, or elution conditions as the wide variety of fractionation needs of researchers prohibits this type of discussion. Instead, for each sorbent I have briefly described the basis by which RNAs are fractionated and the initial conditions that should be tried. I have supplemented this with a wide variety of hints and suggestions, which should be useful in developing the final fractionation strategy. The reader is directed to a number of recent review articles[2-6] for information on HPLC and FPLC methodology.

Ionic Chromatography

Diethylaminoethane

Diethylaminoethyl (DEAE)-cellulose and DEAE-Sephadex are weak anion-exchange columns that have often been used to fractionate and purify RNAs.[7] RNA binds to the column in low-salt buffers through ionic interactions between the phosphodiester backbone and the positively charged DEAE groups. Because the strength of the interactions are dependent on the number of charges present, the RNAs are eluted from the column, with an increasing salt gradient, by increasing chain length. Although these columns have been used successfully to purify tRNAs,[8] 5 S

[1] L. A. Osterman, "Methods of Protein and Nucleic Acid Research," Vol. 3. Springer-Verlag, Berlin and New York, 1984.
[2] T. R. Floyd, J. B. Crowther, and R. A. Hartwick, LC Mag. 3, 508 (1985).
[3] J. A. Thompson, BioChromatography 1, 16 (1986).
[4] R. Dornburg, J. Kruppa, and P. Földi, LC Mag. 4, 22 (1986).
[5] L. W. McLaughlin, Trends Anal. Chem. 5, 215 (1986).
[6] L. W. McLaughlin and R. Bischoff, J. Chromatogr. 418, 51 (1987).
[7] R. M. Bock and J. D. Cherayil, this series, Vol. 12, Part A, p. 638.
[8] Y. Kawade, T. Okamoto, and Y. Yamamoto, Biochem. Biophys. Res. Commun. 10, 200 (1963).

rRNA,[9] and nuclease-generated RNA fragments,[10] the resolution has been generally limited. This problem has been alleviated somewhat by the commercial introduction of silica-based (TSK DEAE-SW) and polymer-based (TSK DEAE-PW) matrices that are designed to be used with HPLC. A TSK DEAE-5PW column has been used recently to separate tRNA, 5 S rRNA, 7 S RNA, and viroid RNA.[4] These and other anion-exchange HPLC columns are described in more detail elsewhere.[2,4,6]

In addition to electrostatic interactions, DEAE-cellulose and DEAE-Sephadex form strong hydrogen-bonding and hydrophobic interactions with nucleic acids. These secondary interactions are minimized by loading and eluting the RNA in the presence of urea, formamide, or ethylene glycol.[11,12] The secondary interactions are less pronounced with DEAE-Sephadex than with DEAE-cellulose, although both columns give similar elution profiles.[8,13] The generalized condition for using these columns is to load the RNA in a buffer containing low salt, 20 mM Tris–HCl, pH 7.5, and 7 M urea. The RNA is eluted with a linearly increasing NaCl gradient in this buffer. The salt concentrations necessary to elute the RNA are dependent on the temperature, RNA, and column matrix. tRNA elutes at 0.3–0.4 M NaCl on DEAE-cellulose and at 0.4–0.65 M NaCl on DEAE-Sephadex.[7] Ideally, the RNA should be loaded on the column at a salt concentration just below that required to elute the material.[7] A small amount of Mg(II) decreases the amount of NaCl required to elute the RNA.[13] Flow rates of ~10 ml/cm²/hr are often used, but this is not critical.

There are numerous variations to this scheme where the pH, salt concentration, and urea concentration are varied to take advantage of differences in the secondary interactions due to the primary, secondary, and tertiary properties of the RNAs. For example, single-stranded, denatured RNA binds stronger than base-paired RNA. These conditions are summarized in tabular form in previous volumes of this series.[7,14] Other examples of ways of fractionating RNA include elevated temperatures with a linear salt gradient,[15] a constant salt concentration with a decreasing temperature gradient,[16] or a constant salt concentration and exponentially decreasing urea gradient.[7]

[9] R. Monier and J. Feunteun, this series, Vol. 20, p. 494.
[10] D. Bell, R. V. Tomlinson, and G. M. Tener, *Biochemistry* **3**, 317 (1964).
[11] R. V. Tomlinson and G. M. Tener, *J. Am. Chem. Soc.* **84**, 2644 (1962).
[12] R. V. Tomlinson and G. M. Tener, *Biochemistry* **2**, 697 (1963).
[13] J. D. Cherayil and R. M. Bock, *Biochemistry* **4**, 1174 (1965).
[14] G. M. Tener, this series, Vol. 12, Part A, p. 398.
[15] B. C. Baguley, P. L. Bergquist, and R. K. Ralph, *Biochim. Biophys. Acta* **95**, 510 (1965).
[16] P. L. Bergquist, B. C. Baguley, and R. K. Ralph, this seris, Vol. 12, Part A, p. 660.

Benzoylated DEAE-Cellulose

Esterification of the hydroxyl groups of cellulose with benzoic acid or with a combination of benzoic and naphthoic acid substantially improves the resolution of DEAE-cellulose by increasing the hydrophobic characteristics of the column.[17,18] Benzoylated DEAE-cellulose (BD-cellulose) and benzoylated, naphthoylated DEAE-cellulose (BND-cellulose) have similar properties to DEAE-cellulose, except that the binding interactions are generally stronger and that fractionation is more dependent on the secondary structure of the nucleic acid.[17,19,20] As with DEAE-cellulose, BD- and BND-cellulose form stronger interactions with more open, denatured structures. The columns have a high capacity and good recovery, although large molecular weight RNAs are sometimes difficult to elute.[20]

The strength of the binding depends on the temperature, the divalent and monovalent cation concentration, and the pH. The initial conditions to try are similar to those used for DEAE-cellulose; the sample is applied to the column in a low-salt buffer containing 20 mM Tris–HCl, pH 7.5, 10 mM MgCl$_2$, at 20°. The RNA is eluted with a linearly increasing gradient of NaCl. Most tRNAs elute from BD-cellulose between 0.5–1.1 M NaCl.[18] BND-cellulose forms stronger interactions and requires higher salt concentrations.[17] The temperature, pH, divalent cation concentration, and chaotropic reagent (urea, formamide, or alcohol) can all be varied to obtain optimal resolution. Strongly bound RNAs are eluted by adding alcohol, formamide, or urea to the high-salt (1–2 M) buffer.[17] Caffeine (2%) is also effective in eluting strongly bound RNA.[20] The concentration of caffeine required has been found to be proportional directly to the length and extent of single strandedness of DNA bound to BND-cellulose[21,22] and to BD-cellulose.[23] The properties of BD-cellulose can vary between commercial batches or can deteriorate during storage for unknown reasons[19,24]; boiling the matrix in water briefly seems to restore the normal chromatographic properties.[24]

[17] I. Gilham, S. Millward, D. Blew, M. von Tigerstrom, E. Wimmer, and G. M. Tener, *Biochemistry* **6**, 3043 (1967).
[18] I. C. Gillam and G. M. Tener, this series, Vol. 20, p. 55.
[19] J. Sedat, A. Lyon, and R. L. Sinsheimer, *J. Mol. Biol.* **44**, 415 (1969).
[20] J. W. Sedat, R. B. Kelly, and R. L. Sinsheimer, *J. Mol. Biol.* **26**, 537 (1967).
[21] V. N. Iyer and W. D. Rupp, *Biochim. Biophys. Acta* **228**, 117 (1971).
[22] M. Haber and B. W. Stewart, *FEBS Lett.* **133**, 72 (1981).
[23] R. A. Schlegel, R. E. Pyeritz, and C. A. Thomas, Jr., *Anal. Biochem.* **50**, 558 (1972).
[24] M. Haber, P. H. T. Huang, and B. W. Stewart, *Anal. Biochem.* **139**, 363 (1984).

Adsorption Chromatography

Hydroxylapatite

Hydroxylapatite $\{[Ca_5(PO_4)_3OH]_2\}$ has been used primarily to separate double-stranded DNA from single-stranded DNA and RNA, but it has also been used to separate aminoacylated and nonaminoacylated tRNAs, synthetic polyribonucleotides, double-stranded RNA, and RNA fragments.[25] A hydroxylapatite column has been used to separate 5 S rRNAs that differed only in their level of 5'-terminal phosphorylation; the 5 S rRNA was homogeneous by gel electrophoresis criteria.[26,27] Altman and Stark were able to separate *Escherichia coli* precursor tRNA[Tyr] from *E. coli* tRNA by hydroxylapatite chromatography, in part, because of the 5'-terminal phosphorylation of the precursor.[28] However, in general, hydroxylapatite discriminates between nucleic acids on the basis of their secondary and tertiary structures rather than on their overall charge. Molecules with rigid, ordered structures have a higher affinity for the sorbent than molecules with flexible, disordered ones.[25] This property has been attributed to the effective charge density of the nucleic acid,[25] to differences in conformational entropy of single-stranded nucleic acids,[29] and to the steric availability of the phosphate groups.[30] Binding is due to electrostatic interactions between the phosphate groups on the polynucleotide and the positively charged calcium groups on the hydroxylapatite.[25]

One of the major advantages of hydroxylapatite is its ability to discriminate between different nucleic acid structures. Hydroxylapatite also has a high binding capacity, and it is resistant to a wide range of temperature, pH values, and organic and inorganic solvents. However, the fragility and slow flow rates of the sorbent necessitates that short, wide columns be used. Several matrices are now commercially available, including one for HPLC, that have greater mechanical strength and can be used at faster flow rates. Hirano *et al.*[31] have also presented a simple method for preparing hydroxylapatite that can be used with 10-fold higher

[25] G. Bernardi, this series, Vol. 21, p. 95.
[26] C. Soave, E. Galank, and G. Torti, *Bull. Soc. Chim. Biol.* **52**, 857 (1970).
[27] C. Soave, R. Nucca, E. Sala, A. Viotti, and E. Galante, *Eur. J. Biochem.* **32**, 392 (1973).
[28] S. Altman and B. C. Stark, *Anal. Biochem.* **59**, 59 (1974).
[29] H. G. Martinson, *Biochemistry* **12**, 2731 (1973).
[30] H. G. Martinson, *Biochemistry* **12**, 2737 (1973).
[31] H. Hirano, T. Nishimura, and J. Iwamura, *Anal. Biochem.* **150**, 228 (1985).

flow rates and that has similar binding properties to those prepared originally by the method of Tiselius *et al.*[32]

Generally, the RNA is loaded on the column in 10–80 mM potassium phosphate buffer, pH 6.8, and is washed with several column volumes of this buffer to remove nonbound material. The RNA is eluted with a shallow linear or step gradient up to 0.20–0.25 M phosphate buffer. Most RNAs elute around 0.10–0.15 M phosphate buffer. The matrix can absorb up to 1 mg of calf thymus DNA/g dry wt of hydroxylapatite. Potassium phosphate is a more effective eluting agent than sodium phosphate.[25] The flow rate can range from 5 to 50 ml/cm^2/hr; slower flow rates give better resolution.[25] Care must be taken at the higher flow rates to avoid crushing the sorbent.

Compounds that have a stronger affinity for calcium than phosphate [ethylenediaminetetraacetic acid (EDTA), citrate, polyphosphates, etc.] decrease the adsorption capacity of hydroxylapatite, while ions with a low affinity for calcium (chlorides) have little affect.[25] The addition of urea, in some instances, can improve the resolution.[33,34] Zimmer and Hartmann found that, at pH 5.4, tRNA sticks more firmly to the sorbent, probably as a result of the increased protonation of the phosphates.[35] However, this did not affect the resolution of the tRNA. At pH 7.8, tRNA was loosely bound. They conclude that pH 6.8 was optimum, although anything that affects the structure of the nucleic acid can alter its affinity for hydroxylapatite. Increasing the structural order of polynucleotides, either through base pairing or base stacking, is correlated with an increased affinity for the sorbent.[29] A temperature gradient has also been used to elute nucleic acids,[25,36] but care should be taken, because hydroxylapatite has been found to catalyze the hydrolysis of RNA at elevated temperatures in the presence of low-ionic-strength buffers.[37] Hydroxylapatite is very soluble in 25% trichloroacetic acid, and this can be used as a way to recover nucleic acids that are strongly adsorbed on the matrix.[38] Since phosphate buffers have a low solubility in ethanol, nucleic acids can be recovered by precipitation with the cationic detergent cetylpyridinium bromide.[39]

[32] A. Tiselius, S. Hjerten, and O. Levin, *Arch. Biochem. Biophys.* **65**, 132 (1956).
[33] T. R. Johnson and J. Ilan, *Anal. Biochem.* **132**, 20 (1983).
[34] K. W. Mundry, *Bull. Soc. Chim. Biol.* **52**, 873 (1970).
[35] A. Zimmer and G. R. Hartmann, *Bull. Soc. Chim. Biol.* **52**, 867 (1970).
[36] T. Sindhuphak, V. Patel, and I. Svensson, *J. Chromatogr.* **314**, 408 (1984).
[37] H. G. Martinson and E. B. Wagenaar, *Biochemistry* **13**, 1641 (1974).
[38] M. Kiper, *Anal. Biochem.* **91**, 70 (1978).
[39] P. Geck and I. Nasz, *Anal. Biochem.* **135**, 264 (1983).

Reversed-Phase Chromatography

Reversed-phase chromatography (RPC) was originally developed as a technique to fractionate tRNAs.[40,41] Early versions of this matrix (RPC-1 through RPC-4) consisted of organic, water immiscible, quaternary ammonium compounds that were thinly coated on diatomaceous earth supports. These were later replaced by RPC-5, which consisted of Adogen 464 (trialkylmethylammonium chloride) coated onto Plaskon (polychlorotrifluoroethylene) beads.[42] RPC-5 provides substantially better resolution than the earlier versions and was widely used to fractionate tRNAs[40,43–46] and ribosomal RNAs.[44–46] The feasibility of using RPC-5 columns to purify mRNAs has also been demonstrated.[47] Most recently, RPC-5 has been a popular alternative to gel electrophoresis for obtaining large quantities of purified plasmid DNA, restriction fragments, and oligonucleotides.[45,48,49] The mild conditions used make it useful for purifying biologically active RNA molecules.[47]

The name "reversed-phase chromatography" is something of a misnomer, because RNA is eluted from the RPC columns, with increasing salt concentrations, by increasing molecular weight. Thus, fractionation seems to be based predominately on ion-exchange interactions. However, the ability of RPC columns to fractionate nucleic acids of similar molecular weights indicates that secondary factors are involved as well.[45] Slight changes in the hydrophobicity of a tRNA can significantly alter its chromatographic behavior on a RPC-5 column.[50]

Despite its popularity, RPC-5 has several disadvantages which have limited its widespread use. The most serious of these is that the Plaskon bead is no longer commercially available. Other problems encountered were the leaching, or bleeding, of the organic layer from the column, the high operating pressures required (up to 500 psi), and the variability be-

[40] A. D. Kelmers, H. O. Weeren, J. F. Weiss, R. L. Pearson, M. P. Stulberg, and G. D. Novelli, this series, Vol. 20, p. 9.
[41] A. D. Kelmers, G. D. Novelli, and M. P. Stulberg, J. Biol. Chem. 240, 3979 (1965).
[42] R. L. Pearson, J. F. Weiss, and A. D. Kelmers, Biochim. Biophys. Acta 228, 770 (1971).
[43] A. E. Hampel and B. Ruefer, this series, Vol. 59, p. 218.
[44] R. M. Kothari and M. W. Taylor, J. Chromatogr. 86, 289 (1973).
[45] J. A. Thompson, R. W. Blakesley, K. Doran, C. J. Hough, and R. D. Wells, this series, Vol. 100, p. 368.
[46] A. D. Kelmers, D. E. Heatherly, and B. Z. Egan, this series, Vol. 29, p. 483.
[47] C. Campbell, S. M. Arfin, and E. Goldman, Anal. Biochem. 102, 153 (1980).
[48] R. D. Wells, J. Chromatogr. 336, 3 (1984).
[49] R. D. Wells, S. C. Hardies, G. T. Horn, B. Klein, J. E. Larson, S. K. Neuendorf, N. Panayotatos, R. K. Patient, and E. Selsing, this series, Vol. 65, p. 327.
[50] M. A. Wosnick and B. N. White, Biochim. Biophys. Acta 561, 194 (1979).

tween batches of the matrix, which limited its reproducibility. A number of alternatives to Plaskon have been developed, which are readily available and involve little preparation. Examples of Adogen 464 coated onto various matrices include Sephadex LH (luteinizing hormone)[51]; siliconized, porous, glass beads[52]; octadecyl silane (ODS)–Hypersil[53]; Kel-F[54]; and Voltalef.[55] These matrices have similar properties and running conditions to RPC-5, but often lack many of its disadvantages.

Another new matrix was developed by Thompson and co-workers called RPC-5 ANALOG.[45] This column consists of a thin film of trioctylmethylammonium chloride on an inert, noncompressible, microparticulate resin. RPC-5 ANALOG is available currently from Bethesda Research Laboratories (Bethesda, MD) under the name NACS (nucleic acid chromatography system). NACS comes in several discrete particle sizes that are suitable for gravity flow, peristaltic pump, and HPLC applications. This material has less bleeding and is more reproducible than the RPC-5 columns. RPC-5 ANALOG can bind up to 0.7 mg of calf thymus DNA/g of sorbent. RPC-5 ANALOG substitutes directly for essentially all the nucleic acid purification procedures that were developed for RPC-5.[45]

Various column sizes and running conditions have been used. Some of these are summarized elsewhere.[40,43–45] In general, tRNAs are loaded on the RPC-5 column (RPC-5 ANALOG or NACS) in a buffer containing 10 mM sodium acetate, pH 4.5, 10 mM MgCl$_2$, and 0.2–0.4 M NaCl. The RNA is eluted with a linear salt gradient up to 1 M NaCl in this buffer. Ribosomal RNAs elute at a higher salt concentration than tRNAs. The resolution is dependent on the column length, which in turn limits the flow rates that can be obtained. Since separation is also dependent on the base sequence and on the secondary and tertiary interactions of the RNA, the pH, divalent cation concentration, and temperature can be varied to obtain optimum resolution. (pH 4.5 was originally used to avoid deacylation of the tRNAs.) Higher temperatures (37°) give better resolution than lower temperatures (4°). Large single-stranded RNAs (>100 nucleotides) do not readily elute from the column, unless they are applied at high-salt concentrations (0.5 M NaCl); smaller RNAs (<100 nucleotides) can be applied to the column with 0.1–0.2 M NaCl.[45] For best separations, care should be taken to remove proteins and organic solvents from the sample prior to loading on the column.

[51] G. Nowak, A. Leonowicz, and J. Trojanowski, *Anal. Biochem.* **94**, 48 (1979).
[52] T. Narihara, Y. Fujita, and T. Mizutani, *J. Chromatogr.* **236**, 513 (1982).
[53] R. Bischoff, E. Graeser, and L. W. McLaughlin, *J. Chromatogr.* **257**, 305 (1983).
[54] B. W.-K. Shum and D. M. Crothers, *Nucleic Acids Res.* **5**, 2297 (1978).
[55] J. M. Flanagan, R. K. Fujimura, and K. B. Jacobson, *Anal. Biochem.* **153**, 299 (1986).

Hydrophobic Interaction Chromatography

High-salt concentrations reduce the solubility of RNAs and favor hydrophobic interactions between the RNA and the sorbent. This is the basis for hydrophobic interaction chromatography (HIC). RNA is loaded onto an unsubstituted agarose column at high $(NH_4)_2SO_4$ concentrations and is eluted with a decreasing salt gradient in order of increasing hydrophobicity. Various tRNA species have been fractionated on a Sepharose 4B column using this technique.[56,57] However, a fractionation mechanism based strictly on hydrophobic interactions is not consistent with the negative temperature dependence observed for tRNA binding.[58] In addition, Spencer and Binns found that the elution profile of tRNAs was dependent both on the gradient slope and on the column length.[59,60] They proposed that these observations could be explained only if two different fractionation mechanisms were involved. The first mechanism is a salt-dependent precipitation of the RNA at the interface of the column. The second mechanism is an adsorptive retardation of the RNA, probably through hydrophobic interactions, once it has become solubilized in the decreasing salt gradient.

Generally, the RNA is loaded on this column with 1.3–2.0 M $(NH_4)_2SO_4$ in a buffer containing 10 mM sodium acetate, pH 4.5, 10 mM $MgCl_2$, and 1 mM EDTA at 4°. The RNA is eluted with a decreasing $(NH_4)_2SO_4$ gradient in this buffer. The concentration of $(NH_4)_2SO_4$ necessary for elution is dependent on the RNA, column length, and temperature.[59] In addition, there are considerable variations between batches of Sepharose 4B.[59,61,62] The interfacial precipitation of the RNA is decreased at high pH values and at higher temperatures; however, good separations are still obtained if longer columns and higher initial concentrations of $(NH_4)_2SO_4$ are used.[58,60]

Other matrices are also suitable for HIC. Ultrogel AcA 44 is a useful alternative to Sepharose 4B.[58] Nitrocellulose columns have been used with a decreasing NaCl gradient to separate 18 and 28 S RNAs[63] and, in some cases, have been found to be superior to agarose.[64] A decreasing

[56] W. H. Holmes, R. E. Hard, B. R. Reid, R. A. Rimerman, and G. W. Hatfield, *Proc. Natl. Acad. Sci. U.S.A.* **72**, 1068 (1975).

[57] V. Colantuoni, L. Guarini, and R. Cortese, *Anal. Biochem.* **93**, 248 (1979).

[58] C. J. O. R. Morris, *J. Chromatogr.* **159**, 33 (1978).

[59] M. Spencer and M. M. Binns, *J. Chromatogr.* **238**, 297 (1982).

[60] M. Spencer, *J. Chromatogr.* **238**, 307 (1982).

[61] M. Spencer, *J. Chromatogr.* **238**, 317 (1982).

[62] V. Patel, U. Hellman, T. Sindhuphak, and I. Svensson, *J. Chromatogr.* **244**, 373 (1982).

[63] D. A. Popovic, *J. Chromatogr.* **236**, 234 (1982).

[64] D. A. Popovic and M. Wintzerith, *J. Chromatogr.* **268**, 285 (1983).

$(NH_4)_2SO_4$ gradient has been used with a TSK Phenyl-5PW column to separate tRNAs and 16 and 23 S rRNAs by HPLC.[65] Other examples of HPLC applications of HIC are reviewed elsewhere.[2,5,66]

$(NH_4)_2SO_4$ is not very soluble in ethanol, so RNA collected by HIC cannot be selectively precipitated. Gillam and Tener have described a simple procedure for removing the $(NH_4)_2SO_4$.[67] tRNA collected from the HIC column is made 2 M in $(NH_4)_2SO_4$ and loaded onto a small column of DEAE-cellulose. The $(NH_4)_2SO_4$ is then washed off the column with 0.3 M NaCl in a buffer containing 50 mM sodium acetate, pH 4.5, and 10 mM $MgCl_2$. The tRNA is eluted with 1.1 M NaCl in this buffer and is ethanol precipitated.

Affinity Chromatography

Oligo(dT)-Cellulose

Affinity chromatography with small columns of oligo(dT)-cellulose has been a popular means of purifying mRNAs containing 3'-terminal poly(A) segments.[68–70] Oligo(dT)-cellulose is commercially available in several grades from several manufactors. T3-grade oligo(dT)-cellulose from Collaborative Research (Bedford, MA) is commonly used for analytical purification of RNAs.[1] It will bind over 4 mg of RNA/g of oligo(dT)-cellulose. It consists of oligomers up to 30 nucleotides long that are attached by their 5' phosphate to a matrix of fibrous cellulose. The matrix-bound oligo(dT) forms stable base pairs with poly(A)-containing RNAs in the presence of high-ionic-strength buffers, while proteins and nonpoly(A)-containing RNAs are washed off the column. The poly(A) RNA is eluted with low-ionic-strength buffers, which destabilize base pairings. The advantages of this column are that it is resistant to nucleases and alkali hydrolysis and that it has a much higher binding capacity than poly(U)-agarose columns.[1] Also, strong denaturants or high temperatures are not required to elute the sample. The disadvantages are that there can be nonspecific adsorption of RNAs and that mRNAs with short (less than 10–20 nucleotides) or missing poly(A) tails will not bind to the column.[69]

Generally, the sample is slowly loaded onto the column in a buffer containing 0.4–0.5 M NaCl, 10 mM Tris, pH 7.5, 0.5% (w/v) sodium

[65] Y. Kato, T. Kitamura, and T. Hashimoto, J. Chromatogr. **292**, 418 (1984).
[66] Z. E. Rassi and C. Horvath, J. Liq. Chromatogr. **9**, 3245 (1986).
[67] I. C. Gillam and G. M. Tener, Anal. Biochem. **105**, 405 (1980).
[68] H. Aviv and P. Leder, Proc. Natl. Acad. Sci. U.S.A. **69**, 1408 (1972).
[69] J. M. Taylor, Annu. Rev. Biochem. **48**, 681 (1979).
[70] H. Nakazato and M. Edmonds, this series, Vol. 29, p. 431.

dodecyl sulfate (SDS), and 1 mM EDTA. The column is washed with this buffer until the unbound RNA and proteins are removed, and the poly(A) RNA is eluted with a buffer containing low (0.1 M) or no salt. Applying the RNA sample slowly to the column (<2 ml/min) allows time for base pairs to form.[71] Even so, it may be necessary to pass the sample through the column several times in order to obtain complete binding. The binding capacity of the column increases with increasing salt concentration (up to 0.4–0.5 M NaCl), but nonspecific binding also increases.[72,73] The column capacity is greater when KCl is used instead of an equivalent concentration of NaCl, but again nonspecific binding is also greater.[72] Slow elution of the sample allows the RNA to be collected in a small volume.[72] Different mRNAs have been resolved or enriched by using an increasing thermal gradient[74,75] or decreasing salt gradient.[72] The resolution is dependent on the amount of RNA bound on the column.[71,72]

Cellulose alone will preferentially bind poly(A) RNAs.[70] This nonspecific binding is dependent on the salt concentration, the salt cation type, the presence of detergents, and on the source of cellulose. Ammonium and potassium cations cause more nonspecific binding than sodium or lithium cations.[72] Washing the column, prior to use, with 0.5 M NaCl saturated with SDS will greatly reduce the nonspecific binding, presumably by removing lignins present in the cellulose.[72] Nonspecifically bound poly(A) RNA will not be eluted completely from the column at low ionic strengths and must be removed by washing the column with 0.1 M NaOH.[76] Fortunately, this nonspecific binding is generally small and it is not of concern in preparative purifications. For analytical work, the nonspecific binding sites can be saturated with carrier RNA prior to loading the sample.[76]

Nonpoly(A)-containing RNAs will form aggregates with mRNAs and will be retained on the column. The formation of these aggregates is dependent on the type and concentration of the other RNAs[76] and on the salt concentration.[72,73] They can be disrupted by heating the sample for 5–10 min at 65°,[71,74] but they may reform prior to loading on the column, especially if high concentrations of rRNA are present. Bantle et al. have developed a two-step protocol for eliminating the RNA aggregates.[76] First, most of the contaminating RNA is removed by an initial passage through the oligo(dT) column. The poly(A) RNA is eluted from the

[71] S. Nadin-Davis and V. A. Mezl, *J. Biochem. Biophys. Methods* **11**, 185 (1985).
[72] J. F. B. Mercer and H. Naora, *J. Chromatogr.* **114**, 115 (1975).
[73] S. Nadin-Davis and V. A. Mezl, *Can. J. Biochem. Cell Biol.* **61**, 353 (1983).
[74] R. E. Rhoads, *J. Biol. Chem.* **250**, 8088 (1975).
[75] C. R. Astell, M. T. Doel, P. A. Jahnke, and M. Smith, *Biochemistry* **12**, 5068 (1973).
[76] J. A. Bantle, I. H. Maxwell, and W. E. Hahn, *Anal. Biochem.* **72**, 413 (1976).

column, ethanol precipitated, and resuspended in Tris buffer. The solution is made to contain 80% formamide, 0.1 M LiCl, 5 mM EDTA, and 0.2% (w/v) SDS. The sample is heated for 5 min at 55°, diluted 10-fold with a buffered solution containing 0.5 M NaCl, and reloaded onto the oligo(dT) column. RNA concentrations less than 2 mg/ml minimize aggregation.[71]

The binding properties of poly(A)-containing RNAs and ribonucleoprotein particles (RNPs) have been investigated as a function of monovalent and divalent ion concentration.[77] Poly(A) RNPs have similar binding properties to poly(A) RNAs in buffers containing 40 mM NaCl and 5 mM MgCl$_2$. However, the poly(A) RNPs remain attached to the column under conditions that completely remove the poly(A) RNAs.

Oligo(dT) has also been attached to glycerylated, porous glass.[78] This matrix has a smaller binding capacity than oligo(dT)-cellulose, but it has less nonspecific binding. It is a better matrix for purifying high-molecular-weight mRNAs. Higher flow rates are possible because the matrix is stronger than cellulose. Oligo(dA)-, oligo(dC)-, oligo(dG)-, and oligo(dI)-cellulose are commercially available for purifying other RNAs. An oligo(dT)-cellulose column has been used to synthesize and immobilize a DNA strand that was complementary to a particular mRNA.[79] The oligo(dT) was used as a primer for a specific, hybridized mRNA and reverse transcriptase. The cDNA made in this way was then used to enrich a crude sample for the complementary mRNA.

Poly(U)-Sepharose

Columns consisting of poly(U)-agarose have also been used to isolate poly(A) RNAs.[69,70,80,81] Poly(U)-Sepharose 4B from Pharmacia (Piscataway, NJ) is the most commonly used matrix. It consists of polyuridylic acid chains, which are ~100 nucleotides long, that have been coupled at multiple points to cyanogen bromide-activated Sepharose.[82,83] The fractionation principle is basically the same as with oligo(dT)-cellulose. The poly(A) RNA is bound to the column in the presence of high-ionic-strength buffers by forming A–U base pairs. The poly(A) RNA is eluted

---

[77] P. De Meyer, E. De Herdt, M. Kondo, and H. Slegers, *J. Biochem. Biophys. Methods* **2**, 311 (1980).
[78] T. Mizutani and Y. Tachibana, *J. Chromatogr.* **356**, 202 (1986).
[79] M. P. Vitek, S. G. Kreissman, and R. H. Gross, *Nucleic Acids Res.* **9**, 1191 (1981).
[80] U. Lindberg and T. Persson, *Eur. J. Biochem.* **31**, 246 (1972).
[81] U. Lindberg and T. Persson, this series, Vol. 34, p. 496.
[82] M. S. Poonian, A. J. Schlabach, and A. Weissbach, *Biochemistry* **10**, 424 (1971).
[83] A. F. Wagner, R. L. Bugianesi, and T. Y. Shen, *Biochem. Biophys. Res. Commun.* **45**, 184 (1971).

with a low-salt buffer containing denaturants, by raising the temperature, or by a combination of the two methods. Poly(U)-Sepharose 4B will bind ~150 μg of mRNA (OD_{260} ~5 U)/ml of gel.

The advantage of this sorbent is that the binding of the RNA is sensitive to the salt concentration, temperature, and presence of denaturants. Thus, the binding and elution conditions of the RNA can be varied readily to minimize the nonspecific adsorption of RNAs and to separate the different types of mRNAs. Furthermore, poly(U)-Sepharose columns are less susceptible to nonspecific adsorption than columns of oligo(dT)-cellulose.[84–86] The long poly(U) chains enable mRNAs with short, terminal poly(A) strands to bind,[1] but as with the oligo(dT)-cellulose column, mRNAs with missing or very short strands are lost.[69] Although the poly(U) is attached at multiple points to the matrix, the column's capacity is reduced with time by the action of nucleases and by hydrolysis, especially when elevated temperatures are used to elute the RNA. The main disadvantage of this column is that strong denaturants or high temperatures are required to elute the poly(A) RNA from the column.[72]

The generalized protocol for using this column is to slowly load the poly(A) RNA in a neutral pH buffer containing 0.2–0.5 M NaCl, 0.1–10 mM EDTA, and 0.2–1% (w/v) SDS. The column is washed with several column volumes of this buffer and then with a low-salt buffer to remove nonspecifically associated RNAs. The poly(A) RNA is eluted with 70–90% deionized formamide in a low-ionic-strength buffer. It may also be necessary to increase the temperature as well. Discontinuous elution has been used to fractionate different mRNAs.[87] Interestingly, Mercer and Naora found there was nonquantitative recovery of the bound mRNA, in buffers of low ionic strength, at temperatures well above the melting point of the poly(A) · poly(U) helix.[72] Quantitative recovery could be obtained only by eluting the RNA with 90% (v/v) formamide or at high temperatures (75°) in the presence of 0.5 M NaCl. These results suggest that there are probably secondary (ionic) interactions between the poly(A) RNA and the Sepharose column.

The poly(A) tails may not always be accessible for good binding to the column. Double-stranded structures and aggregates are generally destroyed, prior to loading the sample on the column, by heating the RNA at 65–70° and then by cooling rapidly.[86] However, it is not clear that this will effectively destroy intrastrand base pairing.

[84] D. J. Shapiro and R. T. Schimke, *J. Biol. Chem.* **250,** 1759 (1975).
[85] M. Atger and E. Migram, *J. Biol. Chem.* **252,** 5412 (1977).
[86] J. M. Taylor and T. P. H. Tse, *J. Biol. Chem.* **251,** 7461 (1976).
[87] J. W. Bynum and E. Volkin, *Anal. Biochem.* **107,** 406 (1980).

Nadin-Davis and Mezl found variable recovery of RNA by ethanol precipitation in the formamide solutions used to elute the RNA from the column.[88] The variability increased with increasing formamide concentrations, and there was an average of a 7% decrease in recovery with every 10% increase in the formamide concentration. The sample should be diluted to at least a 30% (v/v) formamide concentration before ethanol precipitation of the RNA.

Columns of poly(A)-, poly(C)-, poly(G)-, poly(I)-, and poly(I) · poly(C)-agarose are also commercially available. Poly(U)-cellulose has also been used to isolate poly(A) RNAs.[89] In contrast to poly(U)-Sepharose, RNA bound to poly(U)-cellulose elutes readily with low-salt-containing buffers and by increasing temperatures, but it has a smaller binding capacity, and the binding capacity is more rapidly lost.[72]

Boronyl Columns

Dihydroxyboryl affinity columns have been used to separate aminoacylated tRNAs from nonaminoacylated tRNAs,[18,90] to isolate 3'-terminal RNA fragments[91] and 5'-terminal RNA fragments containing 7-methylguanosine caps,[92] to purify small nuclear RNAs with 7-methyl caps,[93] to isolate Q base-containing tRNAs,[94] and to remove DNA from RNA transcription reactions.[95] The boronyl residue on the column forms a stable tetrahedral complex with *cis*-diols, such as the 2',3'-hydroxyls of the ribose sugar ring, at high pH values.[96,97] Molecules lacking a *cis*-diol are not retained on the column. At low pH values, the tetrahedral complex becomes unstable, and the bound RNA is eluted from the column.[97] Boronyl-substituted columns are commercially available on agarose, polyacrylamide, cellulose, and recently acrylic bead supports. Boronic acid-modified silica is also available, and it has been used with HPLC.[98] Each boronyl-substituted matrix has different loading capacities and binding properties that make them suitable for specific applications (see below).

[88] S. Nadin-Davis and V. A. Mezl, *Prep. Biochem.* **12**, 49 (1982).
[89] J. Kates, *Cold Spring Harbor Symp. Quant. Biol.* **35**, 743 (1976).
[90] H. Schott, E. Rudloff, P. Schmitt, R. Roychoudhurg, and H. Kossel, *Biochemistry* **12**, 932 (1973).
[91] M. Rosenberg and P. T. Gilham, *Biochim. Biophys. Acta* **246**, 337 (1971).
[92] R. E. Gelinas and R. J. Roberts, *Cell (Cambridge, Mass.)* **11**, 533 (1977).
[93] H.-E. Wilk, N. Kecskemethy, and K. P. Schafer, *Nucleic Acids Res.* **10**, 7621 (1982).
[94] T. F. McCutchan, P. T. Gilham, and D. Söll, *Nucleic Acids Res.* **2**, 853 (1975).
[95] S. Ackerman, B. Cool, and J. J. Furth, *Anal. Biochem.* **100**, 174 (1979).
[96] J. X. Khym, this series, Vol. 12, Part A, p. 93.
[97] H. L. Weith, J. L. Wiebers, and P. T. Gilham, *Biochemistry* **9**, 4396 (1970).
[98] E. Hagemeier, K.-S. Boos, and E. Schlimme, *J. Chromatogr.* **268**, 291 (1983).

Dihydroxyboryl columns are highly selective for cis-diol-containing molecules, but their use has been generally limited to the purification of tRNAs and other low-molecular-weight RNAs. The 3′-terminal ends of larger RNAs may not be generally accessible for complexation with the boronyl residue.[99] The generalized protocol for using these columns is to load the sample in 50 mM HEPES or N-methylmorpholine–HCl, pH 7.7–8.5, 10 mM MgCl$_2$, 0.2 M NaCl, and 20% (v/v) ethanol. Generally, the sample should be loaded at a flow rate of 1–3 column volumes/hr, or less, but this also varies with the type of RNA being purified.[94,95] The column is washed with several column volumes of this buffer to remove nonbound nucleic acids, and the bound RNA is eluted with 50 mM sodium acetate or sodium 2-(N-morpholine)ethane sulfonate, pH 5.0–5.5, and 0.2 M NaCl. Bound RNA can be eluted at a higher pH with competing cis-diols, like sorbitol or ethylene glycol,[94,100] although the effectiveness of this technique seems to depend on the characteristics of the bound RNA.[93,99] The capacity of the column is higher at 2–4° than at room temperature, but nonspecific binding also tends to be greater.[93,94]

The concentrations of the salts and ethanol are important for binding; no binding occurs in the absence of ethanol.[99] Ethanol has been found to be more effective than dimethyl sulfoxide, which requires higher salt concentrations.[94] Replacing the ethanol with urea or formamide reduces or eliminates binding.[93,99] Magnesium can be omitted, but higher concentrations of monovalent salt (0.5 M) are needed to ensure good binding capacity.[94]

Columns of N-[N'-(m-dihydroxyborylphenyl)succinamyl]aminoethyl cellulose (DBAE-cellulose) are acetylated to reduce the ionic binding of RNAs through residual charged aminoethyl groups on the column. This also lowers the pH required for binding (from pH 8.5 to 7.7).[94] The properties of DBAE-cellulose can vary and deteriorate with time, presumably because of the loss of the acetyl groups.[101] The boronyl group is also photosensitive.[99]

Pace and Pace have made extensive comparisons of the different boronyl matrices.[99] They found that acetylated DBAE-cellulose will bind macromolecular RNA (5 S rRNA and tRNA), but that it is not useful for trace amounts of RNA, because of nonspecific, and irreversible, adsorption. However, this nonspecific adsorption can be eliminated by saturating the sites with carrier RNA prior to using the column for the first time.[102] Pace and Pace were able to fractionate trace amounts of RNA on

[99] B. Pace and N. R. Pace, Anal. Biochem. **107**, 128 (1980).
[100] R. E. Duncan and P. T. Gilham, Anal. Biochem. **66**, 532 (1975).
[101] B. J. B. Johnson, Biochemistry **20**, 6103 (1981).
[102] M. Rosenberg, Nucleic Acids Res. **1**, 653 (1974).

a m-aminophenylboronic acid agarose column, but it was ineffective for mononucleotides. Boronate agarose also was the most effective column tested in distinguishing RNAs with 2',3'-hydroxyls from those with a 2'-hydroxyl and 3'-phosphate. A m-aminophenylboronic acid polyacrylamide column also had low, nonspecific adsorption. It bound cis-diol mononucleotides and short oligonucleotides, but it did not retain polyribonucleotides strong enough for preparative use.

Size-Exclusion Chromatography

Size-exclusion chromatography (SEC) is used to separate molecules according to their size. It is often referred to as gel filtration, gel permeation, or molecular-sieve chromatography. This column is a popular way to remove salts and unincorporated nucleotides and has been used to separate various RNAs.[103] There are various types of SEC matrices that are commercially available. They all work in the same way: the molecules are separated by their ability to permeate the pores of the matrix. Small molecules spend more of their time within the pores than larger molecules, which are more often excluded. Thus smaller molecules are retarded on the column and elute later than the larger molecules. The column matrices are available with different pore sizes so that RNAs of a particular size range can be resolved.

The main advantages of SEC are the mild elution conditions and the large capacity of the columns. However, very long columns (150 cm or more) and slow elution times (up to 80 hr or more) are often required to resolve various types of RNAs.[104] Even so, the resolution is generally poor. These problems have been alleviated somewhat by the recent introduction of several HPLC and FPLC columns. Dornburg et al. were able to resolve 4, 5, 9, 16, and 23 S RNAs on a 60-cm × 7.5-mm TSK G 4000SW HPLC column in less than 3 hr.[4] Unfortunately, these columns are expensive and have generally been limited to analytical work. The properties and running conditions of these columns have been described elsewhere.[2,4,105,106]

The buffers and operating conditions for SEC columns are quite variable and depend on the particular experimental need. Columns used to desalt RNA samples can be as small as 4–5 times the sample volume, and the column height to diameter ratio can be 10:1 or less.[1] Rather high flow rates (~20 ml/cm²/hr) can be used with these columns, but care should be

[103] F. O. Wetlstein and C. H. Noll, J. Mol. Biol. **11**, 35 (1965).
[104] K. M. Lee and A. G. Marshall, Prep. Biochem. **16**, 247 (1986).
[105] T. Andersson, M. Carlsson, L. Hagel, P.-A. Pernemalm, and J.-C. Janson, J. Chromatogr. **326**, 33 (1985).
[106] S. Uchiyama, T. Imamura, S. Nagai, and K. Konishi, J. Biochem. (Tokyo) **90**, 643 (1981).

taken to avoid compressing the gel.[1] A simple and rapid method for desalting a sample is described by Gorray and Quay.[107] Columns used to fractionate RNA samples should, if possible, be 50–100 times larger than the sample volume, and the column height to diameter ratio should be about 100:1.[1] Fast flow rates give poor resolution,[108,109] presumably because the molecules are less likely to enter the pores of the matrix. Also, long columns are particularly susceptible to compression at the faster flow rates. Clearly, a compromise must be made between the elution time and the resolution of the RNA. Flow rates are generally 2–10 ml/cm²/hr.

Resolution of the RNA is strongly dependent on the structural stability of the RNA, which in turn is dependent on the pH and ionic strength of the buffer, the base sequence of the RNA, and the temperature. In addition, the matrix of the column is not inert. At low ionic strengths, dextran columns have been found to separate RNAs by charge as well as by size.[110] At high salt concentrations, there can be hydrophobic interactions with the matrix.[108] Popovic and Leskovac were able to take advantage of the selective retention of rRNAs by hydrophobic interactions to separate four major classes of nucleic acids on a Sepharose CL column[111] (see Hydrophobic Interaction Chromatography). Generally, salt concentrations of 0.2–0.5 M are used to minimize the ionic and hydrophobic interactions. Addition of 6–8 M urea, or other denaturants, to the buffer often improves the resolution,[108,110] although care should be taken to ensure that the matrix is compatible with the reagents.

Conclusions

Unfortunately, there is no single ideal sorbent for purifying RNAs; the purification strategy for a particular RNA can be determined only by experimenting with a number of different sorbents and elution conditions. In this chapter, I have compiled a number of the more widely used chromatographic methods. While it is by no means intended to be inclusive of all the possible chromatographic techniques available, it should provide the basic information needed.

Acknowledgments

I wish to thank Michael Clark and Shawn Westaway for critically reading this manuscript. This work was supported by NIH grant number GM11823.

[107] K. C. Gorray and W. B. Quay, *Anal. Biochem.* **82**, 69 (1977).
[108] B. Oberg and L. Philipson, *Arch. Biochem. Biophys.* **119**, 504 (1967).
[109] S. Hjerten, *Arch. Biochem. Biophys.* **99**, 466 (1962).
[110] T. Hohn and H. Schaller, *Biochim. Biophys. Acta* **138**, 466 (1967).
[111] D. A. Popovic and V. Leskovac, *Anal. Biochem.* **153**, 139 (1986).

[4] Synthesis of Long, Capped Transcripts *in Vitro* by SP6 and T7 RNA Polymerases

By Joel K. Yisraeli and Doug A. Melton

Introduction

The advent of the SP6/T7 *in vitro* transcription system has simplified and increased the detective sensitivity of a number of protocols involving DNA probes and has also enabled a number of experiments not previously possible. RNA probes offer advantages over DNA probes in virtually any hybridization assay, including Southern blots,[1] Nothern blots,[2,3] *in situ* hybridization,[4–7] or RNase mapping,[2,8] with a resultant higher hybrid melting temperature and increased detection of small amounts of the assayed material. The ability to synthesize *in vitro* large quantities of pure RNA transcripts, often highly unstable when isolated *in vivo,* has allowed great advances in our understanding of 3' end generation[9–12] and splicing.[13,14] Capped SP6/T7 transcripts are fully functional in *in vitro* translation systems[15–18]; *in vivo,* injected synthetic RNAs can generate alternative phenotypes as a result of directing ectopic overexpression of their

[1] G. M. Church and W. Gilbert, *Proc. Natl. Acad. Sci. U.S.A.* **81,** 1991 (1984).
[2] D. A. Melton, P. A. Krieg, M. R. Rebagliati, T. Maniatis, K. Zinn, and M. R. Green, *Nucleic Acids Res.* **12,** 7035 (1984).
[3] P. A. Krieg, M. R. Rebagliati, M. R. Green, and D. A. Melton, *in* "Genetic Engineering" (J. K. Setlow and A. Hollaender, eds.), p. 165. Plenum, New York, 1985.
[4] D. V. DeLeon, K. H. Cox, L. M. Angerer, and R. C. Angerer, *Dev. Biol.* **100,** 197 (1983).
[5] K. H. Cox, D. V. DeLeon, L. M. Angerer, and R. C. Angerer, *Dev. Biol.* **101,** 485 (1984).
[6] R. C. Angerer, K. H. Cox, and L. M. Angerer, *in* "Genetic Engineering" (J. K. Setlow and A. Hollaender, eds.), p. 43. Plenum, New York, 1985.
[7] D. A. Melton, *Nature (London)* **328,** 80 (1987).
[8] K. Zinn, D. DiMaio, and T. Maniatis, *Cell (Cambridge, Mass.)* **34,** 865 (1983).
[9] P. A. Krieg and D. A. Melton, *Nature (London)* **308,** 203 (1984).
[10] O. Georgiev, J. Mous, and M. Birnstiel, *Nucleic Acids Res.* **12,** 8589 (1984).
[11] R. Hart, M. McDevitt, and J. Nevins, *Cell (Cambridge, Mass.)* **43,** 677 (1985).
[12] D. Zarkower and M. Wickens, *J. Biol. Chem.* **263,** 5780 (1988).
[13] R. A. Padgett, P. J. Grabowski, M. M. Konarska, S. Seiler, and P. A. Sharp, *Annu. Rev. Biochem.* **55,** 1119 (1986).
[14] T. Maniatis and R. Reed, *Nature (London)* **325,** 673 (1987).
[15] P. A. Krieg and D. A. Melton, *Nucleic Acids Res.* **12,** 7057 (1984).
[16] H. Persson, L. Hennighausen, R. Taub, W. DeGrado, and P. Leder, *Science* **225,** 687 (1984).
[17] D. A. Mead, E. S. Skorupa, and B. Kemper, *Nucleic Acids Res.* **13,** 1103 (1985).
[18] D. R. Drummond, J. Armstrong, and A. Colman, *Nucleic Acids Res.* **13,** 7375 (1985).

METHODS IN ENZYMOLOGY, VOL. 180

protein products.[19,20] Injected antisense RNA transcripts can also produce specific phenotypes,[21,22] and RNAs synthesized with modified nucleotides have been instrumental in isolating certain proteins interacting with the RNA.[23] Recently, SP6 transcripts of a maternal mRNA have been shown to be recognized by the localization machinery of oocytes.[24] In short, the SP6/T7 transcription system provides a fast, simple way to generate hybridization probes and biologically active RNAs.

Many of the experiments mentioned above require full-length, 5'-capped transcripts. In this chapter, we review the basic technique for synthesizing such transcripts *in vitro*. The use of modified nucleotides is discussed as well. Many biochemical parameters of the transcription reaction, such as rNTP and buffer requirements, have been analyzed and discussed previously.[25]

Basic Transcription Reaction

The steps in making *in vitro* RNAs are outlined in Fig. 1. The sequence to be transcribed is cloned into the appropriate vector. The transcription reaction consists of a linear DNA template, ribonucleotide triphosphates, RNA polymerase, and, generally, a RNase inhibitor in a relatively simple buffer. These steps are discussed below.

Preparation of Template

An increasing number of plasmid vectors has been constructed to facilitate the cloning of sequences to be transcribed *in vitro* (Fig. 2). In these vectors, a bacteriophage promoter is located immediately upstream from a sequence containing multiple restriction enzyme sites. This polylinker sequence provides both a convenient site for inserting the DNA sequence to be transcribed as well as sites for linearization of the plasmid. More recently, a number of these vectors contain a second, different bacteriophage promoter flanking the polylinker on the other side, allowing transcription of the opposite strand (in the opposite direction). Linearization of the DNA determines which promoter remains proximal to the

[19] R. P. Harvey and D. A. Melton, *Cell (Cambridge, Mass.)* **53**, 687 (1988).
[20] A. Ruiz i Altaba and D. A. Melton, *Cell (Cambridge, Mass.)* **57**, 317 (1989).
[21] C. V. Cabera, M. C. Honso, P. Johnston, R. G. Phillips, and P. A. Lawrence, *Cell (Cambridge, Mass.)* **50**, 659.
[22] U. B. Rosenberg, A. Preiss, E. Seifert, H. Jackle, and D. C. Knipple, *Nature (London)* **313**, 703 (1985).
[23] P. J. Grabowski and P. A. Sharp, *Science* **233**, 1294 (1986).
[24] J. K. Yisraeli and D. A. Melton, *Nature (London)* **336**, 592 (1988).
[25] P. A. Krieg and D. A. Melton, this series, Vol. 155, p. 397.

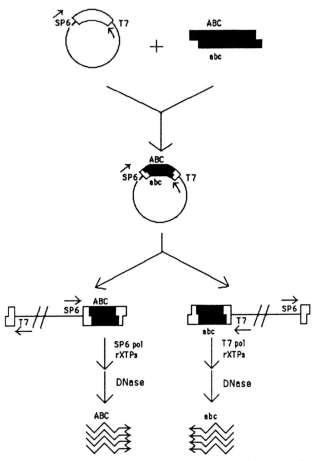

Fig. 1. General scheme for the *in vitro* synthesis of RNA with a bacteriophage RNA polymerase. The DNA sequence to be transcribed (the black box) is cloned into the polylinker region (open box) of a plasmid containing bacteriophage RNA polymerase promoters flanking this region. Generally, the plasmid is linearized by a restriction enzyme which cleaves in the polylinker; the DNA is now a template for the polymerase whose promoter remains proximal to the insert. Transcription of the template by the appropriate polymerase, in the presence of nucleoside triphosphates, yields a large molar excess of RNA transcripts to template. The template is then removed by RNase-free DNase. The orientation of the transcripts relative to the insert is represented by the letters above the transcripts and the insert.

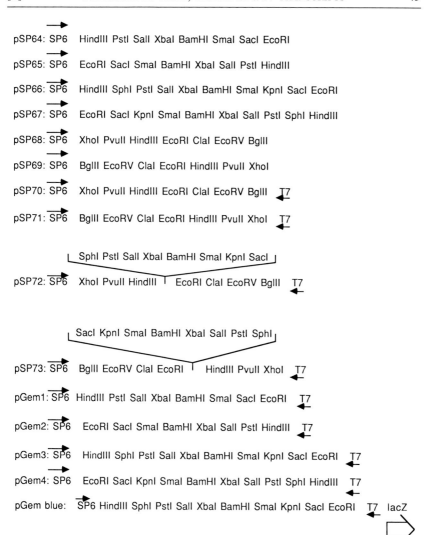

FIG. 2. A sampling of transcription vectors containing SP6 and/or T7 promoters. Restriction sites in the polylinker of a number of transcription vectors containing the SP6 and/or T7 promoters are shown. The direction of transcription is indicated by the arrows. Many of the vectors listed were made in the laboratory, and a number of them, along with others, are commercially available. The pGem blue vector allows color selection of recombinants by detecting the interruption of the *lacZ* gene (indicated by the open arrow).

insertion and is then capable of directing transcription of the cloned DNA in the presence of the appropriate RNA polymerase.

Plasmids containing the cloned DNA sequence can be grown and purified using any of the common techniques.[26–28] Because supercoiled plasmids containing a bacteriophage promoter are very efficiently transcribed, yielding large, often multimeric transcripts, it is very important to cleave all of the template to completion. Although this is usually achieved by linearizing the template plasmid, multiple cuts do not affect the transcription reaction as long as the promoter remains contiguous with the DNA sequence to be transcribed. Any residual RNase activity is removed by stopping the restriction enzyme with 200 μg/ml of proteinase K for 30 min at 37° followed by a phenol–chloroform extraction and ethanol precipitation. Templates are routinely resuspended at 1 μg/μl in diethyl pyrocarbonate (DEPC)-treated water or 10 mM Tris, pH 7.5, 1 mM ethylenediaminetetraacetic acid (EDTA).

RNA Synthesis

The optimal conditions for *in vitro* transcription by the SP6 RNA polymerase of DNA cloned downstream of a SP6 promoter have been determined.[2,29]

40 mM Tris–HCl, pH 7.5
6 mM MgCl$_2$
2 mM Spermidine–HCl
10 mM Dithiothreitol
(Optional: 100 μg/ml bovine serum albumin)
100 μg/ml DNA template
500 μM Each ATP, CTP, GTP, and UTP
(Optional: 1–10 μCi of radiolabeled rNTP, usually [α-^{32}P]UTP at 400 Ci/mmol)
1 U/μl Human placental ribonuclease inhibitor
100 U/ml SP6 RNA polymerase

We have found that the T7 polymerase works well under these conditions as well. In practice, the Tris–HCl, MgCl$_2$, and the spermidine–HCl are combined in a 10× stock, which can be stored frozen for extended periods of time. It should be noted that the polymerases are generally supplied in

[26] T. Maniatis, E. F. Fritsch, and J. Sambrook, "Molecular Cloning: A Laboratory Manual." Cold Spring Harbor Laboratory, Cold Spring Harbor, New York, 1982.
[27] H. C. Birnboim and J. Doly, *Nucleic Acids Res.* **7,** 1513 (1979).
[28] J. T. Lis, this series, Vol. 65, p. 347.
[29] E. T. Butler and M. J. Chamberlin, *J. Biol. Chem.* **257,** 5772 (1982).

a 200 mM KCl buffer, so the final salt concentration in the buffer is 2–10 mM. The nucleotides can also be stored as a frozen cocktail. All the reagents are prepared with DEPC-treated water. Because concentrations of spermidine above 4 mM can precipitate DNA at 0–4°,[30] the reaction is assembled at room temperature, adding the 10× buffer after most of the other components. Although 40° is the optimal temperature for the reaction, incubation at 37° generally yields less incomplete transcripts with only a negligible drop in yield.[25] Normally, the reaction is allowed to proceed for 1 hr, yielding approximately 8–10 mol of RNA transcript/mol of DNA template. Addition of another aliquot of the enzyme after this period and another hour of incubation will approximately double this yield; in general, under optimal conditions, the template and the polymerase are limiting.[2]

It is often convenient to include trace amounts of a radiolabeled ribonucleotide triphosphate in the reaction. This allows for easy quantification of the amount of RNA synthesis or incorporation. Due to the dilution of the radionucleotide by unlabeled nucleotide, however, the specific activity is rather low, on the order of 1×10^6 cpm/μg. Much higher specific activities can be obtained by lowering or eliminating the amount of the unlabeled nucleotide, by using a higher specific activity nucleotide, or by using more than one labeled nucleotide in the reaction. The highest concentration of the labeled nucleotide obtainable in the reaction mix with commercially available radionucleotides is well below the K_m of the reaction; there can be problems, therefore, with incomplete transcripts. In addition, radiolysis can also be quite a severe problem if the RNA is not used within 1 day. Generally, RNA transcripts with a specific activity of up to 3×10^7 cpm/μg pose a minimum problem with breakdown and can be synthesized by reducing the volume of the reaction to 5–10 μl (e.g., 4 μl of [α-^{32}P]UTP at 10 μCi/μl and 400 Ci/mmol in a 10-μl reaction with 75 mM of unlabeled UTP yields transcripts with a specific activity of approximately 3×10^7 cpm/μg).

After the reaction is completed, the DNA template is digested by the addition of RNase-free DNase (generally obtained commercially, e.g., from Promega, Madison, WI) to a final concentration of 20 μg/ml for 10 min at 37°. The reaction is brought up to a volume of at least 100 μl and is extracted with an equal volume of phenol–chloroform. Unincorporated nucleotides are generally removed by passing the reaction over a Sephadex G-50 spin column in 10 mM Tris–HCl, pH 7.5, and 0.1% (w/v) sodium dodecyl sulfate (SDS). The RNA is precipitated with 0.7 M ammonium acetate and 2.5 volumes of ethanol.

[30] B. C. Hoopes and W. R. McLure, *Nucleic Acids Res.* **9,** 5493 (1981).

Synthesis of Capped Transcripts

Both *in vivo* and *in vitro*, many reactions involving RNA require or are more efficient with transcripts that are capped at their 5' ends.[31,32] In addition, a 5' cap is essential for the stability of RNA injected into oocytes.[15,18,33] Although *in vitro* synthesized RNAs can be capped enzymatically by guanylyl transferase,[3,33,34] this reaction can be quite expensive for large amounts of RNA. Alternatively, RNAs can be synthesized in the presence of a cap analog.[32,35] By including the proper concentration of the cap analog in the reaction, one can obtain an essentially pure population of capped molecules with virtually no loss of efficiency.

The basic reaction mix described above is modified by changing the concentration of GTP and including the cap analog in the reaction.

40 mM Tris–HCl, pH 7.5
6 mM MgCl$_2$
2 mM Spermidine–HCl
10 mM Dithiothreitol
  (Optional: 100 μg/ml bovine serum albumin)
100 μg/ml DNA template
500 μM Each ATP, CTP, and UTP
50 μM GTP
500 μM G(5')ppp(5')G
  (Optional: 1–10 μCi of radiolabeled rNTP, usually [α-^{32}P]UTP at 400 Ci/mmol)
1 U/μl Human placental ribonuclease inhibitor
100 U/ml SP6 or T7 RNA polymerase

The cap analog can be obtained commercially (e.g., from PL Biochemicals, Milwaukee, WI) as a dry salt which should be resuspended in 10 mM Tris–HCl, pH 7.5. The SP6 and T7 RNA polymerases normally initiate at a G residue. As shown in Fig. 3, the 10-fold molar excess of G(5')ppp(5')G over GTP results in the incorporation of the cap analog by the SP6 or T7 polymerase in virtually all of the transcripts. Other cap analogs, such as mG(5')ppp(5')G and mG(5')ppp(5')mG, appear to be incorporated equally well. Labeled, capped transcripts of low or high specific activity can be synthesized in an analogous fashion to the method described in the pre-

[31] A. R. Krainer, T. Maniatis, B. Ruskin, and M. R. Green, *Cell (Cambridge, Mass.)* **36**, 993 (1984).

[32] M. M. Konarska, R. A. Padgett, and P. A. Sharp, *Cell (Cambridge, Mass.)* **38**, 731 (1984).

[33] M. R. Green, T. Maniatis, and C. Melton, *Cell (Cambridge, Mass.)* **32**, 681 (1983).

[34] G. Monroy, E. Spenser, and J. Hurwitz, *J. Biol. Chem.* **253**, 4481 (1978).

[35] R. Contreras, H. Cheroutre, W. Degrave, and W. Fiers, *Nucleic Acids Res.* **10**, 6353 (1982).

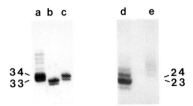

FIG. 3. Incorporation of the cap analog into SP6 and T7 transcripts. Transcription reactions were performed in the presence of labeled nucleotide as described in the text using either the SP6 (lanes a–c) or T7 (d–e) RNA polymerase. In the case of the SP6 polymerase, the template was cut by *Dde*I, which should have produced transcripts that were 33 nucleotides long. In the case of the T7 polymerase, the template was cut by *Rsa*I and should have produced transcripts 23 nucleotides long. When no cap analog was used in the reaction (lanes b and d), the major transcript is the expected size; the additional, larger transcripts detected with the *Rsa*I template (lane d) may be the result of adding nontemplate nucleotides onto the end of a transcript from a blunt-end template (see Ref. 2 for a discussion of how the 3′ end of the template influences the transcripts). The addition of G(5′)ppp(5′)G in a 10-fold molar excess over GTP results in a one nucleotide shift in the size of all the transcripts, both major and minor, as a result of the incorporation of the dinucleotide (lanes c and e). Incubation of the transcripts from lane b, not containing any dinucleotide, with guanylyl transferase also causes the transcripts to increase by one nucleotide as a result of the transfer of GTP to the 5′ end (lane a). Additional, larger transcripts are also detectable, presumably because the transferase can add on more than one G.

vious section (RNA Synthesis) without any additional effects on the amounts or completion of the transcripts.

A curious fact about this reaction is that, although the concentration of GTP has been reduced to well below its optimum (50 μM, as opposed to over 250 μM), full-length transcripts are produced to the same extent and with the same yield as when no cap analog is included and the GTP concentration is higher. In fact, under the conditions described above, the amount of RNA synthesized exceeds the amount which can be synthesized from just the GTP in the reaction alone. Apparently, the dinucleotide cap analog can function as a substrate for chain elongation as well as for transcription intiation.

Synthesis of Modified RNA Transcripts

RNA transcripts containing modified nucleotides have been used in a variety of ways. RNAs containing biotinylated nucleotides can be used as nonradioactive probes in any sort of hybridization assay in place of DNA probes. In addition, biotinylated RNAs will complex with spliceosomes and can subsequently be isolated with avidin or streptavidin.[23] RNAs containing bromo-UTP or thio-UTP nucleotides can be ultraviolet (UV) cross-linked at long wavelengths (300 nm or greater) with high efficiency to proteins which are bound to it.[36]

Synthesis of these modified RNAs is very straightforward. In general, the modified nucleotide is included in the reaction either replacing the unmodified nucleotide or in addition to it. UTP connected to a biotin moiety by an 11-carbon chain (from BRL, Bethesda, MD) is incorporated approximately half as efficiently as unmodified UTP. At high concentrations of biotin, however, the biotinylated RNA will separate to the interphase of a phenol–chloroform extract. This problem can be avoided by diluting the biotinylated UTP with unmodified UTP (~1 : 80).[23] Bromo-UTP or thio-UTP can also be substituted for UTP in the basic synthesis reaction, with approximately comparable yields (M. Hanna, personal communication).

General Comments

Incomplete transcripts, although not common, can occur for a variety of reasons. Often, ensuring that none of the nucleotides is limiting by simply increasing the concentration can yield significantly more full-length transcripts. Reducing the temperature can also increase the yield of full-length RNAs. As mentioned above, the reaction works almost as well at 37° as it does at 40°, with often a significant drop in short transcripts. In fact, reducing the temperature to 30° is often even more efficient at producing full-length transcripts, but the yield of RNA drops by about 50%.[25] A third possible source of incomplete transcripts is the presence of sequences in the transcribed region, which cause premature termination by the RNA polymerase. Because these sequences show some specificity for a particular polymerase, we have had some success in eliminating short stops by recloning the sequence into a vector where it can be transcribed by another polymerase.

Acknowledgments

We would like to thank Michael Green for his suggestion for monitoring the incorporation of the cap analog. This work was supported by a Charles Weizmann Postdoctoral Fellowship to Joel K. Yisraeli and a NIH grant to Doug A. Melton.

[36] N. K. Tanner, M. M. Hanna, and J. Abelson, *Biochemistry* **27**, 8852 (1988).

[5] Synthesis of Small RNAs Using T7 RNA Polymerase

By JOHN F. MILLIGAN and OLKE C. UHLENBECK

Introduction

Bacteriophage RNA polymerases are exceptionally active for *in vitro* transcription.[1] The high level of activity may be due in part to the fact that they are composed of a single polypeptide chain and thus do not employ a dissociating initiation factor. In addition, they initiate at a single site and terminate precisely. Although these polymerases have been shown to be more active on supercoiled templates,[2] they are also very active on linear templates. Perhaps most importantly, the T3, T7, and SP6 RNA polymerase genes have all been cloned, and the enzymes are available in large amounts.[3–5] Altogether, these considerations make this a good system to synthesize RNA *in vitro* for biochemical and biophysical studies.

This chapter will concentrate on using T7 RNA polymerase, although the enzymological properties of these three polymerases are apparently very similar.[1] In addition, the focus will be on making multimilligram amounts of oligoribonucleotides less than 100 nucleotides in length. This means that efficient initiation of transcription (the rate-limiting step) will be critical for a successful reaction. Both synthetic and cloned DNA templates will be discussed.

Materials

T7 RNA Polymerase. Large amounts of T7 RNA polymerase at high concentrations are needed for the many cycles of initiation required for the synthesis of short RNAs. Commercially available enzyme preparations are often too dilute for effective use. Fortunately, the purification of the polymerase from *Escherichia coli* containing the gene on an expression vector is relatively straightforward. In our experience, the *E. coli* strain BL21 containing the plasmid pAR219[4] yields at least 30 mg of puri-

[1] M. Chamberlin and T. Ryan, *in* "The Enzymes" (P. D. Boyer, ed.), 3rd ed., Vol. 15, p. 85. Academic Press, New York, 1982.
[2] S. P. Smeekens and L. J. Romano, *Nucleic Acids Res.* **14**, 2811 (1986).
[3] W. T. McAllister, unpublished results.
[4] P. Davanloo, A. Rosenberg, J. Dunn, and F. W. Studier, *Proc. Natl. Acad. Sci. U.S.A.* **81**, 2035 (1984).
[5] H. Kontani, Y. Ishizaki, N. Hiraoka, and A. Obayashi, *Nucleic Acids Res.* **15**, 2653 (1987).

METHODS IN ENZYMOLOGY, VOL. 180

fied enzyme/liter of cell culture.[6] We follow the early stages of the purification procedure of Davanloo *et al.*[4] including cell lysis in low salt, streptomycin sulfate precipitation of the DNA–T7 polymerase complex, high-salt wash of the precipitate to release the enzyme, and ammonium sulfate precipitation. At this stage, the enzyme is nearly 50% pure, and the major contaminant is fragmented DNA. T7 RNA polymerase absorbs to and elutes from phosphocellulose, diethylaminoethyl (DEAE)-cellulose, and blue dye columns.[1,4] We find that any two of these columns are generally sufficient to remove contaminating DNA and nucleases. Purified enzyme has a specific activity between 300,000 and 500,000 U/mg.[7]

In most transcription reactions, the final enzyme concentration is ~0.1 mg/ml, so a 1 mg/ml stock is convenient. The enzyme is stored at $-20°$ in 100 mM NH$_4$Cl, 50 μM ethylenediaminetetraacetic acid (EDTA), 1 mM dithiothreitol (DTT), 10 mM Tris–HCl (pH 8.6 at 20°), and 50% (v/v) glycerol. The enzyme is stable for long periods of time, provided that additional DTT is added every 6 months to prevent oxidative inactivation.

Synthetic DNA Templates. For a fully active template DNA strand, a fragment, containing the region from -17 to -1 of the class III T7 RNA polymerase promoter followed by the complement of the RNA sequence desired, must be made (Fig. 1A). DNA is synthesized on an automated synthesizer using either methoxy- or β-cyanoethyl-protected phosphoramidites. Since a 0.2-μmol column provides enough DNA to make up to 20 μmol of RNA, a 1.0-μmol column is generally not necessary. We find that, in order to minimize misincorporation and premature termination when transcribing synthetic DNA, it is essential to use more rigorous deblocking procedures than generally recommended. DNA should be incubated in concentrated ammonium hydroxide (14.8 N) at 55° for at least 12 hr to fully remove base-protecting groups. After removing the ammonium hydroxide under vacuum, the DNA is ready for purification.

If the DNA synthesis reaction gives a large fraction of full-length product, the DNA may be purified for transcription by simple ethanol precipitation. The dried DNA is resuspended in 0.25–1 ml of 0.5 M sodium acetate and 1 mM EDTA, pH 5.2, and 2.5–3 volumes of 95% (v/v) ethanol is added. After cooling to $-70°$, the DNA is centrifuged and the pellet is washed with 70% ethanol. The dried pellet is resuspended in 250 μl of TE buffer (10 mM Tris–HCl and 1 mM EDTA, pH 8.0) to give a final DNA concentration around 100 μM.

[6] P. Lowary, J. Sampson, J. Milligan, D. Groebe, and O. Uhlenbeck, *NATO ASI Ser. A* **110**, 69 (1986).
[7] M. Chamberlin and J. Ring, *J. Biol. Chem.* **248**, 2235 (1973).

A

-17   -1
TAATACGACTCACTATA
ATTATGCTGAGTGATATCCTGTACTCCTAATGGGTACA
                    +1 ➝

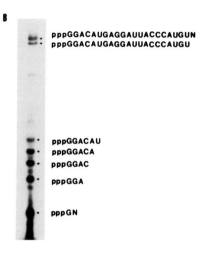

FIG. 1. (A) Synthetic DNA template, promoter region base paired from −17 to −1. Transcription starts at +1. (B) Autoradiograph of [γ-³²P]GTP-labeled transcription products from the template in A. N indicates a mixture of the four nucleotides.

If the DNA synthesis is not efficient or a trial transcription reaction does not give satisfactory results, the DNA can be further purified on a denaturing 20% (w/v) polyacrylamide gel. After electrophoresis to single nucleotide resolution, the DNA is localized by ultraviolet shadowing, and the band is cut out. The DNA is eluted in 0.5 M sodium acetate, pH 5.2, and is recovered by filtration through a paper frit and ethanol precipitation.[8]

In order to create a transcription template, a complementary strand must be annealed to the promoter portion of the template strand (Fig. 1A). Although a variety of lengths can be used,[9,10] the 17-nucleotide fragment corresponding to the −17 to −1 of the promoter is convenient. To prepare

[8] P. G. Sealey and E. M. Southern, in "Gel Electrophoresis of Nucleic Acids" (D. Rickwood and B. D. Hames, eds.), p. 39. IRL Press, Washington, D.C., 1982.
[9] J. F. Milligan, D. R. Groebe, G. W. Witherell, and O. C. Uhlenbeck, Nucleic Acids Res. 15, 8783 (1987).
[10] C. T. Martin and J. E. Coleman, Biochemistry 26, 2690 (1987).

a template, equal molar concentrations of the two strands are mixed in storage buffer, heated to 90° for 3 min, and cooled quickly on ice. When the DNA is not gel purified, a small (1.2-fold) excess of top strand is added to compensate for the lower extinction coefficient of the failure sequences. We find that, although active template often forms without the annealing step, certain sequences required annealing to prevent aberrant transcription products. Annealed templates are stored frozen in TE buffer at 0.5–5 μM and can be thawed and diluted without further treatment.

Plasmid DNA Templates. In some cases, it may be desirable to clone the sequence of interest downstream of a T7 promoter into a multicopy plasmid and use plasmid DNA, linearized with a restriction enzyme, as a transcription template. While this approach is more labor intensive than the direct use of synthetic DNA and does not produce templates of significantly higher quality,[9] it is not restricted by the length of the DNA fragment that can be synthesized.

Plasmids for transcription can be constructed by standard cloning procedures.[11,12] A variety of transcription vectors containing a T7 promoter upstream of a multiple cloning site is available. Alternatively, DNA synthesis can be used to place residues from −17 to −1 of the T7 promoter directly upstream of the desired RNA sequence.[13] This latter strategy prevents the introduction of unwanted 5'-terminal nucleotides, but places some restrictions on the first few nucleotides of the transcript.

Any one of a variety of available procedures for isolating plasmid DNA is adequate for producing transcription template DNA. The major requirements are that the DNA must be free of ribonucleases and pure enough to ensure complete cleavage by restriction enzymes. We prefer methods that use both phenol extraction and one cesium chloride–ethidium bromide density gradient.

Transcription templates are prepared by cleaving the plasmid DNA with a restriction enzyme of choice. Complete cleavage is essential, since small percentages of uncut plasmid will dramatically reduce yields by producing very long "runaround" transcripts that use up much of the available nucleoside triphosphates (NTPs). The type of restriction enzyme will also affect transcription yield and product quality. Enzymes, which leave 3'-overhanging ends, should be avoided, since they are prone to giving end-initiated products.[14] Enzymes, which leave blunt or 5'-over-

[11] D. E. Draper, S. A. White, and J. M. Kean, this series, Vol. 164, p. 221.
[12] D. A. Melton, P. A. Krieg, M. R. Rebagliati, T. Maniatis, R. Zinn, and M. R. Green, *Nucleic Acids Res.* **12**, 7035 (1984).
[13] J. R. Sampson and O. C. Uhlenbeck, *Proc. Natl. Acad. Sci. U.S.A.* **85**, 1033 (1988).
[14] E. T. Schenborn and R. C. Mierendorf, *Nucleic Acids Res.* **13**, 6223 (1985).

hanging ends, are both suitable, but show varying 3'-terminal heterogeneity of products.[9,11,12]

Nucleoside Triphosphates. Nucleoside 5'-triphosphates are available commercially as either sodium or lithium salts and are stored dry below $-20°$ until needed. Stock solutions are prepared by first dissolving each NTP individually in sterile, ribonuclease-free H_2O and adjusting the pH to 8.1 with NaOH. This step is critical, since poorly titrated NTPs can alter the pH of the transcription reaction and thereby lower yields. The individual NTPs are diluted to desired concentrations and are mixed in equimolar ratios to make a stock solution. Convenient stock solutions vary from 5 to 25 mM in each NTP and are stored frozen at $-20°$. Although stocks can be frozen and thawed without loss of activity, they slowly degrade, so fresh stocks should be made every several months.

[α-^{32}P]NTPs and [α-^{35}S]NTPs can be purchased as aqueous solutions from a variety of sources and used directly. If they are supplied in ethanol and will be used in large amounts, the ethanol should be removed prior to use. [γ-^{32}P]GTP can be used conveniently to 5' end label transcripts and can easily be prepared from inorganic [^{32}P]-phosphate.[15]

Buffers. The basic transcription buffer is made as a 10× stock solution that gives the following buffer when diluted.

40	mM	Tris–HCl (pH 8.1 at 37°)
6	mM	MgCl$_2$
5	mM	DTT
1	mM	Spermidine
0.01% (v/v)		Triton X-100
(50	μg/ml	Bovine serum albumin)

The stock solution is prepared by mixing autoclaved MgCl$_2$ and Tris–HCl to the next three components and sterilizing with a Nalgene 0.25-μm filter. Bovine serum albumin (Fraction V powder) is then added, and the stock is brought to the correct final volume. The addition of serum albumin to the reaction is optional and is generally omitted in large-scale reactions. The stock solution is stored at $-20°$ and can be thawed repeatedly, but eventually becomes oxidized. Since NTP concentrations are often varied, additional MgCl$_2$ must be added in order to keep the magnesium concentration greater than the NTP concentration. An autoclaved solution of 0.5 M MgCl$_2$ is useful for this purpose.

For large-scale transcription reactions, the addition of polyethylene glycol (PEG 8000, Sigma, St. Louis, MO) to a final concentration of 80 mg/ml can improve reaction yields. A 400 mg/ml of stock solution pre-

[15] R. A. Johnson and T. F. Walseth, *Adv. Cyclic Nucleotide Res.* **10**, 135 (1979).

pared with sterile water and stored at $-20°$ can be used (although it is difficult to pipet accurately).

Methods

Transcription Reactions. A set of transcription conditions that is generally successful for the preparation of short RNAs contains the following components.

Transcription buffer (see *Buffers*)	1×
NTPs (each)	1 mM
Synthetic DNA template	100 nM
(or plasmid DNA template	50 nM)
T7 RNA polymerase	0.1 mg/ml

All the stock solutions, except the polymerase, are brought to room temperature prior to mixing to avoid precipitating the DNA template by spermidine and MgCl$_2$. After the addition of enzyme, reactions are incubated at 37°. After 1–2 hr, no additional product is made.

Transcription products labeled with $[\gamma$-$^{32}P]GTP$ and analyzed by gel electrophoresis are shown in Fig. 1B. By using this radioisotope, the amount of label is proportional to the molar yield. Two or sometimes three major products are observed at the size expected for a full-length runoff transcript. In addition, substantial amounts of shorter products, ranging from two to six nucleotides, are inevitably present. All the products appear in constant proportion during the course of the reaction and do not appear to interconvert, suggesting that they are independent outcomes of transcription[9] and not degradation products.

Sequencing the longer products in Fig. 1B reveals that the shorter of the two corresponds to the expected full-length transcript, whereas the longer one is the same molecule with an extra 3′-terminal nucleotide of variable identity. Sequencing of products from a variety of other templates reveals that, while the situation in Fig. 1B is generally the case, one occasionally obtains products with two or more nontemplate-encoded 3′ nucleotides. This situation is potentially even more complicated if the DNA template contains a substantial amount of shorter products. It is therefore recommended to determine the 3′ terminal nucleotide of transcription products[9] if its identity is important for the application.

The 3′-terminal heterogeneity observed with synthetic DNA templates has also been reported with plasmid templates. We have attempted, without success, to find reaction conditions that would minimize the heterogeneity. Comparison of a variety of different 3′-terminal sequences, generated either by restriction enzymes or with synthetic templates, did not reveal any useful regularities. In fact, it appears that the extent of 3'-terminal heterogeneity may depend on the sequence (or structure) of the entire template. For example, a *Bst*NI-cut plasmid that contains the yeast tRNA[Phe] sequence gives a nearly homogeneous, correct 3' terminus,[13] whereas multiple 3' termini are observed with transcripts terminating at other *Bst*NI sites.[16]

The smaller products observed in Fig. 1B correspond to the series of oligonucleotides that initiate at +1 and terminate prematurely. Although such abortive initiation products have been described with *E. coli* RNA polymerase on certain promoters,[17,18] they are much more prevalent with T7 RNA polymerase on every promoter tested.[19] While their amount and length distribution varies from promoter to promoter, it is quite common for less than one initiation event in ten to result in a full-length transcript. For example, a 20-μl reaction (Fig. 1B) yielded 140 pmol of the product doublet and about 1000 pmol of the six abortive products. As a result, nearly half of the incorporated triphosphates appears in the abortive products. Attempts to alter the fraction of abortive products by varying reaction conditions have not been successful.

Optimizing Incorporation. Differences in template sequence and length can result in substantial differences in the optimal enzyme and template concentrations required for the production of RNA. In addition, one or more of the components may be more valuable than the others. It is, therefore, usually worthwhile to carry out trial reactions designed to find optimal reaction conditions for a given synthesis.

For the production of the largest amount of RNA in a given reaction volume, it is usually possible to raise the concentration of each NTP to as high as 4 mM. The $MgCl_2$ concentration should be adjusted to 6 mM above the total NTP concentration. Increasing the template concentration will result in a proportional increase in yield until a maximum is reached between 5 and 500 nM of template. Since additional DNA is not inhibitory and the amount of available template is rarely limiting, it is generally possible to find an optimal template concentration with only a few trial reactions. Reaction yield will also increase with increasing enzyme concentration until an optimum is reached between 0.01 and 0.2 mg/ml. However, when the enzyme concentration exceeds the optimum, much lower yields are encountered. Thus the enzyme concentration should be carefully varied to ensure that the optimum has been found. An optimized

[16] J. Sampson, personal communication.
[17] A. J. Carpousis and J. D. Gralla, *Biochemistry* **19**, 3245 (1980).
[18] J. R. Levin, B. Krummel, and M. J. Chamberlin, *J. Mol. Biol.* **196**, 85 (1987).
[19] C. T. Martin, D. K. Muller, and J. E. Coleman, *Biochemistry* **27**, 3966 (1988).

TABLE I
OPTIMAL REACTION CONDITIONS FOR LARGE-SCALE REACTIONS[a]

Reaction	Transcript length (nucleotides)	Synthetic or plasmid DNA	Concentration		Reaction		Reference
			Polymerase (U/μl)	DNA (nM)	Volume (ml)	Yield (mg)	
1	13	Synthetic	30	400	30	8.5	b
2	19	Synthetic	30	400	10	2.1	b
3	24	Synthetic	30	200	40	23.8	9
4	24	Synthetic	10	50	10	1.5	9
5	24	Synthetic	10	50	10	0.6	9
6	24	Synthetic	10	50	10	0.7	9
7	24	Synthetic	2	50	10	0.6	9
8	26	Synthetic	30	400	40	3.8	b
9	31	Plasmid	2	50	25	1.2	c
10	76	Plasmid	30	50	10	15.2	13
11	81	Plasmid	30	50	10	21.3	d

[a] All reactions were 4 mM in each NTP, 20 mM MgCl$_2$, and incubated at 37° for 4 hr, except reaction 9, which was 1 mM in each NTP, 6 mM MgCl$_2$. All reactions were purified by denaturing polyacrylamide gel electrophoresis (PAGE), except for reactions 3, 10, and 11, which were purified by high-performance liquid chromatography (HPLC), and reaction 9, which was purified by gel filtration chromatography. The final product of reaction 3 was heterogeneous, consisting of a 24- and 25-mer. All other products were homogeneous.

[b] J. R. Wyatt, personal communication; J. D. Puglisi, J. R. Wyatt, and I. Tinoco, *Nature* (*London*) **331**, 283 (1987).

[c] H. A. Heus, Ph.D. Thesis, Department of Biochemistry, State University of Leiden, Leiden, The Netherlands, 1987.

[d] J. Sampson and K. Hall, personal communication.

reaction is generally complete after 4 hr of incubation. Reducing the enzyme concentration and incubating longer is rarely successful.

Although performing trial reactions to optimize reaction conditions can be tedious, reaction yields can be substantially higher than those obtained using the general conditions given previously, and the reactions scale up readily. Table I gives the optimized reaction conditions and reaction yields for preparative transcription reactions from several different templates. Although the yield per milliliter of reaction volume can vary substantially among templates, large amounts of RNA can still be produced.

When a radiolabeled or modified NTP is used, it is usually desirable to modify the reaction conditions to optimize the incorporation of this valuable component. This can be done by reducing the concentration of the valuable NTP to as low as 50–100 μM. Although the K_m of most NTPs are estimated to be in the 100-μM range,[20] a high level of incorporation of these components occurs before the concentration drops too low to bind the enzyme. This strategy is, however, less effective if the rare nucleotide is one of the first nucleotides in the transcript because of the large number of abortive initiation products. For example, the transcript in Fig. 1B, which starts pppGGACAU, can be used to efficiently incorporate low concentrations of UTP and CTP, but not ATP and GTP, because so much of the latter two triphosphates are converted to short oligomers. When radiolabeled GTP or a modified GTP is to be used, it is a good idea to add GMP as a primer if low concentrations of GTP are to be used.

Purification of RNA Transcripts. Products of small-scale transcription reactions can be purified most conveniently by polyacrylamide–urea gel electrophoresis.[8] RNAs are located either by shadowing, ethidium bromide staining, or autoradiography.[21] After excision from the gel, the RNA is recovered by crush and soak or by electroelution methods.[21]

Large-scale transcription reactions often generate considerable amounts of a white precipitate (probably a magnesium–pyrophosphate complex) which first must be removed by centrifugation or by the addition of EDTA. Large-scale reactions are generally then concentrated by ethanol precipitation prior to further purification. Since the transcripts are contaminated with considerable concentrations of G-rich abortive initiation products, which can bind RNA, it is advisable to use denaturing conditions for purification.

We have found that DEAE-Fractogel (Bio-Rad, Richmond, CA) is a suitable resin for high-performance liquid chromatography (HPLC) purification of the transcript from these short RNAs. The crude sample is completely denatured by heating in 7 M urea, 10 mM Tris, 1 mM EDTA (pH 8.0). The sample is loaded onto the column, and the products are fractionated, using a gradient from 0.1 to 0.5 M NaCl, in the TE buffer containing 7 M urea. This resin will give satisfactory separation of the main products (N and $N + 1$) away from the small abortive initiation products. The main advantage of this system is the high capacity of the resin; 10 mg of full-length product can be separated successfully from the smaller products on a 25-ml column in a single run.

[20] J. L. Oakley, R. E. Strothkamp, A. H. Sarris, and J. E. Coleman, *Biochemistry* **18,** 528 (1979).
[21] T. Maniatis, E. F. Fritsch, and J. Sambrook, "Molecular Cloning: A Laboratory Manual," p. 149. Cold Spring Harbor Laboratory, Cold Spring Harbor, New York, 1982.

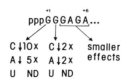

FIG. 2. Effect of sequence variations in the +1 to +6 region on reaction yield (shown here as the RNA sequence). ND, Not determined.

Discussion

The protocols outlined thus far have been successful in making hundreds of different small RNAs. During the course of our investigations, we have uncovered several variations in the protocol, which allow greater flexibility in the sequences that can be made. Further, many issues have arisen on specific details of the synthesis. In this section, we will attempt to comment on both topics.

Allowed Sequences in the +1 to +6 Region. The T7 consensus class III promoters are absolutely conserved from +1 to +6. However, many of the class II promoters differ in this region. Many nucleotide changes were made from +1 to +6 to determine their effect upon the final yield. In these cases, the sequences downstream of +6 were different, so templates of approximately the same length were compared. While not exact, this gives us a good estimate of the relative effect of base changes in the +1 to +6 region (Fig. 2).[9] These mutations have been assessed only with respect to the final yield and still await kinetic analysis.

The effects appear to be independent and additive. Furthermore, changes that place many AU base pairs in this region work poorly as well. For example, the sequence pppGGUU is approximately 5-fold down in a 76-nucleotide transcript of a tRNA.[16] Nevertheless, there are relatively few sequence constraints so that nearly any RNA can be made.

Priming of Transcription. Further sequence changes at the 5' end can be accomplished through the use of small primers, which will initiate transcription. Dimers of the type NpG work well to create a template with one extra nucleotide on the 5' end. The dimer ApG works the best, as it is complementary to the template strand.[22] Alternatively, if a 5'-triphosphate is undesirable, then either GDP or GMP can be used to initiate transcription. Typically, the primer is used at ~4 mM and the four triphosphates are at 1 mM. Under these conditions, >95% of the transcripts

---

[22] V. D. Axelrod and F. D. Kramer, *Biochemistry* **24**, 5716 (1985).

TABLE II
NTP SUBSTRATE SPECIFICITY OF T7 RNA POLYMERASE

NTP	Activity		Reference
[α-S]NTP (Sp)	+		24
[α-S]NTP (Rp)	−		24
5-BrUTP	+		—
5-FUTP	+		—
5-Hexamethyleneamino-UTP	+		—
6-Aza-UTP	+		—
4-Thio-UTP	+		—
Pseudo-UTP	+		—
8-BrATP	+		—
7-MeGTP	−		—
ITP + primer	+		22
ITP − primer	−		22
2'-dNTP	+/−		23
3'-dNTP	+	(Terminator)	22
3'-O-Me-NTP	+/−	(Terminator)	22

begin with the primer.[13] Presumably, this high level of primer incorporation is due to the rapid depletion of GTP (the normal initiating nucleotide)
by the abortive initiation process. Priming by trimers has been attempted
on a limited scale and has been only moderately successful. Tetramers
will not prime at all.

Modified NTPs as Substrates. An additional advantage of synthesizing RNAs with T7 RNA polymerase is that many modified NTPs are
substrates and can be incorporated. Table II lists those NTPs that have
been tested and whether they are active. Modified NTPs, which are non-
aromatic or block functional groups involved in base pairing, have not
been tested, since base pairing is essential for transcription. Inosine triphosphate (ITP) is a special case, since it cannot be used to initiate transcription, but will work in elongation. If a primer is used (ApG or GMP,
for example) to initiate transcription, then ITP works as a substitute for
GTP.[22] Deoxynucleotides will incorporate at a lower efficiency than the
corresponding ribonucleotides.[23] When a dNTP is used in place of a
rNTP, the polymerase will terminate ~30% of the time and read through
and incorporate the dNTP about 70% of the time. Thus, this technique
can be used to incorporate dNTPs if there is only one or two of a given
nucleotide present in the transcript. Since the preference for rNTPs is

[23] J. R. Wyatt, personal communication.

much higher than the dNTPs, it is essential to use very pure preparations of dNTPs. The Sp isomer of the [α-S]NTPs [nucleoside 5'-O-(1-thio-triphosphate)] is also a substrate for T7 polymerase and allows incorporation of modified phosphates. They are incorporated with an inversion of configuration such that the RNA contains the Rp isomer.[24]

Accuracy of Transcription. The accuracy of T7 RNA polymerase has not been carefully measured, but is likely to be high. Misincorporation is expected to be low on transcripts from cloned templates, based on aminoacylation levels of tRNAs.[13] There is some evidence that synthetic templates may generate RNA with a slight amount of misincorporated nucleotides. Self-cleaving RNAs (hammerhead structures) never cleave above 95%.[25] This may be due to misincorporation by the polymerase or to low levels of heterogeneity in the DNA templates. The subsequent RNA heterogeneity cannot be detected by any of the sequencing techniques we have used, and therefore, the incomplete cutting of the hammerhead is not understood.

T7 RNA polymerase has no trouble transcribing stretches of adenosines or uridines less than eight residues. However, when a region of greater than eight adenosines or uridines is encountered, the polymerase begins to slip, creating a series of molecules with N, $N + 1$, $N + 2$ (etc.) adenosines or uridines within the molecule.[26] This phenomenon has been seen with both cloned and synthetic DNA templates. This phenomenon does not occur through stretches of guanosines or cytidines and apparently is due to instability due to the presence of AU or UA base pairs, possibly in conjunction with pausing of transcription due to localized depletion of ATP and UTP. Thus, these regions should be avoided if possible.

Acknowledgment

This work was supported by the National Institutes of Health Grant GM 36944.

[24] A. D. Griffiths, B. V. L. Potter, and I. C. Eperon, *Nucleic Acids Res.* **15,** 4145 (1987).
[25] O. C. Uhlenbeck, *Nature* (*London*) **328,** 596 (1987).
[26] D. R. Groebe and O. C. Uhlenbeck, *Nucleic Acids Res.* **16,** 11725 (1988).

[6] *In Vitro* Synthesis of End-Mature, Intron-Containing Transfer RNAs

By VICENTE M. REYES and JOHN N. ABELSON

Systematic study of transfer RNA (tRNA) splicing depends on the investigator's ability to introduce mutations at will anywhere in the pre-tRNA (tRNA precursor) gene, transcribe the corresponding mutant pre-tRNA, and subject it to an *in vitro* splicing assay using purified tRNA splicing enzymes. A highly efficient and accurate *in vitro* splicing assay system for yeast tRNA precursors is currently available, consisting of highly purified yeast endonuclease and ligase enzymes.[1,2] However, there has been a lack of a system for the efficient synthesis of pre-tRNA molecules containing any predesigned mutation. Previously, end-mature pre-tRNA substrates were either isolated from [32P]RNA synthesized *in vivo* at the nonpermissive temperature in a temperature-sensitive yeast strain, RNA1 *ts* 136,[3] or from RNA transcribed *in vitro* with the appropriate DNA template, nucleotide triphosphates, and a crude wild type yeast nuclear extract (YNE). The *in vivo* method precludes design and construction of site-specific mutations in the pre-tRNA gene, is laborious, and involves working with high levels of radioactivity. In the YNE method, transcription and end processing take place simultaneously in the test tube. But certain sequences in the mature domain of the pre-tRNA are known to affect these steps as well as the stability of the product. Hence, mutations in these regions cannot be tested for their effects on splicing.

In this chapter, we present a method by which end-mature, intron-containing pre-tRNA is transcribed from a synthetic DNA template containing a bacteriophage T7 promoter. Since mutations in the pre-tRNA gene cannot affect transcription by T7 RNA polymerase, pre-tRNA molecules containing any predesigned mutation can be synthesized. Moreover, this method allows large quantities of pre-tRNA (in the order of milligrams) to be synthesized for structural studies.

[1] C. L. Peebles, P. Gegenheimer, and J. Abelson, *Cell* (*Cambridge, Mass.*) **32,** 525 (1983).

[2] C. L. Greer, C. L. Peebles, P. Gegenheimer, and J. Abelson, *Cell* (*Cambridge, Mass.*) **32,** 537 (1983).

[3] A. K. Hopper, F. Banks, and V. Evangelidis, *Cell* (*Cambridge, Mass.*) **14,** 211 (1978).

Construction of Pre-tRNA Gene Linked to a Bacteriophage T7 Promoter

A *Saccharomyces cerevisiae* pre-tRNA[Phe] gene, containing a 19-base intron[4] and under the control of a bacteriophage T7 promoter, was constructed by self-assembly of eight synthetic oligodeoxynucleotides. The resulting DNA fragment was cloned into the *Eco*RI/*Bam*HI site of the phage vector M13mp10 RF. The resulting construct is depicted in Fig. 1A, and the sequences of the oligodeoxynucleotides used are shown in Table I; the cloverleaf structure of the pre-tRNA transcribed from this system is shown in Fig. 1B. To ensure that the transcript would have a mature 5' end, the construct was designed in such a way that transcription initiation by T7 RNA polymerase coincided with the 5' end of the gene. To provide the correct 3' end, a *Bst*NI restriction site was placed at the 3' end of the gene; cleavage of the plasmid with *Bst*NI produced a template DNA coding strand ending in GGT-5', which gave a transcript terminating in CCA-3'.

Procedure

1. Purify each oligodeoxynucleotide by electrophoresis in a 20% polyacrylamide–8.0 M urea gel. Elute the DNA from the gel, extract the solution once with chlorophane (a 1 : 1 mixture of chloroform and phenol), and then ethanol precipitate. Dissolve the resulting DNA pellet in TE [10 mM Tris, 1 mM ethylenediaminetetraacetic acid (EDTA), pH 7.4] to a concentration of 1–5 μg/μl.

2. Phosphorylate the 5' end of each oligonucleotide with (unlabeled) ATP and polynucleotide kinase using standard procedures (see Ref. 5, p. 122). After the reaction, add approximately 10 μg of glycogen carrier, extract once with chlorophane, and ethanol precipitate. Dissolve each 5'-phosphorylated oligonucleotide in TE to a concentration of 0.1–0.5 μg/μl.

3. Mix equimolar amounts of each oligonucleotide (see Table I) as follows. Divide the length (in number of bases) of each oligonucleotide with that of the shortest oligonucleotide. Then add this amount, in micrograms, of each oligonucleotide. For example, if the longest oligonucleotide is 43 bases long and the shortest is 19 bases, then 43/19 = 2.2 μg of the longest oligonucleotide and 19/19 = 1.0 μg of the shortest oligonucleotide that should be included in the oligonucleotide cocktail.

4. Add NaCl to the oligonucleotide cocktail to a final concentration of 50 mM in a total volume of 100 μl.

5. Cover the solution with Paraffin Oil and heat to 95° for 5 min.

[4] P. Valenzuela, A. Venegas, F. Weinberg, R. Bishop, and W. J. Rutter, *Proc. Natl. Acad. Sci. U.S.A.* **75**, 190 (1978).

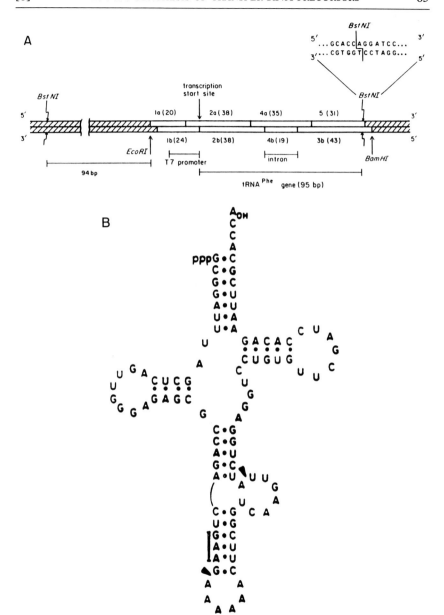

FIG. 1. (A) Bacteriophage T7 promoter/pre-tRNA^Phe gene system. The eight synthetic oligodeoxynucleotides (see Table I for sequences) are indicated with their corresponding lengths (in parentheses). The *Bst*NI restriction site at the 3′ end of the gene is detailed. (▨), M13mp10. (▭), Synthetic gene insert. (B) Cloverleaf structure of the synthetic pre-tRNA^Phe molecule. The intron is demarcated by two arrowheads; the anticodon is indicated by the bar.

TABLE I

CONSTRUCTION OF T7 PROMOTER/PRE-tRNA^{Phe} GENE SYSTEM: SEQUENCES OF SYNTHETIC OLIGODEOXYNUCLEOTIDES EMPLOYED

Name	Length (number of bases)	Sequence (5' → 3')
1a	20	AAT TGC TGC AGT AAT ACG AC
2a	38	TCA CTA TAG CGG ATT TAG CTC AGT TGG GAG AGC GCC AG
4a	35	ACT GAA GAA AAA ACT TCG GTC AAG TTA TCT GGA GG
5	31	TCC TGT GTT CGA TCC ACA GAA TTC GCA CCA G
1b	24	TAT AGT GAG TCG TAT TAC TGC AGC A
2b	38	TCT TCA GTC TGG CGC TCT CCC AAC TGA GCT AAA TCC GC
4b	19	TAA CTT GAC CGA AGT TTT T
3b	43	GAT CCT GGT GCG AAT TCT GTG GAT CGA ACA CAG GAC CTC CAG A

6. Let the sample cool slowly to room temperature (this takes about 1 hr).

7. Add 100 ng of M13mp10 RF vector cut with EcoRI and BamHI for annealing and add a 1/10 volume of 10× ligation buffer [500 mM Tris, pH 7.4, 100 mM MgCl$_2$, 100 mM dithiothreitol (DTT), 10 mM spermidine, and 10 mM ATP].

8. Add 50 U of T4 DNA ligase and incubate at 15° overnight.

9. Transform competent JM101 cells as described (see Ref. 5, p. 250) and screen the resulting plaques with a radioactively labeled oligonucleotide from the coding strand of the DNA ["b" oligonucleotides (see Table I)].[5]

10. Preferably, rescreen with a second b oligonucleotide.

11. Pick appropriate plaques, amplify in *Escherichia coli,* and sequence each isolate.

To create mutations in the pre-tRNA gene, either we isolate single-stranded M13mp10 phage DNA for site-specific mutagenesis using synthetic oligonucleotides,[6] or for more extensive mutations, we re-synthesize the entire gene construct by replacing the appropriate oligonucleotides (see Fig. 1) with ones containing the desired alterations.[7]

Synthesis of Pre-tRNA by T7 RNA Polymerase

Double-stranded plasmid DNA containing the pre-tRNA gene is prepared in the usual way (see Ref. 5, p. 90), except that CsCl ultracentrifugation is always done twice to ensure cleanliness of the DNA template. The DNA is dissolved in TE.

Procedure

1. Mix 200 μg of supercoiled plasmid DNA with 500–600 U of *Bst*NI restriction endonuclease and the appropriate restriction buffer. Cover the surface of the solution with Paraffin Oil and incubate at 65° for 2 hr.

2. Carefully transfer the aqueous phase into another Eppendorf tube and extract once with chlorophane (do not add any RNA or glycogen carrier).

3. Ethanol precipitate, wash, and dry. Dissolve the DNA pellet in TE at a final concentration of 1.0 μg/μl.

[5] T. Maniatis, E. F. Fritsch, and J. Sambrook, "Molecular Cloning: A Laboratory Manual." Cold Spring Harbor Laboratory, Cold Spring Harbor, New York, 1982.
[6] M. J. Zoller and M. Smith, *Nucleic Acids Res.* **10,** 6487 (1982).
[7] V. M. Reyes and J. Abelson, *Cell (Cambridge, Mass.)* **55,** 719 (1988).

4. Depending on the amount of pre-tRNA needed, one can conveniently use 5, 10, or 20 μg of template DNA/transcription. A typical medium-sized transcription reaction using 10 μg of template DNA, would be as follows.

BstNI-restricted DNA (1.0 μg/μl)	10 μl
10× NTP mix	5 μl
10× T7 RNA polymerase buffer	5 μl
RNasin (40 units/μl)	2 μl
5'-GMP (100 mM)	1 μl
[α-^{32}P]NTP label (3000 Ci/mmol)	10 μl
Distilled H$_2$O	to 50 μl

The 10× NTP mix is 0.5 mM in the NTP being used as label and 2.5 mM in the other three. The 10× T7 RNA polymerase buffer is 250 mM NaCl, 250 mM MgCl$_2$, 50 mM Tris (pH 7.8), 5 mM DTT, and 1% (v/v) Triton X-100.

5. Incubate at 40° for 1–1½ hr.

6. Bring the volume to 100 μl with TE and extract once with 100 μl of chlorophane.

7. Retrieve the aqueous phase, ethanol precipitate, and dissolve the pellet in 20–25 μl of formamide loading buffer [95% (w/v) deionized formamide and 0.5% (w/v) each of xylene cyanol (XC) and bromphenol blue].

8. Load onto a 10% (w/v) polyacrylamide–4.0 M urea gel (12 × 18 cm in length, 0.3 mm thick) and electrophorese at 250–300 V until the XC dye is midway through the gel (the pre-tRNA comigrates roughly with the XC dye at these conditions).

9. Visualize the transcripts by autoradiography and slice out the pre-tRNA band from the gel.

10. Place the gel slice in an Eppendorf tube containing 0.4 ml of gel elution buffer [0.5 M ammonium acetate, 50 mM magnesium acetate, 1 mM EDTA, and 0.1% (w/v) sodium dodecyle sulfate (SDS)] and passively elute the RNA by incubation at 37° with shaking for at least 4 hr.

11. Take the gel slice from the tube by means of sterile forceps.

12. Add 10–20 μg of glycogen carrier to the tube (do not use RNA carrier as it inhibits splicing), vortex briefly, extract once with chlorophane, and then ethanol precipitate.

In our hands, a single transcription reaction using 20 μg of template DNA and 250 U of T7 RNA polymerase in a total reaction volume of 80–100 μl yields approximately 10^6–10^7 cpm of pre-tRNA with a specific activity of 8–9 × 10^4 dpm/pmol. This corresponds roughly to 4–5 μg of pre-tRNA (the pre-tRNA^{Phe} synthesized from this system is 95 bases long and has a molecular weight of 32,500 Da). 5'-GMP is included in the transcription reaction to ensure that the pre-tRNA is initiated with a 5'-monophosphate,[8] just as a native pre-tRNA.

The biochemical properties of this pre-tRNA have been characterized thoroughly.[9] It is efficiently cleaved by the purified yeast endonuclease with a K_m of 25–30 nM and a V_{max} of 10^{-3} to 10^{-2} pmol/min at an enzyme concentration of approximately 10^{-11} to 10^{-10} M. The resulting half-tRNAs are also efficiently joined by the yeast ligase, yielding mature tRNA^{Phe}.

---

[8] J. Sampson and O. Uhlenbeck, *Proc. Natl. Acad. Sci. U.S.A.* **85**, 1033 (1988).
[9] V. M. Reyes and J. Abelson, *Anal. Biochem.* **166**, 90 (1987).

[7] Preparation of Precursors to mRNA from Mammalian Cell Nuclei

By SELINA CHEN-KIANG and DANIEL J. LAVERY

The processing of primary RNA transcripts in the nucleus of mammalian cells is a complex event essential for the biogenesis of cytoplasmic mRNAs. The development of *in vitro* site-specific polyadenylation,[1,2] splicing,[3–6] and methylation[7] systems has allowed advances in the understanding of the *cis*-acting sequences and biochemical components important for these reactions. The kinetics and regulation of nuclear RNA processing, however, have not been recapitulated *in vitro*.

Isolation and analysis of steady-state and radiolabeled nuclear RNAs permits insight into the kinetics and intermediates of polyadenylation and

---

[1] S. Chen-Kiang, D. J. Wolgemuth, M.-T. Hsu, and J. E. Darnell, Jr., *Cell* (*Cambridge, Mass.*) **28**, 575 (1982).
[2] C. Moore and P. Sharp, *Cell* (*Cambridge, Mass.*) **36**, 581 (1984).
[3] N. Hernandez and W. Keller, *Cell* (*Cambridge, Mass.*) **35**, 89 (1983).
[4] R. A. Padgett, S. F. Hardy, and P. A. Sharp, *Proc. Natl. Acad. Sci. U.S.A.* **80**, 5230 (1983).
[5] S. F. Hardy, P. J. Grabowski, R. A. Padgett, and P. A. Sharp, *Nature* (*London*) **308**, 375 (1984).
[6] A. R. Krainer, T. Maniatis, B. Ruskin, and M. R. Green, *Cell* (*Cambridge, Mass.*) **36**, 993 (1984).
[7] P. Narayan and F. Rottman, *Science* **242**, 1159 (1988).

splicing[8] and internal methylation[9] *in vivo*. Separation and analysis of polyadenylated nuclear RNAs from those without poly(A) at the 3' end allow an estimate of the utilization of primary transcripts for mRNA and for turnover in the nucleus. Furthermore, analysis of poly(A)⁻ nuclear RNAs have revealed the existence of prematurely terminated RNAs, which constitute yet another level of regulation of RNA synthesis for viral and cellular genes.[10-13] An understanding of the nuclear dwell time and exiting of spliced mRNAs from the nucleus to the cytoplasm is also a prerequisite for the investigation of mRNA transport, which is presently poorly understood. This chapter will describe comprehensive methods for labeling of pre-mRNAs by radioisotopes in whole cells and the separation and isolation of nuclear and cytoplasmic RNAs.

Solutions

20× phosphate-buffered saline (PBS)
  Mix 160 g/liter of NaCl, 4 g/liter of KCl, 43 g/liter of $Na_2HPO_4 \cdot 7 H_2O$ and 4 g/liter of KH_2PO_4. Autoclave and store as a 20× solution. Add $MgCl_2$ to 1 mM upon dilution to 1×.
4 M Guanidine thiocyanate
  To a 100-g bottle of guanidine thiocyanate (Kodak, Rochester, NY), add 100 ml of H_2O, 7 ml of 3 M sodium acetate, pH 5.0, and 2.1 ml of 0.5 M ethylenediaminetetraacetic acid (EDTA), pH 8.0; mix until dissolved (warm to 37° if required). Add 14 ml of 30% (v/v) sarkosyl and 2.1 ml of 14.4 M 2-mercaptoethanol and bring the total volume to 212 ml with H_2O. Filter through a 0.45-μm
6 M Guanidine thiocyanate
  Add 40 ml of H_2O to a 100-g bottle of guanidine thiocyanate. Add 4.6 ml of 3 M sodium acetate, pH 5.0, 1.4 ml of 0.5 M EDTA, and dissolve at 37°. When completely dissolved, add 9.3 ml of 30% (v/v) sarkosyl and 1.4 ml of 14.4 M 2-mercaptoethanol. Bring the total volume to 141 ml with H_2O, filter, and store as the 4 M solution.
CsCl solution
  Mix 0.83 g/ml of optical grade CsCl (Boehringer Mannheim, Indianapolis, IN) in 0.1 M sodium acetate, pH 5.2, and 5 mM EDTA.

[8] J. R. Nevins, *J. Mol. Biol.* **130**, 493 (1979).
[9] S. Chen-Kiang, J. R. Nevins, and J. E. Darnell, Jr., *J. Mol. Biol.* **135**, 733 (1979).
[10] A. Maderious and S. Chen-Kiang, *Proc. Natl. Acad. Sci. U.S.A.* **81**, 5931 (1984).
[11] N. Hay, H. Scolnik-David, and Y. Aloni, *Cell (Cambridge, Mass.)* **29**, 183 (1982).
[12] D. M. Bentley and M. Groudine, *Nature (London)* **321**, 702 (1986).
[13] S.-Y. Kao, A. F. Calman, P. A. Luciw, and B. M. Peterlin, *Nature (London)* **330**, 489 (1987).

Filter through a 0.45-μm nitrocellulose filter and store at room temperature.

Iso-Hi solution

Mix 0.14 M NaCl, 10 mM Tris base, pH 8.4, and 1.5 mM MgCl$_2$. Autoclave and store at 4°.

High-salt buffer (HSB)

Mix 0.5 M NaCl, 10 mM Tris–HCl, pH 7.4, 50 mM MgCl$_2$, and 2 mM CaCl$_2$. Autoclave and store at 4°.

DNase I

The ribonuclease-free grade is purchased from Worthington (Freehold, NJ) or BRL and is kept frozen in aliquots at a concentration of ~4000 U/ml.

10× Magic Wash (see Refs. 14 and 15)

Mix 425 mM Tris–OH, pH 8.3, 85 mM NaCl, 26 mM MgCl$_2$, 40 mM vanadyladenosine, 12 mM phenylmethylsulfonyl fluoride (PMSF), 6% (v/v) Tween 40, and 3% (w/v) sodium deoxycholate.

ETS solution

Mix 10 mM Tris–HCl, pH 7.4, 10 mM EDTA, and 0.2% (w/v) sodium dodecyl sulfate (SDS).

NETS solution

Mix 10 mM Tris–HCl, pH 7.4, 10 mM EDTA, 0.2% (w/v) SDS, and 0.3 M NaCl.

5% (v/v) Nonidet P-40 (NP-40)

Combine 5% (v/v) nonionic detergent NP-40 in Iso-Hi solution.

Acidic phenol

Equilibrate a freshly thawed bottle of phenol repeatedly with equal volumes of 50 mM sodium acetate, pH 5.2, 10 mM EDTA, and 0.2% (w/v) SDS until the pH is 5.2. Store in a brown bottle at 4° for no longer than 1 month or at −20°. Be careful to check the pH before each use.

Acidic phenol–chloroform

Equilibrate a mixture of phenol, chloroform, and isoamyl alcohol (25 : 24 : 1) exactly as above and store at 4°.

Dimethyl sulfoxide (DMSO)

To a fresh bottle of DMSO, add 10 mM Tris–HCl, pH 7.4, and 10 mM EDTA. Store in a brown bottle sealed with Parafilm in the dark at room temperature for up to 6 months.

[14] S. Penman, *J. Mol. Biol.* **17,** 117 (1966).
[15] Z. Zhai, J. A. Nickerson, G. Krochmalnic, and S. Penman, *J. Virol.* **61,** 1007 (1987).

Procedure

Step 1: In Vivo Labeling of Newly Synthesized RNas with Ortho[^{32}P]phosphate

The polyadenylated, spliced, mature cytoplasmic mRNAs represent the major species of RNAs of most cellular and viral genes at steady state. Due to the rapid kinetics of RNA processing in vivo, mRNA precursors, processing intermediates, and metabolic by-products comprise a very small proportion of the steady-state nuclear RNAs in mammalian cells. Their low abundance often renders them undetectable by Northern blot or nuclease protection analysis of steady-state RNAs with radiolabeled probes.

A convenient method for following the processing of nuclear RNAs in vivo is by labeling newly synthesized RNAs with ortho[^{32}P]phosphate or [^{14}C]uridine. Labeling with ortho[^{32}P]phosphate has the advantage of high specific activity. After isolation of nuclear RNAs labeled in vivo in whole cells for designated periods of time, abundant RNAs such as adenoviral RNAs can be analyzed directly by gel electrophoresis (compare newly synthesized RNAs in Figs. 2 and 3 with steady-state RNAs in Fig. 2), while less abundant RNAs must first undergo selection by appropriate probes in order to distinguish them from the background.

To increase efficiency of in vivo labeling of cultured cells, phosphate in serum used during the labeling period is first removed by dialysis. Approximately 100 ml of serum in dialysis tubing (molecular weight cutoff of <1500) is dialyzed, with at least three changes of dialysis buffer, against >10 volumes of 0.15 M NaCl buffered with 10 mM HEPES, pH 7.4, for 3–5 hr at 4°. The dialyzed serum is filter sterilized and stored in 10-ml aliquots at −20°. Due to the large amount of radioactivity to be used, isolate an area for the labeling procedure, preferably a shaking water bath surrounded by Lucite shields, away from other workers and use disposable materials only. Be sure to monitor the area and yourself for radioactivity throughout the procedure.

To achieve equilibrium labeling of nuclear RNAs in approximately 3 hr, ~2–4 mCi of ortho[^{32}P]phosphate is used for labeling 1 × 10^8 cells. Cells of suspension culture are sedimented and resuspended three times in serum-free, phosphate-free culture media. For labeling, the cells are resuspended in phosphate-free media supplemented with dialyzed serum at 2 × 10^6 cells/ml. Add ortho[^{32}P]phosphate ["carrier-free" (New England Nuclear, Boston, MA)] dropwise. In a sealed culture flask lacking CO$_2$ and with reduced amount of phosphate, the pH of the media may drop with time during the 3-hr labeling period. Add HEPES, pH 7.4, as needed, but do not exceed 20 mM final concentration, as higher concentrations may be toxic to the cells. To label monolayer cultures, the growth media are decanted. The monolayer is rinsed three times with serum-free, phosphate-free culture media and is labeled finally with ortho[^{32}P]phosphate in the presence of dialyzed serum, usually at the same specific activity per cell as the suspension culture.

Labeled cells (monolayer cells are scraped) are sedimented by centrifugation and are resuspended in ice-cold PBS three times, changing the centrifugation bottle each time to prevent adherence of free radioactive phosphate to the flasks.

Step 2: Separation of RNA into Nuclear, Postnuclear ("Magic Wash"), and Cytoplasmic Fractions

The partitioning of nuclei and cytoplasm with buffer containing the nonionic detergent NP-40 and the subsequent cleaning of nuclei with Magic Wash,[14,15] as well as the isolation of RNAs from each fraction, are outlined in Fig. 1. At appropriate concentrations (such as 0.5%), NP-40 disrupts the plasma membrane while leaving the nuclear membrane in-

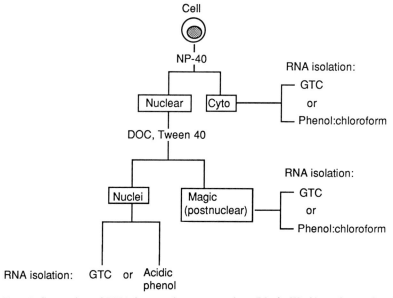

FIG. 1. Separation of RNA into nuclear, postnuclear (Magic Wash), and cytoplasmic fractions. Flow diagram of detergent treatment procedures resulting in the separation of subcellular fractions, indicated within boxes. See the text for details.

tact, allowing separation of nuclei from the cytoplasmic fraction by centrifugation. Treatment of the nuclear fraction with Magic Wash, which contains deoxycholate, a steroid emulsifier, and Tween 40, an oleate ester emulsifier, removes the outer layer of the nuclear membrane, as well as residual cytoplasmic materials. As can be seen in Figs. 2 and 3, the newly synthesized 45 and 32 S ribosomal RNA precursors partition in the nuclear fraction, but not in the postnuclear (Magic Wash) or cytoplasmic fractions. Furthermore, the poly(A)$^+$ nuclear precursors of adenoviral mRNA partition in the nuclear fraction, while the poly(A)$^+$ mRNAs are found in the cytoplasmic fraction, indicating the effectiveness of the fractionation.

The nature of these methods, however, calls for special caution in their use and interpretation. First, the limits of the separation procedures must be kept in mind, as well as the functional definitions of the subcellular compartments themselves (e.g., does the "nuclear" fraction contain some endoplasmic reticulum material?). Second, different cell lines, and even the same cell line under different conditions, may exhibit altered behavior in response to the detergents. For example, while nuclei from one clone of a human T-cell line are completely lysed by 0.5% (v/v) NP-40 concentration, nuclei can be isolated efficiently at this concentration from another clone of the same cell line (K. Park and S. Chen-Kiang, unpublished observations). Therefore, care should be taken to define empirically the detergent concentrations required, using the conditions given in this chapter as a reference. These conditions have proved successful for the isolation of nuclei from HeLa cells, KB cells, 293 cells, and monolayer fibroblasts, as well as many human lymphoid cell lines.

Two methods are presented for the isolation of RNA after subcellular fractionation. These are the guanidine thiocyanate (GTC) method[16] and RNA extraction by acidic phenol.[17] We have found the former to be widely applicable in the isolation of intact RNAs. Cell fractionation ruptures endosomes, causing the release of various enzymes, including ribonucleases (RNases). The speed with which nucleases are inactivated by GTC makes its use preferable when extracting RNAs from sources, such as lymphocytes or hepatocytes, which may be rich in endogenous RNases. However, the acidic phenol method has the advantages of isolating RNAs of all sizes with equal efficiency and quick recovery; there is no requirement for an overnight ultracentrifugation step. A native minigel can be used conveniently to monitor the intactness of RNAs isolated by

[16] J. M. Chirgwin, A. E. Przybyla, R. J. MacDonald, and W. J. Rutter, *Biochemistry* **18,** 5294 (1979).
[17] R. Soeiro and J. E. Darnell, *J. Mol. Biol.* **44,** 551 (1969).

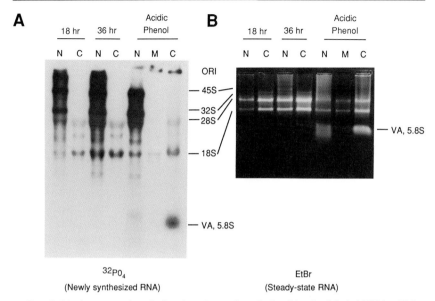

FIG. 2. Nuclear–cytoplasmic fractionation and analysis of *in vivo* labeled RNAs. HeLa cells (1.5×10^8) in the late phase of infection by adenovirus type 2 were labeled with 1 mCi of ortho[^{32}P]phosphate for 2 hr, as described in the text. After labeling, cells were divided into three equal aliquots. Nuclear (N), cytoplasmic (C), and postnuclear [Magic Wash (M)] fraction RNAs were isolated by acidic phenol extraction (Acidic Phenol), or nuclear and cytoplasmic RNAs were isolated by guanidine thiocyanate homogenization, followed by centrifugation at 33,000 rpm for 18 (18 hr) or 36 hr (36 hr). RNAs were precipitated twice in ethanol, and 0.2% of each sample (1×10^5 cell equivalents) was used for ethidium bromide–native gel analysis. The samples were loaded onto a 25-ml TBE–1% (w/v) agarose minigel containing 0.5 μg/ml of ethidium bromide (EtBr). Samples were electrophoresed at 100 V for 1 hr at constant current. The gel was visualized and photographed under ultraviolet (UV) light, as shown in B. RNAs from 1×10^6 cells were also separated on a 250-ml (20×40 cm) denaturing formaldehyde–agarose gel as described.[1] After electrophoresis (14 hr at 40 V), the gel was dried under vacuum and was exposed to Cronex X-ray film for <1 hr. The resulting autoradiogram is shown in A (^{32}PO$_4$). Note that while adenovirus-infected HeLa cells show accumulation of 28 and 18 S ribosomal RNAs (B), greater labeling is found in 45 and 32 S ribosomal RNA precursors (A). Also, 36-hr centrifugation, rather than 18-hr centrifugation, improved recovery of RNAs of 2 kb (18 S) and smaller, but did not recover very small RNAs, such as 5.8 S rRNA (~120 nucleotides) or the adenovirus type 2 *VA* gene products (~160 nucleotides), which are recovered by acidic phenol RNA isolation. ORI, Sample origin.

either method. As shown in Fig. 2, the ethidium bromide (EtBr)-stained 28 and 18 S ribosomal RNA bands, after just 1 hr of electrophoresis, serve to indicate that RNAs are not degraded.

Step 2a: Guanidine Thiocyanate Isolation. Pellet cells (~5 × 10^7 to 1 × 10^8) at 275 *g* for 5 min and wash twice with ice-cold PBS. Resuspend

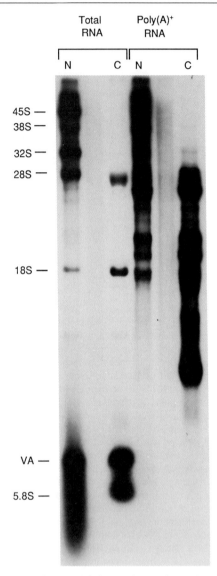

FIG. 3. Poly(A) selection of *in vivo* labeled nuclear and cytoplasmic RNAs. Nuclear (N) and cytoplasmic (C) RNAs were extracted by acidic phenol from 1×10^8 adenovirus-infected HeLa cells labeled with 4 mCi of ortho[^{32}P]phosphate for 3 hr and were subjected to poly(A) selection by poly(U)–agarose chromatography, as described.[1] Poly(A)$^+$ RNAs from 1×10^6 cells and unselected RNAs (total RNA) from 1×10^4 cells were electrophoresed on denaturing formaldehyde–agarose gels. The dried gel was exposed to X-ray film, and the resulting autoradiogram is presented. Note that nuclear poly(A)$^+$ RNA precursors to mRNAs are generally larger in size than those in the cytoplasmic poly(A)$^+$ fraction.

the pellet, after gentle tapping, in 0.45 ml of Iso-Hi solution in a 15-ml conical centrifuge tube and add 50 μl of 5% NP-40 (v/v in Iso-Hi) dropwise; mix well by swirling and tapping. If desired, quickly withdraw a small volume (10–20 μl) with a Pasteur pipet to monitor the disruption of intact cells and the release of nuclei by phase-contrast microscopy. Centrifuge immediately at 550 g for 5 min at 4°. Carefully transfer the supernatant cytoplasmic fraction with a Pasteur pipet directly to 4.5 ml of 4 M GTC solution. Immediately homogenize the suspension using a 10-ml syringe with a $1\frac{1}{2}$-inch 20-gauge needle until the viscosity appears even.

The nuclei, after gentle tapping, are resuspended in 0.9 ml of Iso-Hi buffer. The nuclei should be a loose pellet, easily dispersed by tapping. A viscous pellet indicates lysis of the nuclei. One hundred microliters of 10× Magic Wash solution is added. The cells are tapped to ensure even resuspension and are centrifuged again at 550 g for 5 min at 4°. The supernatant postnuclear fraction is carefully transferred directly into 4 ml of 4 M GTC solution with a Pasteur pipet and is homogenized as described for the cytoplasmic fraction. The postnuclear fraction contains the 28 and 18 S rRNAs, as well as poly(A)$^+$ RNAs (compare newly synthesized RNAs and steady-state RNAs in Fig. 2). Researchers may choose to pool the cytoplasmic and postnuclear fractions as one cytoplasmic fraction. To accommodate the total volume (0.5 ml of cytoplasmic fraction and 1 ml of postnuclear fraction) in one CsCl gradient, 3.5 ml of 6 M GTC is used in place of 4.5 ml of 4 M GTC.

The remaining nuclear pellet is homogenized in 5 ml of 4 M GTC as described above. This homogenization will be more difficult, however, due to the presence of chromatin, and thus, will require more cycles of syringing (typically 20–25 for 10^8 HeLa cells).

The GTC solutions are layered onto 4 ml of CsCl solution in a SW41-type ultracentrifuge tube and are overlayed with paraffin oil. The samples are centrifuged for 18–36 hr at 33,000 rpm. We have found that centrifugation for 36 rather than 18 hr results in more efficient recovery of RNAs of 2 kb in length or smaller (Fig. 2). RNAs smaller than 300 nucleotides are not recovered in the CsCl gradient, but are recovered by acidic phenol extraction (Figs. 2 and 3).

To recover RNA in the pellet, first remove the supernatant with a pipet. DNA in the nuclear fraction will appear as an opaque band at the CsCl interface. Be certain to completely remove this layer first to prevent it from contaminating the RNA pellet. Wipe the outside of the tube with 70% (v/v) ethanol and mark the bottom of the tube. Carefully cut the tube 2 cm from the bottom using a clean scalpel or razor blade. Remove any residual CsCl gradient with an Eppendorf pipet. The RNA should be visible as a tight, clear pellet at the bottom of the tube. Pipet 200 μl of sterile ETS solution onto the pellet and let it sit for a few minutes. Pipet the ETS solution up and down in an Eppendorf pipet to dissolve the RNAs and then transfer the ETS–RNA solution to a siliconized 15-ml Corex tube. Repeat this step four more times to ensure complete recovery of RNAs. Bring the final volume of ETS in the Corex tube to 4 ml, add LiCl to a concentration of 0.2 M, and precipitate with 2 volumes of ethanol. After sufficient precipitation time, pellet the RNAs from ethanol and reprecipitate in the same volume to remove residual CsCl. Store the RNAs in ethanol at −20°.

Step 2b: Acidic Phenol Extraction. Because this method is not restricted by volume for CsCl gradients, as is the GTC method, larger volumes are used to increase the efficacy of organic extractions.

Cells are pelleted and washed twice with ice-cold PBS. The cells are resuspended in 2.25 ml of ice-cold Iso-Hi solution in a 15-ml conical centrifuge tube, tapping the pellet to aid resuspension. The volumes of Iso-Hi solution and 5% NP-40 solution used for 5 × 10^8 to 10^8 cells are 2.25 ml and 250 µl, respectively, thus maintaining the final concentration of NP-40 at 0.5%. Following addition of 250 µl of 5% (v/v) NP-40 and sedimentation of the nuclei, the 2.5-ml supernatant cytoplasmic fraction is transferred to a new tube. The volume of the cytoplasmic fraction is increased to 5 ml with ETS solution; 25 µl of 20% (w/v) SDS, 25 µl of 0.5 M EDTA, and 125 µl of 8 M LiCl are added. The aqueous phase is extracted twice with 2 volumes of neutral pH phenol–chloroform and twice with chloroform. The cytoplasmic RNAs are in the aqueous phase and are precipitated twice with 2.5 volumes of ethanol.

The nuclear pellet from above is resuspended in 4.5 ml of Iso-Hi. To this suspension, 0.5 ml of 10× Magic Wash solution is added. The postnuclear supernatant fraction, after sedimenting nuclei at 550 g for 5 min, is processed identically as the cytoplasmic fraction. Alternatively, the postnuclear and cytoplasmic fractions may be combined before organic extraction.

To the remaining nuclear pellet, add 2.5 ml of HSB, which lyses the nuclei and makes the solution very viscous. To relieve the viscosity and to aid in RNA extraction, nick the chromatin with DNase I. Add ~100 U of RNase-free DNase I in 25 µl and triturate in a Pasteur pipet at room temperature. The viscosity should be greatly reduced within 30 sec. If not, add more DNase I. (The efficacy of each DNase I lot should be tested empirically before use.) Add 2.5 ml of 50 mM sodium acetate, pH 5.2, 25 µl of 20% (w/v) SDS, 25 µl of 0.5 M EDTA, and 25 µl of 1 M sodium acetate, pH 5.2. Extract RNAs for 5 min by intermittent vigorous shaking and mixing in a 65° water bath with 2 volumes of acidic phenol, prewarmed to 65°. Extraction of nucleic acid in acidic pH results in denaturation of nucleic acids and preferential partition of double-stranded DNA in the interphase. Repeat the acidic phenol extraction, followed by extraction of the aqueous phase twice with acidic phenol–chloroform and once with chloroform. Precipitate the RNAs from the aqueous phase twice with 2.5 volumes of ethanol in a new tube.

Step 3: Selection of Poly(A)$^+$ RNA

The use of poly(U)–Sepharose (or agarose) chromatography for selection of polyadenylated RNAs is preferred over the more commonly used oligo(dT) column chromatography for the following reasons: (1) A–U pairing is more specific and stable than A–dT pairing and (2) elution of poly(A) from poly(U) columns is temperature and formamide concentration dependent, allowing selective elution of stretches of As of various lengths.[18] The combination of these two features allows the poly(A)$^+$ fractions, representing only 0.1–0.3% of total RNAs, to be virtually free of contamination by ribosomal RNA, which constitutes the majority of RNAs in both cytoplasmic and nuclear fractions (Fig. 2B). For example, compare the poly(A)$^+$ RNA and total RNA fractions in Fig. 3.

Step 3a: Preparation of Poly(U) Columns. The poly(U)–agarose is supplied by the manufacturer (Pharmacia, Piscataway, NJ) in 50% (v/v) glycerol, which can be reduced by repeated dilution of poly(U)–agarose slurry in 20 volumes of ETS and by aspiration of supernatant solution after the beads are settled. As outlined in Fig. 4, for selection of poly(A)$^+$ RNAs from 5×10^7 to 1×10^8 cells, a 0.5-ml bed volume of poly(U)–agarose is packed in a siliconized, large-bore Pasteur pipet plugged with siliconized glass wool. The pipet is connected to a 10-cm piece of Tygon tubing ($\frac{1}{16}$-inch i.d.) at the bottom. Attached to the top of the pipet is a feeder line, comprising a 20-gauge needle inserted into a rubber cap (Kontes, Vineland, NJ), which fits snugly on top of the Pasteur pipet. The blunt end of the needle is inserted into one end of a 25-cm-length piece of Tygon tubing ($\frac{1}{32}$-inch i.d.). The other end of the tubing is connected to a 50-μl glass micropipet. When the area above the poly(U)–agarose in the column is filled with buffer, and when the micropipet is immersed in the reservoir placed above the column with a pressure head of about 35 cm, gravity will allow continuous flow of the buffer from the reservoir.

The entire column, including feeder line and the inside of the rubber cap, is washed with at least 40 bed volumes (20 ml) of ETS, followed by washing with 10 ml of 70% (v/v) formamide (recrystallized three times) in ETS. The agarose will be compressed and will appear transparent at this formamide concentration. The column is washed immediately with at

[18] S. G. Sawicki, W. Jelinek, and J. E. Darnell, *J. Mol. Biol.* **113**, 219 (1977).

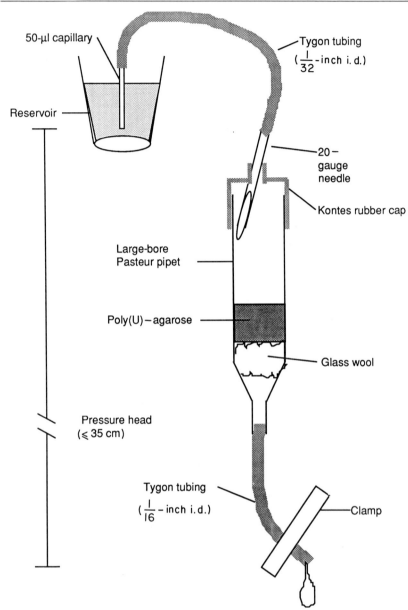

FIG. 4. Assembly of poly(U)–agarose columns. Schematic diagram for the assembly of columns for use in poly(A)$^+$ selection. See the text for details.

least 40 volumes (20 ml) of ETS and, finally, is equilibrated with 40 volumes (20 ml) of NETS. The column is now ready for RNA loading.

Step 3b: Loading of RNA Sample. Selection of poly(A)$^+$ RNA is performed at room temperature (20°). The RNA sample, precipitated twice in ethanol, is dissolved completely in 100 μl of ETS. To maximize binding efficiency, the RNAs are denatured in 75% (v/v) DMSO by adding 300 μl of DMSO and incubating at 65° for 5 min, followed by immediate chilling on ice. Since DMSO at high concentrations prevents base pairing, the denatured RNA solution is then diluted by adding 2.6 ml of ETS. Sodium chloride (180 μl of 5 M) is then added, such that the final NaCl concentration is 0.3 M.

The RNAs in NETS are loaded onto the column in 1-ml increments at a very slow flow rate (~1 drop/3 sec) to facilitate binding. The loading is recycled three times. The column is then further washed two times with 1 ml of ETS; this 2-ml wash and the 3-ml flow-through from the loading steps are collected as the poly(A)$^-$ fraction. Lithium chloride is added to a concentration of 0.2 M, and RNAs are precipitated with 2 volumes of ethanol.

Step 3c: Elution of Poly(A)$^+$ RNAs. Wash the column with 20 ml of ETS. Elute RNAs containing short stretches of As (<20 As), including rRNAs, in 20 ml of 10% formamide in ETS. Elute poly(A)$^+$ RNAs in 2.5 ml (5 bed volumes) of 70% formamide in ETS in 0.8-ml increments at a slow flow rate of 1 drop/3 sec. Wash the column with 2.5 ml of ETS. Collect the 2.5-ml 70% (v/v) formamide and the 2.5-ml ETS eluates together as the poly(A)$^+$ RNA fraction. Add 10 μg of yeast tRNA as carrier and LiCl to a concentration of 0.25 M and precipitate RNAs with 2 volumes of ethanol. It is important to sediment the RNAs rapidly after precipitation (overnight at −20° or 2 hr at −70°) to prevent the formation of sodium formate salts. Reprecipitate the poly(A)$^+$ RNAs at least once more to ensure complete elimination of formamide.

Poly(U)–Sepharose columns can be regenerated and reused at least 10 times. Equilibrate the column thoroughly with 40 ml of ETS and then with 40 ml of NETS, in which the columns can be stored at room temperature. Before reuse, equilibrate the column with 20 ml of ETS and repeat the 10-ml wash with 70% (v/v) formamide as above. Before use each time, prepare fresh 70% (v/v) formamide.

The methods described apply to isolation of steady-state RNAs (Fig. 2B), which include the radiolabeled newly synthesized RNAs (Fig. 2A). For abundant RNAs, such as viral RNAs, radiolabeling may not be necessary. In this case, the unlabeled poly(A)$^+$ and poly(A)$^-$ RNAs from nuclear and cytoplasmic fractions are then subjected to Northern blot analysis or nuclease protection assays. We have found that, in Northern blot analysis, inclusion of DNA (10 μg/ml) of a 28 S mouse ribosomal RNA clone,[19] in addition to yeast tRNA (100 μg/ml) in the prehybridization and hybridization mixture, reduces the background substantially and allows the successive reprobing of the nitrocellulose blots up to 13 times with various probes.

[19] D. C. Tiemeier, S. M. Tilghman, and P. Leder, *Gene* **2**, 173 (1977).

[8] Pulse Labeling of Heterogeneous Nuclear RNA in Isolated Nuclei

By SELINA CHEN-KIANG and DANIEL J. LAVERY

In the analysis of gene regulation, it is essential to distinguish the relative contribution of transcriptional and posttranscriptional events to the accumulation of mRNAs. Investigators have thus sought to develop an assay system that adequately and reproducibly represents *in vivo* transcriptional events. While the radiolabeling of RNA in intact cells would represent true *in vivo* conditions, the relative impermeability of cells to charged ribonucleoside triphosphates requires the use of radiolabeled precursors, such as uridine, in these experiments. However, the time required for equilibration of cellular triphosphate pools, for example, 20 min for UTP in HeLa cells, makes this method impractical for the study of rapid events, such as RNA transcription.

Nuclei isolated from various cells or tissues have been shown to be capable of supporting elongation of transcripts initiated *in vivo* by RNA polymerases I, II, and III prior to isolation of nuclei.[1-4] The relative permeability of nuclei to triphosphates permits rapid incorporation of radiolabeled triphosphates in pulse-labeling times as short as 2 min[5] (Fig. 3). Derman *et al.*[6] optimized this system to incorporate large amounts of radiolabeled ribonucleotides, allowing the transcriptional analysis of single-copy genes expressed in specific tissues. Pulse labeling consistent with yielding reproducible results is important also because RNA pro-

[1] R. F. Cox, *Cell (Cambridge, Mass.)* **7**, 455 (1976).
[2] R. H. Reeder and R. G. Roeder, *J. Mol. Biol.* **67**, 433 (1972).
[3] W. F. Marzluff, Jr., E. C. Murphy, Jr., and R. C. C. Huang, *Biochemistry* **13**, 3689 (1974).
[4] R. Weinmann and R. G. Roeder, *Proc. Natl. Acad. Sci. U.S.A.* **71**, 1790 (1974).
[5] J. Weber, W. Jelinek, and J. E. Darnell, Jr., *Cell (Cambridge, Mass.)* **10**, 611 (1977).
[6] E. Derman, K. Krauter, L. Walling, C. Weinberger, M. Ray, and J. E. Darnell, Jr., *Cell (Cambridge, Mass.)* **23**, 731 (1981).

cessing events, such as polyadenylation, which occurs in isolated nu-clei,[7,8] as well as less-defined events, such as nuclear turnover, may inter-fere with transcriptional analysis by long labeling times. Thus, an isolated nuclei system optimized for rapid incorporation of radiolabeled ribonu-cleotides is essential. We present in this chapter procedures for the ra-diolabeling within 3–5 min in nuclei from various human cells, including human lymphoid cells, and for the hybridization analysis of *in vivo* initi-ated RNA transcripts, which permit the analysis of relative transcription of single-copy genes. This system is also useful in the analysis of tran-scription pausing–termination, which is emerging as an important elonga-tion control of gene expression.[9-11]

Solutions

20× Phosphate-buffered saline (PBS)
   Mix 160 g/liter of NaCl, 4 g/liter of KCl, 43 g/liter of $Na_2HPO_4 \cdot 7H_2O$, and 4 g/liter of KH_2PO_4. Autoclave as a 20× solution. Add $MgCl_2$ to 1 mM upon dilution and store at 4°.
TITE buffer
   Mix 100 mM NaCl, 20 mM Tris–HCl, pH 7.4, 2 mM ethylene-diaminetetraacetic acid (EDTA), disodium salt, and 0.02% (v/v) Triton X-100 (diluted from 10% (v/v) stock in H_2O). Store at 4°.
10% (w/v) Sucrose in TITE
   RNase-free sucrose (Schwarz-Mann, Cleveland, OH) is weighed out in a sterile tube; the appropriate volume of sterile TITE buffer is then added.
ETS solution
   Mix 10 mM EDTA, 10 mM Tris–HCl, pH 7.4, and 0.2% (w/v) sodium dodecyl sulfate (SDS).
Transcription buffer for pulse labeling with [^{32}P]UTP
   Mix 10 mM Tris, pH 7.9, 140 mM KCl, 14 mM 2-mercaptoethanol, 5 mM $MgCl_2$, 1 mM $MnCl_2$, 1 mM each of ATP, CTP, and GTP, 1 mM *S*-adenosylmethionine, 10% (v/v) glycerol, and H_2O to 10 ml. (*Notes*) Be sure that the mix is pH 7.4. All nucleoside triphosphate stock solutions (0.1 *M*) should be adjusted to pH 7.0, aliquoted, and stored at −70°. *S*-Adenosylmethionine

[7] J. L. Manley, P. A. Sharp, and M. L. Gefter, *J. Mol. Biol.* **135,** 171 (1979).
[8] S. Chen-Kiang, D. J. Wolgemuth, M.-T. Hsu, and J. E. Darnell, Jr., *Cell (Cambridge, Mass.)* **28,** 575 (1982).
[9] A. Maderious and S. Chen-Kiang, *Proc. Natl. Acad. Sci. U.S.A.* **81,** 5931 (1984).
[10] D. M. Bentley and M. Groudine, *Nature (London)* **321,** 702 (1986).
[11] M.-C. Raynal, Z. Liu, T. Hirano, L. Mayer, T. Kishimoto, and S. Chen-Kiang, *Proc. Natl. Acad. Sci. U.S.A.* (submitted for publication).

(Boehringer Mannheim, Indianapolis, IN), 0.1 M stock, is dissolved in ice-cold 0.1 N acetic acid and stored at $-70°$. The presence of S-adenosylmethionine during *in vitro* transcription of isolated nuclei reproducibly yields higher incorporation of [^{32}P]UTP,[8] though the mechanism is not known.

UTP chase solution

    Use 3 mM UTP in 1× PBS. Dissolve 106 mg of UTP in 60 ml of 1× PBS. Aliquot 6 ml for each transcription reaction into 15-ml plastic conical centrifuge tubes; chill to a slurry at $-20°$ immediately before use.

CsCl solution

    Mix 0.83 g/ml of optical grade CsCl (Boehringer Mannheim) in 0.1 M sodium acetate, pH 5.2, and 5 mM EDTA. Filter through a 0.45-μm nitrocellulose filter and store at room temperature.

Hybridization solution

    Recrystallize formamide (BRL, Bethesda, MD, nucleic-acid grade) three times. Formamide is recrystallized by packing in ice overnight in a 4° room, seeding with a formamide seed crystal, and allowing the crystals to grow in an ice bucket at 4° overnight. Uncrystallized liquid is decanted, and crystals are melted at room temperature and are recrystallized twice more. The 3× recrystallized formamide is stored in 50-ml aliquots at $-20°$. Mix formamide, 50% (v/v), with 5× SSC (1× = 0.15 M NaCl, 15 mM sodium citrate, pH 7.0), 50 mM Na$_x$PO$_4$, pH 7.4, 0.1% (w/v) SDS, 5% (w/v) dextran sulfate, 1× Denhardt's solution [0.2 mg/ml of Ficoll, 0.2 mg/ml of poly(vinylpyrrolidone), and 0.2 mg/ml of bovine serum albumin Fraction V], 0.1 mg/ml of yeast tRNA, and 20 μg/ml of mouse 28 S ribosomal DNA clone I–19,[12] restriction digested and boiled. The 28 S rDNA clone is very high in G–C content and is used to reduce nonspecific hybridization.

RNase A

    Dissolve lyophilized RNase A (Calbiochem, San Diego, CA) in 10 mM sodium acetate, pH 5.2, to a concentration of 10 mg/ml. Heat to 85° for 30 min, then allow to cool to room temperature. This step will inactivate contaminating DNases. Aliquot and store at $-20°$. Each preparation must be titrated to assess optimal concentration and incubation times for removal of noncomplementary single-stranded RNAs from filters.

---

[12] D. C. Tiemeier, S. M. Tilghman, and P. Leder, *Gene* **2**, 173 (1977).

DNase I
  From Worthington (Freehold, NJ) or BRL as ribonuclease-free grade, stored at $-20°$ at a concentration of ~4000 U/ml.
Proteinase K
  Dissolve lyophilized proteinase K (Boehringer Mannheim) in ETS solution to 10 mg/ml. The solution is self-digested at 37° for 30 min. Store at 4°.
High-salt buffer (HSB)
  Mix 0.5 M NaCl, 10 mM Tris–HCl, pH 7.4, 50 mM MgCl$_2$, and 2 mM CaCl$_2$. Autoclave and store at 4°.
20× SSC
  Mix 3 M NaCl and 0.3 M sodium citrate. The pH is adjusted to 7.0 with concentrated HCl, and the solution is autoclaved and stored at room temperature.
4 M Guanidine thiocyanate
  To a 100-g bottle of guanidine thiocyanate (Kodak, Rochester, NY), add 100 ml of H$_2$O, 7 ml of 3 M sodium acetate, pH 5.0, and 2.1 ml of 0.5 M EDTA; mix until dissolved (warm to 37° if required). Add 14 ml of 30% (v/v) sarkosyl and 2.1 ml of 14.4 M 2-mercaptoethanol and bring the volume to 212 ml with H$_2$O. Filter through a 0.45-μm nitrocellulose filter and store at 4° in a dark bottle.
10× TBE buffer for native minigels
  Mix 108 g/liter of Tris base, 55 g/liter of boric acid, and 40 ml/liter of 0.5 M EDTA, pH 8.0. Autoclave and store at room temperature.

Procedure

  Transcriptional elongation, but not initiation, occurs in isolated nuclei; therefore, polymerase elongation during isolation of nuclei before pulse labeling can be a problem, resulting in the depletion of polymerases at the 5' end of a transcription unit. To minimize this problem, (1) perform the experiment in a cold room with everything chilled or (2) freeze all plastic and glassware until use.
  All plastic and glassware must be RNase-free. The Dounce homogenizer [tight-fitting, B pestle (Bellco, Vineland, NJ)] should not be autoclaved, because expansion and contraction of glass during autoclaving alters the calibrated clearance. Rather, the homogenizer can be rinsed with 30% (v/v) hydrogen peroxide, followed by TITE buffer, and finally with 50% (v/v) ethanol–H$_2$O, and store in 50% (v/v) ethanol–H$_2$O until

use. The Dounce homogenizer should be rinsed with TITE buffer before each use, and the same Dounce homogenizer should be used for each set of samples.

Step 1: Isolation of Nuclei

Carry out all steps on ice or at 4°, as outlined in Fig. 1A. Pellet cells of the suspension culture (usually 5×10^7–1×10^8/reaction) at 275 g for 5 min. Harvest monolayer cultures by scraping with an RNase-free rubber policeman. Tap the pellet gently before resuspension to prevent clumping. Resuspend and sediment the pellet two times with ice-cold PBS and then resuspend in TITE buffer at 10^7 cells/ml; leave on ice for 1–3 min, depending on the cell type. Disrupt cells by homogenization in a tight-fitting Dounce homogenizer (B pestle) and monitor under phase-contrast microscopy.

Whole cells appear bright and refractile, while nuclei appear dark, flat, and scruffy. (*Important*) Be efficient and keep every pipet ice-cold. Dounce homogenizer and all tubes should be packed *in* ice rather than sitting *on* ice. Douncing is considered complete when 90% of the cells are disrupted. The number of strokes necessary for disrupting cells varies among cell lines and other parameters (virus infection, etc.). Typically, we find that HeLa cells require 15–20 strokes, human B myeloma cells require 15 strokes, human T cells require 5 strokes when uninfected, but 35 strokes when adenoviral infected (K. Park, D. Lavery, and S. Chen-Kiang, unpublished observations). Alternatively, disruption of cells may be achieved by suspending cells in TITE buffer with the Triton X-100 concentration elevated to 0.1% or with the use of 0.5% (v/v) Nonidet P-40 (NP-40). These detergent concentrations neither dissociate polymerase II from the DNA templates, nor do they increase transcription rates significantly (Fig. 2). Whether they will eliminate regulatory proteins from the template is an open question. For this reason, the Dounce homogenization method is preferred.

After douncing, nuclei are separated from cytoplasm by careful layering the dounced suspension (10 ml for 10^8 cells) onto 4 ml of 10% (w/v) sucrose in TITE buffer in a chilled 15-ml conical centrifuge tube, using a chilled Pasteur pipet, and the suspension is pelleted by centrifugation for 5 min at 750 g (for example, in the Sorvall RT 6000B). The supernatant is removed by first eliminating cell debris at the interphase using a Pasteur pipet and by removing as much supernatant as possible without disturbing the nuclear pellet.

Step 2: Transcription in Vitro

Use nuclei without delay. Equilibrate the nuclei in transcription reaction buffer by resuspending the nuclei in 3 nuclei volumes of ice-cold

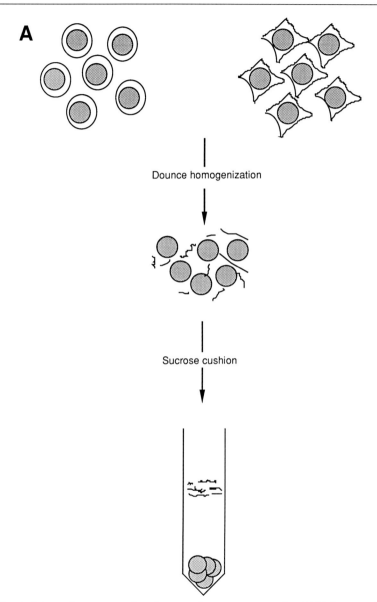

FIG. 1. Schematic representation of nascent heterogeneous nuclear RNA (hnRNA) labeling in isolated nuclei, as presented in the text. (A) Isolation of nuclei. Cells from various sources, both suspension and monolayer, are lysed by douncing in the presence of low concentrations of nonionic detergent. The resultant suspension is layered on a sucrose cushion and is centrifuged. Intact nuclei will pellet, while membrane and cytosolic material, including nucleases and proteases, will remain at the sucrose interface. (B) Pulse labeling of

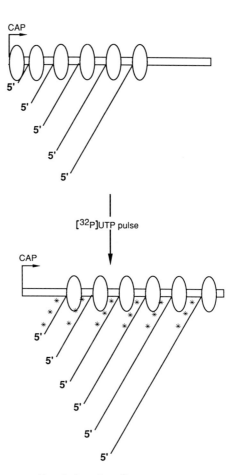

FIG. 1. (*continued*)

nascent RNA molecules. The addition of [³²P]UTP results in its incorporation into the growing 3' end of nascent RNA chains. Because initiation does not occur in isolated nuclei, only polymerase complexes engaged ( ) to the DNA template before nuclear isolation will incorporate [³²P]UTP. Thus, incorporation reflects relative polymerase density on a locus at the time of isolation. (∗) [³²P]UTP incorporation. (C) Isolation, limited alkali digestion, and hybridization of radiolabeled RNA to DNA probes. RNAs are purified away from DNA and proteins by acidic phenol extraction or by the guanidine thiocyanate method. The RNA molecules are then limitedly hydrolyzed by NaOH to an average size of 300–400 nucleotides. This aids in standardizing hybridization kinetics and prevents cross-hybridization to intragenic probes. After hybridization to DNA probes immobilized on filters, nonhomologous regions are trimmed from the RNA molecule by treatment with RNase A to eliminate the contribution of nonhomologous counts to a signal and to reduce background.

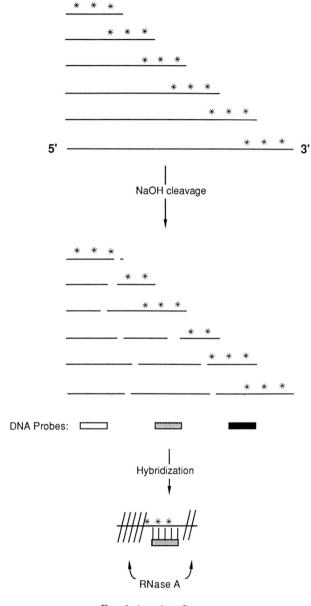

FIG. 1. (*continued*)

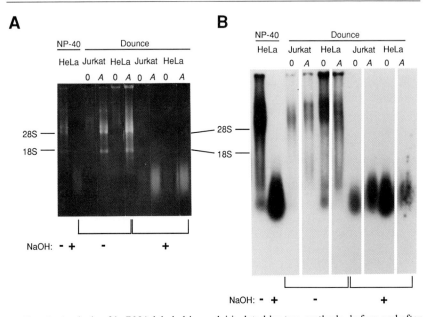

FIG. 2. Analysis of hnRNA labeled in nuclei isolated by two methods, before and after limited alkali hydrolysis. Nuclei were isolated from HeLa cells (HeLa) and human Jurkat T cells (Jurkat), both uninfected (0) or adenovirus infected (A), by Dounce homogenization (Dounce). Nuclei were isolated from another sample of uninfected HeLa cells by treatment with 0.5% (v/v) Nonidet P-40 (NP-40). Nuclei (5 × 10[7]) were pulse labeled with 300 μCi of [32P]UTP at 31° for 5 min in 300 μl of transcription buffer. RNAs were isolated by the guanidine thiocyanate method as described (−) and were subjected to limited alkali digestion (+). RNAs from 1 × 10[6] cells were separated by electrophoresis in a 75 ml of "mini-Northern" denaturing formaldehyde gel,[8] containing 0.5 μg/ml of ethidium bromide (EtBr), for 8 hr at 20 V/cm (A). Note the accumulation of 28 and 18 S ribosomal RNAs in the nuclei of adenovirus-infected cells (A), a documented result of infection. (B) The autoradiographic exposure of the [32P]UTP-labeled RNAs of A after drying of the gel. Note that the 28 and 18 S rRNA bands are not as prominent in the newly transcribed, radioactive RNA population. The heterogeneity in nascent RNA transcripts is seen as a broad smear from greater than 5 kb to 300 nucleotides. However, limited alkali digestion (NaOH: +) reduces all RNAs to 300–400 nucleotides.

transcription buffer (without UTP) and then pellet the nuclei by centrifugation at 750 g for 3 min. (The equilibration step is optional; omit if the nuclei are very fragile.) Resuspend the nuclei for the transcription reaction in a minimal volume of transcription buffer, usually a 300-μl final volume/10[8] nuclei, depending on the size of nuclear pellet. Typically, the nuclei volume of 1 × 10[8] cells is ~100 μl. Start pulse labeling by adding [32P]UTP (100–200 μCi/10[8] cells; 3000-Ci/mmol specific activity) and incubate in a 31° water bath. For continuous labeling, add unlabeled UTP to a final concentration of 30 μM. For pulse–chase experiments, add unlabeled UTP to a final concentration of 3 mM during chase. Stop transcription by adding 20 reaction volumes of ice-cold PBS containing 3 mM unlabeled UTP. Typically, the elongation rate is ~150 nucleotides/min.[5] Pulse labeling for 3–5 min by this method results in the incorporation of 1–2 × 10^7 cpm into 10^8 HeLa cell nuclei or into other human lymphoid cell nuclei. Immediately centrifuge 750 g for 3 min at 4° to separate nuclei from unincorporated [^{32}P]UTP in chase solution.

Step 3: Isolation of RNA from Nuclei

RNAs can be isolated with the acidic phenol method or with the guanidine thiocyanate method. The acid phenol method is rapid. However, for isolation of RNA from nuclei rich in nucleases, the guanidine thiocyanate method is preferred due to its nearly immediate inactivation of nucleases.[13] Furthermore, this method, though time-consuming, also allows for efficient separation of RNA from DNA.

Step 3a: Acidic Phenol Method. NICKING OF CHROMATIN BY DNase I (OPTIONAL). Resuspend nuclei in HSB (0.5 ml/10^8 cells), in which nuclei lyse immediately. Add DNase (Worthington DPFF), 10 U/10^8 cells. Triturate 1–2 min at room temperature. The suspension should become less viscous with time. This step nicks chromatin and facilitates separation of the DNA into the interphase during acid–phenol extraction.

DIGESTION WITH PROTEINASE K. Add 5 volumes of 10 mM Tris–HCl, pH 7.4, and 10 mM EDTA to the above solution or to the nuclear pellet if the optional step above is omitted. Add SDS to a final concentration of 0.2%. Add proteinase K (self-digested for 30 min at 37°) to final concentration of 200 μg/ml. Digest for 30 min–1 hr at 37°. Add LiCl to a final concentration of 0.2 M. Extract with 2 volumes of acidic phenol (phenol saturated with 20 mM sodium acetate and 10 mM EDTA, pH 5.1) two times, followed by extraction with 2 volumes of acidic phenol : chloroform : isoamyl alcohol (25 : 24 : 1) one time, and extraction with chloroform : isoamyl alcohol (24 : 1) one time. Precipitate the aqueous phase in a new tube with 2.5 volumes of ethanol.

Step 3b: Guanidine Thiocyanate Method. Tap the nuclear pellet to prevent clumping and add 5 ml of guanidine thiocyanate solution in a 5-ml syringe with a 1.5-inch, 20-gauge needle. Syringe up and down to reduce viscosity (no clumps), which requires ~20 strokes. Avoid foaming during syringing, which can interfere with homogenization and may cause overflow of the suspension. Place on ice for a few minutes after syringing if

[13] J. M. Chirgwin, A. E. Przybyla, R. J. MacDonald, and W. J. Rutter, *Biochemistry* **18**, 5294 (1979).

foam persists or spin at low speed ($\sim$275 g) to remove bubbles. Add yeast tRNA in ETS to 10 μg/10^8 cells as a carrier for the recovery of small RNAs. Carefully layer the homogenized suspension onto a 4-ml cushion of CsCl solution in a SW41 centrifuge tube or equivalent and overlay the tube with paraffin oil. Pellet RNAs at 33,000 rpm for 36 hr. We have found that extended centrifugation time (36 vs 18–24 hr) is necessary for efficient recovery of small RNAs. After centrifugation, remove the CsCl gradient, the DNA interface first, with a Pasteur pipet. Wipe the outside of the tube with 70% (v/v) ethanol, label the bottom of the tube, and cut off the tube above the bottom seam with a razor blade or scalpel. Use caution, as the tube is slippery and can be difficult to cut. Remove the residual supernatant with a 200-μl Eppendorf pipettor and add 200 μl of ETS on top of the RNA pellet, which should be visible as a clear, tight pellet at the bottom of the tube. Allow the ETS solution to sit for a few minutes, then pipet up and down to dissolve the RNAs. Transfer the RNAs dissolved in ETS solution to a siliconized 15-ml Corex tube and repeat this step four more times to ensure complete recovery of RNAs. Add ETS to a final volume of 4 ml in the Corex tube, add LiCl to a concentration of 0.2 M, and precipitate with 10 ml of ice-cold ethanol. After recovery of the RNAs by centrifugation, it may be necessary to repeat the precipitation step in the same volume to remove any residual salt. Store RNAs in ethanol at all times until use.

Step 4: RNA–DNA Hybridization

Step 4a: Limited Alkaline Hydrolysis. Nascent RNAs at the 5' and 3' ends of a transcription unit can be very different in length, which may influence their hybridization efficiency to DNA templates, as shown in Figs. 1B and C. To minimize this potential problem, RNAs are subjected to limited hydrolysis with NaOH to a size of 300–400 nucleotides.

To accomplish this, spin down RNAs precipitated in ethanol. Dissolve RNAs in ETS solution (200 μl for 10^8 cell equivalents) and leave on ice until SDS precipitates out, indicating that the RNA solution is very cold. Add 50 μl of freshly prepared, ice-cold 1 N NaOH dropwise. Mix well and leave on ice for 15 min to allow limited hydrolysis. Neutralize the pH by adding 50 μl of unequilibrated 2 M HEPES buffer. The RNAs are now $\sim$400 nucleotides in length and ready for hybridization (Fig. 2). Optionally, one may choose to reduce the volume by ethanol precipitation.

Step 4b: Filter Hybridization. DNA probes, in at least 20-fold molar excess, are immobilized onto nitrocellulose filters using a slot–blot apparatus (Schleicher and Schuell, Keene, NH), according to the manufacturer's specifications. Filters are prehybridized in hybridization solution at 42° overnight. This solution is then replaced with a minimal volume of fresh hybridization solution (blot area × 0.067 ml). Labeled RNAs after alkali cleavage are added to the hybridization solution and are allowed to hybridize for 24 hr at 42°. The hybridization solution is collected carefully from each sample and is hybridized to duplicate filters, which have been prehybridized as before. This second round of hybridization is performed to confirm that, with DNA probe in excess, the first hybridization has gone to near completion. Following hybridization, these filters are washed and processed identically to the first set of filters. Typically, one does not detect any signals in a third round of hybridization.

Perform washing of the filters essentially according to the method of Thomas.[14] Remove the filters from the hybridization bags and rinse in a small volume of 2× SSC and 0.2% (w/v) SDS ("2×"). Swirl this gently to remove residual hybridization mix from the filters and dispose of the wash. Wash the filters four times for 10 min each at 42° with approximately 1 liter/wash of 2×, prewarmed to 42°. Next wash twice for 20 min each at 55° with approximately 1 liter/wash of 0.1× SSC and 0.2% (w/v) SDS ("0.1×"), prewarmed to 55°.

The filters are then treated with RNase A at 10 μg/ml in 2× for 30 min at 37° for a length of time determined for each RNase A preparation (5–20 min) in a container reserved for this purpose (to discourage the spread of RNases). This will eliminate labeled RNAs, which are not complementary to a given probe and which may otherwise be included as signals given by the probe (Fig. 1C). The RNase A treatment is followed by proteinase K digestion, 100 μg/ml in 2× for 30 min at 37°, which will inactivate the RNase A and which will reduce background on the filter. The filters are rinsed well with 2× and are carefully blotted on Whatman 3MM paper, but care must be taken not to let them dry out. The filters are wrapped immediately in Saran wrap and are exposed to an X-ray film with an intensifying screen at −70° overnight.

Step 5: Analysis of Data

The hybridization signals are quantified in the following two ways: (1) densitometric scanning of autoradiograms within the linear range of sensitivity and (2) scintillation counting of probes cut out of the filter after hybridization. The latter method is more accurate. The former has the advantage that the data can be compared to results obtained from steady-state mRNAs by densitometer tracing of Northern blot analysis or by

[14] P. Thomas, *Proc. Natl. Acad. Sci. U.S.A.* **77**, 5201 (1980).

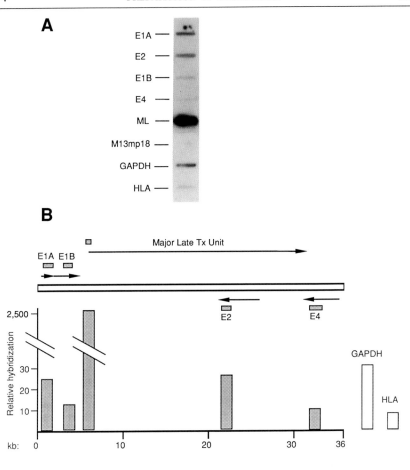

FIG. 3. Hybridization and analysis of hnRNA pulse labeled in nuclei from adenovirus-infected Jurkat T cells. Alkali-digested, pulse-labeled hnRNAs from adenovirus-infected Jurkat cell nuclei (Fig. 2) were hybridized to DNA probes specific for adenoviral and cellular genes in order to assess relative rates of transcription. Probes for adenoviral genes *E1A* (nucleotides 0–1,574; 1.57 kb), *E2* (nucleotides 22,896–21,672, 1.22 kb), *E4* (nucleotides 33,660–32,076, 1.59 kb), and the major late transcription unit (ML; nucleotides 5,983–6,542, 0.56 kb) were single-stranded DNA complementary to the mRNA, isolated from M13 phage vectors. Probes for adenovirus gene *E1B* (nucleotides 2,804–3,933, 1.13 kb) and cellular genes *HLA* class I antigen (human cDNA clone, 1.4 kb) and glyceraldehyde-3-phosphate dehydrogenase (*GAPDH;* rat cDNA clone, 1.3 kb) were double-stranded DNA inserts isolated from plasmid vectors. M13 vector mp18 (M13mp18) single-stranded DNA was used as control DNA. DNA probes were in 20-fold molar excess. Hybridization was performed as described, and filters were exposed to X-ray film overnight. (A) The autoradiogram of the hybridized filters. (B) The size, location, and orientation of adenovirus type 2 transcription

nuclease protection analysis. As shown in Fig. 3, the relative hybridization per base = (intensity of the band)/[(length of the probe) × (percentage G–C content of the probe)]. In cases in which two rounds of hybridizations are necessary to exhaustively hybridize the nascent RNAs, the intensity of the band is the sum of the two hybridizations. Comparison of transcription among cell lines is accomplished through the use of the relative hybridization signal for a housekeeping gene, such as glyceraldehyde-3-phosphate dehydrogenase (GAPDH; Fig. 3), as a standard. Transcription rates expressed as a ratio of this standard hybridization signal will help to correct for intrinsic differences in ploidy, efficiency of nuclei isolation, and incorporation of radionucleotides between cell lines.

Discussion

The procedure described in this chapter represents one of the currently available methods for assessing transcription *in vivo*.[15] Transcripts initiated prior to isolation of nuclei are elongated in the presence of [^{32}P]UTP. The signals hybridized to specific probes represent incorporation of [^{32}P]UTP into the growing 3' end of nascent RNA chains and hence the loading of polymerase. Since these signals are biochemical averages of transcription complementary to the probes used, it is important to (1) use short probes (preferably less than or equal to 400 nucleotides in size) representing various regions of a transcription unit to assess whether transcription is equimolar and to (2) use single-stranded probes to ascertain the polarity of transcription. Indeed, transcriptional pausing and termination, two events not distinguishable by pulse labeling of isolated nuclei, are found to occur in viral and cellular genes.[9-11]

While this method is widely used, it must be noted that whether the alteration of the nuclear organization due to Dounce homogenization or the leakage of small molecules, such as small nuclear ribonucleoproteins (snRNPs), from nuclei during preparation of nuclei influence transcription is still an open question. The importance of pulse labeling in as short a period of time as possible should also be emphasized. The nuclear dwell

[15] U. Schibler, O. Hagenbuchle, P. K. Wellauer, and A. C. Pittet, *Cell (Cambridge, Mass.)* **33**, 501 (1983).

units on the viral genome, represented by arrows. The DNA probes specific for each transcription (Tx) unit are also shown, represented by dotted bars. Beneath each probe is a histogram expressing the relative hybridization for each probe, calculated as described in the text. Results are presented graphically as arbitrary hybridization units. Relative hybridization values for *GAPDH* and *HLA* are also shown, represented by open bars. The relative hybridization values for each probe are as follows: *E1A*, 24.32; *E2*, 25.65; *E1B*, 11.93; major late promoter (ML), 2,645.8; *GAPDH*, 30.76; *HLA*, 7.86.

time of viral RNAs in human cells is ~30 min, and the predominant RNA species in the nucleus are spliced, mature RNAs ready for export,[16] implying that many nuclear processing events take place rapidly. Postinitiation events, such as nuclear turnover, which is currently poorly understood, may complicate the transcriptional analysis in experiments with prolonged labeling of nascent transcripts.

[16] S. Chen-Kiang, J. R. Nevins, and J. E. Darnell, Jr., *J. Mol. Biol.* **135**, 733 (1979).

[9] Isolation and Characterization of Yeast Ribosomal RNA Precursors and Preribosomes

By Jacobus Klootwijk and Rudi J. Planta

Introduction

The 80 S ribosomes from *Saccharomyces cerevisiae* contain four ribosomal RNAs, namely, 18 S rRNA in the 40 S subunit and 25, 5.8, and 5 S rRNAs in the 60 S subunit. All four species are encoded by a 9.1-kb (kilobase) rDNA unit, which is 100–200 times tandemly repeated on chromosome XII.[1-4] One such repeat (see Fig. 1) contains two transcription units, namely, a 35 S pre-rRNA (rRNA precursor) operon, which is transcribed by RNA polymerase I, and a 5 S rRNA gene, which is transcribed by RNA polymerase III. This chapter deals with the identification and characterization of the precursor RNAs and RNPs (ribonucleoproteins) from the 35 S pre-rRNA operon.

Figure 1 shows the major precursor rRNA species in wild-type *S. cerevisiae*. The 35 S pre-rRNA is the largest detectable transcript from the rRNA operon, but is unlikely to be the primary transcript, since very rapid processing has been observed at the 3' end.[5] In addition to 35 S pre-rRNA, 20 and 27 S pre-rRNAs are the predominant intermediates. The 32 S pre-rRNA is found in variable yields depending on the genetic background of the particular strain and the conditions of growth and harvest-

[1] T. D. Petes, *Proc. Natl. Acad. Sci. U.S.A.* **76**, 410 (1979).
[2] E. Schweizer, C. MacKechnie, and H. O. Halvorson, *J. Mol. Biol.* **40**, 261 (1969).
[3] J. Retèl and R. J. Planta, *Biochim. Biophys. Acta* **169**, 416 (1968).
[4] T. B. Øyen, G. Saelid, and G. V. Skuladottir, *Biochim. Biophys. Acta* **520**, 88 (1978).
[5] A. E. Kempers-Veenstra, J. Oliemans, H. Offenberg, A. F. Dekker, P. W. Piper, R. J. Planta, and J. Klootwijk, *EMBO J.* **5**, 2703 (1986).

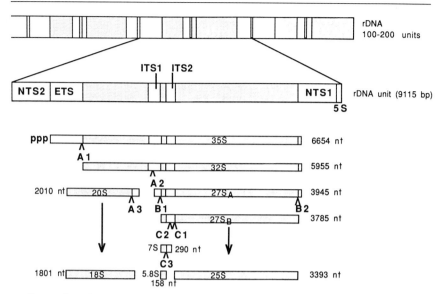

FIG. 1. Genetic organization of the rRNA operon in yeast and the processing pathway of the largest detectable transcript derived from it. More than one nomenclature is used for yeast (pre-)rRNAs; the one given in the figure is the one most frequently used. It relates to that employed by other laboratories as follows: 35 = 37 S, 27 = 29 S, 25 = 26 S, 20 = 18 S, 18 = 17 S. It should be noted that 35 S pre-rRNA is the largest detectable, but not the primary transcript. There is no obligatory temporal order for the cuts at A1, A2, and B1, but in most cases, those at A1 and A2 occur virtually simultaneously, followed by that at B1. The actual timing of the cut at B2 is unknown.

ing. The 27 S pre-rRNA consists of a mixture of two species, which have different 5' ends (either at A2 or at B1). Finally, 5.8 S rRNA is generated via a 7 S intermediate.

The major part of the ribosome biogenesis pathway is located in the nucleolus. Shortly after, or possibly already during transcription, 35 S pre-rRNA is subjected to a large number of modifications, including methylation and pseudouridylation. Furthermore, a large number of ribosomal and nonribosomal proteins associates with the 35 S pre-rRNA, resulting in the formation of a 90 S pre-rRNP. This particle is then split into separate precursors for the 60 and 40 S subunits, namely, a 66 and a 43 S pre-rRNP, respectively.[6] The 66 S particle matures completely within the nucleus, whereas the 43 S particle undergoes its final maturation steps in the cytoplasm.[6] Identification and isolation methods for the pre-rRNAs

[6] J. Trapman, J. Retèl, and R. J. Planta, *Exp. Cell Res.* **90**, 95 (1975).

and rRNPs depend on both their biochemical properties and the intracellular localization of these components.

Identification of Pre-rRNAs

Pre-rRNAs can be identified directly in total RNA extracted from cells or spheroplasts. The methods of detection are either Northern blot hybridization, using pre-rRNA-specific probes, or pulse labeling *in vivo*, using either [³H]uracil or [*methyl*-³H]methionine. Several protocols are available for the extractions of total RNA from yeast. However, to achieve quantitative yields of pre-rRNA, the following precautions have to be taken. (1) Exponentially growing cells or metabolically active spheroplasts should be used as a source for the RNA. (2) The cells or spheroplasts have to be chilled rapidly to 0–4° and processed further as quickly as possible. (3) RNase inhibitors [e.g., vanadyl–ribonucleoside complex (BRL, Gaithersburg, MD, 10 mM) or ATA (aurintricarboxylic acid; 50 μM)] should be included in the media used for breaking the cells or for lysis of the spheroplasts. (4) Stringent extraction procedures have to be used for a full recovery of 35 S pre-rRNA (see the following section).

Extraction from Cells

All glassware should be heated at 120° for 2 hr to inactivate nucleases. Twenty to forty milliliters of an exponentially growing culture is poured into a chilled glass tube containing sterile crushed ice. Cells are collected by centrifugation and are washed once with 20 ml of ice-cold EB [extraction buffer: 0.1 M Tris–HCl (pH 7.4), 0.1 M LiCl, 0.1 mM EDTA (ethylenediaminetetraacetic acid)]. Cells are suspended in 2 ml of EB, are transferred to a 2.2-ml Eppendorf tube, and are centrifuged, and the pellet is resuspended in 0.5 ml of EB. Then 1 g of glass beads (0.45–0.50 mm in diameter; washed in 1 N HCl and rinsed with H$_2$O several times) and ATA are added. The suspension is shaken vigorously by vortexing six times for 1 min, taking care to keep the sample chilled. Subsequently, SDS (sodium dodecyl sulfate), to a final concentration of 0.5% (w/v), and 0.5 ml of phenol : chloroform : isoamyl alcohol [50 : 50 : 1 (v/v/v)], containing 0.1% (w/v) 8-hydroxyquinoline and saturated with EB, are added. The mixture is vortexed vigorously for 15 min, while maintaining the mixture at 60°. After centrifugation for 5 min, the aqueous phase is transferred to a fresh tube and the phenol : chloroform layer *plus* the interphase is reextracted twice for 1 min using 0.15 ml of EB. The aqueous phases are pooled and extracted a second time with the phenol : chloroform mixture at room temperature for 5 min. RNA is precipitated by adding 0.1 volume of 3 M sodium acetate and 2 volumes of 100% (v/v) ethanol.

Extraction from Spheroplasts

When subcellular fractionation is desired, yeast cells have to be converted into spheroplasts, e.g., by suspending them in 1 M sorbitol, 20 mM K_2HPO_4 (pH 6.5), and 10 mM DTT (dithiothreitol), containing Zymolyase 100 T (1 mg/g of cells), and by shaking them gently at 20–25°. The time needed for complete spheroplast formation is variable (usually in the range of 5–15 min) and has to be judged by microscopic examination. After two washes in 1 M sorbitol, the spheroplasts are resuspended in an appropriate medium, containing 1 M sorbitol, and are allowed to return to full metabolic activity by gently shaking at 30° for at least 45 min, before the spheroplasts are labeled and processed. For isolation of total RNA, spheroplasts can be lysed in 0.1 M NaCl, 0.01 M EDTA, 50 mM sodium acetate (pH 5.0), and 1% (w/v) SDS and are extracted with phenol : chloroform as described in Extraction from Cells.

Fractionation of Pre-rRNAs

High-molecular-weight pre-rRNAs (20–35 S) are fractionated on agarose gels after glyoxylation[7] or on standard formamide–formaldehyde gels and then are blotted onto a solid support. Filters are hybridized with probes containing either ETS, ITS1, or ITS2 sequences (see Fig. 1), but lacking sequences corresponding to the mature 25 and 18 S rRNAs. A recombinant, containing ITS1 *plus* ITS2 sequences, is suitable for visualizing all regular pre-rRNAs in one hybridization.

Pulse Labeling

Both [³H]uracil and [*methyl*-³H]methionine can be used for pulse labeling of pre-rRNAs *in vivo*. [*methyl*-³H]Methionine is very useful for routine labeling of the high-molecular-weight pre-rRNAs, since the bulk of the methyl groups are introduced at the level of the 35 S pre-rRNA (see below) shortly after or during transcription. Furthermore, pre-mRNAs are labeled very poorly by this compound, since pre-RNAs usually contain only one methyl group per molecule.[8] However, for trustworthy kinetic analysis, [³H]uracil is recommended. Since 5.8 and 5 S rRNAs are

---

[7] P. Thomas, *Proc. Natl. Acad. Sci. U.S.A.* **77**, 5201 (1980).
[8] W. H. Mager, J. Klootwijk, and I. Klein, *Mol. Biol. Rep.* **3**, 9 (1976).

devoid of methyl groups, [³H]uracil has to be used for the kinetic analysis of these low-molecular-weight RNAs.

For studying the details of the processing, labeling of the culture at lower temperatures can be considered, e.g., to achieve preferential labeling of the 35 S pre-rRNA. Short pulses (e.g., 2 min) at 30° cause already significant labeling of the 27 and 20 S pre-rRNA species. Lowering the temperature to 15° slows down the rate of ribosome formation by a factor of three.[9] Pulse labeling of cells can be terminated by transfer to a tube containing sterile crushed ice together with an excess of unlabeled uracil (or methionine). Substituting a solution containing glycerol for plain ice allows an even more drastic chilling. Spheroplasts have to be poured into a 10-fold excess of ice-cold 1 M sorbitol containing unlabeled uracil or methionine. For visualization of the pre-rRNAs, total RNA is extracted and fractionated on agarose–formaldehyde gels, is transferred to GeneScreen membrane (NEN, Boston, MA), and is subjected to fluorography. Alternatively, gels can be sliced, and radioactivity in the slices can be quantitated by scintillation counting.

Identification and Isolation of Pre-rRNPs

Pulse labeling of spheroplasts with [³H]uracil and subsequent extraction of RNPs reveal three major peaks of radioactivity, sedimenting at 90, 66, and 43 S, respectively.[6] These pre-rRNPs contain 35, 27, and 20 S pre-rRNAs, respectively. Subcellular fractionation has revealed that 90 and 66 S particles are exclusively nuclear constituents, while the majority of the 43 S particles is found in the cytoplasm. The sedimentation values, in particular of the 90 and 66 S particles, are considerably affected by the ionic strength of the media used in the extraction and in the subsequent sucrose gradient centrifugation.[6] Furthermore, these particles probably contain a set of small nuclear RNAs.[10]

Pre-rRNPs, as originally described by Trapman et al.,[6] were detected in the following way. Spheroplasts or crude nuclei (see below) were suspended after an appropriate time of labeling in 10 mM Tris–HCl (pH 7.4), containing 5 mM MgCl₂, 10 mM DTT, 0.2% (w/v) Brij-58, and 0.01 or 0.2 M KCl. The inclusion of a RNase inhibitor such as RNasin to block further processing and degradation of the pre-rRNA proved to be crucial. After stirring for 5 min at 0°, the suspension was centrifuged for 10 min at 4°. The supernatant was loaded on a 15–30% linear sucrose gradient in 10 mM Tris–HCl (pH 7.4), containing 5 mM MgCl₂, 1 mM DTT, and 0.01 or

[9] J. Trapman and R. J. Planta, Biochim. Biophys. Acta 442, 265 (1976).
[10] D. Tollervey, EMBO J. 6, 4169 (1987).

0.2 M KCl. RNP fractions from this gradient can now be subjected to RNA analysis and to determination of the buoyant density. Labeling with ^{3}H-labeled amino acids allows the analysis of the protein composition of the 66 and 43 S particles.[11]

Buoyant density analysis of the 90 and 66 S particles revealed a considerably higher protein–RNA ratio than observed for the 60 S subunits. The 43 S particles and the 40 S subunits, on the other hand, were found to have identical buoyant densities. Buoyant densities were measured after fixation of the particles with formaldehyde. Sucrose gradient fractions were pooled, and concentrated formaldehyde (neutralized with $NaHCO_3$) in 50 mM triethanolamine–HCl (pH 7.8), containing 5 mM $MgCl_2$ and 0.01 or 0.2 M KCl, was added to a final concentration of 6%. The suspension was incubated for 24 hr at 4° and was dialyzed against the same triethanolamine buffer, but lacking formaldehyde. Fixed particles were subjected to CsCl equilibrium centrifugation at 4° in a SW65 rotor at 30,000 rpm for 48 hr. Fractions obtained after puncturing the bottom of the tube were assayed for ^{3}H radioactivity and for refractive index.

Preparation of Labeled Pre-rRNAs for Structural Analysis

Structural analysis of pre-rRNAs requires the isolation of radiochemically pure preparations of a sufficiently high specific activity. Using these samples, classical RNA fingerprinting procedures can be used to analyze end groups, modifications, and other structural features. The study of the methylation characteristics of the pre-rRNAs requires labeling with [*methyl*-^{3}H]methionine. For obtaining an appropriate specific activity, subcellular fractionation is essential, and therefore, labeling has to be performed in spheroplasts. The 35, 27, and 7 S pre-rRNAs require the isolation of a crude nuclear pellet. The 20 S pre-rRNA can be obtained from 43 S particles isolated under conditions which retain the bulk of the 40 S subunits in polyribosomes.

In all cases, cells can be grown in a rich medium (e.g., 20 g of glucose, 1 g of bactopeptone, 2 g of yeast extract, and 3 g of malt extract/liter) and can be converted into spheroplasts as rapidly as possible (see above). After washing the spheroplasts twice in prewarmed medium, the spheroplasts are incubated at 30° at an $OD_{550 nm}$ of 1.0. Spheroplasts should be suspended carefully without vigorous shaking. A minimal medium should be used for labeling. A YNB (yeast nitrogen base) medium without amino acids, containing 1 M sorbitol, is suitable for labeling with [*methyl*-^{3}H]methionine. Adenosine and guanosine (10 mg/liter) and 0.01 M sodium

---

[11] T. Kruiswijk, R. J. Planta, and J. M. Krop, *Biochim. Biophys. Acta* **517**, 378 (1978).

formate were routinely included to suppress randomization of the label. For labeling with ortho[^{32}P]phosphate, we have used a phosphate-free medium containing 10 g of glucose, 1 g of $MgSO_4 \cdot 7H_2O$, 4 g of $(NH_4)_2SO_4$, 0.8 g of KCl, 2 g of sodium citrate $\cdot 2H_2O$, 20 mg of methionine, 10 mg of glutamic acid, 10 mg of tryptophan, 50 mg of inositol, 6 ml of 70% sodium lactate, 120 g of mannitol/liter, brought to pH 5.8 by the addition of 1 M citric acid. After 45 min, a nutritional shift up is created by adding a 0.01 volume of a mixture of all amino acids and ribonucleosides (15 mg of each amino acid and 25 mg of each nucleoside dissolved in 20 ml of H_2O). Ten minutes later, 200 ml of the culture is shifted to 15°. After labeling (see below), spheroplasts are poured into 3 volumes of ice-cold 1 M sorbitol and are collected by centrifugation. The supernatant is removed carefully, and the pellets are suspended in 20 ml of 1 mM $MgCl_2 \cdot 6H_2O$, 20 mM K_2HPO_4 (pH 6.5), 8% (w/v) poly(vinylpyrrolidone), 0.03% (v/v) Triton X-100, 10 mM DTT, 50 mg/liter of poly-(vinyl sulfate), and 10 mM VRC (vanadium–ribonucleoside complex) using a glass pipet; the suspension is kept at 0° for 10 min and is homogenized subsequently in a Potter–Elvehjem-type homogenizer with a tight-fitting pestle and is diluted after homogenization with an equal volume of the same solution containing 20% (w/v) sucrose. Intact cells and debris are removed by centrifugation at 2000 g for 5 min. Nuclei are pelleted at 20,000 g for 15 min, and RNA is extracted (see above). For labeling 35 and 27 S pre-RNAs, 5 mCi of [*methyl*-^{3}H]methionine or 50 mCi of ortho[^{32}P]-phosphate is added for 13–15 and 20 min, respectively. Pre-rRNAs are purified by two cycles of sucrose gradient centrifugation [15–30% (w/v) sucrose in 0.1 M NaCl and 0.01 M EDTA]. For the isolation of 7 S pre-rRNA, the period of labeling with ortho[^{32}P]phosphate at 15° can be extended for 45–60 min. The 66 S particles are then isolated as described above. After extraction of the RNA, the RNA is dissolved in 50 mM Tris–HCl (pH 7.4) containing 1 mM EDTA. The 7 S pre-rRNA is released from the high-molecular-weight rRNA by incubation for 5 min at 60° and is purified on 8% (w/v) polyacrylamide gels. After radioautography, 7 S pre-rRNA is eluted from gel slices using 1 M NaCl. For the isolation of cytoplasmic 20 S pre-rRNA, spheroplasts can be labeled for a period up to 35 min.

Pre-RNPs are then extracted from lysed spheroplasts as described above. ^{32}P-Labeled 20 S pre-rRNA is isolated by pooling the leading edge of the 43–40 S peak in the sucrose gradient and by subsequent extraction with phenol–SDS and by precipitation with ethanol. The specific activity is usually about 10^5 dpm/μg. This value is about 20-fold higher than that obtained when extracting RNA directly from spheroplasts. The reason for this is that, using the above method, the bulk of the mature 40 S subunits is in the polyribosomes, and therefore, contamination of the 43 S particles with mature 40 S subunits is reduced appreciably.

Finally, any part of the 35 S pre-rRNA operon can be cloned behind either the T7 or SP6 promotor to synthesize (labeled) pre-rRNA or fragments thereof *in vitro*. These transcripts have proved to be very useful for studies of rRNA–ribosomal protein interactions[12] and will also provide suitable substrates for *in vitro* processing and modification studies.

Characterization of Yeast Pre-rRNAs

Mapping of Sites for Transcription Initiation and Processing

Analysis of the 5'-terminal sequence of 35 S pre-rRNA revealed the presence of a triphosphate group (pppA-Up).[13] This result allowed the precise mapping of the transcription initiation site on the rDNA unit. This mapping was based on primer extension and S1 nuclease analysis[14–16] using either total or nuclear unlabeled RNA. A fragment, spanning the region from positions −149 to +15 relative to the site of initiation, has been shown to promote transcription of a ribosomal minigene.[17,18]

Eight processing sites have been identified in the 35 S pre-rRNA (see Fig. 1). Six of these are located at the junctions between the ends of the three mature rRNAs and the adjoining transcribed spacers. The two remaining sites lie within ITS1 and ITS2, respectively. These eight sites are sufficient to account for the formation of all major pre-rRNA intermediates.

Analysis of the 5' end of cytoplasmic 20 S pre-rRNA revealed a sequence identical to that present at the 5' end of 18 S rRNA.[19] This shows that the nonconserved nucleotides present in 20 S pre-rRNA are exclusively located at the 3' end. These ITS1 sequences are removed in the cytoplasm in the last maturation step detectable before the 40 S subunits

---

[12] T. T. A. L. El-Baradi, V. C. H. F. de Regt, S. W. C. Einerhand, J. Teixido, R. J. Planta, J. P. G. Ballesta, and H. A. Raué, *J. Mol. Biol.* **195,** 909 (1987).

[13] J. Klootwijk, P. de Jonge, and R. J. Planta, *Nucleic Acids Res.* **6,** 27 (1979).

[14] R. Klemenz and P. Geiduschek, *Nucleic Acids Res.* **8,** 2679 (1980).

[15] A. A. Bayev, O. I. Georgiev, A. A. Hadjiolov, M. B. Kermekchiev, N. Nikolaev, K. G. Skryabin, and V. M. Zakharyev, *Nucleic Acids Res.* **8,** 4919 (1980).

[16] J. Klootwijk, M. P. Verbeet, G. M. Veldman, V. C. H. F. de Regt, H. van Heerikhuizen, J. Bogerd, and R. J. Planta, *Nucleic Acids Res.* **12,** 1377 (1984).

[17] A. E. Kempers-Veenstra, H. van Heerikhuizen, W. Musters, J. Klootwijk, and R. J. Planta, *EMBO J.* **3,** 1377 (1984).

[18] A. E. Kempers-Veenstra, W. Musters, A. F. Dekker, J. Klootwijk, and R. J. Planta, *Curr. Genet.* **10,** 253 (1985).

[19] P. de Jonge, J. Klootwijk, and R. J. Planta, *Eur. J. Biochem.* **72,** 361 (1977).

become active in protein synthesis. The position of the 3'-terminal sequence of 20 S pre-rRNA has been traced to a location near the middle of ITS1.[20] Analysis of 7 S pre-rRNA revealed a similar picture: its 5'-terminal sequence is identical to that of 5.8 S rRNA, and thus, all sequences in 7 S pre-rRNA, not found in the mature 5.8 S rRNA, are present at the 3' end.[21] The cut separating the 7 and 26 S rRNA sequences (site C2, see Fig. 1) is made near the middle of ITS2.[22] End-group analyses of 27 S pre-rRNA did not give unambiguous results.

Primary Structure of Pre-rRNAs

The sequence of the pre-rRNAs can be deduced from the sequence analysis of the rDNA unit, taking into account the location of the established sites for transcription initiation and for processing. The sequence obtained in this way has been confirmed by cataloging oligonucleotides obtained by two-dimensional fractionation of RNase T1 or RNase A digests of ^{32}P-labeled 27, 20, and 7 S pre-rRNAs[20-22] and by comparing these catalogs with those of the corresponding mature rRNAs. The lengths of the various ribosomal RNA species are given in Fig. 1. To facilitate the compilation of the scattered sequence data on yeast rDNA, the various sources are summarized below. For this, one has to take into account that *Saccharomyces carlsbergensis* is now classified as a synonym of *S. cerevisiae,* and therefore, the *S. carlsbergensis* rDNA unit has to be considered as one of the *S. cerevisiae* alleles.

Nontranscribed Spacers Including 5 S rRNA Gene. Since no site for transcription termination has yet been identified unambiguously, the entire so-called nontranscribed spacers may still turn out to be actually part of the RNA polymerase I transcription unit. The complete sequences of the nontranscribed spacers of two alleles can be found in Skryabin *et al.*[23] This part of the rDNA unit exhibits extended sequence polymorphism, which is further illustrated by partial sequences of other alleles.[24-26]

Transcribed Spacer Sequences. Sequence data for the ETS have been

[20] G. M. Veldman, R. C. Brand, J. Klootwijk, and R. J. Planta, *Nucleic Acids Res.* **8,** 2907 (1980).

[21] P. de Jonge, R. A. Kastelein, and R. J. Planta, *Eur. J. Biochem.* **83,** 537 (1978).

[22] G. M. Veldman, J. Klootwijk, H. van Heerikhuizen, and R. J. Planta, *Nucleic Acids Res.* **9,** 4847 (1981).

[23] K. G. Skryabin, M. A. Eldarov, V. L. Larionov, A. A. Bayev, J. Klootwijk, V. C. H. F. de Regt, G. M. Veldman, R. J. Planta, O. I. Georgiev, and A. A. Hadjiolov, *Nucleic Acids Res.* **12,** 2955 (1984).

[24] R. Jemtland, E. Maehlum, O. S. Gabrielsen, and T. B. Øyen, *Nucleic Acids Res.* **14,** 5145 (1986).

[25] M. E. Swanson and M. J. Holland, *J. Biol. Chem.* **258,** 3242 (1984).

[26] M. E. Swanson, M. Yip, and M. J. Holland, *J. Biol. Chem.* **260,** 9905 (1985).

published for three different alleles.[15,16,27,28] They reveal only very limited sequence heterogeneity. The length of the ETS varies from 695 to 699 bp (base pairs).

Complete sequences for ITS1[20,29] set its length at 360–362 bp. The processing site A2 leaves 209 nucleotides of ITS1 at the 3′ end of 20 S pre-rRNA and 153 at the 5′ end of 27 S_A pre-rRNA. The length of ITS2 is 234 bp.[22,30] The cut at C2 (see Fig. 1) leaves 132–135 nucleotides of ITS2 at the 3′ end of 7 S pre-rRNA.

18, 5.8, and 25 S rRNA. The original sequence published for 18 S rRNA[31] has recently been updated.[32] In addition, Hogan *et al.*[33] predict the insertion of another 3 bp on the basis of phylogenetic comparison resulting in a molecule of 1801 nt (nucleotides).

The 5.8 S rRNA was sequenced by Rubin.[34] Ten percent of the 5.8 S rRNA population consists of molecules having a 5′-terminal extension of 6 or 7 nt.[35] This extension is also present in a minor portion of the 7 S pre-rRNA molecules.[21]

Two independent sequences of 25 S rRNA have been published.[36,37] The two sequences differ in 16 positions, which could reflect either experimental errors or sequence heterogeneity. The inferred length of the 25 S rRNA molecule is 3390–3393 nt.

35 S Pre-rRNA. The combined sequence data discussed above allow us to calculate the length of the 35 S pre-rRNA to be 6645–6654 nt, taking into account a 3′-terminal extension of 7 nt beyond the 3′ end of 25 S pre-rRNA[38]; 19.6% of the molecule is removed during processing.

[27] P. Valenzuela, G. I. Bell, A. Venegas, E. T. Sewell, F. R. Masiarz, L. J. DeGennaro, F. Weinberg, and W. J. Rutter, *J. Biol. Chem.* **252**, 8126 (1977).

[28] K. G. Skryabin, V. M. Zakharyev, P. M. Rubtsov, and A. A. Bayev, *Dokl. Akad. Nauk SSSR* **247**, 1275 (1979).

[29] K. G. Skryabin, A. S. Krayev, P. M. Rubtsov, and A. A. Bayev, *Dokl. Akad. Nauk SSSR* **247**, 761 (1979).

[30] A. A. Bayev, O. I. Georgiev, A. A. Hadjiolov, N. Nikolaev, K. G. Skryabin, and V. M. Zakharyev, *Nucleic Acids Res.* **9**, 789 (1981).

[31] P. M. Rubtsov, M. M. Musakhanov, V. M. Zakharyev, A. S. Krayev, K. G. Skryabin, and A. A. Bayev, *Nucleic Acids Res.* **8**, 5779 (1980).

[32] A. S. Mankin, K. G. Skryabin, and P. M. Rubtsov, *Gene* **44**, 143 (1986).

[33] J. J. Hogan, R. R. Gutell, and H. F. Noller, *Biochemistry* **23**, 3322 (1984).

[34] G. M. Rubin, *J. Biol. Chem.* **248**, 3860 (1973).

[35] G. M. Rubin, *Eur. J. Biochem.* **41**, 197 (1974).

[36] G. M. Veldman, J. Klootwijk, V. C. H. F. de Regt, R. J. Planta, C. Branlant, A. Krol, and J. P. Ebel, *Nucleic Acids Res.* **9**, 6935 (1981).

[37] O. I. Georgiev, N. Nikolaev, A. A. Hadjiolov, K. G. Skryabin, V. M. Zakharyev, and A. A. Bayev, *Nucleic Acids Res.* **9**, 6953 (1981).

[38] G. M. Veldman, J. Klootwijk, P. de Jonge, R. J. Leer, and R. J. Planta, *Nucleic Acids Res.* **8**, 5179 (1980).

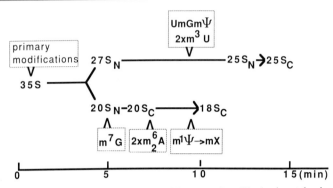

Fig. 2. Timing of modifications during rRNA maturation. The horizontal axis represents a time scale indicating the lifetime of the various pre-rRNAs at 15°. The primary modifications involve most, if not all, pseudouridylations, most, except one or two, 2'-O-methylations, introduction of four-base methyl groups into the 25 S rRNA sequence, and finally, the formation of $m^1\Psi$, which is the precursor of mX.

Modification of Yeast Pre-rRNAs

The 25 and 18 S rRNAs contain 43 and 24 methyl groups, respectively.[39] In each molecule, six of these methyl groups are attached to bases: two m^1A, two m^5C, and two m^3U residues in 25 S rRNA and two m_2^6A, one m^7G, and one methylated, hypermodified nucleotide (mX) in 18 S rRNA. The latter nucleotide has been identified as 1-methyl-3-γ-(α-amino-α-carboxypropyl)pseudouridine.[40] The remaining methyl groups are all linked to the 2'-O of ribose moieties. Analysis of methyl-^{3}H-labeled 35 S pre-rRNA has revealed that all but eight or nine methyl groups are introduced at the level of 35 S pre-rRNA.[41] As illustrated in Fig. 2, the remaining methyl groups are introduced at the level of the 27 and 20 S pre-rRNAs. For 27 S pre-rRNA, these modifications involve the formation of two m^3U residues and the introduction of one or both methyl groups at the highly conserved Um-Gm-Ψ sequence shortly before the conversion into 25 S rRNA. These secondary modifications are located in domain V.[36] For 20 S pre-rRNA, the late methylations consist of formation of a m^7G, probably introduced in the nucleus shortly before transport to the cytoplasm. Furthermore, two adjacent m_2^6As are formed immediately after the

[39] J. Klootwijk and R. J. Planta, Eur. J. Biochem. 39, 325 (1973).
[40] B. E. H. Maden, J. Forbes, P. de Jonge, and J. Klootwijk, FEBS Lett. 59, 60 (1975).
[41] R. C. Brand, J. Klootwijk, T. J. M. van Steenbergen, A. J. de Kok, and R. J. Planta, Eur. J. Biochem. 75, 311 (1977).

pre-rRNA reaches the cytoplasm. Finally, shortly before the conversion of the 20 S pre-rRNA to the mature 18 S rRNA, the formation of mX is completed by the attachment of the 3-γ-(α-amino-α-carboxypropyl) group to $m^1\Psi$.[42] The 25 and 18 S rRNAs contain about 36 Ψ-residues.[43] Analysis of ^{32}P-labeled 35 S pre-rRNA showed that most, if not all, of these pseudouridines are formed at an early stage of ribosome biogenesis. Only a few of them, however, have been located in the primary structure of 25 S rRNA.[36]

Methylation of Pre-rRNA in Isolated Nuclei

Methylation of pre-rRNA in isolated nuclei can be studied by using SAM (*S*-[*methyl*-^{3}H]adenosylmethionine). Several protocols have been described for the isolation of transcriptionally active nuclei from yeast.[44,45] These nuclei can incorporate 0.3–0.5 nmol of NTPs (nucleoside triphosphates)/μg of DNA within 15 min. The inclusion of SAM in the medium during transcription leads to efficient 2'-O-methylation of pre-rRNA at sites also modified *in vivo*.[44] A significant amount of 2'-O-methylation occurs even in the absence of transcription, indicating a time lag between transcription and primary methylation at the time of isolation of the nuclei.

For methylation studies, nuclei can be suspended (20–50 μg DNA/ml) in 0.2 M Tris–HCl (pH 8.0), 2 mM MgCl$_2$, 3 mM ATP, 1.2 mM each of GTP, CTP, and UTP, 0.5 mM DTT, 0.1 mM EDTA, 60 mM KCl, 0.16 mM SAM, 20% (w/v) sucrose, 10% (v/v) glycerol, and 20 μg/μl of α-amanitin to inhibit RNA polymerase II. Nuclei are incubated for 20 min at 24° to complete the methylation. 2'-O-Methylation can be assayed by adding [*methyl*-^{3}H]SAM at 40 μCi/ml. Roughly 25% of the *methyl*-^{3}H label is incorporated into tRNA and less than 5% in 5' cap structures of pre-mRNAs and/or snRNAs (small nuclear RNAs). The remainder of the label is present at the 2'-O of ribose residues in pre-rRNA. In the absence of CTP and GTP, the DNA-dependent RNA synthesis drops to 4%, whereas the overall methylation is reduced by only 35%.[44] Analysis of specific methylation sites is performed by mixing the *in vitro* [*methyl*-^{3}H]RNA with *methyl*-^{14}C-labeled mature 25 and 18 S rRNAs,[39] followed by ribonuclease T1 digestion and by subsequent two-dimensional fractionation of the oligonucleotides. The labeled oligonucleotides are located

[42] R. C. Brand, J. Klootwijk, R. J. Planta, and B. E. H. Maden, *Biochem. J.* **169**, 71 (1978).
[43] R. C. Brand, J. Klootwijk, C. P. Sibum, and R. J. Planta, *Nucleic Acids Res.* **7**, 121 (1979).
[44] J. Klootwijk, R. J. Planta, and J. P. Bakker, *J. Microsc. Biol. Cell* **26**, 91 (1976).
[45] J. F. Jerome and J. A. Jaehning, *Mol. Cell. Biol.* **6**, 1633 (1986).

by autoradiography, are cut from the paper, and are counted for both ^3H and ^{14}C radioactivity.

New Approaches for Study of Ribosome Biogenesis

As ribosome biogenesis appears to comprise a large number of interrelated steps, involving many different RNA–protein interactions, it will be extremely difficult to design systems that allow *in vitro* reconstitution of part(s) of the overall pathway. A more feasible approach, therefore, is to develop systems for mutational analysis of yeast rDNA *in vivo*. These systems will enable us to identify sequences involved in the processing, modification, and assembly of 35 S pre-rRNA and to unravel the function of specific regions of interest, such as transcribed spacers and expansion segments. The first of these systems has been developed recently in our laboratory.[46] The basic idea is to supply the yeast cell with a small proportion of tagged rDNA units, in which the tag behaves as a neutral mutation, that, nevertheless, can be used to distinguish the transcripts produced by these units from the bulk of the cellular rRNA. Thus, the tagged rDNA unit can be subjected to *in vitro* mutagenesis, and the fate of its rRNA products can be monitored *in vivo*. For this purpose, we have constructed a plasmid called pORCS (**o**ligonucleotide-tagged **r**ibosomal **c**entromeric DNA of *S. cerevisiae*), which contains 9.7 kb of yeast rDNA in the *Bam*HI–*Hin*dIII sites of YCp7. The insert extends from the *Bcl*I site in the 25 S rRNA gene (at 3065 nt) to the *Hin*dIII site at 270 bp downstream of the 3' end of the 25 S rRNA gene of the next 35 S pre-rRNA operon. The insert includes two enhancers for RNA polymerase I to ensure its optimal transcription. The inclusion of the centromere is crucial to achieve stable yeast transformants, since the centromere selects against inadvertent integration into the chromosome which might lead to uncoupling of the mutation from the tag.

To tag the plasmid-borne 35 S pre-rRNA operon, we inserted an 18-mer oligonucleotide into the unique *Kpn*I site, which is present in the first expansion segment of 25 S rRNA (see Fig. 3).[36] The insertion slightly extends the stem, and a larger loop will be present at the tip. After transformation of the tagged operon (pORCS) into yeast, mature, tagged 25 S rRNA can be easily detected by either Northern blot hybridization or primer extension, using the 18-mer oligonucleotide as a probe. Furthermore, we have found tagged 25 S rRNA to be present in polyribosomes in

[46] W. Musters, J. Venema, H. van Heerikhuizen, J. Klootwijk, and R. J. Planta, *Mol. Cell. Biol.* **9**, 551 (1989).

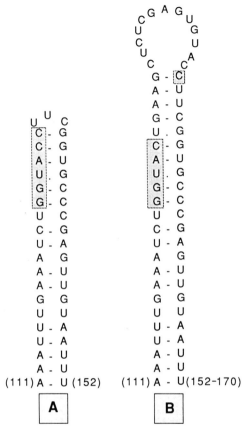

FIG. 3. (A) The first expansion segment in 25 S rRNA used for tagging. The coordinates are from Veldman et al.[36] The KpnI site used for the insertion of the 18-mer oligonucleotide is shaded. (B) The resulting secondary structure.

a distribution indistinguishable from that of untagged 25 S rRNA. This allows us to conclude that the tag behaves as a neutral mutation with respect to ribosome biogenesis and function. The tagged operon opens the way for a systematic study of rDNA mutations created *in vitro* that will ultimately delimit critical sequences involved in processing, modification, and assembly within 35 S pre-rRNA.

[10] Preparation of Yeast Transfer RNA Precursors *in Vivo*

By GAYLE KNAPP

The ability to isolate the substrates for RNA processing enzymes is a major step in the characterization of RNA biosynthetic pathways. Currently, we have available the *in vitro* transcription methods that allow the synthesis of specific substrates.[1] However, these methods are limited to the synthesis of those substrates for which genes have been cloned or for which complete sequences (and synthetic genes) are available. In addition, the *in vitro* synthesized substrates, e.g., tRNA precursors (pre-tRNAs), do not always contain the full set of base modifications that are present when the precursors are isolated *in vivo*. Lack of base modifications could potentially affect the structure and activity of the precursors; therefore, the *in vivo* isolation procedure provides an alternative source for tRNA splicing substrates.

This chapter deals with the preparation of pre-tRNAs from the yeast *Saccharomyces cerevisiae*. The procedure is adaptable for the preparation of both unlabeled and radioisotopically labeled pre-tRNA and utilizes a RNA biosynthesis mutant that accumulates tRNA precursors. Precursor-accumulating mutants are essential for the isolation of biochemically useful amounts of pre-tRNAs, although low levels of precursors in RNA from wild-type cells can be detected on Northern blots using high-specific-activity probes.[2]

Media, Solutions, and Reagents

YEPD media
    1% (w/v) Bactoyeast extract
    2% (w/v) Bactopeptone
    2% (w/v) Glucose

    Autoclave yeast extract and peptone in nine-tenths of a final volume of distilled H_2O for 30 min. Cool slowly to room temperature. Autoclave for another 30 min. Cool. Add one-tenth of a final volume of sterile 20% (w/v) glucose.

$10\times$ Lo-Phos
    1. Dissolve 20 g of peptone, 10 g of yeast extract, and 13.4 g of yeast nitrogen base with amino acids in 75 ml of water with heating.

---

[1] V. M. Reyes and J. N. Abelson, this volume [6].

[2] J. P. O'Connor, Master's Thesis, University of Pittsburgh, Pittsburgh, Pennsylvania, 1986.

METHODS IN ENZYMOLOGY, VOL. 180

2. Add concentrated NH$_4$OH to pH 10.0 (~20 ml).
3. Add 10 ml of 1 *M* MgCl$_2$ and stir in a cold room overnight.
4. Add another 10 ml of 1 *M* MgCl$_2$. Leave in a cold room *at least* 2 hr.
5. Filter media using a vacuum flask and a Büchner funnel with Whatman #1 paper.
6. Assay for PO$_4$ concentration (see Phosphate Determination). If the concentration is less than 20 μg/ml, skip to step 10.
7. Add 5 ml more of 1 *M* MgCl$_2$. Stir in a cold room for 2 hr.
8. Filter media through a sterile Nalgene filter unit or as in step 5.
9. Reassay for PO$_4$ concentration. If the concentration is *not* less than 20 μg/ml, return to step 7.
10. Adjust the pH to 7.0 with HCl (**Caution:** Ammonia and HCl are relatively volatile; care should be used not to inhale the fumes.)
11. Adjust to a final volume of 200 ml with H$_2$O.
12. Autoclave 30 min.
13. Cool slowly to room temperature.
14. Autoclave 30 min.

Phosphate determination

The phosphate content of the media is measured using a modification of the Fiske–Subbarow method.[3]

1. Reagent A is a mixture containing 6 *N* sulfuric acid : distilled H$_2$O : 2.5% (w/v) ammonium molybdate (1 : 2 : 1).
2. Standard and unknown samples for phosphate determination are each contained in a final volume of 1.0 ml (adjusted using glass-distilled H$_2$O).
3. The 1-ml sample and 4.0 ml of Reagent A are mixed thoroughly and are incubated at 37° for 2 hr.
4. After cooling to room temperature, the absorbance at 820 nm is determined for each tube.
5. The phosphate content of each unknown is determined from a standard curve (1.0–40 μg of K$_2$HPO$_4$).

10× Yeast succinate
10% (w/v) Succinic acid
6% (w/v) NaOH

Autoclave for 30 min.

Lo-Phos media
25 ml 10× Yeast succinate

---

[3] C. H. Fiske and Y. Subbarow, *J. Biol. Chem.* **66**, 375 (1925).

25 ml 10× Lo-Phos
25 ml 20% (w/v) Glucose
175 ml Sterile H_2O
SEA buffer
    1% (w/v) Sodium lauryl (or dodecyl) sulfate (SDS)
    2 mM Ethylenediaminetetraacetic acid (EDTA)
    0.1 M Sodium acetate, pH 5.0
0.3 M NaCl TE buffer
    10 mM Tris–HCl, pH 7.4
    1 mM EDTA
    0.3 M NaCl
1.0 M NaCl TE buffer
    10 mM Tris–HCl, pH 7.4
    1 mM EDTA
    1 M NaCl
RNA gel dye buffer
    7 M Urea
    10% (w/v) Glycerol
    0.05% (w/v) Xylene cyanol
    0.05% (w/v) Bromphenol blue
TBE
    90 mM Tris–borate, pH 8.3
    1 mM EDTA
RNA elution buffer
    0.3 M NaCl TE buffer containing 1% (v/v) redistilled phenol; keep refrigerated and away from light.
Glycogen
    20 mg/ml of glycogen in diethyl pyrocarbonate (DEPC)-treated H_2O; the glycogen must be RNase-free, e.g., molecular biology grade, from Boehringer Mannheim (Indianapolis, IN).

Growth of Yeast

This procedure is presented for the isolation of [^{32}P]RNA from a temperature-sensitive yeast mutant, but is easily adapted for preparation of unlabeled or other isotopically labeled RNA. Variations will be explained at the end of this section. When working with yeast, it is important to remember that the capacity of this organism to produce large amounts of degradative enzymes increases with higher cell densities. A good rule of thumb is to harvest the cells before the midpoint of the log phase.

The diploid strain, M304, carrying the *rna1* mutation,[4] is grown in 200 ml of YEPD media at 23° to a density of 3 × 10⁷ cells/ml ($OD_{660} \sim 2.0$). The cells are harvested in sterile bottles by centrifugation at 4200 g at 4° for 5 min. The cell pellet is washed sterilely with 50 ml of Lo-Phos media and is repelleted at 4200 g for 5 min. The washed cells are resuspended in 200 ml of Lo-Phos media, are transferred sterilely to a 500-ml culture flask, and are shaken at 23° for 3–5 hr to deplete the cells of PO_4. At the end of the depletion time, up to 50 mCi of ³²PO₄ [HCl-free, carrier-free (Amersham, Arlington Heights, IL, or ICN, Irvine, CA)] is added to the culture. The culture is shaken briefly (2 min) at 23° to allow incorporation of the PO_4 into the ribonucleotides and then is placed at 37° for 30 min to allow synthesis and accumulation of the pre-tRNAs.

RNA Isolation and Characterization

Isolation from Yeast Cells

It is essential to chill and keep the cells on ice during the work-up. This limits the extent of *in vivo* processing of the pre-tRNAs. After the labeling time at the nonpermissive temperature has elapsed, the cells are pelleted immediately by centrifugation at 4200 g for 5 min at 4°. The spent media are discarded to liquid radioactive waste, and the pellet is quickly resuspended in 4 ml of SEA buffer. The resuspended cells are transferred to a 50-ml polypropylene centrifuge tube (Corning, disposable), containing 5 ml of SEA buffer-saturated phenol (redistilled). The phases are mixed thoroughly and are incubated at 55° for 30 min. The RNA-containing aqueous phase (top), separated by centrifugation at 1500 rpm (Dupont clinical centrifuge) for 10 min at room temperature, is transferred to another 50-ml tube, containing 5 ml of distilled phenol (this is the *second phenol*), and is reextracted at 23° for 15 min. The first phenol phase is reextracted twice with 5 ml of SEA buffer. Each of these additional aqueous phases is reextracted successively using the *second phenol*. The three aqueous phases are pooled, and the combined volume is estimated. The RNA in this pooled solution is precipitated by the addition of 2 volumes of ethanol, followed by storage at −20° for 2–4 hr and by centrifugation at 12,000 g for 40 min at 4°.

The precipitated RNA is dissolved in 2–4 ml of 0.3 M NaCl TE buffer and is applied to a 0.5- to 1-ml diethylaminoethyl (DEAE)-cellulose

[4] G. Knapp, J. S. Beckmann, P. F. Johnson, S. A. Fuhrman, and J. Abelson, *Cell (Cambridge, Mass.)* **14**, 221 (1978).

column that has been equilibrated in 0.3 M NaCl TE buffer. Small RNAs and residual unincorporated $^{32}PO_4$ are removed by washing the column with 5 column volumes of 0.3 M NaCl TE buffer. The [^{32}P]RNA is then eluted using 5 column volumes of 1 M NaCl TE buffer. Following standard ethanol precipitation, the RNAs are separated by two-dimensional polyacrylamide gel electrophoresis (PAGE).

Polyacrylamide Gel Electrophoresis

The [^{32}P]RNA is dissolved in 100 μl of RNA gel dye buffer and is divided equally among four wells (12 × 1.5 mm) of a 10% (w/v) polyacrylamide gel ($\sim$40 × 28 cm), containing 4 M urea and TBE. During electrophoresis at 300 V, additional dye is loaded when the xylene cyanol marker dye has migrated to the bottom of the gel. Electrophoresis continues until the second xylene cyanol marker dye has migrated 6–8 cm from the top of the gel. Following autoradiography, the strips of gel, containing the pre-tRNAs and tRNAs, are excised, are placed at the bottom between a pair of glass plates, and are sealed in a plug of the 10% gel to which xylene cyanol has been added prior to polymerization. A 20% (w/v) polyacrylamide gel, containing 7 M urea and TBE, is polymerized above the plug. RNAs are separated in the second gel by electrophoresis at 400 V for the time that it takes the xylene cyanol dye to migrate, the equivalent of twice through the gel.

The separation pattern of the yeast intron-containing pre-tRNAs is shown in Fig. 1.[5] The identities of the various RNAs have been confirmed using a variety of techniques, including colony filter hybridization to known tRNA gene-containing clones, RNA sequence analysis, including both fingerprinting of *in vivo* labeled RNA and enzymatic ladder sequencing of *in vitro* 3' end-labeled RNA, hybrid selection using cloned genes, and splicing cleavage patterns.

Following autoradiography, each gel is aligned with its film using ^{32}P-labeled ink labels, and gel slices containing the pre-tRNAs are excised. The pre-tRNAs are eluted by diffusion from crushed gel into RNA elution buffer. The buffer is exchanged after elution overnight, and small fragments of gel carried in the eluant are removed by filtration through QuikSep filters (Isolab, Akron, OH) into 15- or 30-ml Corex centrifuge tubes. Elutions (3–4 hr each) continue until at least 80–90% of the [^{32}P]RNA has been removed from the gel. The RNA is precipitated by adjusting the combined eluants to 0.2 M sodium acetate, pH 5.0, and 20 μg/ml of glycogen, by adding 3 volumes of ethanol, and by storing at −70°

---

[5] R. C. Ogden, M. C. Lee, and G. Knapp, *Nucleic Acids Res.* **12**, 9367 (1984).

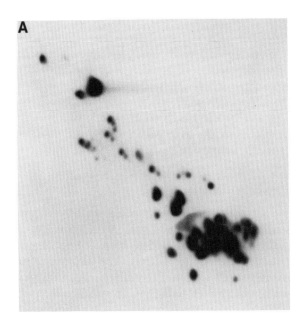

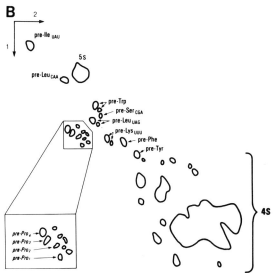

FIG. 1. Separation of tRNA precursors. Two-dimensional polyacrylamide gel electrophoretic separation is shown. Uniformly labeled [^{32}P]RNA was isolated as described from a temperature-sensitive diploid yeast, M304, that is homozygous at the *rna1* locus. (A) Autoradiography. (B) Schematic representation and identification of the RNA species. The two dimensions indicated in B at the top left are as follows: first, 10% (w/v) polyacrylamide [acrylamide : bisacrylamide (30 : 1)], 4 M urea; second, 20% (w/v) polyacrylamide [acrylamide : bisacrylamide (30 : 1)], 7 M urea. Both dimensions contained 90 mM Tris–borate, pH 8.3, and 1 mM EDTA. (Reprinted with permission from IRL Press, Washington, D.C.[5])

for 1 hr. It is important to use glycogen as the carrier instead of carrier RNA in this final precipitation, because high concentrations of random RNAs can inhibit the RNA splicing reaction. The RNA precipitate is collected by centrifugation at 12,000 g at $-5°$ for 40 min, is washed with 85% (v/v) ethanol ($-20°$) and is dried *in vacuo*. The pre-tRNAs are now ready for use in splicing assays; however, for quantitative work, e.g., kinetics, the specific activity of each pre-tRNA must be determined.

Specific Activity Determination

The specific activity of *in vivo* ^{32}P-labeled pre-tRNAs can be calculated using an experimentally determined value for the nucleotide specific activity. To do this, bulk [32P]RNA will be used to determine the disintegrations per minute per microgram of RNA. The distintegrations per minute in a sample are measured by liquid scintillation counting; the micrograms in the sample are determined using the absorbance measured at 260 nm and the specific extinction coefficient $E_{260 \text{ nm}}$ of 20 ml/mg. It is assumed that this RNA reflects the average of RNA in the cell and, as such, should contain equimolar quantities of the four nucleotides. Equal nucleotide specific activity will be verified by complete digestion of the [32P]RNA and by separation of the nucleotides using thin-layer chromatography[6] or acid gel electrophoresis (G. Knapp, unpublished data).

Notes and Variations

Care should be taken in [32]PO$_4$ labeling of yeast cultures. The cells are sufficiently depleted of PO$_4$ within 3–5 hr. After a 3-hr depletion, the cells will take up >90% of added [32]PO$_4$ in 5 min. Depletion for periods much longer than 5 hr will result in the synthesis of large amounts of polyphosphates which reduce yields of [32P]RNA and which complicate purification of the RNA. The chain lengths of the polyphosphate can vary from three to several hundred[7] and can comigrate with tRNAs and pre-tRNAs on certain polyacrylamide gels (e.g., see Fig. 3 in Knapp et al.[4]).

Cell death by irradiation is significant at [32]PO$_4$ concentrations greater than 50 mCi/200 ml of media. This concentration is used in this protocol, since the yeast is grown in the presence of the label for less than half a generation time and since no regrowth of the cells is required. Lower

[6] G. P. Mazzara, J. G. Seidman, W. H. McClain, H. Yesian, J. Abelson, and C. Guthrie, *J. Biol. Chem.* **252**, 8245 (1977).
[7] D. G. Fraenkel, *in* "The Molecular Biology of the Yeast Saccharomyces" (J. N. Strathern, E. W. Jones, and J. R. Broach, eds.), p. 1. Cold Spring Harbor Laboratory, Cold Spring Harbor, New York, 1982.

$^{32}PO_4$ concentrations should be used in protocols that require multiple generations.

If unlabeled RNA is required for *in vitro* labeling procedures, the phosphate-depletion step should be omitted, and the remaining protocol should be followed for precursor accumulation and for RNA isolation. Radioactive labeling of yeast RNA using ^{3}H- or ^{14}C-labeled purines, pyrimidines, or methionine (methyl moiety) is discussed in Rubin (see Ref. 8 and references therein).

[8] G. M. Rubin, *Methods in Cell Biol.* **12,** 45 (1975).

Section II

Characterization of RNAs

A. Primary Structure
Articles 11 through 15

B. Secondary Structure
Articles 16 through 22

C. RNA Functions
Articles 23 through 26

[11] Dideoxy Sequencing of RNA Using Reverse Transcriptase

By CHANG S. HAHN, ELLEN G. STRAUSS, and JAMES H. STRAUSS

Introduction

Shortly after the dideoxy (dd) chain-termination method for DNA sequencing was described by Sanger *et al.*,[1] Zimmern and Kaesberg[2] modified this method to sequence RNA directly, using purified encephalomyocarditis viral RNA as a template, pdT$_7$rC as a specific primer, and reverse transcriptase as the elongation enzyme. Bina-Stein *et al.*[3] further developed the method to use total cellular mRNA as a template and complementary DNA restriction fragments, which anneal to a particular mRNA at specific locations, as primers for reverse transcriptase. Recently, sequence-specific oligonucleotides, synthesized with an automated DNA synthesizer, have been used as primers on both DNA[4] and RNA templates.[5-8]

There are several advantages to sequencing RNA directly. First, the method can be very fast, since the multiple manipulations involved in cDNA cloning are omitted. In fact, unfractionated RNA can be used, since use of a specific primer dictates the sequence to be obtained, although the sequence ladder will not be as clean as when highly purified RNA is used. Even somewhat degraded RNA preparations that would not be suitable for cloning can be used, since the average length of the RNA need only be greater than the length of the sequence to be obtained in the reaction. Second, the sequence obtained automatically represents the majority nucleotide at any position and reduces the risk of sequencing errors

[1] F. Sanger, S. Nicklen, and A. R. Coulsen, *Proc. Natl. Acad. Sci. U.S.A.* **74**, 5463 (1977).
[2] D. Zimmern and P. Kaesberg, *Proc. Natl. Acad. Sci. U.S.A.* **75**, 4257 (1978).
[3] M. Bina-Stein, M. Thoren, N. Salzman, and J. A. Thompson, *Proc. Natl. Acad. Sci. U.S.A.* **76**, 731 (1979).
[4] E. C. Strauss, J. A. Kobori, J. Siu, and L. E. Hood, *Anal. Biochem.* **154**, 353 (1986).
[5] S. H. Clarke, K. Huppi, D. Ruezinsky, L. Staudt, W. Gerhard, and M. Weigert, *J. Exp. Med.* **161**, 687 (1985).
[6] D. J. Lane, B. Pace, G. J. Olsen, D. A. Stahl, M. L. Sogin, and N. R. Pace, *Proc. Natl. Acad. Sci. U.S.A.* **82**, 6955 (1985).
[7] J. Geliebter, R. Zeff, R. W. Melvold, and S. G. Nathenson, *Proc. Natl. Acad. Sci. U.S.A.* **83**, 3371 (1986).
[8] A. Baroin, R. Perasso, L.-H. Qu, G. Brugerolle, J.-P. Bachellerie, and A. Adoutte, *Proc. Natl. Acad. Sci. U.S.A.* **85**, 3474 (1988).

due either to cloning of a minority RNA or to errors introduced during transcription; because reverse transcriptase has a high error frequency,[9,10] two or more independent cDNA clones must normally be sequenced to be sure of obtaining the correct sequence. The drawbacks of the method are the relatively large amounts of RNA needed and the unavoidable presence of ambiguities in the sequence ladder. These ambiguities arise from two sources. One is the familiar compression artifacts due to secondary structure in the cDNA product; because only one strand is sequenced, it is not possible to completely resolve all of the compression problems, although every effort can be taken to run the sequencing gels as hot as possible. The second source is the falloff of the enzyme during reverse transcription, caused by secondary structure in the RNA template. The lack of proofreading activity in reverse transcriptase means that the concentration of dideoxynucleoside triphosphates required for proper ladder formation is approximately 1% of that needed with the large fragment of DNA polymerase I, which in turn means it is possible to increase the concentration of deoxynucleoside triphosphates and eliminate many of these nonspecific bands. Other methods designed to minimize these ambiguities are presented in this chapter. However, it has not proved feasible to eliminate ambiguities completely.

This method is thus particularly well suited to comparative sequence analysis of abundant RNA molecules, for example, ribosomal RNAs or viral RNAs, as it can be done rapidly and efficiently. In comparative sequencing, ambiguities are also less of a problem, because a change of sequence in the region will lead to a change in the pattern.

Preparation of RNA Template

Virtually any RNA purification method[11-15] that removes most of the protein and DNA can be used. We have sequenced alphavirus RNA isolated both from purified virions and from infected cells. To prepare virions, the medium over infected cells is harvested, and the virus is precipitated with polyethylene glycol (PEG) ($\frac{1}{3}$ volume of 40% PEG in 2 M NaCl

[9] K. P. Gopinathan, L. A. Weymouth, T. A. Kunkel, and L. A. Loeb, *Nature (London)* **278**, 857 (1979).
[10] J. M. Leider, P. Palese, and F. I. Smith, *J. Virol.* **62**, 3084 (1988).
[11] S. Penman, *J. Mol. Biol.* **17**, 117 (1966).
[12] D. T. Simmons and J. H. Strauss, *J. Mol. Biol.* **71**, 599 (1972).
[13] C. Auffray and F. Rougeon, *Eur. J. Biochem.* **107**, 303 (1980).
[14] J. M. Chirgwin, A. E. Przybyla, R. J. MacDonald, and W. J. Rutter, *Biochemistry* **18**, 5294 (1979).
[15] J.-H. Ou, E. G. Strauss, and J. H. Strauss, *Virology* **109**, 281 (1981).

is added on ice, and the precipitate is collected by low-speed centrifugation). To obtain a good sequencing ladder, it is very important to remove the supernatant completely, but further purification is usually not necessary. Alternatively, the virus can be pelleted from the medium by centrifugation at high speed. In either case, the viral pellet is resuspended in Tris–NaCl–ethylenediaminetetraacetic acid (EDTA) buffer (50 mM Tris–Cl, pH 7.4, 100 mM NaCl, 1 mM EDTA) containing 0.5–1% (w/v) sodium dodecyl sulfate (SDS), is extracted several times with phenol and phenol–chloroform (containing 1–2.5% (v/v) isoamyl alcohol), and the RNA is precipitated with ethanol in 200 mM final concentration sodium acetate.[11,12] The RNA precipitate is resuspended in 5 mM Tris–Cl, pH 7.4, and 0.05 mM EDTA.

To isolate intracellular RNA, cells are treated with Nonidet P-40 (NP-40) in PIPES buffer (500 mM NaCl, 15 mM MgCl$_2$, 5 mM EGTA, 50 mM PIPES, pH 6.5, and 1% (v/v) NP-40) to disrupt cytoplasmic membranes only and are centrifuged to remove intact nuclei and membrane debris. Sodium dodecyl sulfate is added to 1%, and the solution is extracted and precipitated as described above for RNA extraction from virions.[15]

Another method to extract total cellular RNA uses guanidine thiocyanate.[14] Either cells or tissue are treated with 5.5 M guanidine thiocyanate and are centrifuged at low speed to remove cell debris. The RNA is then pelleted onto either a cesium chloride (density 1.71 g/ml) or a cesium (trifluoroacetate) cushion (density 1.51 g/ml) in a SW27 rotor at 25,000 rpm for 24 hr at 15°.[16,17] The RNA pellet is resuspended in 4 M guanidine thiocyanate and is precipitated twice with 200 mM sodium acetate and 70% (v/v) ethanol. If the RNA of interest is a mRNA, a cleaner sequence ladder will be obtained by purifying poly(A)-containing RNA from total intracellular RNA by oligo(dT)-cellulose column chromatography.[18]

Reagents and Buffers

Reverse Transcriptase

Two forms of the enzyme [RNA-dependent DNA polymerase (EC 2.7.7.49)] are commercially available: the avian form [isolated from avian myeloblastosis virus (AMV)] and a murine enzyme [expressed from a cloned polymerase gene from mouse mammary tumor virus (MMTV)]. Avian reverse transcriptase is a better enzyme for direct RNA sequenc-

[16] V. Glisin, R. Crkvenjakov, and C. Byus, *Biochemistry* **13**, 2633 (1974).
[17] D. S. Zarlenga and H. R. Gamble, *Anal. Biochem.* **162**, 569 (1987).
[18] H. Aviv and P. Leder, *Proc. Natl. Acad. Sci. U.S.A.* **69**, 1408 (1972).

ing, because the optimum temperature for polymerization is 42.5°, and it is even possible to incubate the reaction at 50°. On the other hand, murine reverse transcriptase has a temperature optimum of 37° and is unstable at higher temperatures. The higher temperatures used with the avian enzyme result in many fewer ambiguous crossbands, because of decreased RNA secondary structure during reverse transcription. Therefore, all the conditions and procedures described in this chapter pertain to avian reverse transcriptase.

Preparing Stock Solutions

Since the concentrations of reagents are quite critical, we recommend making a significant volume of concentrated stock solutions and optimizing the reaction using these stocks. In particular, the conditions must be readjusted for each new batch of dideoxynucleotides. We also emphasize that the reaction is extraordinarily sensitive to pH, and in our hands, even a change of 0.1 pH unit from the optimum leads to inferior results. The essential stock solution list is as follows.

>   1 M Tris–Cl, adjusted to be pH 8.30 at 42.5° when diluted to 50 mM
>   0.5 M MgCl$_2$
>   1 M KCl
>   400 mM Dithiothreitol (DTT)
>   25 mM dATP, 25 mM dCTP, 25 mM dGTP, and 25 mM dTTP
>   5 mM ddATP, 5 mM ddCTP, 5 mM ddGTP, and 5 mM ddTTP

Reverse Transcriptase Reaction Conditions

The buffer for the reaction (RDDP buffer) is as follows.[19]

>   50 mM Tris–HCl, pH 8.30 at 42.5°
>   8 mM MgCl$_2$
>   50 mM KCl
>   10 mM DTT

We usually make a 5× buffer stock solution without DTT (5× RDDP buffer) and store it in small aliquots at −70°. Since the pH of Tris buffers changes with both temperature (−0.03 pH unit/degree increase) and concentration (+0.05 pH unit/10-fold dilution) and since the pH of the reaction is critical, it is important to make the solution at 42.5° or to adjust the pH by calculation. Dithiothreitol is added to the reaction from the 400 mM stock solution.

---

[19] C. M. Rice and J. H. Strauss, *J. Mol. Biol.* **150**, 315 (1981).

Deoxynucleoside triphosphates (dNTPs) and dideoxynucleoside triphosphates (ddNTPs) are added to the reaction from 10× solutions as follows.

10× dNTPs (for labeling with [α-^{32}P]dATP)
Solution A
   25 μM dATP, 0.5 mM dCTP, 0.5 mM dGTP, and 0.5 mM dTTP
Solution B
   75 μM dATP, 1.5 mM dCTP, 1.5 mM dGTP, and 1.5 mM dTTP
10× dNTPs (for sequencing with labeled primer)
Solution A
   0.5 mM dATP, 0.5 mM dCTP, 0.5 mM dGTP, and 0.5 mM dTTP
10× ddATP for labeling with [α-^{32}P]dATP
Solution A
   50 μM ddATP
Solution B
   18 μM ddATP
10× ddATP for sequencing with labeled primer
Solution A
   0.5 mM ddATP
10× ddCTP
Solution A
   0.5 mM ddCTP
Solution B
   0.18 mM ddCTP
10× ddGTP
Solution A
   0.25 mM ddGTP
Solution B
   0.18 mM ddGTP
10× ddTTP
Solution A
   1 mM ddTTP
Solution B
   0.27 mM ddTTP

Two concentrations of dNTPs and ddNTPs are given that cover two different ranges. The A conditions will give readable sequence from about 30 to 350 nucleotides from the primer, whereas the B conditions cover the interval from about 75 to 450 nucleotides. If sequence information further from the primer or closer to the primer is desired, it will be necessary to decrease or increase the ddNTP concentrations, respectively, relative to the dNTP concentrations. In any event, the optimal ratio of dideoxynucleotide to deoxynucleotide varies depending on the exact conditions and the particular lot of each ddNTP. Thus, the above ddNTP concentrations should be used as a guideline only, and the concentration of each ddNTP should be adjusted after performing a test reaction and examining the ladder formed after gel electrophoresis.

The concentrations of dNTPs (and consequently of ddNTPs) are much higher than those normally used in the literature. We attempt to use the highest concentrations of dNTPs feasible in any experiment in order to reduce the number of ambiguous crossbands present in the sequence ladder.

Selection of Primers

Length of Primers

The length of the primer is not critical. We have found oligodeoxynucleotides between 15 and 20 residues in length to work well.

GC Content and Position Effects

The GC content of an oligonucleotide primer is more important than its length. Primers with low GC content (25–45% GC) work better than primers with higher GC content. The position of the primer is also important, and ideally, it should bind to a region of the RNA template that is free of strong secondary structure. However, in practice, it is usually difficult to predict the secondary structure of a RNA.

Reaction Protocols

There are two ways of sequencing using reverse transcriptase. The first method uses a primer, which has been radiolabeled with T4 polynucleotide kinase and [γ-^{32}P]ATP, and uses unlabeled dNTPs in the elongation reaction. One advantage of this method is that, since all four dNTPs are present at the same (high) concentration, crossband artifacts can be reduced. However, this method also has disadvantages. First, there is the extra step of labeling the primer. Second, since it is not feasible to use an excess of labeled primer, the annealing conditions become more critical. Third, since not as much ^{32}P is incorporated, it is difficult to read more than 300 nucleotides of sequence from a given reaction.

The second protocol uses unlabeled primer; [α-^{32}P]dATP is incorporated during the chain-elongation reaction. Alternatively, one can perform a short labeling reaction, followed by chain-termination reactions.[3]

This method has several advantages. Since primer is present in large excess, the annealing step can be virtually omitted. Also, it is possible to substitute [^{35}S]dATP for [^{32}P]dATP to obtain sharper bands on the autoradiograms. Even though the band intensities may not be as uniform and more ambiguous crossbands may be present, these problems can be minimized by using an optimum ratio of dideoxy- to deoxynucleotides and by following the chain-termination reaction with a chase period using a high concentration of all four dNTPs. Use of terminal transferase during such a chase period has also been reported to eliminate many of the remaining crossbands.[20]

Purification of Oligonucleotide Primers

The products from an automated oligonucleotide synthesizer, which have been purified by high-performance liquid chromatography (HPLC), can be used as primers directly, but additional purification of the primers can improve the sequencing results, and purification is recommended after labeling a primer with polynucleotide kinase.

Labeling Primers with [γ^{32}P]ATP and T4 Polynucleotide Kinase. One to ten nanomoles per milliliter of oligonucleotide is recommended for the labeling reaction. The reaction contains 1–10 pmol oligonucleotide, 0.5- to 1.2-fold molar ratio of [γ-^{32}P]ATP (5000 Ci/mmol), 50 mM Tris–HCl, pH 9.0, 10 mM MgCl$_2$, 10 mM DTT, 50 μg/ml bovine serum albumin (BSA), and 5 U of polynucleotide kinase. The reaction is incubated at 37° for 30–60 min. An equal volume of 99% (w/v) formamide containing 0.1% (w/v) xylene cyanol (XC) is added to stop the reaction, and the mixture is loaded directly onto a denaturing polyacrylamide gel for purification of the labeled primer. If the gel purification is to be omitted, the reaction is stopped by heating to 70° for 10 min, and the primer is used directly.

Purification by Polyacrylamide Gel Electrophoresis. For purification by polyacrylamide gel electrophoresis (PAGE), either the labeled primer in the mixture described above or the unlabeled oligonucleotide dissolved in 10 μl of 80% (w/v) formamide, containing 0.1% (w/v) XC and ½× TBE (TBE is 90 mM Tris, 100 mM boric acid, and 1 mM EDTA, pH 8.3), is electrophoresed at 30 mA on a 20% (w/v) polyacrylamide [acrylamide : bisacrylamide (19 : 1)] gel 0.5–1.5 mm thick containing 8 M urea, using TBE as the electrode buffer. After electrophoresis, the gel is wrapped in Saran wrap and is placed on a plate with a fluorescent indicator [usually Merck silica gel 60F 254, but most thin-layer chromatography (TLC) plates can be used]. The gels are illuminated from above with ultraviolet

---

[20] D. C. DeBorde, C. W. Naeve, M. L. Herlocher, and H. F. Maassab *Anal. Biochem.* **157,** 275 (1986).

(UV) irradiation (254 nm) in order to visualize the bands, and the bands are cut from the gel and are eluted overnight either at room temperature or at 37° with 500 mM ammonium acetate 10 mM magnesium acetate, and 0.1% (w/v) SDS. The eluate containing the oligonucleotide is centrifuged through siliconized glass wool in an Eppendorf tube which has been punctured at the bottom and is placed in a second tube, in order to remove pieces of polyacrylamide. The purified oligonucleotides are desalted and concentrated either by ethanol precipitation with 0.2 M sodium acetate and 70% (v/v) ethanol, by drop dialysis, or by binding to diethylaminoethyl (DEAE)-Sephadex A-25. The purified oligonucleotide is resuspended in TE buffer (10 mM Tris, pH 7.6, and 0.1 mM EDTA) at a final concentration of 1–10 × 10^{-6} M.

Annealing Template and Primer

Annealing of the primer to the template is not necessary when the primer is in excess, and we often omit the annealing step when sequencing. However, if the template is a clean RNA preparation, careful annealing can give improved results. In addition, when primer is not in excess (as when sequencing using a labeled primer), the annealing step is rather important. For a set of four reactions, the annealing procedure is as follows.

One to four micrograms of RNA (in 5 μl of 5 mM Tris–HCl and 0.05 mM EDTA, pH 7.6) is mixed with 5 μl of primer (5–50 pmol) and 5 μl of 5× RDDP buffer and is heated at 70–90° for 2–5 min. The temperature of heating depends on the template; the higher temperature is used for RNAs suspected of having complex secondary structures. The primer and the template are then allowed to anneal for 0–60 min at a temperature determined by the composition of the primer. Although there are many ways to calculate the optimum temperature for annealing of a DNA–RNA duplex, a good estimate is 5–10° below the denaturation temperature (T_d) calculated by

$$T_d(°C) = 4(G + C) + 2(A + T)$$

where G + C or A + T refer to the composition of the oligonucleotide.[21] In most cases, it is not necessary to anneal for more than 5 min, and if the RNA preparation is not clean, the time of annealing should be kept to a minimum. For RNA molecules with a lot of secondary structure, a shorter

[21] S. V. Suggs, T. Hirose, T. Miyake, E. H. Kawashima, M. J. Johnson, K. Itakura, and R. B. Wallace, in "Developmental Biology Using Purified Genes" (D. Brown and C. F. Fox, eds.), p. 683. Academic Press, New York, 1981.

primer (15- to 18-mer) works better, and after heating at 90° for 3 min, the mixture is put directly on ice.

Chain-Termination Reaction

Chain-Termination Reactions Using Labeled Primer. To each of four small (0.5-ml) Eppendorf tubes labeled C, T, A, or G, add 0.5 μl of 10× dNTPs, 0.5 μl of the proper 10× ddNTP (ddCTP into the tube labeled C, etc.), 0.25 μl of 400 mM DTT, 1–3 U of AMV reverse transcriptase, and make up to 2 μl with sterile, deionized water. To each tube, then add 3 μl of RNA–primer–buffer mix and incubate at 42.5° for 30 min. Stop the reaction by adding 5 μl of 99% (w/v) deionized formamide containing 0.2% (w/v) XC (stop solution). Heat the samples at 90° for 2 min just before loading onto the gel. Load 2–3 μl/lane.

Chain-Termination Reactions Using Unlabeled Primer and Low Specific Activity [α-^{32}P]dATP or [α-^{35}S]dATP. To each of four small (0.5-ml) Eppendorf tubes labeled C, T, A, or G, add 0.5 μl of 10× dNTPs, 0.5 μl of the appropriate 10× ddNTP, 0.1 μl of radioactive dATP (400 Ci/mmol, either [^{32}P]dATP or [^{35}S]dATP), 0.25 μl of 400 mM DTT, 1–3 U of AMV reverse transcriptase, and make up to 2 μl with sterile, deionized water. To each tube, add 3 μl of RNA–primer–buffer mix and incubate for 30 min at 42.5°. Add 1 μl of chase mix (a mix of all four dNTPs, each at 3 mM) and incubate for an additional 15 min at 42.5°. Add 5 μl of stop solution to terminate the reaction. Heat the samples at 90° for 2 min just before loading onto the gel, loading 2–3 μl in each lane.

Prelabeling before Chain-Termination Reaction. To 13 μl of the annealed RNA–primer–buffer mix, add 4 μl of labeling mix (2 μl of 0.5 mM dNTPs minus dATP, 0.4 μl of radioactive dATP as above, 4–12 U of AMV reverse transcriptase, and 1 μl of 400 mM DTT, made up to 4 μl with deionized water). Incubate at 37° for 2–5 min. Then distribute 4 μl into each of four small Eppendorf tubes containing 1 μl of termination reaction mix (each termination reaction mix is 0.5 μl of 0.5 mM dATP and 0.5 μl of one ddNTP). Incubate at 42.5° for 30 min. Terminate the reaction by adding 5 μl of stop solution. Just before loading, heat the samples at 90° for 2 min and quick chill. Load 2–3 μl in each lane.

Analysis of Reaction Products on Sequencing Gels

As stated earlier, the conditions described previously, especially the concentrations of dNTPs and ddNTPs, are optimized for determining the nucleotide sequence in the region either from 30 to 350 nucleotides from the primer or from 75 to 450 nucleotides from the primer. To sequence the region closer to the primer or more distant requires altering the ddNTP–dNTP ratios during the chain-termination reactions. For the conditions described in this chapter, three sequential loadings on 80-cm-long sequencing gels, running a XC marker to 35–40, 90–100, and 150 cm, can cover the complete range from 45 to 450 nucleotides. Alternatively, by using either field-strength gradient gels,[22,23] or buffer gradient gels,[24] it is possible to reduce the number of loadings required.

Acknowledgments

This work was supported by Grants AI 10793 and AI 20612 from the National Institutes of Health and by Grant DMB 86-17372 from the National Science Foundation.

[22] A. Olsson, T. Moks, M. Uhlen, and A. B. Gaal, *J. Biochem. Biophys. Methods* **10**, 83 (1984).
[23] W. Ansorge and S. Labeit, *J. Biochem. Biophys. Methods* **10**, 237 (1984).
[24] M. D. Biggin, T. J. Gibson, and G. F. Hong, *Proc. Natl. Acad. Sci. U.S.A.* **80**, 3963 (1983).

[12] RNA Fingerprinting

By Andrea D. Branch, Bonnie J. Benenfeld, and Hugh D. Robertson

Role of RNA Fingerprinting in RNA Processing Studies

RNA fingerprinting analysis is often the fastest way to obtain accurate information about a RNA processing reaction. Figure 1 provides an example of RNA fingerprinting used to analyze the products of RNase III cleavage. Figure 1A shows a fingerprint of a 131-base-long transcript which contains a RNase III cleavage site. The oligonucleotide CCUUUAUGp is missing from the fingerprints of the two RNase III cleavage products (Fig. 1B and C). The presence of a new spot, pGp, in one of the products identifies the exact phosphodiester bond cleaved during the processing reaction and indicates the final deposition of the phosphate group.

Unlike many alternatives, RNA fingerprinting (one of the techniques for direct RNA analysis developed by F. Sanger and colleagues in the 1960s) permits all parts of a RNA molecule to be examined in a single assay. Largely because the base-specific ribonucleases used for RNA

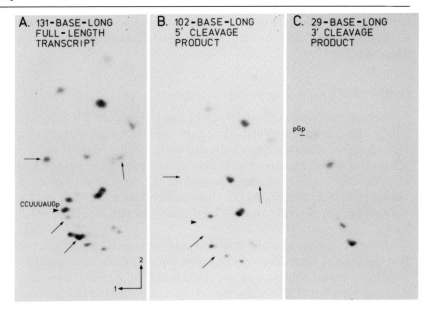

A. 131-BASE-LONG FULL-LENGTH TRANSCRIPT   B. 102-BASE-LONG 5' CLEAVAGE PRODUCT   C. 29-BASE-LONG 3' CLEAVAGE PRODUCT

FIG. 1. RNase T1 fingerprints of a RNase III substrate and two products of RNase III cleavage. A 131-base-long RNA containing a RNase III cleavage site (derived from the bacteriophage T7 early mRNA precursor) was transcribed *in vitro* from a plasmid generously supplied by Dr. John J. Dunn (Brookhaven National Laboratories, Upton, NY). One aliquot was fingerprinted directly (A), while a second aliquot was first cleaved by RNase III. The resulting cleavage products, one from the 5' end of the transcript and the other from the 3' end, were purified by gel electrophoresis and fingerprinted (B and C, respectively). The oligonucleotide which contained the RNase III-sensitive phosphodiester bond in the substrate RNA is identified in A by an arrowhead, and the oligonucleotides from the region beyond the RNase III-sensitive bond are marked by arrows. This latter set of oligonucleotides is present in the fingerprint of the RNase III product from the 3' end of the transcript (C), along with pGp, a nucleoside diphosphate whose 5'-phosphate terminus was created by RNase III cleavage. None of the oligonucleotides marked in A appear in the fingerprint of the cleavage product from the 5' end of the substrate RNA; however, the expected positions of these spots are indicated in B. The pair of arrows in the lower right-hand corner of A indicates the directions of high-voltage electrophoresis on cellulose acetate strips (the first dimension) and RNA homochromatography on diethylaminoethyl (DEAE)-cellulose thin-layer plates (the second dimension).

fingerprinting were discovered before the DNA restriction enzymes, approaches for determining RNA sequences—RNA fingerprinting and secondary analysis—were worked out before procedures for DNA sequencing. While most nucleic acid sequencing is now carried out at the DNA level, direct RNA analysis is still the most rigorous approach to use when it is important to obtain certain precise information about RNA biochemistry (e.g., the phosphate polarity of a cleavage site, the positions of modifying groups) and the structure of lariats and other intermediates of RNA splicing.

Outline of RNA Fingerprinting Procedure

Several different methods[1-4] can be used to prepare RNA fingerprints, which are two-dimensional arrays of ribonuclease-resistant oligonucleotides. All methods begin with ribonuclease digestion. Usually either RNase T1, which cleaves after G residues, or pancreatic RNase A, which cleaves after C and U residues, is used for this step. One commonly used approach is then to separate the oligonucleotides on the basis of charge in the first dimension and (roughly) on the basis of size in the second dimension.

Either cellulose acetate strips or acid–urea gels can be used for the first dimension. Cellulose acetate strips are finicky and are usually run in expensive high-voltage tanks, as diagrammed by Barrell[1]; however, Furdon and Kole[5] have devised a method that circumvents this requirement. Strip fingerprinting is fast [20 min for electrophoresis and 20 min for transfer to the diethylaminoethyl (DEAE) thin-layer plate used for the second dimension]. Under ideal conditions, eight RNAs can be processed in one set. We generally use cellulose acetate strips when fingerprinting pancreatic RNase-resistant oligonucleotides. Alternatively, gel first dimensions can be run in almost any laboratory and give more reproducible patterns. However, first-dimension gels are time-consuming (2 hr for electrophoresis and 2 hr for transfer), and only four RNAs can be fractionated in each 20 × 40-cm gel. First-dimension gels yield excellent RNase T1 fingerprints of ^{32}P-labeled oligonucleotides (both internally labeled and kinase labeled), but can sometimes be less reliable with iodinated RNAs and RNAs digested by pancreatic RNase A.

We usually transfer oligonucleotides to DEAE-cellulose thin-layer plates and use ascending RNA homochromatography for the second dimension. Polyethyleneimine (PEI) plates can be used to obtain second-dimension separations which are similar to those obtained with DEAE-cellulose.[3] Alternatively, although polyacrylamide gels can produce very good second-dimension separations and are used by a number of

[1] B. G. Barrell, in "Procedures in Nucleic Acid Research" (G. L. Cantoni and D. R. Davies, eds.), Vol. 2, p. 751. Harper, New York, 1971.
[2] G. G. Brownlee and F. Sanger, Eur. J. Biochem. 11, 395 (1969).
[3] T. Platt and C. Yanofsky, Proc. Natl. Acad. Sci. U.S.A. 72, 2399 (1975).
[4] D. V. Faller and N. Hopkins, J. Virol. 23, 188 (1977).
[5] P. Furdon and R. Kole, Anal. Biochem. 162, 74 (1987).

groups,[4,6] homochromatography has a greater ability to separate oligonucleotides evenly over a wide range of sizes. Also, if it becomes necessary to analyze the oligonucleotides further, the oligonucleotides can be eluted from the DEAE plate with relative ease using triethylamine carbonate, as described before.[1] The original RNA fingerprints were made using high-voltage electrophoresis on DEAE paper for the second dimension.[7] This fractionation procedure provides considerable information about the base composition of the oligonucleotides. However, because the procedure cannot be used to resolve oligonucleotides of similar base composition that contain more than about 10 residues, this method is usually reserved for small RNAs.

Ribonuclease Digestion

Materials and Methods for Ribonuclease Digestion

The protocol chosen for ribonuclease digestion depends on whether the RNA is already radiolabeled (see below). In all cases, RNAs are dried down in small silicon-treated glass tubes. Glass tubes are used because their clarity permits even minute quantities of RNA to be seen. On request, 13 × 54-mm glass tubes can be purchased with a partial band of colored paint around their midsection (Dynalabs, Rochester, NY). This band provides a space for indelibly labeling the tube with a felt pen. To prepare the tubes for use, the glass tubes are first filled with 5% (v/v) dichlorodimethylsilane (Pierce Chemical Co., Rockford IL) in chloroform, then are emptied, washed twice with distilled water, drained, and baked in a 150° oven for 2 hr.

Radiolabeled RNAs are dried down in the presence of tRNA (usually 10 μg). The *Escherichia coli* tRNA carrier is prepared by extracting 0.5-g aliquots (Sigma, St. Louis, MO) with BDH phenol (BDH Biochemicals, Poole, England) [saturated with 0.01 *M* Tris, pH 7.6, and 0.001 *M* ethylenediaminetetraacetic acid (EDTA)] three times followed by two rounds of chromatography on Whatman cellulose CF11.[8]

For ribonuclease digestion, the dried RNA is resuspended in 2 μl of either RNase T1 [Sankyo Co., Calbiochem (San Diego, CA)] or pancreatic RNase A (Worthington, Freehold, NJ). Individual, disposable Drummond microcaps are used for measuring the ribonucleases. Ribonuclease stocks are made up in 0.010 *M* Tris–HCl, pH 7.6, and 0.001 *M* EDTA and

[6] P. Palukaitis and M. Zaitlin, *Virology* **132,** 426 (1984).
[7] F. Sanger, G. G. Brownlee, and B. G. Barrell, *J. Mol. Biol.* **13,** 373 (1965).
[8] R. M. Franklin, *Proc. Natl. Acad. Sci. U.S.A.* **55,** 1504 (1966).

are stored in a freezer designated for this purpose at $-20°$. Procedures involving ribonucleases are conducted in a specified part of the laboratory. Digestions are carried out in the tips of drawn-out capillary tubes [0.8–1.10 $\times$ 100 mm (Kimax)] in 37° incubators (one for RNase T1 and a separate one for pancreatic RNase).

RNase T1 Digestion of Internally Labeled RNA

Radiolabeled RNAs are dried down in the presence of 10 μg of cold *E. coli* tRNA (to prevent overdigestion) and then are resuspended in 2 μl of 1-mg/ml RNase T1. The 2-μl samples, each containing 1 mg/ml of RNase T1, 0.010 M Tris–HCl, pH 7.6, 0.001 M EDTA, the radiolabeled RNA, and 10 μg of tRNA carrier, are incubated at 37° for 45 min. Under these conditions, RNase T1 cleaves primarily after G residues and resolves almost all cyclic phosphate intermediates to 3'-phosphate groups. However, certain A residues are nearly as susceptible to cleavage by RNase T1 as standard G residues, leading to fingerprint spots ending with A rather than G. Furthermore, stable structural elements may inhibit cleavage after some of the G residues contained within them. This latter problem can be overcome by carrying out the digestion at 65° for 1 hr in a sealed drawn-out capillary. Similarly, RNase T1 will not cleave RNA in RNA : DNA hybrids or in RNA : RNA duplexes. If either of these double-stranded structures is likely to be present (as in RNAs purified by affinity hybridization or RNA from a replication complex), the dried sample is resuspended in 0.010 M Tris–HCl, pH 7.6, and 0.001 M EDTA, sealed into a drawn-out capillary, submerged in a boiling water bath for 3 min, and plunged into a water–ice bath. The capillary is then opened and the sample is mixed immediately with the ribonuclease.

Pancreatic RNase Digests of Internally Labeled RNA

To prepare pancreatic RNase digests of radiolabeled RNAs, samples are dried down in the presence of 10 μg of *E. coli* tRNA carrier and then are resuspended in 2 μl of the ribonuclease. The 2-μl samples, containing 0.25 mg/ml of pancreatic RNase A, 0.010 M Tris–HCl, pH 7.6, 0.001 M EDTA, the radiolabeled RNA, and 10 μg of tRNA, are incubated at 37° for 10 min. These conditions are a compromise; they minimize cleavage after A residues, while permitting efficient cleavage after C and U residues. However, some long A-rich oligonucleotides may suffer nicking and thus appear in less than their actual molar ratio. In addition, many of the oligonucleotides may split into doublets with a 3'-phosphate on one component and a 2',3'-cyclic phosphate on the other, due to incomplete resolution of the cyclic phosphate intermediate. If either of these problems is acute, it may be necessary to raise or lower the ribonuclease concentration or to vary the digestion time.

Preparation of Digests for First-Dimension Separation

At the end of digestion, the samples are either spotted directly onto cellulose acetate strips or are expelled into 4 μl of a dye mixture, containing 9 M urea, 0.0375 M citric acid, pH 3.5, 1% (w/v) xylene cyanol FF, 0.5% (w/v) acid fuchsin, and 1% (w/v) methyl orange, in preparation for separation in an acid–urea gel.

Digestion and Kinase Labeling

Special conditions are needed for nonradioactive RNAs whose oligonucleotides are to be labeled after digestion by incubation in the presence of polynucleotide kinase and [γ-^{32}P]ATP. Such RNAs are often available in only limited amounts (far less than 1 μg) and thus require a reduced concentration of ribonuclease to avoid overdigestion. For example, 2 μl of 30-μg/ml RNase T1, 0.010 M Tris–HCl, pH 7.6, and 0.001 M EDTA are used to digest 5–500 ng of RNA. Incubation is carried out in the tip of a drawn-out capillary tube at 37° for 30 min. After digestion, samples in the 200- to 500-ng range are treated with 0.15 U of calf alkaline phosphatase [(Boehringer Mannheim, Indianapolis, IN) which had been purified by chromatography on Sephadex G-75 prior to use] for 15 min at room temperature and then for 15 min at 37°. The phosphatase treatment removes the 3'-phosphate groups produced by ribonuclease digestion. Samples are then sealed into drawn-out capillary tubes and are heated in a bath of boiling water for 3 min to inactivate the phosphatase. Reactions to label the 5' termini are carried out in 25-μl volumes containing the RNA digest, 15–20 U of polynucleotide kinase (Pharmacia, Piscataway, NJ), 0.01 M Tris–HCl, pH 8.0, 0.010 M MgCl$_2$, 0.015 M 2-mercaptoethanol, and 1–10 μCi of [γ-^{32}P]ATP (Amersham, Arlington Heights, IL), and are incubated in a 37° waterbath for 45 min.

Samples in the 5- to 200-ng range do not require treatment with calf alkaline phosphatase because even the relatively weak phosphatase activity provided by the kinase preparation is sufficient to completely remove the 3'-phosphate groups. The calf alkaline phosphatase treatment is needed for samples containing larger amounts of RNA to avoid a situation in which the weak phosphatase associated with the kinase preparation would partially remove the 3'-phosphatase groups, causing the fingerprint spots to appear as doublets.

Best results require compatible amounts of RNase T1, RNA and [^{32}P]ATP. When it is not possible to obtain an accurate estimate of the RNA concentration, it may be necessary to test a variety of conditions.

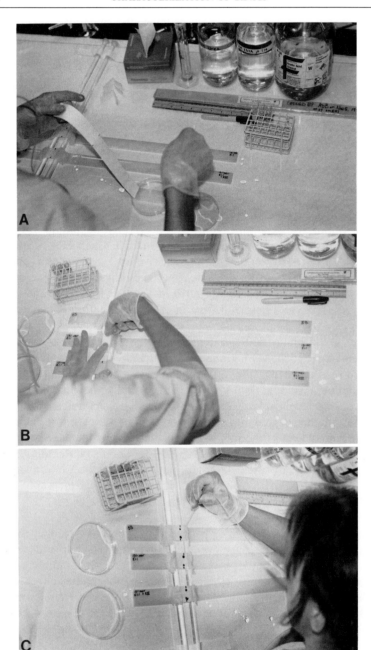

For a quick check of the labeled oligonucleotides from a large number of test reactions, we first fractionate an aliquot of each reaction in a 7 M urea–20% (w/v) polyacrylamide gel and fingerprint only the products of successful reactions. Much specific information about the fingerprinting of kinase-labeled oligonucleotides appears in Silberklang et al.[9]

First-Dimension Separation in RNA Fingerprinting and Transfer to DEAE Thin-Layer Plates

Fractionation on Cellulose Acetate Strips

Once this procedure, developed by Sanger and colleagues,[7] is underway, fast action is required. To aid in following our description of this technique, the individual steps have been set off by asterisks (*). Several steps are illustrated in Figs. 2–4. It may help to look at these figures before reading further.

Preliminaries. During the time needed for the ribonuclease digestions to be carried out, the room must be set up for strip fingerprinting. *To provide a cushiony and absorbent surface, several layers of paper toweling are taped around the edge of a tabletop or other overhang in such a way that a cellulose acetate strip, dangling vertically, could be pressed against these towels and not touch anything else. These towels will be used to remove excess buffer from the strips just prior to electrophoresis. *One cellulose acetate strip is needed for each RNA digest. A maximum of eight strips [3 × 55 cm (Schleicher and Schuell, Keene, NH)] is laid out on a freshly cleaned glass surface, taking care not to bend the strips as they are removed from their casing. These strips can accommodate up to 100 μg of a RNA digest. The strips can vary in quality from lot to lot. We only purchase a large quantity of a new lot after testing a representative

[9] M. Silberklang, A. M. Gillum, and U. L. RajBhandary, this series, Vol. 59, Part G, p. 58.

FIG. 2. Application of RNA digests to cellulose acetate strips. (A) A cellulose acetate strip is wetted with first-dimension buffer. Felt-tipped pen marks on the strip identify the approximate location of the future origin. Two previously wetted strips are draped over Lucite support rods and have buffer-soaked half-Kimwipes in place to maintain moisture. Three V-shaped blotters lie near the box of Kimwipes. (B) Excess buffer is removed from the top and bottom surfaces of each cellulose acetate strip with an individual V-shaped blotter. A rack holds small glass tubes, with the drawn-out capillaries containing the RNA digests inside the tubes. (C) RNA digests are applied to the cellulose acetate strips in small aliquots. Dye markers have already been applied with a drawn-out capillary tube and can be seen inside the felt pen marks. Some first-dimension buffer remains in the petri dish. As necessary, one of the half-Kimwipes will be used to spread a thin film of buffer over the long and short arms of the cellulose acetate strips, on the regions which lie outside the Lucite rods.

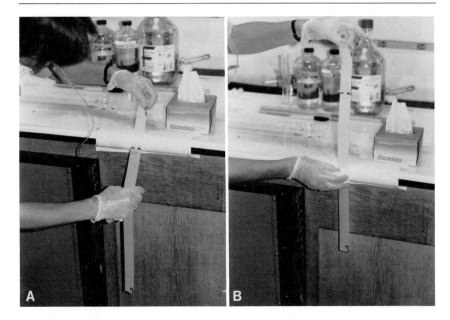

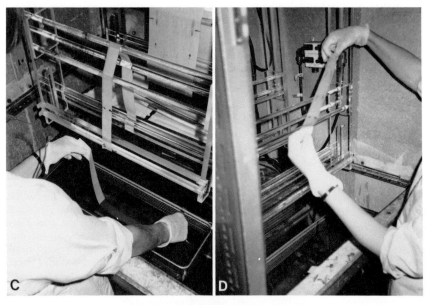

FIG. 3. High-voltage electrophoresis on cellulose acetate strips. (A) To prepare the strips to be positioned on the high-voltage rack used for electrophoresis, excess buffer must be removed. In the first step of this blotting process, several layers of paper toweling (taped to a benchtop overhang) are used to dry the back side of the short arm of the cellulose acetate

package. Eventually, the digests will be applied 12 cm from one end of each strip (at what will become the cathode end). A felt pen is used to mark this distance. Very close to the edge of the strip, a short line is drawn. The sample will be spotted between these marks, at the center of the strip. A felt pen is also used to label both ends of the strip.

*A twenty milliliter volume of first-dimension buffer is needed for each two strips. First-dimension buffer is made by combining the following ingredients in the specified order: 19 ml of 7 M urea, 0.2 ml of 0.5 M EDTA, pH 8.0, and 1 ml of glacial acetic acid. The 7 M urea used in this buffer is produced by dissolving 420.4 g of urea in 500 ml of distilled water in the presence of heat. Once the urea is in solution, it is topped to 1 liter with distilled water. Then 50 g of mixed bed resin (Bio-Rad, Richmond, CA) is added to the urea solution, and the solution is stirred for 1 hr. The urea is filtered through Whatman 3MM paper prior to use and then is stored at room temperature. The first-dimension buffer should have a pH between 3.5 and 3.7. If a fresh solution of urea (one less than 3 months old) is used, the buffer may be too acidic, and pyridine (Fisher Scientific, Pittsburgh, PA) is added dropwise with constant stirring, while the pH is monitored. After the desired pH is obtained, the buffer is poured into a petri dish and is covered by the top of the dish.

*For each strip, a blotter (two layers thick, about 1.5 cm across and 8 cm long) is cut from a paper towel and is bent in half so that the blotter is V shaped. *In addition, for each strip, one Kimwipe is torn in half; the Kimwipe halves are soaked in buffer and are set aside in their own covered petri dish. *In the location where the digests are to be applied to the

---

strip. Starting at a position very close to the sample, the strip is gently pulled downward so that the part of the strip making contact with the toweling moves from a site just above the sample toward the end of the short arm of the strip. At no time does the portion of the strip containing the sample touch the paper toweling. (B) For the second blotting step, the strip is repositioned so that the back side of the long arm contacts the paper toweling just beyond the area where the sample was applied. Several dry Kimwipes are pressed onto the front of the strip. A gentle upward pull allows both surfaces of the long arm to be blotted at the same time. (C) After blotting, the cellulose acetate strips are immersed in Varsol, a nonpolar liquid, which will temporarily halt evaporation of the small amount of first-dimension buffer remaining on the strips, but which will not dissolve the RNA oligonucleotides. The immersion is carried out as shown, by passing a strip from one end to the other (from the end of the short arm toward the long arm) through the Varsol. A cellulose acetate strip, already wetted with Varsol, rests on the high-voltage rack hanging inside the Savant enclosure. For electrophoresis, the rack will be lowered into the tank below, through the deep layer of Varsol, which also serves as an electrophoresis coolant, so that the two ends of the strip are submerged in the buffer reservoir lying at the bottom of the tank beneath the Varsol. (D) At the end of electrophoresis, the high-voltage rack is raised. From the end of the strip nearer to the origin, the buffer-soaked tip is torn off. The separation of the dye markers is clearly visible on the strip being removed from the rack.

FIG. 4. Transfer of oligonucleotides from cellulose acetate to DEAE-cellulose thin-layer plates. Using the dye markers as guides, a cellulose acetate strip is positioned on a DEAE plate. The blue dye is about 3 cm in from the right edge of the plate. As illustrated by the DEAE plate at the right, which is already set up for the transfer of oligonucleotides, water-soaked strips of Whatman 3MM paper are placed on top of the cellulose acetate strip and then a small rectangular glass plate is put on top of the Whatman 3MM paper.

strips, two plastic rods, each about 1 cm in diameter and 55 cm in length, are placed parallel to each other, separated by about 2 cm. The distance separating the rods must be slightly greater than the width of the blotters, which must be able to fit between them. The rods are taped to the glass surface.

Preparing the Strips to Receive the Digests. *When the ribonuclease digestions are completed, each strip (in turn) is saturated with buffer by moving it from one end to the other through the buffer in the petri dish (Fig. 2A). The remaining buffer is saved and used to rewet the strips later. *Each buffer-soaked strip is draped over the two rods (perpendicular to the long axis of the rods) so that the future origin is centered above the space separating the rods and thus is raised slightly above the surface. *A pair of buffer-soaked half-Kimwipes is used to spread additional buffer over the two arms of the cellulose acetate which lie on either side of the rods. The half-Kimwipes are left sitting on the cellulose acetate at the inflection point, where the strip bends over the rods and rests on the glass surface. *Successive strips are spaced about 6 cm apart from each other. *Once all of the strips have been prepared, the V-shaped blotters are carefully used to remove excess buffer from the portion of the strips which lie between the rods, leaving this part of the strip moist, but with no visible drops of buffer, thus readying the origins for the RNA oligonucleotides (Fig. 2B). The V shape of the blotters makes it relatively easy to quickly remove excess buffer from both the bottom and top sides of the strip. During this procedure, it is important to consider that cellulose acetate has very little affinity for RNA oligonucleotides. Any excess buffer, or accidental movement of buffer into the region of the sample, can cause the sample to spread out, thus spoiling the analysis.

Spotting Digests onto Cellulose Acetate. *Because the strips have a limited capacity to absorb liquid, the digests are applied in small (0.5-μl) aliquots (Fig. 2C). The aliquots are spotted onto the center of the strip at the distance from the end specified by the two felt pen marks. Every effort is made to spot all aliquots of a given sample onto exactly the same site. When the entire sample has been applied, the digest will form a barely visible dot about 3 mm in diameter. *While the first round of aliquots is sinking into the strip, 1-μl aliquots of dye [2% (w/v) xylene cyanol FF, 1% (w/v) acid fuchsin, and 2% (w/v) methyl orange (Searle Scientific Services, Chicago, IL)] are applied with a drawn-out capillary tube just inside each of the two felt pen marks. *If the strips appear to be drying out during the time it takes for the digests to be applied (watch out for urea crystals), the strips can be rewetted by dipping one of the half-Kimwipes in the remaining buffer and spreading an even film of fresh buffer over the strips—taking care not to get buffer too close to the samples.

Preparing for Electrophoresis. *As soon as the final aliquot of one of the samples has been absorbed, the strip is rewetted and then its buffer-saturated half-Kimwipes are discarded. *Immediately thereafter, fresh, dry Kimwipes are used to blot the top side of the short arm of the strip. Any buffer which escapes this blotting has the potential for flowing into the sample and ruining it as the next step of the procedure is undertaken. *The strip is grasped at both ends and lifted carefully from the glass surface. It is still very wet throughout most of its length and especially along the long end beyond the sample. Thus, the strip must be carefully managed while it is carried over to the table-edge, paper-towel blotter. *The back side of the short arm is dried first by pressing the strip against the paper towels at the point on the strip that is as close to the sample as prudent (Fig. 3A). A downward motion is then used to dry the short arm, moving from very close to the sample toward the beginning of the strip. *The strip is quickly repositioned so that the back side of the strip touches the paper towels just beyond the sample, on the long arm. Several dry Kimwipes are pressed against the front of the strip. An upward motion lifts the strip, which is simultaneously blotted on its front and back surfaces (Fig. 3B). The objective is to end up with a strip that is moist all over, but has no freely flowing drops of buffer. Time is of the essence at this point; it is critical that the strips do not dry out. The Kimwipes are quickly discarded, freeing one hand.

*The strip is now ready to be wetted with Varsol (Savant Instruments, Inc., Hicksville, NY), which will temporarily prevent further drying. If a number of samples are being fingerprinted as a set, it can be difficult to complete this part of the procedure quickly enough to keep the strips from drying out. Thus, it is wise to enlist a partner, a wide receiver, who can dip one strip in Varsol and position the strip for electrophoresis while the next strip is being readied. The strip to be submerged in Varsol is grasped at both ends. The strip is wetted by passing it through Varsol with a smooth motion, starting with the end closer to the sample (Fig. 3C). *The strip is then draped over the top rung of the high-voltage sample rack (Savant) so that the long arm is in back and the sample is in front. *After all strips have been positioned on the rack, the strips are clamped into place with Lucite rods and are lowered slowly into the tank, taking care not to displace the strips.

Electrophoresis. *Electrophoresis is begun at "low voltage" (2000 V). After the dyes have begun to separate visibly, 6000 V is applied. Electrophoresis is carried out in an interlocked enclosure (Savant) using an 8000-V power supply. The electrophoresis tanks contain 12 liters of pH 3.5 buffer [5% (v/v) acetic acid and 0.5% (v/v) pyridine[1]]. This buffer lies under 24 liters of Varsol, an electrophoresis coolant, which is chilled by coils containing circulating cold water. Electrophoresis continues until the blue dye has moved 8.5–10 cm from the origin.

Positioning Cellulose Acetate on DEAE Thin-Layer Plate. *While the electrophoresis is going on, a dull pencil is used to label the 20 × 40 cm DEAE thin-layer plates [Machery-Nagel DEAE-cellulose 300 (Sybron/Brinkman, Westbury, CT)]. Like the cellulose acetate strips, a representative package is tested before quantities from a given lot of DEAE plates are purchased. *At the end of electrophoresis, the sample rack is raised and allowed to drain briefly. *As will be evident, both ends of the strips will be wet with buffer (from the reservoirs at the bottom of the tank). While the strip is still on the rack, the buffer-soaked end of the short arm is torn off. *The strip is carefully removed from the rack, keeping an eye on the buffer-soaked tip of the long arm of the strip (Figure 3D). *The strip is placed across the short axis of a 20 × 40-cm-DEAE thin-layer plate so that the bottom edge of the strip is about $\frac{1}{2}$ cm above the bottom of the plate and so that the blue dye is about 3 cm from the right edge of the plate (Fig. 4).

The Transfer to DEAE Thin-Layer Plates. *Without delay, four water-soaked strips of Whatman 3MM paper, 2.5 cm wide and 22 cm long, are centered on the cellulose acetate strip. It is essential that the strips of Whatman 3MM paper be narrower than the cellulose acetate, otherwise, the Whatman 3MM paper may contact the DEAE plate and may create a hole by removing portions of the DEAE-cellulose. *A glass plate (20 × 5 cm) is put on top of the wet Whatman 3MM papers. Pressure is applied to the glass to ensure a firm seal and to encourage water to flow from the Whatman 3MM papers through the cellulose acetate, carrying the oligonucleotides to the DEAE-cellulose thin-layer plate, which binds the oligonucleotides avidly. *After 5 min, the glass plate is gently lifted in a motion that leaves the Whatman 3MM papers and the cellulose acetate strip undisturbed. More water is applied to the Whatman 3MM papers and is allowed to sink in briefly. The glass plate is replaced. This rewetting is repeated three times. *When the transfer has been completed, the glass plate is set aside and the Whatman 3MM papers are gently removed as a group. The cellulose acetate strip is carefully lifted and is sent to the radioactive waste because the transfer is only about 75–90% efficient. *Each DEAE plate is moved to a large tray containing 95% (v/v) ethanol and is washed with gentle agitation for 5 min and then is air-dried on paper towels. *The plates can be stored at this point between layers of paper towels or can be prepared for homochromatography by attaching them to supportive pieces of glass (20 × 40 × 0.3 cm) with double-sided tape (see below).

Use of Acid–Urea Gels for First-Dimension Separation of RNA Fingerprinting

When we use polyacrylamide gels for the first-dimension separation of oligonucleotides, the ribonuclease digests are fractionated in an ultrathin version of the 10% (w/v) polyacrylamide gel with 6 M urea introduced by De Wachter and Fiers.[10]

Preparing First-Dimension Gel. Note: We have found that urea is the component most commonly at fault when poor resolution occurs. Smeary spots signal the need to purchase a new lot. The recipe for a first-dimension gel is as follows:

10% Polyacrylamide gel with 6 M urea and citric acid (50 ml)
    18 g of urea
    5 ml of 0.25 M citric acid, pH 1.85 (Fisher Scientific)

[10] R. De Wachter and W. Fiers, this series, Vol. 21, p. 167.

12.5 ml of 40% (w/v) acrylamide/1.3% (w/v) bisacrylamide (Kodak, Rochester, NY/Bio-Rad)
(The acrylamide/bisacrylamide solution is filtered through Whatman 3MM paper when made up as a stock solution)

The ingredients listed above are stirred until the urea dissolves. The volume is raised to 50 ml. The following are added just prior to pouring:

0.2 ml of $FeSO_4 \cdot 7H_2O$ (2.5 g/liter; Sigma)
0.2 ml of 10% (w/v) ascorbic acid (Sigma)
0.050 ml of 30% (v/v) hydrogen peroxide (Fisher Scientific)

As with all ultrathin gels, to ease pouring and to reduce later sticking of the gel to the glass plates, one of the plates must be treated with 5% (v/v) dichlorodimethylsilane in chloroform (siliconized) before the glass plates and side rails are assembled and sealed with waterproof tape (Scotch 3M). The silicone coating is produced by squirting one of the scrupulously cleaned glass plates with about 1 ml of a solution containing 5% (v/v) dichlorodimethylsilane and chloroform in a fume hood. This fluid is rapidly spread around with a Kimwipe. The glass plate is then rinsed with distilled water and air-dried. Gels are poured between glass plates measuring 20 × 40 × 0.5 cm, which are separated by 0.3-mm-thick side rails. The comb and spacers are cut from plastic sheets [Slater's (Plastikard) Ltd., Matlock Bath, Derbyshire, England]. We make 16-slot combs.

Gel Electrophoresis. As mentioned in Preparation of Digests for First-Dimension Separation, the 2-μl ribonuclease digests are combined with 4 μl of 1.5× citrate dye [9 M urea, 0.0375 M citric acid, pH 3.5, 1% (w/v) xylene cyanol FF, 0.5% (w/v) acid fuchsin, and 1% (w/v) methyl orange (Searle Scientific Services)]. Samples are loaded onto the gels through drawn-out glass capillary tubes with especially long tips. Since the gels will eventually be cut into strips (for transfer to the second dimension), it is important to leave at least 2 cm (two slots) between samples when loading the gel. In addition, for best resolution, we leave three blank lanes at each side of the gels.

These ultrathin gels are run at 1000 V using a citric acid–sodium citrate running buffer (see below) until the blue and yellow (methyl orange) dyes are 10.5 cm apart. A 10× stock of citric acid–sodium citrate running buffer is made by titrating a 0.25 M citric acid solution to pH 3.5 by the addition of 0.25 M sodium citrate. This buffer is filter sterilized, stored at 4°, and diluted to a 1× solution just prior to use as the running buffer for the gel. Note that the citric acid solution used to make up the gel itself has a pH of 1.85; when this solution is combined with the urea and acrylamide, the final mixture should have a pH of 3.5 without adjustment. If the pH of the urea–citric acid–acrylamide solution is much above 3.5, resolution will be diminished.

Transfer of Oligonucleotides from Gel to DEAE-Cellulose Thin-Layer Plate. At the end of electrophoresis, the top glass plate is removed from the gel. The gel is covered with Saran Wrap. A clear plastic straight edge is placed on the Saran Wrap and used as a guide for cutting the gel lengthwise into strips, each about 3.0 cm wide, with a set of dye markers running down the center of each strip. To make it easier to pick the gel strip up from the glass plate, the teeth are trimmed off of the gel, and the lower part of the gel is cut away, leaving a strip about 3.0 × 25 cm. After carefully removing the Saran wrap and the excess gel from the first lane to be transferred, a strip of Whatman 3MM paper about 28 cm long and 2.5 cm wide (i.e., narrower than the gel strip) is centered on the gel strip and is pressed gently against the gel strip. The paper should extend beyond the top and bottom of the gel strip.

The only difficult part of the transfer comes next, when the paper is used to lift the gel from the glass. This step is demonstrated in Fig. 5. The objective is to arc the paper gently upward from the bottom of the gel strip toward the origin, bringing the gel strip along. If the gel seems reluctant to leave the glass, the bottom 2–3 cm of the paper is curled upward, then a single-edge razor blade is used to raise the bottom edge of the gel from the glass onto the Whatman 3MM paper. The bottom of the gel strip is then held in contact with the Whatman 3MM paper by pinching them gently together with the thumb and first finger. As one hand rolls the gel and paper assembly backward, the thumb of the opposite hand runs ahead of the rising portion to maintain the bond between the gel and the paper (see Fig. 5). When this maneuver has been completed, the free hand is used to secure the origin of the gel strip to the paper.

As a result of the motion used for lifting, an inversion takes place. The gel strip ends up on top of the paper, and thus gravity helps to hold the gel strip in place as it is moved in the direction of the 20 × 40-cm DEAE thin-layer plate [Machery-Nagel DEAE-cellulose 300 (Sybron/Brinkman)]. The gel and paper are inverted so that the gel comes to face the prelabeled DEAE plate and then the gel strip is placed directly across the short axis of the DEAE plate, about 0.5 cm from the plate's bottom edge. The origin of the gel should fall just beyond the right outside edge of the DEAE thin-layer plate.

The strip of Whatman 3MM paper used for the transfer is left on top of the gel and is overlain by four water-soaked Whatman 3MM strips (2.5 × 22 cm). The stack of Whatman paper is covered with a glass plate (5 × 20

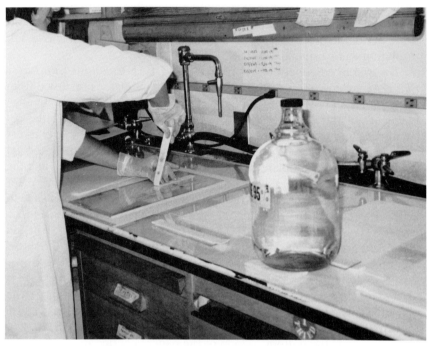

Fɪɢ. 5. Lift-off of first-dimension gel strips containing RNA oligonucleotides. At the end of electrophoresis, the polyacrylamide gel used for the first-dimension separation of oligonucleotides is sliced into strips containing the digestion products of individual RNA samples. A piece of Whatman 3MM paper, slightly narrower than the gel strip, is used to remove the gel strip from the electrophoresis glass plate. The gel strip will be placed in direct contact with the nearby DEAE-cellulose thin-layer plate. As illustrated by the DEAE plate at the right, which is already set up for the transfer of oligonucleotides, water-soaked strips of Whatman 3MM paper are placed on top of the cellulose acetate strips and then a small rectangular glass plate is put on top of the Whatman 3MM paper. Finally, a jug is placed on the small glass plate to provide a weight. A pad of paper toweling, used to stabilize the jug, is just visible between the bottom outside edge of the jug and the surface of the laboratory bench.

cm). Pressure is applied by hand to create a good seal, and then a glass 1-gal jug, filled with liquid (usually ethanol), is placed on top to act as a weight. Blotting takes a total of 2 hr. The glass plate is lifted and water is added to the top of the Whatman papers every 30 min, taking care not to disturb the gel or to allow the Whatman papers to contact the DEAE thin-layer plate.

By the end of the transfer, water has usually moved about 15 cm up from the bottom of the DEAE-cellulose thin-layer plate. The glass plate and Whatman papers are removed. The gel strip is carefully lifted from the DEAE plate. The DEAE plate is then washed in a large tray containing 95% (v/v) ethanol for 5 min with gentle agitation. The plates are air-dried on paper towels. The plates may be stored at this point between layers of paper towels or can be prepared for homochromatography by affixing them to glass support plates (20 × 40 × 0.3 cm) with double-sided tape (Scotch 3M), as described below.

Ascending RNA Homochromatography

The version of this technique we use was developed by Brownlee and Sanger.[2] This second dimension of RNA fingerprinting is carried out inside a 65° oven within a set of chromatography tanks (one placed on top of the other) which are large enough to hold three 20 × 40 × 0.3-cm glass plates. The chromatography tanks have outside dimensions of 11.4 × 21.6 × 21.6 cm (Desaga, FRG). It is important to test a new tank to make sure that the inside partitions will securely hold the glass plates to be used. Each tank requires 140 ml of homomix.

Homomix

Homomix is a mixture of urea and partially hydrolyzed RNA fragments which provides the moving phase.[2] The homomix is saved at the end of each run, stored at room temperature, and recycled. Taking into account the complexity of the RNA to be fingerprinted, the strength of this "from tank" homomix is adjusted by the addition of 3% and 5% stocks, which differ considerably in their ability to move large oligonucleotides up the DEAE plate. A typical homomix for 5 S rRNA (120 bases) is 110 ml of "from tank," 20 ml of 3%, and 10 ml of 5%. The 2000 base-long 18 S rRNA might require 100 ml of "from tank," 20 ml of 3%, and 20 ml of 5%. For the 10,000 base-long 45 S rRNA precursor, the recipe might be 90 ml of "from tank," 10 ml of 3%, and 40 ml of 5%. The recent history and hence composition of the "from tank" will determine the exact percentages to use. Know your homomix. If you have no homomix to recycle, begin with a 1 : 1 mixture of virgin 3% and 5% homomix, but cook this homomix overnight at 65° before use. Poor resolution in the second dimension is often due to a homomix that is too strong. An undesirably strong "from tank" can be improved by incubating the entire bottle at 65° for 12–18 hr.

One-half liter of 3% homomix is prepared by adding 15 g of yeast RNA (BDH, product number 42045 from Gallard-Schlessinger, Carle Place, NY) to 150 ml containing 8.4 g of potassium hydroxide and stirring vigorously. Once most of the RNA is in solution (90% or more), the mixture is stirred for 12 min. Then, while continuing the stirring, the pH is quickly adjusted to 7.5 by the addition of concentrated hydrochloric acid. The hydrolyzate is later poured into a 1.2 m length of wide (6.6 cm in circumference) prewashed dialysis tubing (VWR Scientific, Piscataway, NJ) and is dialyzed against 2 liters of tap water for 2 hr. The water is changed and dialysis is repeated. The dialysis tubing is moved to the largest available beaker and is slit, and then the hydrolyzate is transferred to a beaker of more normal proportions. After the addition of 210 g of urea, the volume is adjusted to 500 ml with water.

The 5% homomix is prepared in the same way, except that 25 g of yeast RNA are treated with the potassium hydroxide solution for 5 min. The need for the oversized beaker is much more acute due to the high pressure which builds up inside the dialysis tubing when 5% homomix is made. Occasionally, the 5% hydrolyzate picks up so much water during dialysis that its volume is greater than the desired 500 ml by the time 210 g of urea has been added. The hydrolyzate can still be used, but will be somewhat weakened by the dilution. Stocks of 3% and 5% homomix are stored at 4°.

Setting Up Second Dimension of RNA Fingerprinting

DEAE thin-layer plates are prepared for homochromatography by attaching them to supportive glass plates (20 × 40 × 0.3 cm) with double-sided tape (Scotch 3M). With the plate upright, propped in a nearly vertical position, four small pieces of double-sided tape are placed along each long side of the glass, on both sides of the glass plate if the glass is to be used for two DEAE plates. A DEAE plate is attached to the glass with its bottom edge about 0.3 cm above the bottom edge of the glass plate. A wick, consisting of four thicknesses of 2.5 × 20-cm Whatman 3MM paper and bearing the name of the appropriate digest (sometimes the label inscribed on the DEAE plate becomes unreadable during homochromatography), is fastened across the top of each DEAE plate with four jumbo paper clips.

Like all other components to be used in the chromatography run, the glass and attached DEAE plates are preheated in a 65° oven for at least one-half an hour before beginning the second-dimension separation. At least half an hour before the run, 140 ml of the appropriate homomix is made up, and the homomix and tanks are preheated to 65°. If the chromatography tanks are at room temperature, the homomix can be poured into the tanks for the preheating. If the tank is still warm from a previous run, the homomix should be placed in the oven in its own vessel to prevent a thermal shock to the glass tanks. Before preheating the tanks, Dow Corning high-vacuum grease is applied to the top lip of the bottom tank. The second tank is inverted, placed on top of the bottom tank, and the two tanks are put into the oven.

When all components have equilibrated to 65° and it is time to begin the run, tanks are removed from the oven and copious stopcock grease is again applied to the lip of the bottom tank. The homomix is poured into the tank, if necessary. The DEAE plates (attached to glass plates) are removed from the oven and approximately the bottom one-third of each plate is thoroughly wetted with a mist of water (Fig. 6). We employ a hand-powered sprayer such as the ones used to moisten the leaves of house plants. The part of the plate that gets misted is the part which absorbed water and urea during the transfer of oligonucleotides. From a favorable angle, it is possible to see the discontinuity in the plate that was

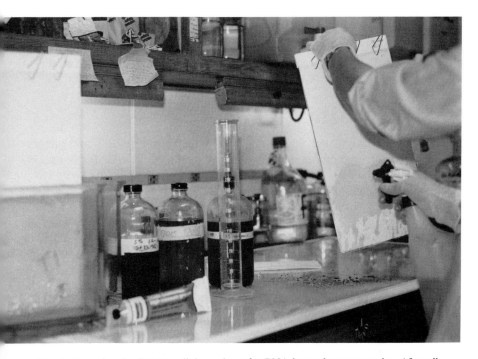

FIG. 6. Preparing the DEAE-cellulose plates for RNA homochromatography. After all components to be used for homochromatography have been preheated to 65°, a spray bottle is used to mist the lower portion of the DEAE thin-layer plates with water. During the misting procedure, a DEAE plate may be held by the wick clipped to its top, without damaging the surface of the DEAE-cellulose. Excess water is drained from the bottom of the plate before the plate is placed into the chromatography tank.

created during the oligonucleotide transfer: the plate should be misted from this point downward. If the misting leaves the plate dripping, excess water is drained briefly from the bottom. The plates are carefully lowered into the tank. It is important that the fingerprints do not touch each other or the stopcock grease while going in or out of the tank. We leave a blank slot between adjacent glass plates. When all the fingerprints have been lowered into the bottom tank, the top tank is replaced. Its position on top of the bottom tank is adjusted to establish an airtight seal. The chromatography tanks are placed well back in the oven (Fig. 7). For even heating, we then put a large sheet (28 × 44 cm) of Styrofoam just inside the oven door. To obtain a matched set of fingerprints, the tank is rotated halfway around at about the midpoint of the run.

Ascending RNA homochromatography requires about 8–16 hr, depending on the batch of plates. It is finished when the blue dye has migrated about 30 cm up the DEAE-cellulose thin-layer plate. The fingerprints are removed from the tank. After the fingerprints have dried thoroughly and have been marked with radioactive ink (both for identification and to provide guides for the elution of any spots to be studied by secondary analysis), the fingerprints can be mounted and later placed in direct contact with X-ray film.

Interpreting RNA Fingerprint

Fingerprints of the RNase T1-resistant oligonucleotides and pancreatic RNase-resistant oligonucleotides of a representative RNA (a 131-base-long RNase III substrate) are shown in Fig. 8.

Considerable information can often be assembled from simple inspection of a fingerprint. While secondary analysis[1] is used for final identification, an initial reading can be made by following a few guidelines. As exemplified by the fingerprints shown in Fig. 8, in which the spots have been identified, the longer and more characteristic oligonucleotides migrate slowly in the second dimension and make up the bottom portion of the pattern.

In the fingerprints of internally labeled RNAs, the RNase T1-resistant oligonucleotides derived from the interior of the molecule have the structure $_{OH}$NNNNNGp (the number of "N" nucleotides is variable). U-rich fragments migrate the fastest in the first dimension, C-rich oligonucleotides are the slowest, and A-rich oligonucleotides fall into the middle. The RNase T1 fingerprint shown in Fig. 8A was made using a 10% (w/v) polyacrylamide gel for the first dimension. To see the difference between a fingerprint made using a gel for this step and one made using a cellulose

FIG. 7. Placement of the chromatography tank within the oven. An air-tight seal connects the bottom and top parts of the chromatography tanks. Second-dimension separations take place in an oven set to 65°.

acetate strip, the fingerprint in Fig. 8A should be compared to the finger-print of the same RNA shown in Fig. 1A. RNase T1-resistant oligonu-cleotides in kinase fingerprints have the structure $pNNNNNG_{OH}$, but follow the same mobility rules as internally labeled fragments. In the second dimension, U- and C-rich oligonucleotides migrate rapidly; A-rich oligonucleotides are much slower. Because second-dimension mobility is

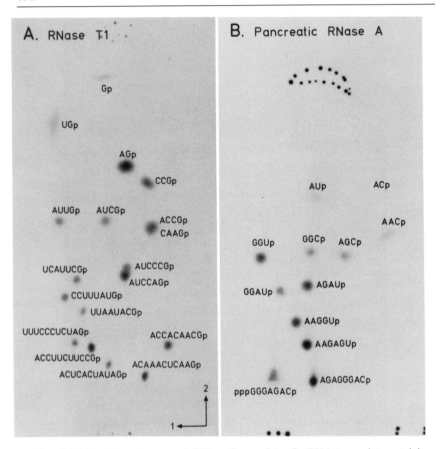

FIG. 8. RNase T1 and pancreatic RNase fingerprints of a RNA transcript containing a RNase III processing site. A 131-base-long RNA transcript was digested with either RNase T1 (A) of pancreatic RNase A (B) and was fingerprinted. Unless otherwise specified, oligonucleotides have hydroxyl groups at their 5' termini. The pair of arrows in the lower right-hand corner of A indicates the directions of first- and second-dimension separations.

strongly influenced by base composition, it is not possible to make a very accurate estimate of an oligonucleotide's length by its position in a RNase T1 fingerprint.

Fingerprints of pancreatic RNase-resistant oligonucleotides are usually simpler than those of RNase T1-resistant oligonucleotides and easier to read (compare Fig. 8A and B). The oligonucleotides from the interior portions are either $_{OH}$NNNNNUp or $_{OH}$NNNNNCp. Since G and A residues have about the same impact on mobility in the second dimension as each other (although G residues slow mobility somewhat more than A residues) and since C and U residues are roughly equivalent in their impact on second-dimension mobility, conventional pancreatic RNase oligonucleotides of the same length fall into a (bumpy) horizontal line. In pancreatic RNase fingerprints, the impact of G residues on the first-dimension mobility is a significant factor because unlike RNase T1 spots, pancreatic RNase spots contain a variable number of G residues: G residues speed mobility in the first dimension, but not as much as U residues.

In general, oligonucleotides of the same length and base composition comigrate in fingerprints. However, a 5'-terminal A residue sometimes causes the separation of oligonucleotides of the same length and base composition and the comigration of oligonucleotides of the same length, but slightly different base compositions.

RNA homochromatography is said to "count phosphate groups." Thus, oligonucleotides with an unusually high number of phosphate groups for their length, such as oligonucleotides from the 5' ends of primary transcripts, certain products of RNA processing reactions, capped oligonucleotides, or fragments containing a branch point or any other polyphosphorylated structure, will be slower in the second dimension than the number of residues and base composition alone would indicate.

Mobility in both the first and second dimension is especially sensitive to the number and configuration of *terminal* phosphate groups. Terminal phosphates speed mobility in the first dimension and slow mobility in the second dimension. As a result, oligonucleotides with other than the single terminal phosphate present on conventional fragments will often appear in surprising places in the fingerprint. Oligonucleotides with no terminal phosphates, typically those from the 3' terminus of the RNA, are shifted dramatically toward the origin in the first dimension and move much faster in the second dimension than their conventional counterparts. Furthermore, such oligonucleotides can often be identified by their irregular shapes. Oligonucleotides with two or more terminal phosphates will move exceptionally rapidly in the first dimension and markedly slowly in the second dimension.

To define the products of a RNA processing reaction, it is often necessary to study the terminal phosphate groups present on the newly created ends. By carrying out a few simple manipulations, it is often possible to determine the phosphate polarity of fingerprint spots and thus to classify the cleavage reaction. For example, prior to fingerprinting, phosphatase treatment can be used to remove any exposed phosphate groups from the products of a processing reaction. Fingerprinting of phosphatase pretreated and control samples will then reveal which oligonucleotides contain terminal phosphates. With knowledge of the sequence of the RNA, it is often possible to determine the type of cleavage which must have taken place. This approach is particularly powerful when coupled with "nearest neighbor" analysis, a technique that relies on the ability to synthesize RNA from each of the four ^{32}P-labeled nucleoside triphosphates and later to digest the RNA with nucleases, such as RNase T1 and pancreatic RNase, which transfer the labeled phosphate onto the preceding nucleotide.

One phosphate group that requires special techniques for analysis is the 2',3'-cyclic phosphate created during certain RNA processing reactions. The cyclic phosphate groups produced by RNA processing may or may not still be present on fingerprint spots, depending on the ribonuclease used for digestion and on the nature of the residue bearing the 2',3'-cyclic intermediate; hence, RNase T1 could be expected to resolve a cyclic phosphate if it fell after G, but not if it were present on a C, U, or A residue. These cyclic phosphate groups can often be identified by their distinctive resistance to both calf alkaline phosphatase and the 3'-phosphatase activity present in nuclease P1 coupled with the fact that they can be resolved and rendered phosphatase sensitive by either acid[1] or 2',3'-cyclic-nucleotide 3'-phosphodiesterase (Sigma). The unusual properties of the 2',3'-cyclic phosphate group often allow them to be mapped directly by devising the appropriate fingerprinting strategy. Additional techniques for studying 2',3'-cyclic phosphate groups appear in the chapter by Symons and colleagues.[11]

[11] A. C. Forster, C. Davies, C. J. Hutchins, and R. H. Symons, this series, Vol. 181.

[13] Enzymatic RNA Sequencing

By Y. KUCHINO and S. NISHIMURA

Several postlabeling techniques for RNA sequencing have been developed. These techniques involve enzymatic digestion,[1-3] chemical degradation,[4] formamide degradation,[5-8] or the wandering-spot method.[3,9,10]

[1] H. Donis-Keller, A. M. Maxam, and W. Gilbert, *Nucleic Acids Res.* **4**, 2527 (1977).
[2] A. Simoncsits, G. G. Brownlee, R. S. Brown, J. R. Rubin, and H. Guilley, *Nature (London)* **269**, 833 (1977).
[3] R. E. Lockard, B. Alzner-Deweerd, J. E. Heckman, J. MacGee, M. W. Tabor, and U. L. RajBhandary, *Nucleic Acids Res.* **5**, 37 (1978).
[4] D. A. Peattie, *Proc. Natl. Acad. Sci. U.S.A.* **76**, 1760 (1979).
[5] J. Stanley and S. Vassilenko, *Nature (London)* **274**, 87 (1978).

These methods are fast and highly sensitive, because the methods allow direct reading of the nucleotide sequence of RNA on acrylamide gel. Moreover, with these methods, the nucleotide sequences of very small amounts of RNA (a few micrograms of RNA) can be determined. Among these postlabeling techniques, the enzymatic RNA sequencing method has the advantage of being available for direct determination of the 5'- and 3'-terminal nucleotide sequences of RNA. Moreover, this method has recently become more widely used, because a variety of ribonucleases, especially pyrimidine-specific enzymes, have been found to be useful for distinguishing C and U residues.

The classical enzymatic method for sequencing RNA is sequence analysis of the RNA fragments generated by complete digestion with RNase T1 or RNase A, followed by study of the overlapping of fragments in partial digests of RNA for their arrangement in order and for connection of nucleotide sequences at junction points of RNA sequences. Techniques, such as fingerprint analysis and homochromatographic analysis, have been used for the determination of overlapping. However, these techniques require great skill and much time.

To overcome the complications of the overlapping technique, Donis-Keller et al.[1] developed a gel readout method based on the procedure of Maxam and Gilbert[11] for DNA sequence analysis. By using this method, the nucleotide sequence of RNA can be read out directly and rapidly on a polyacrylamide gel without overlapping of RNA digests.

The principle of the enzymatic RNA-sequencing method described in this article is based on definition of specific cleavage sites on partial enzymatic digestion of RNA labeled with ^{32}P at its 5' or 3' terminus. Table I lists the ribonucleases now available for RNA sequence analysis by this postlabeling technique.[12-18]

The positions of G and A residues in the polynucleotide chains are identified without difficulty with RNase T1 and RNase U2, which cleave

---

[6] Y. Tanaka, T. A. Dyer, and G. G. Brownlee, *Nucleic Acids Res.* **8,** 1259 (1980).

[7] K. Randerath, R. C. Gupta, and E. Randerath, this series, Vol. 65, p. 638.

[8] Y. Kuchino, N. Hanyu, and S. Nishimura, this series, Vol. 155, p. 379.

[9] A. Nomoto and N. Imura, *Nucleic Acids Res.* **7,** 1233 (1979).

[10] Y. Kuchino, S. Nishimura, R. E. Smith, and Y. Furuichi, *J. Virol.* **44,** 538 (1982).

[11] A. M. Maxam and W. Gilbert, this series, Vol. 65, p. 499.

[12] T. Uchida, *CRC Handb. Microbiol.* **8,** 369 (1987).

[13] D. Pilly, A. Niemeter, M. Schmidt, and J. P. Bargetzi, *J. Biol. Chem.* **253,** 437 (1978).

[14] H. Donis-Keller, *Nucleic Acids Res.* **8,** 3133 (1980).

[15] M. S. Boguski, P. A. Hieter, and C. C. Levy, *J. Biol. Chem.* **255,** 2160 (1980).

[16] M. Kunitz, *J. Gen. Physiol.* **24,** 15 (1940).

[17] G. Krupp and H. J. Gross, *Nucleic Acids Res.* **6,** 3481 (1979).

[18] Pharmacia P-L Biochemicals, Piscataway, New Jersey, unpublished data.

TABLE I
RNases for RNA Sequencing

Enzyme	Source	pH optimum	Specific cleavage	Reference(s)
RNase T1	*Aspergillus oryzae*	7.5	Gp ↓ Np	12
RNase T2	*Aspergillus oryzae*	4.5	Np ↓ Mp	12
RNase U2	*Ustilago sphaerogena*	4.5	Ap ↓ Np	12
RNase *Phy*I	*Physarum polycephalum*	4.5	Ap ↓ Np	12, 13
			Gp ↓ Np	
			Up ↓ Np	
RNase *Phy*M	*Physarum polycephalum*	4.5		12, 14
		With urea	Ap ↓ Np	
			Up ↓ Np	
		4.5		
		Without urea	Ap ↓ Np	
			Gp ↓ Np	
			Up ↓ Np	
RNase *Bc*	*Bacillus cereus*	7.5	Up ↓ Np	3
			Cp ↓ Np	
RNase CL3	Chicken liver	6.5	Cp ↓ Np	15
RNase A	Bovine pancreas	7.5	Cp ↓ Np	16
			Up ↓ Np	
RNase S7	*Staphylococcus aureus*	7.5	Np ↓ Ap	17
		With Ca^{2+}	Np ↓ Ap	
			Np ↓ Up	
		3.5		
		Without Ca^{2+}	Cp ↓ Np	
			Up ↓ Np	
	Neurospora crassa	7.5		
		With urea	Np ↓ Mp	17
		7.5		
		Without urea	Np ↓ Mp	
			(except C–U)	
Nuclease M1	*Cucumis mels*		Np ↓ Mp	18
			(except N–C)	

bonds specifically next to G and A residues, respectively. But the assignment of pyrimidine residues is sometimes difficult, due to the absence of highly specific ribonucleases that hydrolyze phosphodiester bonds next to C or U residues. Therefore, for distinction of C and U residues, certain pyrimidine-specific enzymes such as pancreatic ribonuclease (RNase A) are used generally in combination with other enzymes. For example, chicken liver ribonuclease (RNase CL3), which preferentially cleaves phosphodiester bonds next to C residues, has been used in combination with RNase *Phy*I from *Physarum polycephalum* or RNase S7 from *Staphylococcus aureus*. RNase *Phy*I cleaves RNA at all phosphodiester bonds except C–N bonds, where N represents any of the four constitutive nucleosides. RNase S7 cleaves all pyrimidine–N bonds more uniformly and efficiently than RNase A at pH 3.5 in the absence of Ca^{2+}, but hydrolyzes N–U and N–A bonds rapidly at pH 7.5 in 10 mM Ca^{2+}. RNase *Phy*I has also been used in combination with RNase A or RNase *Bc* from *Bacillus cereus*. Like RNase A, RNase *Bc* cleaves RNA mainly at U–N and C–N bonds, but in a more sequence-specific manner. Other methods for distinguishing U and C residues are use of RNase *Phy*M from *P. polycephalum*, nuclease M1 from *Cucumis melo*, and *Neurospora crassa* endonuclease in combination with RNase *Bc* or RNase A. Like RNase *Phy*I, *N. crassa* endonuclease, reported by Krupp and Gross,[17] can cleave all phosphodiester bonds except C–U bonds in the presence of 7 M urea. RNase *Phy*M cleaves U–N and A–N bonds exclusively. However, in the absence of urea, RNase *Phy*M cleaves A–N, G–N, and U–N bonds, but not C–N bonds like RNase *Phy*I. Nuclease M1 can hydrolyze all phosphodiester bonds, except those adjacent to C (N–C).

Preparation of Terminally Labeled RNA

3′-Terminal Labeling

As RNA molecules contain a free hydroxy group at their 3′ terminus, the procedure described by Bruce and Uhlenbeck[19] using 5′-[^{32}P]pCp as a donor molecule is used for 3′-terminal labeling of RNA. The reaction mixture (50 μl) contains 100 pmol of 5′-[^{32}P]pCp (specific activity, 3000 Ci/mmol), RNA equivalent to 33 pmol of 3′ terminus, 300 pmol of ATP, 6.3 U of T4 RNA ligase, and 0.5 μg of bovine serum albumin in 50 mM N-2-hydroxyethylpiperazine-N'-2-ethanesulfonic acid (HEPES) buffer, pH 7.5, with 20 mM MgCl$_2$ and 10% (v/v) dimethyl sulfoxide (DMSO). After incubation at 4° overnight, an equal volume of a dye solution, containing 0.02% (w/v) xylene cyanole (XC), 0.02% (w/v) bromphenol blue (BPB), and 50% (v/v) glycerol, is added, and then the sample is loaded on a 15% (w/v) polyacrylamide–7 M urea gel to purify the terminally labeled RNA. The 15% gel (height, 20 cm; width, 20 cm; thickness, 0.5 mm) is polymerized in 1× TBE buffer, containing 14.25% (w/v) acrylamide, 0.75% (w/v) N,N'-methylenebisacrylamide, 0.13% (v/v) 3-dimethylaminopropionitrile (DMAPN), 0.1% (w/v) ammonium persulfate, 10% (v/v) glycerol, and 7 M urea, at room temperature. Electrophoresis is carried out at 500 V for 5–7

[19] A. G. Bruce and O. C. Uhlenbeck, *Nucleic Acids Res.* **5**, 3665 (1978).

hr in 1× TBE electrophoresis buffer. TBE buffer (1×) consists of 90 mM Tris base, 90 mM boric acid, and 1 mM disodium salt ethylenediaminetetraacetic acid (Na₂EDTA). For purification of large labeled RNAs, such as viral RNA and mRNA, the RNAs are recovered by phenol extraction after incubation and are filtrated through a column of Sephadex G-50 to remove residual [5′-³²P]pCp. Then the labeled RNAs, precipitated with carrier tRNA at a final concentration of 0.3 A_{260}/ml, are applied to 5% (w/v) polyacrylamide gel (height, 20 cm; width, 20 cm; thickness, 1 mm). Electrophoresis is performed at 4° for 10–15 hr at 20–25 mA with circulation buffer, containing 37 mM Tris–phosphate, pH 7.4, and 5 mM ethylenediaminetetraacetic acid (EDTA). After electrophoresis, the gel is placed on an autoradiogram, and the portion of the gel corresponding to labeled RNA is cut out with a razor. The gel strip is placed in a 1.5-ml conical plastic centrifuge tube and is shaken with 0.4 ml of MG buffer, containing 0.5 M ammonium acetate, 0.1 mM EDTA, and 0.1% (w/v) sodium dodecyl sulfate (SDS) for 6 hr at room temperature to extract the [³²P]RNA. The extracted solution is transferred to a fresh tube, and appropriate amounts of carrier tRNA are added with 2.5 volumes of cold ethanol to precipitate the labeled RNA.

5′-Terminal Labeling

Most eukaryotic and viral mRNAs contain a cap structure, m⁷GpppN-, at their 5′ terminus. For 5′-terminal labeling of these mRNAs with ³²P, the cap structure must be removed before phosphorylation with T4 polynucleotide kinase. The methods available for removal of the cap structure are a chemical method, reported by Fraenkel-Conrat and Steinschneider,[3,20] and an enzymatic method with tobacco acid pyrophosphatase, described by Efstradiatis et al.[21] In both procedures, the initial decapping step is followed by dephosphorylation with alkaline phosphatase and labeling with T4 polynucleotide kinase and [γ-³²P]ATP. In this step, if selective 5′-terminal labeling of the decapped mRNA molecule is desired, a recapping reaction with mRNA guanylyltransferase and [α-³²P]GTP can be used.[22,23] RNA molecules, having a polyphosphate group at the 5′ terminus (ppN-RNA and pppN-RNA), are also good substrates for the capping reaction, while RNAs, containing a monophosphate group and a free hydroxy group at the 5′ terminus (pN-RNA and HO-N-RNA), are completely inert. Therefore, the recapping reaction with mRNA guan-

[20] H. Fraenkel-Conrat and A. Steinschneider, this series, Vol. 12, Part B, p. 243.
[21] A. Efstradiatis, J. N. Vournakis, H. Donis-Keller, G. Chaconas, D. K. Dougall, and F. C. Kafatos, *Nucleic Acids Res.* **4**, 4165 (1977).
[22] K. Mizumoto and Y. Kaziro, *Prog. Nucleic Acid Res. Mol. Biol.* **34**, 1 (1987).
[23] I. Financsek, K. Mizumoto, and M. Muramatsu, *Gene* **18**, 115 (1982).

ylyltransferase and [α-^{32}P]GTP is useful not only for specific 5′-terminal labeling of decapped eukaryotic mRNAs, but also for other terminal labeling of decapped eukaryotic mRNAs, and also for other large RNAs, such as primary transcripts of eukaryotic rRNAs and bacterial mRNAs.

The reaction mixture for capping (50 μl) contains 50 mM Tris–HCl, pH 7.9, 5 mM magnesium acetate, 10 mM dithiothreitol (DTT), 0.05 μg of yeast inorganic pyrophosphatase, 0.6–2 μM [α-^{32}P]GTP (specific activity, 400–2000 Ci/mmol), 0.3–4 μg of acceptor RNA, and 2.5 μg of mRNA guanylyltransferase (rat liver). After incubation for 60 min at 25°, the RNAs are extracted with phenol and are passed through a Sephadex G-50 column to remove residual [α-^{32}P]GTP.

For labeling of RNAs with a monophosphate group at their 5′ terminus, this group must first be removed. Dephosphorylation of the 5′-terminal phosphate of RNA molecules (5 μg, chain length < 150) is performed in 5 μl of solution, containing 50 mM Tris–HCl, pH 8.0, 0.1 mM EDTA, and 0.2 μg/μl of calf intestinal alkaline phosphatase, by incubation for 30 min at 55°. After incubation, 1 μl of 100 mM nitrilotriacetic acid is added to inactivate the phosphatase, and then the mixture is incubated again for 20 min at 23°. Then, for labeling of the 5′ terminus with ^{32}P, a solution (4 μl) of 150 mM MgCl$_2$, 150 mM 2-mercaptoethanol, 5 U of polynucleotide kinase, and 5 μl of [γ-^{32}P]ATP (specific activity, ~7000 Ci/mmol) is added. After incubation for 30 min at 37°, the end-labeled RNA is purified by gel electrophoresis on 15% (w/v) denaturing polyacrylamide gel as described previously.

Partial Digestion of Terminally Labeled RNA

Partial digestion of terminally labeled RNAs with base-specific ribonucleases is usually performed at elevated temperature (50–55°) and in the presence of 7 M urea to avoid interference of the secondary–tertiary structure of RNA with enzymatic hydrolysis steps.

RNase T1 Digestion

The labeled RNA (1–3 × 10^4 cpm), added as carrier tRNA to give a total of 5 μg of RNA, is digested by incubation with 0.02 U of RNase T1 in 10 μl of 20 mM sodium acetate buffer, pH 5, containing 7 M urea and 1 mM EDTA. The incubation is carried out at 50° for 15 min.

RNase U2 Digestion

The same amount of RNA as above is digested by incubation with 0.05 U of RNase U2 in 10 μl of 20 mM of 20 mM sodium citrate buffer, pH 5, containing 7 M urea and 1 mM EDTA. The incubation is carried out at 50° for 15 min.

RNase PhyI Digestion

The same amount of RNA as above is digested by incubation with 0.07 U of RNase PhyI in 10 μl of sodium citrate buffer, pH 5, containing 7 M urea and 1 mM EDTA. The incubation is carried out at 50° for 10 min.

RNase PhyM Digestion

The same amount of RNA as above is digested by incubation with 5 U of RNase PhyM in 10 μl of 20 mM sodium citrate, pH 5, containing 7 M urea and 1 mM EDTA. The incubation is carried out at 50° for 10 min.

If the same amount of RNA is digested by incubation at 24° for 15 min with 1 U of RNase PhyM in 10 mM sodium acetate buffer, pH 5, containing 1 mM EDTA, the digestion pattern of the RNA should be the same as that with RNase PhyI.

RNase Bc Digestion

The same amount of RNA as above is digested by incubation with 2 U of RNase Bc in 10 μl of 25 mM sodium citrate buffer, pH 5, containing 1 mM EDTA. The incubation is carried out at 50° for 15 min.

RNase CL3 Digestion

The same amount of RNA as above is digested by incubation for 5 min at 37° with 0.005 U of RNase CL3 in 10 μl of 20 mM sodium phosphate buffer, pH 6.5.

RNase A Digestion

The same amount of RNA as above is digested by incubation with 0.002 μg of RNase A in 10 μl of 30 mM Tris–HCl buffer, pH 7.8, containing 70% (v/v) DMSO. The incubation is carried out at 37° for 20 min.

Alkaline Hydrolysis

The same amount of RNA as above is hydrolyzed by incubation at 90° for 15 min in 10 μl of 50 mM sodium bicarbonate, pH 9, containing 1 mM EDTA.

After incubation, the reaction is terminated by addition of 10 μl of dye solution, and then the digests are immediately loaded side-by-side onto 20% (w/v) denaturing polyacrylamide gel. The 12% (w/v) denaturing gel can be used for the separation of longer RNA fragments of 50–120 nucleotides.

If the conditions for partial digestion of RNA given here are not optimal, the conditions should be estimated by changes of the enzyme–substrate ratio in the reaction mixture. Recently, RNA sequencing kits containing RNase T1, U2, *Phy*M, *Bc,* and CL3, containing RNase T1, U2, CL3, and S7, and containing RNase T1, U2, CL3, *Phy*M, *Bc,* T2, and P1 have become available from Pharmacia P-L Biochemicals Inc. (Piscataway, NJ), Boehringer Mannheim (Indianapolis, IN), and Bethesda Research Laboratories Inc. (Bethesda, MD), respectively. RNase *Phy*I, *N. crassa* endonuclease, and nuclease M1 are commercially available from Bethesda Research Laboratories Inc., Boehringer Mannheim GmbH, and Pharmacia P-L Biochemicals Inc., respectively.

Identification of Capped Nucleotides by Terminal Nucleotide Analysis

RNA, labeled by the capping reaction (2000–3000 cpm), is dissolved in 10 μl of 30 mM ammonium acetate buffer, pH 5.8, containing 5 μg of nuclease P1 and then is incubated for 60 min at 37°. After incubation, the solution (10 μl) containing 0.2 M Tris–HCl, pH 9.0, 10 mM magnesium acetate, and 0.8 μg of calf intestinal alkaline phosphatase is added. Then the reaction mixture is incubated for 60 min at 37°, and the RNA digest is applied to a diethylaminoethyl (DEAE)-cellulose paper (Whatman DE-81) and is subjected to electrophoresis to analyze the capped nucleotide.[22,23]

Analysis of 5'-³²P-Labeled Nucleotide

RNA, labeled by the kination reaction (2000–3000 cpm), is digested in 10 μl of 20 mM ammonium acetate buffer, pH 5.3, by incubation with 1 μg of nuclease P1 for 3 hr at 37°. After incubation, the RNA digest is applied to a thin-layer cellulose plate and is subjected to two-dimensional thin-layer chromatography.[8]

Analysis of 3'-Terminal Nucleotide

RNA, labeled by the ligation reaction (2000–3000 cpm), is digested in 10 μl of 50 mM sodium acetate buffer, pH 4.5, by incubation with 0.01 U of RNase T2 for 3 hr at 37°. After incubation, the labeled nucleotide is identified by two-dimensional thin-layer chromatography by the same procedure for identification of 5'-terminal nucleotide.[8]

Reading

The labeled RNA fragments, generated by partial enzymatic digestion of 5'- or 3'-terminal [^{32}P]RNA, are separated according to their chain length by polyacrylamide gel electrophoresis (PAGE). After digestion of labeled RNAs, each reaction mixture is mixed with an equal volume of dye solution and is fractionated on 20% (w/v) polyacrylamide–7 M urea gel (height, 40 cm; width, 20 cm; thickness, 0.5 mm; width of slot, 1 cm). The 20% denaturing gel is polymerized in 1× TBE buffer, containing 19% (w/v) acrylamide, 1% (w/v) N,N'-methylenebisacrylamide, 0.13% (v/v) DMAPN, 0.1% (w/v) ammonium persulfate, 7 M urea, and 10% (v/v) glycerol, at room temperature. Electrophoresis is carried out at 1000 V in 1× TBE electrophoresis buffer, until the BPB marker has moved to 15 cm from the bottom edge of the gel. After electrophoresis, the glass plate on one side of the gel is removed, and then the surface of the gel is attached to a transparent film (usually an X-ray film). The exposed gel surface is then covered with a thin polyethylene sheet. Fluorescent or radioactive markers are applied to the gel at appropriate places for later alignment, and an autoradiogram is taken with an exposure time of 1–2 nights.

The chain length of each fragment in the ladders extends from the labeled terminus to an internal G, A, C, or U residue. A series of nested fragments, generated by limited alkaline hydrolysis or partial RNase T2 digestion, is used for determination of the size of RNA fragments on the acrylamide gel. Each RNA fragment has an identical terminus labeled with ^{32}P, and its size is one nucleotide longer than the next shorter fragment at its variable terminus. As cleavage products with ribonucleases are ordered by size, the position of each A and G in the polynucleotide chain can be read directly and easily from the ladder obtained with RNase U2 or RNase T1 in the autoradiogram. However, the assignments of the positions of C and U residues are sometimes difficult. When RNase CL3 is used in combination with RNase PhyI or RNase S7 under partial digestion conditions, CL3 preferentially cleaves C–N bonds. PhyI hydrolyzes all phosphodiester bonds, except C–N bonds, and RNase S7 cleaves all pyrimidine–N bonds, like pancreatic RNase. Hence, C residues can be identified by the presence of a band in the ladders of both RNase CL3 and S7 digests and by the absence of a band at the same level in the RNase PhyI digest. U residues are identified by the presence of a band in both RNase PhyI and S7 digests and by the absence of a band at the same level in the RNase CL3 digest. Nuclease M1, specific for A, G, and U, and RNase Bc, specific for U and C, may be used for further determination of the RNA sequence and especially for the distinctions of C and U.

Comments

The enzymatic RNA sequencing method with terminally labeled RNA has the advantage that it is available for sequence determination of both terminal regions of the RNA. However, the partial digestion of labeled RNAs with base-specific ribonucleases for sequence analysis is affected by not only the presence of modified nucleotides in the RNA molecules, but also the secondary structure even at high temperature and in the presence of 7 M urea. These conditions induce gaps and compressions in the sequence ladder, resulting in the missing of sequences in these regions. For overcoming difficulties in sequencing of ambiguous regions, the chemical modification–degradation method[4] and formamide degradation method,[5–8] which give reasonably regular patterns of products even in the presence of strong secondary interactions, can be used.

Most small RNAs such as tRNAs contain a variety of modified nucleotides. The identifications of these modified nucleotides in RNAs is essential in sequence analysis of these RNAs. However, with the enzymatic sequencing method, and even with chemical methods, it is impossible to assign any modified nucleotide. This difficulty in the assignment of modified nucleotides during sequence analysis is overcome by use of the formamide degradation method using a two-dimensional thin-layer chromatographic system described previously.[8]

With large RNAs, the 3'-terminal sequence of the RNA is first determined by the enzymatic method or by another method. Then the remaining RNA sequence is analyzed by the primer extension method with reverse transcriptase and with the DNA complementary to the 3'-terminal sequence of the RNA as a primer.[24–26] Sequence analysis of eukaryotic mRNA is carried out by use of oligo(dT)$_n$ that is complementary to a stretch of poly(A)$_n$ residues at their 3'-terminus as a primer.

[24] P. H. Hamlyn, G. G. Brownlee, C. C. Cheng, M. J. Gait, and C. Milstein, Cell (Cambridge, Mass.) **15,** 1067 (1978).
[25] J. D. Parvin and L. H. Wang, Virology **138,** 236 (1984).
[26] L. H. Qu, B. Michot, and J. P. Bachellerie, Nucleic Acids Res. **11,** 5903 (1983).

[14] Characterization of Cap Structures

By Yasuhiro Furuichi and Aaron J. Shatkin

Introduction

A cap of the general structure $m^7G(5')ppp(5')N^{(m)}pN^{(m)}p^1$ (Fig. 1) is present at the 5' terminus of almost all eukaryotic mRNAs.[1a,2] During cellular gene expression, formation of the cap on mRNA precursors occurs in the nucleus at the initial phase of transcription before several RNA-processing steps, including splicing, 3'-poly(A) addition, and internal methylation to form N^6-methyladenosine. Cap structures with multiple methyl groups on the terminal G residue, i.e., 2,2-dimethyl-m^7GpppN^m, are also present on the small nuclear RNA species, which apparently provide scaffolding for the various steps in RNA splicing.[3] With one possible exception,[4] all capped cellular and viral mRNAs so far examined contain only a single methyl group on the first G residue, while the adjacent nucleotides are methylated to different extents, providing a basis for the following cap nomenclature: m^7GpppN (cap 0), m^7GpppN^m (cap 1), $m^7GpppN^m_pN^m$ (cap 2), $m^7GpppN^m_pN^m$ (cap 3), and $m^7GpppN^m_pN^m_pN^m_pN^m$ (cap 4).[5] mRNAs in many lower eukaryotes, including yeast, contain mainly cap 0, while higher organisms usually have more extensively methylated caps.

Despite these variations on the methylation theme, the important biological consequences of a cap structure appear to correlate only with the N-7-methyl group on the 5'-terminal G. For example, caps increase mRNA stability by protecting against 5'–3' exonucleolytic degradation.[6] Unmethylated caps can also stabilize mRNA, but, unless the GpppN is converted to m^7GpppN, the unmethylated cap is hydrolyzed to (p)pN, resulting in destabilization of the decapped mRNA. Splicing accuracy and efficiency are both increased by the presence of the 5'-terminal m^7GpppN (Refs. 7–10), and at least in some mRNAs, 3'-terminal processing also is

---

[1] N, Nucleoside; N^m, 2'-O-methyl nucleoside.
[1a] A. J. Shatkin, Cell (Cambridge, Mass.) 9, 645 (1976).
[2] A. K. Banerjee, Microbiol. Rev. 44, 175 (1980).
[3] T. Maniatis and R. Reed, Nature (London) 325, 673 (1987).
[4] C.-C. HsuChen and D. T. Dubin, Nature (London) 264, 190 (1976).
[5] K. K. Perry, K. P. Watkins, and N. Agabian, Proc. Natl. Acad. Sci. U.S.A. 84, 8190 (1987).
[6] Y. Furuichi, A. LaFiandra, and A. J. Shatkin, Nature (London) 266, 235 (1977).
[7] M. R. Green, T. Maniatis, and D. A. Melton, Cell (Cambridge, Mass.) 32, 681 (1983).

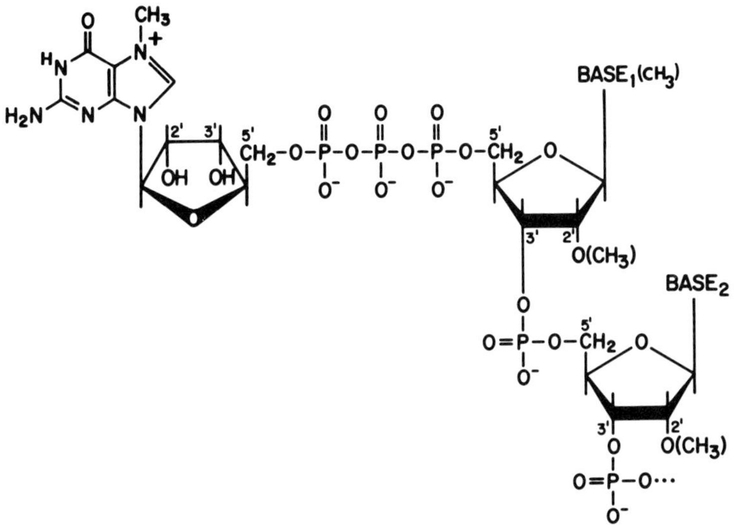

FIG. 1. Structure of the 5' cap.

cap-dependent.[11] In addition, the presence of the cap stimulates mRNA translation at the level of initiation; GpppN cannot substitute for m^7GpppN in the enhancement.[1,2] This may be due to a strict structural requirement by a cap-binding protein, which serves as one of the initiation factors in eukaryotic protein synthesis.[12]

The donor of the methyl group(s) in the cap is an activated form of L-methionine, S-adenosylmethionine (Adomet). It is interesting to note that protein synthesis is initiated with an activated L-methionine intermediate, initiator methionyl-tRNA, and that mRNA is functional as a translational template only after its 5' end is modified by methyl group transfer from Adomet.

The key to complete characterization of mRNA caps is efficient isolation of these unique 5'-blocking structures and separation of the methylated nucleoside components after cleavage of the 5'–5' pyrophosphate linkage of m^7G and $N^{(m)}$. In this chapter, we review the methods, which were developed for the preparation of capped mRNAs, and the isolation and characterization of different cap molecules. These methods were

[8] M. M. Konarska, R. A. Padgett, and P. A. Sharp, Cell (Cambridge, Mass.) **38**, 731 (1984).

[9] I. Edery and N. Sonenberg, Proc. Natl. Acad. Sci. U.S.A. **82**, 7590 (1985).

[10] M. Ohno, H. Sakamoto, and Y. Shimura, Proc. Natl. Acad. Sci. U.S.A. **84**, 5187 (1987).

[11] O. Georgiev, J. Moss, and M. L. Birnstiel, Nucleic Acids Res. **12**, 8539 (1984).

[12] A. J. Shatkin, Cell (Cambridge, Mass.) **40**, 223 (1985).

$$\text{pppG} + \text{pppC} \xrightarrow{\text{RNA polymerase}} \text{pppGpC} + \text{PP}_i \tag{1}$$

$$\text{pppGpC} \xrightarrow{\text{Nucleotide phosphohydrolase}} \text{ppGpC} + \text{P}_i \tag{2}$$

$$\text{pppG} + \text{ppGpC} \xrightarrow{\text{Guanylyltransferase}} \text{GpppGpC} + \text{PP}_i \tag{3}$$

$$\text{GpppGpC} + \text{Adomet} \xrightarrow{\text{Methyltransferase 1}} \text{m}^7\text{GpppGpC} + \text{Adohcy} \tag{4}$$

$$\text{m}^7\text{GpppGpC} + \text{Adomet} \xrightarrow{\text{Methyltransferase 2}} \text{m}^7\text{GpppG}^m\text{pC} + \text{Adohcy} \tag{5}$$

FIG. 2. Mechanism of cap formation. Adohcy, S-Adenosylhomocysteine.

used initially for discovery of the cap structure[13,14] and later for studies on various aspects of caps. For additional details, readers should consult the original publications.

Synthesis in Vitro of Capped mRNAs by Viral Transcription

Several eukaryotic viruses, such as human reovirus, vaccinia virus, and insect cytoplasmic polyhedrosis virus (CPV), contain as components of the purified virus particle enzymes that catalyze the synthesis of capped mRNAs, including RNA polymerase, guanylyltransferase, and methyltransferase(s). Consequently, in vitro viral transcription reactions have proved invaluable for determining the mechanism of cap formation, and it has been shown that cellular mRNA capping proceeds by essentially identical reactions. Figure 2 outlines the mechanism of cap formation during reovirus transcription,[15] but the same general scheme applies for CPV and vaccinia virus. Since these viral mRNAs and most cellular mRNAs initiate with A or G, they provide a convenient source of [3H]methyl- or 32P-labeled forms of cap-A (m^7GpppA^m) and cap-G (m^7GpppG^m) for use as authentic standard markers. Table I summarizes the types of cap structures that can be obtained from the viral transcription reactions. Standard methods for propagation and purification of these viruses have been described.[16]

The selective radiolabeling of specific sites in caps can be planned on the basis of the mechanism of synthesis. For example, reaction mixtures incubated with [3H]Adomet yield [3H]methyl-labeled caps, while use of

[13] Y. Furuichi and K.-I. Miura, Nature (London) 253, 374 (1975).

[14] Y. Furuichi, M. Morgan, S. Muthukrishnan, and A. J. Shatkin, Proc. Natl. Acad. Sci. U.S.A. 72, 362 (1975).

[15] Y. Furuichi, S. Muthukrishnan, J. Tomasz, and A. J. Shatkin, J. Biol. Chem. 251, 5043 (1976).

[16] B. N. Fields, D. M. Knipe, R. M. Chanock, J. L. Melnick, B. Roizman, and R. E. Shope, in "Fundamental Virology." Raven, New York, 1986.

TABLE I
mRNA CAPS SYNTHESIZED *in Vitro* BY VIRAL TRANSCRIPTION

	Transcription conditions		
Virus	+Adomet	Limiting Adomet	−Adomet or +Adohcy
CPV	m^{7}GpppA^m	m^{7}GpppA	GpppA
Reovirus	m^{7}GpppG^m	m^{7}GpppG	GpppG
Vaccinia	m^{7}GpppA^m, m^{7}GpppG^m	m^{7}GpppA, m^{7}GpppG	GpppA, GpppG

[α-^{32}P]GTP or [α-^{32}P]ATP labels both 5′ phosphates in m^{7}GpppG^m vs only the pA^m position in m^{7}GpppA^m, respectively. Addition of inorganic pyrophosphate to reovirus mRNA reaction mixtures diminishes guanylylation and increases the proportion of ppG 5′ ends, while the presence of pyrophosphatase yields mostly 5′-terminal GpppG, which is converted to m^{7}GpppG^m when 0.5 mM Adomet is also present (Table I, Fig. 2).[17] Nonradiolabeled caps and cap analogs have also been chemically synthesized and are commercially available from Pharmacia (Piscataway, NJ).

Cytoplasmic Polyhedrosis Virus

[^{3}H]Methyl-labeled CPV mRNAs are synthesized in a mixture consisting of 0.1 M Tris–HCl, pH 8.0, 10 mM MgCl$_2$, 2 mM each of ATP, CTP, GTP, and UTP, 40 μCi of [^{3}H]Adomet [specific activity, ~70 Ci/mmol (Amersham-Searle, Arlington Heights, IL)], and 10–20 μg of purified virions in a reaction volume of 0.1 ml. After incubation at 31° for 2 hr, followed by phenol extraction, RNAs are isolated by Sephadex G-100 column chromatography and by ethanol precipitation.[18] To obtain [^{3}H]methyl-labeled m^{7}GpppA^m, the mRNA products are digested in 10 mM sodium acetate buffer, pH 6.0, at 37° for 30 min with 50 μg/ml of P1 nuclease [Yamasa Shoyu (Tokyo) and Calbiochem (San Diego, CA)], and caps are purified by high-voltage paper electrophoresis or by chromatography on polyethyleneimine (PEI)-cellulose or paper (see below). Under these conditions, about 10% of the [^{3}H]methyl groups are incorporated into m^{7}GpppA^m from the [^{3}H]Adomet precursor. To prepare ^{32}P-labeled unmethylated GpppA, RNA synthesis is carried out in a modified reaction mixture containing 2 mM Adohcy (Sigma, St. Louis, MO) in place of Adomet and 0.4 mM [β-^{32}P]ATP, prepared from [γ-^{32}P]ATP according to the method of Furuichi and Shatkin,[19] in place of ATP.

[17] Y. Furuichi and A. J. Shatkin, *Proc. Natl. Acad. Sci. U.S.A.* **73,** 3448 (1976).
[18] Y. Furuichi, *Proc. Natl. Acad. Sci. U.S.A.* **75,** 1086 (1978).
[19] Y. Furuichi and A. J. Shatkin, *Nucleic Acids Res.* **4,** 3341 (1977).

Reovirus

Capped reovirus mRNAs are synthesized in a reaction mixture containing 70 mM Tris–HCl, pH 8.0, 7 mM MgCl$_2$, 50 mM KCl, 2 mM each of ATP, CTP, and UTP, 0.5 mM GTP, 24 μCi of [α-^{32}P]GTP [specific activity, 83 Ci/mmol (New England Nuclear, Boston, MA)], 40 μCi of [^{3}H]-Adomet, 5 mM phosphoenol pyruvate, 0.6 U of pyruvate kinase, and 600 μg of washed reovirus cores in a total volume of 0.5 ml. The washed cores are prepared by treating purified reovirions for 30 min at 45° with 1 mg/ml of chymotrypsin (Calbiochem) in 50 mM Tris–HCl buffer, pH 8.0, containing 50 mM KCl.[20] The resulting cores are collected by centrifugation (10,000 g, 15 min, 4°), are resuspended in the same buffer, and are recentrifuged. Washed cores are incubated in the transcription reaction mixture at 45° for 2–4 hr, and capped radiolabeled mRNAs are isolated as described for CPV. To obtain m^{7}GpppG^m, transcripts are digested with P1 nuclease and isolated as for the CPV-generated m^{7}GpppA^m.

Vaccinia Virus

Poly(A)-containing vaccinia mRNAs containing m^{7}GpppA^m and m^{7}GpppG^m are obtained in a 2 : 1 ratio by *in vitro* transcription.[21] The two cap forms released from transcripts by P1 nuclease digestion are readily resolved by paper chromatography. The vaccinia transcription mixture of 0.2 ml contains 0.1 M Tris–HCl buffer, pH 8.0, 10 mM MgCl$_2$, 0.5% (v/v) Nonidet P-40 (NP-40), 10 mM 2-mercaptoethanol, 2 mM each of ATP, GTP, CTP, and UTP, 100 μCi of [^{3}H]Adomet, and 100 μg of purified vaccinia virus. At 37°, this mixture supports RNA synthesis for about 4 hr. After incubation, virions are removed by centrifugation in an Eppendorf microcentrifuge (10,000 rpm, 15 min), and the RNAs in the supernatant fraction are isolated by phenol extraction and gel filtration on a column of Sephadex G-100 or by standard affinity chromatography on oligo(dT)-cellulose.

Characterization of Caps

Isolation

Diethylaminoethyl (DEAE)-Cellulose Column Chromatography. m^{7}GpppN and GpppN are resistant to nucleases that hydrolyze DNA and RNA phosphodiester bonds. In addition, 2'-O-methylation of the ribose

[20] A. J. Shatkin and A. J. LaFiandra, *J. Virol.* **10**, 698 (1972).
[21] C. M. Wei and B. Moss, *Proc. Natl. Acad. Sci. U.S.A.* **72**, 318 (1975).

moiety in an oligonucleotide renders the adjacent 3'–5' phosphodiester linkage resistant to ribonucleases that cleave RNA via 2'–3' phosphate cyclization. Consequently, cap 1 and cap 2 structures, i.e., m⁷GpppNᵐpNp and m⁷GpppNᵐpNᵐpNp that are common in mRNAs of higher eukaryotic cells, are resistant to ribonuclease T2, a nuclease that hydrolyzes RNA without base specificity. The net negative charges of cap 1 and cap 2 are approximately −4.5 and −5.5, as determined by DEAE-cellulose column chromatography in the presence of 7 M urea at pH 8.0[22] (Fig. 3B). These values are based on theoretical charges of +1 at the N-7 position of m⁷G, −3 for GpppNᵐ, and approximately −3 and −4 for -pNp and -pNᵐpNp residues, respectively. Caps obtained by enzymatic digestion of mRNAs elute from DEAE-cellulose columns between 0.2 and 0.25 M NaCl, and for cap isolation, fractions from this region of a linear salt gradient are combined. For concentrating and desalting the caps, the sample is diluted 10-fold with H_2O and is absorbed to a small DEAE-cellulose column (0.6 × 1 cm), which is washed successively with 5 ml each of 50, 75, and 100 mM of freshly prepared NH_4HCO_3 solution. Caps are then eluted with 1 ml of 1 M NH_4HCO_3, which is removed by lyophilization.

Dihydroborylaminoethyl (DBAE)-Cellulose Column Chromatography. DBAE-cellulose is a derivative of DEAE-cellulose containing covalently bound dihydroxyboryl groups. Its utility depends on the selective affinity of 2',3'-*cis*-diols for dihydroboryl groups. DBAE-cellulose was developed originally for the isolation of 3'-terminal oligonucleotides from high-molecular-weight RNA.[23] However, it has also been used to isolate the 5'-terminal capped oligonucleotide from RNA digests, since the cap possesses a 2',3'-*cis*-diol on the ribose moiety of m⁷G.

In one example, avian sarcoma virus (ASV) RNA (1.1 × 10⁷ cpm; specific activity, 1.5–3.0 × 10⁷ cpm/μg) was isolated from ³²P-labeled purified virus, was mixed with [³H]methyl-labeled reovirus mRNA synthesized *in vitro* (10 μg, 10⁵ cpm), and was digested with 50 U of ribonuclease T2 in 0.25 ml of 10 mM sodium acetate buffer, pH 4.5, containing 1 mM ethylenediaminetetraacetic acid (EDTA).[22] After incubation for 6 hr at 37°, the reaction mixture was adjusted to starting buffer A [0.05 M morpholine, pH 8.7, 1.0 M NaCl, 0.1 M $MgCl_2$, 20% (v/v) dimethyl sulfoxide (DMSO)] in a total volume of 1 ml. The mixture was applied to a DBAE-cellulose column (0.6 × 15 cm) that had been equilibrated with buffer A. After removing most of the radioactivity by washing the column

[22] Y. Furuichi, A. J. Shatkin, E. Stavnezer, and J. M. Bishop, *Nature (London)* **257,** 618 (1975).
[23] M. Rosenberg, *Nucleic Acids Res.* **1,** 653 (1974).

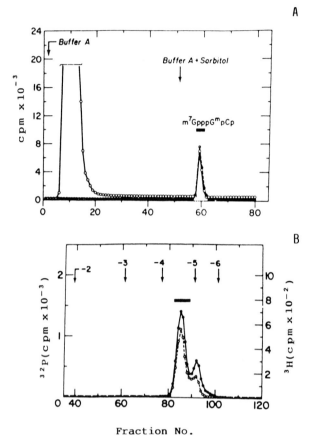

FIG. 3. Separation of capped oligonucleotides by column chromatography on dihydro-borylaminoethyl (DBAE)-cellulose (A) and DEAE-cellulose (B) as detailed in Ref. 22. (○) ^{32}P-labeled avian sarcoma virus (ASV) RNA cap and (●) [^{3}H]methyl-labeled reovirus mRNA cap, m^{7}GpppG^{m}pCp.

with buffer A at a flow rate of 5 ml/hr, the buffer was changed to buffer A plus 1 M sorbitol, and elution was continued at a flow rate of 6–7 ml/hr. As shown in Fig. 3A, ^{32}P-labeled caps eluted immediately after the buffer change in the same fractions as the [^{3}H]methyl-labeled reovirus mRNA caps. The combined cap-containing fractions were reanalyzed by DEAE-cellulose chromatography. A major peak of radioactivity with a net charge of approximately −4.5, which corresponds to the cap 1 structure, m^{7}GpppG^{m}pCp, was resolved by this procedure (Fig. 3B). Capped mRNAs and caps have also been isolated by affinity chromatography on *m*-aminophenylboronate agarose[24] and by m⁷G-specific antibody immobilized on Sepharose,[25] respectively.

Paper Electrophoresis. P1 nuclease digests of capped mRNAs are applied to Whatman 3MM paper, and electrophoresis is performed at 50 V/cm in 10% (v/v) acetic acid–pyridine buffer, pH 3.5. Usually the electrophoretic run is stopped when the control blue marker dye, xylene cyanol (XC), has migrated 10–15 cm from the origin. Nucleotide markers are then located by examination under ultraviolet (UV) light, and the positions of P1 nuclease- and alkaline phosphatase-resistant, ^{32}P-labeled caps are identified by cutting the paper into 1-cm strips and by counting in toluene-based scintillation fluid (Fig. 4A). Caps can be recovered from the paper strips by extraction with water and by lyophilization. The electrophoretic migrations of the different caps relative to 5′pU are as follows: m⁷GpppA^m, 0.38; m⁷GpppG^m, 0.40; GpppA, 0.50; GpppG, 0.52; m⁷GpppA^mpG, 0.26; and m⁷GpppGpA^mpC, 0.24.

Paper and PEI Chromatography. Cap structures and derivatives can be resolved by descending paper chromatography on Whatman 3MM paper in isobutyric acid : 0.5 *M* NH₄OH [10 : 6 (v/v)] with standard markers (Fig. 4B). R_f values in this system are as follows: m⁷GpppA^m, 0.40; m⁷GpppG^m, 0.36; GpppA, 0.25; GpppC, 0.16; GpppG, 0.08; and GpppU, 0.08. It is important to note that the m⁷G residue of the cap is sensitive to alkaline conditions and is degraded to products that include the "ring-opened" form in which the 8–9 bond of the m⁷G imidazole ring is cleaved with accompanying loss of the positive charge. Therefore, it is not appropriate to use alkaline solvents for characterization of caps.

Perry *et al.*[5] have described in detail the separation of a mixture of m⁷GpppAp and m⁷GpppGp, prepared from ribonuclease T2 digests of ^{32}P-labeled *Trypanosoma brucei* poly(A)⁺ RNA, by PEI-cellulose thin-layer chromatography using 2 *M* pyridinium formate, pH 3.4, as solvent.

Enzymatic Cleavage

Nucleotide Pyrophosphatases. As mentioned above, caps are resistant to digestion by ribonuclease T2 and P1 nuclease. After P1 nuclease digestion of mRNA, m⁷GpppN^(m) cap core is released and can be isolated by one of the previous methods. For confirmation of structure and further characterization of the penultimate nucleotide, N^(m), the specific cleavage of the cap 5′–5′ pyrophosphate linkage and the analysis of the resulting products are necessary. Several enzymes are known that can cleave the

[24] H.-E. Wilk, N. Kecskemethy, and K. P. Schäfer, *Nucleic Acids Res.* **10**, 7621 (1982).
[25] T. W. Munns, R. J. Oberst, H. F. Sims, and M. K. Liszewski, *J. Biol. Chem.* **254**, 4327 (1979).

A

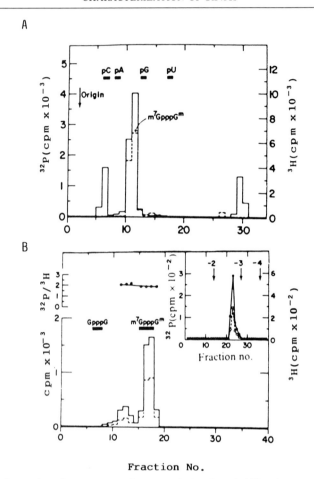

Fraction No.

FIG. 4. Separation of cap structures by paper electrophoresis (A) and paper chromatography (B) as described in Ref. 22. (Inset) DEAE-cellulose column chromatography of ^{32}P-labeled ASV cap and [^{3}H]methyl-labeled reovirus cap. (——) ^{32}P-Labeled ASV RNA cap and (---) [^{3}H]methyl-labeled reovirus mRNA cap.

cap pyrophosphate linkage, including snake venom nucleotide pyrophosphatase, tobacco and potato nucleotide pyrophosphatases, and m^{7}G-specific HeLa cell pyrophosphatase. The choice of enzyme is usually dictated by purpose and availability. Snake venom nucleotide pyrophosphatase (Sigma) catalyzes the conversion of m^{7}GpppN^m to m^{7}Gp + P$_i$ + pN^m. Snake venom nucleotide pyrophosphatase is often used for complete digestion of caps.[14] However, because this pyrophosphatase is always contaminated with a trace amount of phosphodiesterase, venom pyrophosphatase cannot be used to obtain "decapped" mRNA chains.

In contrast, nucleotide pyrophosphatases purified from tobacco tissue culture cells[26] and potato[27] are nuclease-free, which makes them convenient enzymes for removing the cap from mRNA without damaging the polynucleotide strand, i.e.,

$$m^7GpppN^m\text{-----}3' \rightarrow m^7Gp + P_i + pN^mp\text{-----}3'$$

Tobacco pyrophosphatase digestion (followed by phosphatase treatment) is particularly valuable·for preparing the 5' end of mRNA for subsequent radiolabeling by incubation with polynucleotide kinase and [γ-^{32}P]ATP.

m^{7}G-Specific HeLa Cell Pyrophosphatase. An enzyme, which recognizes m^{7}G and which cleaves the pyrophosphate linkage in capped, short oligomers, has been purified from HeLa cells.[28] This enzyme, m^{7}G-specific HeLa nucleotide pyrophosphatase, catalyzes the following reaction.

$$m^7GpppN^mpNp \ldots \rightarrow m^7Gp + ppN^mpNp \ldots$$

The HeLa enzyme preferentially cleaves the pyrophosphate linkage adjacent to the m^{7}G residue and leaves a 5'-diphosphate on the residual oligonucleotide.[29] Trimethyl-G-containing caps, derived from U2 RNA, and unmethylated GpppN structures are not substrates for the HeLa pyrophosphatase.[28,29] In addition and unlike tobacco pyrophosphatase, the HeLa enzyme apparently does not attack m^{7}G-capped polyribonucleotides that are longer than ~10 residues. Presumably, this enzyme, which is mainly cytoplasmic, scavenges capped oligonucleotides, resulting from mRNA turnover, which could otherwise inhibit translation initiation.

A flow chart for analysis of caps by enzymatic digestion is shown as follows.

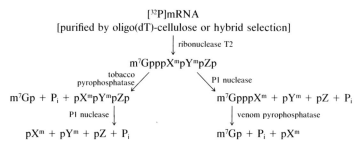

(P$_i$ is released from pZp because P1 nuclease contains a 3'-phosphatase activity)

[26] H. Shinshi, M. Miwa, T. Sugimura, K. Shimotohno, and K. Miura, *FEBS Lett.* **65**, 254 (1976).

[27] M. Zan-Kowalczewska, M. Bretner, H. Sierakowska, E. Szczesna, W. Filipowicz, and A. J. Shatkin, *Nucleic Acids Res.* **4**, 3065 (1977).

[28] D. L. Nuss, Y. Furuichi, G. Koch, and A. J. Shatkin, *Cell (Cambridge, Mass.)* **6**, 21 (1975).

[29] D. L. Nuss and Y. Furuichi, *J. Biol. Chem.* **252**, 2815 (1977).

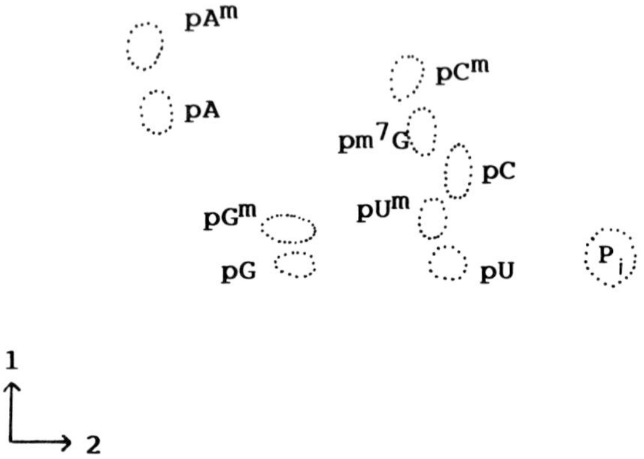

$$pA^m$$

$$pC^m$$

$$pA$$

$$pm^7G$$

$$pC$$

$$pG^m \qquad pU^m$$

$$pG \qquad pU \qquad P_i$$

1

↑

└→ 2

FIG. 5. Two-dimensional thin-layer chromatography of cap constituents. Solvent in the first dimension is isobutyric acid : 0.5 M NH$_4$OH [10 : 6 (v/v)], and solvent in the second dimension is 0.1 M sodium phosphate, pH 6.8 : (NH$_4$)$_2$SO$_4$: 1-propanol [100 : 60 : 2 (v/w/v)].

The constituent nucleotides in the capped oligonucleotide can be identified at various stages by two-dimensional thin-layer cellulose chromatography as diagrammed in Fig. 5.

Yeast mRNA-Decapping Enzyme. A pyrophosphatase that cleaves m^{7}GpppN and GpppN, but not pppN, end groups in RNA has been partially purified from yeast.[30] The nuclease-free preparation catalyzes release of m^{7}GDP and GDP from mRNA and capped synthetic polyribonucleotides at rates directly dependent on chain length; e.g., a 540-mer was 10-fold better as substrate than a 50-mer, and pretreatment with RNase A or nuclease P1 rendered capped RNAs inactive as substrates.

Radiolabeling of mRNA and Caps

With ³H by NaB³H₄ Reduction after Periodate Oxidation

Caps contain a free 2′,3′-*cis*-diol on the ribose ring of the 5′-linked m^{7}G. These hydroxyl groups can be oxidized to the dialdehyde form by treatment with sodium metaperiodate (NaIO$_4$) under conditions that leave the RNA chain intact (with the exception of the 3′-terminal nucleoside, which also contains an oxidized *cis*-diol). The dialdehydes can then be radiolabeled by reduction to the dialcohol form with NaB³H₄:

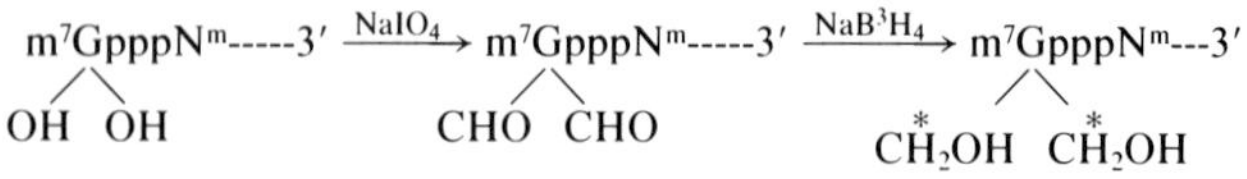

$$
\text{m}^7\text{GpppN}^m\text{-----}3' \xrightarrow{\text{NaIO}_4} \text{m}^7\text{GpppN}^m\text{-----}3' \xrightarrow{\text{NaB}^3\text{H}_4} \text{m}^7\text{GpppN}^m\text{---}3'
$$

OH OH   CHO CHO   CH$_2^*$OH CH$_2^*$OH

[30] A. Stevens, *Mol. Cell. Biol.* **8,** 2005 (1988).

For radiolabeling by this chemical approach, RNA (2–5 μg) is dissolved in 0.1 ml of 20 mM sodium acetate, pH 5.0. NaIO$_4$ (10 mM, 10 μl) is added, and the mixture is kept at 0° in the dark. After 60 min, 20 μl of a 10% solution of propylene glycol is added for another 10 min to destroy excess periodate. RNA is recovered by precipitation with 2.5 volumes of cold ethanol in the presence of 0.3 M sodium acetate. The RNA pellet is dissolved in 0.3 M sodium acetate, is reprecipitated with ethanol to remove propylene aldehyde, is washed with ether, is dried, and is redissolved in 30 μl of 0.5 M sodium phosphate buffer, pH 7.0. NaB^{3}H$_4$ [10 mCi in 10 μl of 0.05 M NaOH; specific activity, 8.8 Ci/mmol (New England Nuclear)] is added to the solution of oxidized RNA, and the mixture is kept on ice for 60 min. To isolate the labeled RNA, the reaction mixture is loaded onto a Sephadex G-100 column (0.6 × 20 cm), equilibrated with 50 mM Tris–HCl buffer, pH 8.0, and RNAs eluted in the excluded fractions are collected by ethanol precipitation.[31,32]

With ^{32}P by Phosphorylation after Decapping

The 5′ termini of decapped mRNAs can be ^{32}P labeled by incubation with polynucleotide kinase and [γ-^{32}P]ATP after phosphatase treatment. The cap is removed by enzymatic digestion with tobacco pyrophosphatase or by chemical reaction. Nuclease-free tobacco pyrophosphatase was used to cleave the cap structure of rabbit globin mRNA, and the resulting decapped mRNA was end labeled and sequenced.[33] Nonenzymatic removal of cap is achieved by a series of two reactions: periodate oxidation followed by selective removal of the resulting dialdehyde by β-elimination using aniline.[32,34]

$$m^7GpppN^m\text{----}3' \xrightarrow{\text{aniline}} m^7\text{'}G + pppN^m\text{-----}3'$$
$$\underset{\text{CHO} \quad \text{CHO}}{\diagdown\diagup}$$

The resulting mRNA retains the 5′-terminal triphosphate, which must be removed by phosphatase treatment before 5′ labeling by polynucleotide kinase. As described in the next section, 5′-triphosphorylated mRNAs can be used as substrates for recapping.

With ^{32}P and/or ^3H by Guanylylation

mRNA guanylyltransferase and mRNA (guanine-7-)methyltransferase catalyze reactions (6) and (7), respectively.

[31] M. Imai, K. Akatani, N. Ikegami, and Y. Furuichi, *J. Virol.* **47**, 125 (1983).
[32] Y. Furuichi, S. Muthukrishnan, and A. J. Shatkin, *Proc. Natl. Acad. Sci. U.S.A.* **72**, 742 (1975).
[33] A. Efstratiadis, J. N. Vournakis, H. Donis-Keller, G. Chaconas, D. K. Dougall, and F. C. Kafatos, *Nucleic Acids Res.* **4**, 4165 (1977).
[34] A. Steinschneider and H. Fraenkel-Conrat, *Biochemistry* **5**, 2735 (1966).

$$\text{ppNpNp} \ldots + \text{GTP} \rightleftarrows \text{GpppNpNp} \ldots + \text{PP}_i \qquad (6)$$

$$\text{GpppNpNp} \ldots + \text{Adomet} \rightarrow \text{m}^7\text{GpppNpNp} \ldots + \text{Adohcy} \qquad (7)$$

These enzymes, purified from vaccinia virus as a heterodimeric complex, can carry out the reactions posttranscriptionally on RNA with diphosphorylated 5′ ends (ppN-) or blocked, unmethylated 5′ termini (GpppN-).[35,36] The enzyme complex also contains RNA triphosphatase, which removes the γ-phosphate from 5′-triphosphorylated mRNA, converting the 5′ end to a diphosphorylated form.[37] Consequently, triphosphate-terminated mRNAs, obtained from capped mRNAs by periodate oxidation and β-elimination, can be recapped and radiolabeled by incubation with the vaccinia enzyme complex plus [α-^{32}P]GTP and/or [*methyl*-^{3}H]Adomet.

$$\text{pppN---3′} \rightarrow \text{ppN----3′} \xrightarrow{\text{GTP}} \text{G}\overset{*}{\text{p}}\text{ppN----3′} \xrightarrow{\text{Adomet}} \text{m}^7\text{G}\overset{*}{\text{p}}\text{ppN----3′}$$

($\overset{*}{\text{p}}$ and $\underset{.}{\text{m}}$ denote ^{32}P- and ^{3}H-labeled positions)

Similar enzyme activities in HeLa cell nuclei have been used to cap transcripts of λc17DNA containing 5′-terminal ppC.[38]

With ^{3}H by 2′-O-Methylation

2′-O-Methyltransferase catalyzes the transfer of the methyl group from Adomet to the 2′-OH of the penultimate nucleoside in capped mRNAs.

$$\text{m}^7\text{GpppN---3′} \xrightarrow{\text{Adomet}} \text{m}^7\text{GpppA}^m\text{-----3′}$$

This enzyme has been purified from vaccinia virus and has been used for radiolabeling brome mosaic and tobacco mosaic virus RNAs by conversion of 5′-terminal m^{7}GpppG to m^{7}GpppG^m.[39] For efficient methyl group transfer, vaccinia 2′-O-methyltransferase requires a m^{7}GpppN-ended polyribonucleotide as substrate. mRNAs from lower eukaryotes and plants both contain mainly cap 0 and thus are good substrates for this enzyme.

Acknowledgments

We are indebted to Mariko Shida and Janet Hansen for excellent secretarial assistance.

[35] S. A. Martin and B. Moss, *J. Biol. Chem.* **250,** 9330 (1975).
[36] S. A. Martin and B. Moss, *J. Biol. Chem.* **251,** 7313 (1976).
[37] S. Venkatesan, A. Gershowitz, and B. Moss, *J. Biol. Chem.* **255,** 903 (1980).
[38] M. Rosenberg and B. M. Paterson, *Nature (London)* **279,** 696 (1979).
[39] E. Barbosa and B. Moss, *J. Biol. Chem.* **253,** 7698 (1978).

[15] Isolation and Characterization of Branched Oligonucleotides from RNA

By J. DAVID REILLY, JOHN C. WALLACE, RANDA F. MELHEM,
DAVID W. KOPP, and MARY EDMONDS

Introduction

The branched RNAs first detected in the polyadenylated nuclear RNA of HeLa cells[1] are now known to be splicing intermediates, the so-called RNA lariats. RNA lariats were first identified in nuclear extracts, where the lariats accumulated in the course of intron removal from specific precursor messenger RNAs (pre-mRNA).[2,3] The striking similarity in composition and sequence of the branch points in these lariats to those recovered from the polyadenylated nuclear RNAs of the cell provided evidence that lariat intermediates exist in the cell nucleus. The branch points isolated from the nonpolyadenylated nuclear RNA fraction presumably are excised lariat introns, although a branch point composition was not determined for this fraction.[4]

It should not be assumed, even in the case of HeLa cells, that lariats are the only form of RNAs that are branched. More recently, a forked RNA branch or Y structure has been implicated in a trans-splicing reaction invoked to account for the transfer of a 39-nucleotide leader sequence from a separately encoded 140-nucleotide transcript to the 5′ ends of all mRNAs of trypanosomes.[5,6] Evidence that the Y branch is formed through a specific A2′p5′G phosphodiester bond has been reported recently.[7] Analogous trans-splicing reaction have now been described for several of the mRNAs of the metazoan *Caenorhabditis elegans*.[8]

Branched RNAs, synthesized in nuclear-splicing extracts, were first detected by their anomalous migration during gel electrophoresis. The subsequent discovery of a specific debranching enzyme in these extracts that converts lariats to their linear forms without loss of nucleotides has

[1] J. C. Wallace and M. Edmonds, *Proc. Natl. Acad. Sci. U.S.A.* **80**, 950 (1983).
[2] R. A. Padgett, M. M. Konarska, P. J. Grabowski, S. F. Hardy, and P. A. Sharp, *Science* **225**, 898 (1984).
[3] B. Ruskin, A. R. Krainer, T. Maniatis, and M. R. Green, *Cell (Cambridge, Mass.)* **38**, 317 (1984).
[4] J. D. Reilly, J. C. Wallace, and M. Edmonds, *Nucleic Acids Res.* **15**, 7103 (1987).
[5] W. J. Murphy, K. P. Watkins, and N. Agabian, *Cell (Cambridge, Mass.)* **47**, 517 (1986).
[6] R. E. Sutton and J. C. Boothroyd, *Cell (Cambridge, Mass.)* **47**, 527 (1986).
[7] R. E. Sutton and J. C. Boothroyd, *EMBO J.* **7**, 1431 (1988).
[8] M. Krause and D. Hirsh, *Cell (Cambridge, Mass.)* **49**, 753 (1987).

METHODS IN ENZYMOLOGY, VOL. 180

provided an important reagent for detecting branched RNAs.[9] A specific alteration in electrophoretic mobility on treatment with the debranching enzyme has provided critical evidence for branched intermediates not only for nuclear pre-mRNA splicing,[9] but also for self-splicing of group II mitochondrial introns[10] and for trans splicing in trypanosomes.[5,6]

A primer extension analysis can be used to map the location of the branch point within an intron, since the branch point blocks reverse transcriptase.[2,3] Sequence analysis of the 5' end of the truncated cDNA product recovered from such reactions can define the branch point within one or two nucleotides. However, precise identification of the branching nucleotide requires the analysis of the ribonuclease-resistant branched oligonucleotide purified from digests of branched RNA.

Procedures are described in this chapter for the isolation, purification, and analysis of these branched oligonucleotides obtained either from RNA labeled *in vivo* or from intermediates formed in splicing extracts from a specific ^{32}P-labeled pre-mRNA.

Isolation of Branch Points[11] from RNAs Labeled *in Vivo* with Inorganic [^{32}P]Phosphate

The procedures described in this section can in principle be applied to any RNA preparation that can be labeled *in vivo* with [^{32}P]phosphate. Success will depend in part on the concentration of branched intermediates in the sample and on the levels of other unidentified components that may cochromatograph with branch points. Branch points were readily detected in the poly(A)$^+$ nuclear RNA of HeLa cells that have one branch per 40,000 nucleotides (Fig. 1B). Branch points were less easily detected in the poly(A)$^-$ nuclear RNA, where additional purification steps were needed, presumably because of the lower concentration of branch points and the higher levels of unknown labeled components that tended to cochromatograph with them.[4]

In addition to HeLa cells, similar procedures have been used for the isolation of branch points from adenovirus RNAs from infected HeLa cells[4] and, more recently, from the poly(A)$^-$ RNA fraction of *Trypanosoma brucei*.[7] A procedure is described for removing 5'-terminal cap structures, specifically the cap 1 that is characteristic of nuclear RNAs

---

[9] B. Ruskin and M. R. Green, *Science* **229**, 135 (1985).
[10] C. L. Peebles, P. S. Perlman, K. L. Mecklenburg, M. L. Petrillo, J.C. Tabor, K. A. Jarell, and H. L. Cheng, *Cell (Cambridge, Mass.)* **44**, 213 (1986).
[11] Branch point is used throughout the text to designate an oligonucleotide that consists of the branching nucleotide and the two nucleotides (or nucleosides) esterified at its 2'- and 3'-hydroxyl groups,

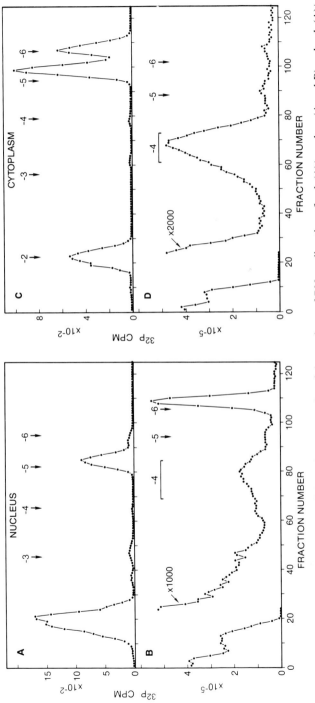

Fig. 1. Diethylaminoethyl (DEAE)-cellulose chromatography of the products of RNase digestion of poly(A)$^+$ nuclear (A and B) and poly(A)$^+$ cytoplasmic (C and D) RNA of HeLa cells.[1] Digests were passed over dihydroxyboronylaminoethyl (DBAE)-cellulose. The bound (A and C) and the unbound (B and D) material was then chromatographed on DEAE-cellulose as described in the text.

that may cochromatograph with branch points released from RNA by ribonuclease T2. Obviously, this step can be omitted if caps have already been eliminated. For example, 5' caps were eliminated by the nuclease treatment that followed hybrid selection of a group of trypanosome pre-mRNAs sharing a common sequence that included the branch point.[7]

Cell Growth and Labeling

HeLa Cells. Approximately 8×10^7 cells in late logarithmic growth were labeled with 800 μCi/ml of [^{32}P]phosphate at a concentration of 2×10^6 cells/ml for 4 hr at 37°. Cells were harvested, washed, and fractionated into cytoplasm and nuclei as described.[1]

Cell Fractionation

Isolation of Nuclei. Nuclei were prepared from HeLa cells by the method of Penman, except that the DNase step was omitted.[12]

RNA Isolation and Fractionation

Total RNA. RNA was extracted from the nuclear pellet by the hot phenol–sodium dodecyl sulfate (SDS) method.[13]
Poly(A)$^+$ RNA. Poly(A)$^+$ was isolated by three cycles of binding of RNA to oligo(dT)-cellulose as described,[14] except for a 2-min heating at 60° before each binding to minimize contamination with nonpolyadenylated RNA.[1] The first unbound RNA fraction is designated as the poly(A)$^-$ RNA.

Isolation and Purification of Branch Points from Nuclear RNAs

RNase Digestion: Poly(A)$^+$ RNA. RNA (about 15 μg RNA that included 10 μg of purified *Escherichia coli* tRNA as carrier) in 200 μl of a solution, containing 5 mM Tris–HCl, pH 7.4, 5 mM disodium salt ethylenediaminetetraacetic acid (Na$_2$EDTA), pH 7.4, 0.3 μg of RNase A (Sigma Type XII-A, St. Louis, MO), 50 U of RNase T1 (Calbiochem, San Diego, CA), and 20 U of RNase T2 (Sigma), was incubated for at least 4 hr at 37°, before the pH was reduced by the addition of sodium acetate, pH 4.5, to a final concentration of 20 mM. An additional 20 U of RNase T2 was added, and incubation was continued for at least 4 hr. Prolonging the incubation to 18 hr did not affect branch recovery. This digest is referred to as the RNase T2 digest.

[12] S. Penman, *J. Mol. Biol.* **17**, 117 (1966).
[13] M. Girard, this series, Vol. 12, p. 581.
[14] H. Nakazato and M. Edmonds, this series, Vol. 29, p. 431.

For a P1 nuclease digestion, the same poly(A)$^+$ nuclear RNA with 10 μg of tRNA carrier is suspended in 100 μl of 10 mM sodium acetate, pH 5.3, before addition of 20 μl of P1 nuclease [1 mg/ml (Sigma)]. After a 1-hr or longer incubation at 37°, the digest is prepared for subsequent chromatography.

Poly(A)$^-$ RNA. About 100 μg of the nuclear RNA, not bound to oligo(dT)-cellulose, was digested as described above for the poly(A)$^+$ RNA, except that, after lowering the pH to 4.5, 100 U of RNase T2 was added. Eighteen hours later, another 100 U was added, and incubation was continued for 4 hr at 37°.

Removal of 5' Caps from RNase Digests. Acetylated dihydroxyboronylaminoethyl (DBAE)-cellulose [no longer commercially available (Collaborative Research)] was prepared for use as follows. A suitable replacement for DBAE-cellulose can be prepared by linking *m*-aminophenylboronic acid to carboxymethyl (CM)-Sepharose by the method of Pace and Pace.[15] The cellulose powder was suspended in buffer A (1 M NaCl and 0.05 M sodium acetate, pH 5.0) and was allowed to swell at 4° for at least 8 hr. The suspension was then pelleted at low centrifugal force and was resuspended in sorbitol buffer (0.5 M morpholine–HCl, pH 8.7, 0.01 M MgCl$_2$, 0.2 M NaCl and 1.0 M sorbitol) at 4°. The sorbitol buffer was removed by centrifugation, the cellulose washed four times with buffer B [1.0 M NaCl and 0.05 M sodium (2-N-morpholino)ethane sulfonate, pH 5.5], before resuspending in buffer A, plus 1 mM NaN$_3$ for storage at 4°. Digests of poly(A)$^+$ RNA, after appropriate dilution, can be applied directly to DBAE columns. However, a preliminary purification that reduces the vast excess of ^{32}P-labeled mononucleotide can simplify subsequent chromatography. This extra step was always used for the poly(A)$^-$ RNA digests, which were diluted 20-fold with 50 mM morpholine, pH 8.0, before passing over a 2.5 × 1-cm-diameter diethylaminoethyl (DEAE) column, equilibrated with 50 mM morpholine–HCl, pH 8.0. The bulk of the ^{32}P-labeled mononucleotides (>90%) was recovered in a 20-ml wash of the column. Bound material was eluted with 300 mM NaCl in 50 mM morpholine, pH 8.0. The eluate was adjusted to the composition of buffer C (500 mM NaCl, 100 mM MgCl$_2$, 50 mM morpholine–HCl, pH 7.0, and 20% (v/v) ethanol), before the eluate was applied to a 7 × 0.7-cm column of DBAE-cellulose, equilibrated with buffer C at 4°. After 10–15 min to allow binding, buffer C was passed over the column at 4° until 50 1-ml fractions were collected. Bound materials that include caps were eluted from the column with 50 mM sodium acetate, pH 4.5, at 23°. The cap 1 structure of nuclear RNA,

[15] B. Pace and N. R. Pace, *Anal. Biochem.* **107**, 128 (1980).

$m^7G_{ppp}N^m_pN_p$, subsequently purified on the DBAE column described below, elutes with charge markers of about -5 (Fig. 1A). Cap 1 and cap 2, $m^7G_{ppp}N^m_pN^m_pN_p$ (-6 charge), found only in cytoplasmic RNA, are separated on this column (Fig. 1C).

DEAE-Cellulose Chromatography. The ^{32}P-labeled fractions, not bound to DBAE-cellulose, were pooled and diluted with 11 volumes of 7 M urea in 25 mM Tris–HCl, pH 7.4. Afterward, ^3H-labeled oligonucleotides, with net charges ranging from -2 to -7, were added. The 7 M urea protects them from digestion by residual RNases. The digest was then pumped over a 25 × 1-cm DEAE-Sephacel column that had been equilibrated with 50 mM NaCl in 7 M urea and 25 mM Tris, pH 7.4. The column was washed with 40 ml of this buffer, before the digest was developed with a 250-ml linear gradient of 50–225 mM NaCl in this same urea–Tris buffer. In contrast to RNase digests of poly(A)$^+$ RNAs (Fig. 1), the ^{32}P elution profile in the region of the -6 charge marker for the poly(A)$^-$ RNA was heterodispersed. The fractions pooled from this region were, therefore, diluted to 50 mM NaCl with the 7 M urea–Tris buffer, pH 7.4, containing fresh ^3H-labeled oligonucleotide markers, before reapplication to another identical DEAE column that was developed as before. Although more homogeneous, the ^{32}P eluting with the -6 charge marker did not elute as the symmetrical peak observed for the branch points from a poly(A)$^+$ RNA digest (Fig. 1B). To quantitate branch points in this poly(A)$^-$ RNA,[4] the ^{32}P-labeled material was desalted as described, was dried at low temperature in a Bio Dryer (Savant), and was electrophoresed in a 20% (w/v) polyacrylamide : 7 M urea vertical slab gel that separates oligonucleotides on the basis of size and charge.[16]

Isolation of Branch Points from RNA Lariats Synthesized *in Vitro*

Branch points, recovered from digests of RNA lariats purified from nuclear-splicing extracts, do not require the extensive purification described in the previous section for those recovered from complex RNA populations, although secondary analysis is simplified by removal of the excess mononucleotides generated by nuclease digestion.

The possibility of obtaining differentially labeled branch points using different ^{32}P-labeled nucleoside triphosphates for the synthesis of pre-mRNA transcripts can provide additional data not only for structural analysis of the branch point, but for localizing the site(s) of the branch within an intron.[17] This approach revealed the presence of multiple

[16] F. Sanger and A. R. Coulson, *FEBS Lett.* **87**, 107 (1978).
[17] M. M. Konarska, P. J. Grabowski, R. A. Padgett, and P. A. Sharp, *Nature (London)* **313**, 552 (1985).

branch points involving both CMP and UMP, but not AMP, in the first intron of the human growth hormone pre-mRNA[18] (see Fig. 4).

Preparation of ^{32}P-Labeled Pre-mRNA Transcripts

A DNA template is prepared by inserting, at appropriate restriction sites of a transcriptional cloning vector, a complete intron flanked by exons. The construction of this vector, designated as Bluescript, has been described[19] and is supplied by Stratagene (3770 Tansy Street, San Diego, CA 92121). In addition to a large collection of restriction sites serving as a polylinker, the template contains promoters for T7 and T3 RNA polymerases in opposite orientation. The plasmid vector containing the insert is linearized by cutting at a suitable restriction site prior to transcription. The linearized DNA is precipitated directly from the restriction endonuclease reaction buffer with 2.5 volumes of ethanol after adjustment to an appropriate salt concentration. Additional purification of the DNAs used in our studies was not required to obtain maximal yields of transcript. The ethanol-precipitated DNA is dried *in vacuo* and is taken up in sterile water prior to the removal of aliquots for transcription.

The DNA is transcribed essentially as described by Carter *et al.*[20] with some modifications designed to enhance the specific radioactivity and to cap the 5' ends of the transcript to stabilize them in nuclear-splicing extracts. All components in transcription and splicing reactions are from sterile solutions either supplied commercially or in solutions passed through a Millex GS 0.22-μm filter, except for $G_{5'ppp5'}G$, DNA, and the T7 polymerase.

The reaction is for 1 hr at 37° in 100 μl containing 40 mM Tris, pH 8.0, 8 mM MgCl$_2$, 25 mM NaCl$_2$, 2 mM spermidine–HCl, 5 mM dithiothreitol, 1 mM $G_{5'ppp5'}G$, 1 mM each of ATP and CTP, 100 μM GTP, and 70 μM [α-^{32}P]UTP with specific activity of 100 Ci/mmol, 2 pmol of purified, linearized DNA, and 100 U of T7 polymerase. When T3 RNA polymerase is used, the sodium chloride concentration of the reaction is raised to 100 mM to suppress transcription from the T7 promoter. The reaction is terminated by the adition of 10 μl of 0.5 M ethylenediaminetetraacetic acid (EDTA) and is extracted once with an equal volume of water-saturated phenol : chloroform : isoamyl alcohol [25 : 24 : 1 (v/v/v)] containing 0.1% hydroxyquinoline, once with an equal volume of chloroform : isoamyl alcohol [24 : 1 (v/v)], and is precipitated with 2.5 volumes of

[18] K. Hartmuth and A. Barta, *Mol. Cell. Biol.* **8**, 2011 (1988).
[19] J. M. Short, J. M. Fernandez, J. A. Sorge, and W. D. Huse, *Nucleic Acids Res.* **18**, 7583 (1988).
[20] A. D. Carter, C. E. Morris, and W. T. McAllister, *J. Virol.* **37**, 636 (1981).

ethanol after adjusting the salt concentration to 0.3 M with 3 M sodium acetate.

Purification of Transcript

The transcript is purified from unincorporated nucleotides, the DNA template, and abortive transcription products by electrophoresis on 4% (w/v) polyacrylamide gels, containing 89 mM Tris–borate, pH 8.3, 2 mM EDTA, and 8.3 M urea.[16] The gel band, revealed by autoradiography that corresponds to the size of the transcript, is excised and crushed with a siliconized glass rod in a siliconized glass tube. The crushed gel, resuspended in 2 ml of elution buffer (0.05 M ammonium acetate, 0.2% (w/v) SDS, 1 mM EDTA and 5 μg/ml tRNA carrier), is incubated for 2 hr (on a shaker) or overnight (not shaken) at 37°. Afterward, the acrylamide fragments are removed by filtering through a syringe plugged with siliconized glass wool. The acrylamide fragments were washed with 2 ml of elution buffer. The transcript is precipitated with 2.5 volumes of ethanol. It is further purified after dissolving in 0.1 M Tris–HCl, pH 7.9, 1 mM EDTA, and 0.4% (v/v) triethylamine by binding to a NEN-Sorb Column (New England Nuclear). After binding, the column is washed with 3 ml of the binding buffer and 3 ml of water. Bound transcript is eluted with 0.5 ml of 50% (v/v) ethanol. The average yield of purified transcript is approximately 6 pmol with a specific activity of 33 × 10^6 cpm/pmol. Greater than 80% of the transcript is capped.

Synthesis of Lariats in Nuclear Extracts

Nuclear extracts are prepared as described by Dignam et al.[21] Splicing reactions, essentially as described by Krainer et al.,[22] are done in 100-μl reaction volumes containing 60% (v/v) nuclear extract and gel-purified transcript at a concentration of 5 nM. The final concentration of the components of the processing reaction is as follows: 12 mM HEPES, pH 7.9, 60 mM KCl, 0.12 mM EDTA, 0.7 mM dithiothreitol, 20 mM phosphocreatine, 1.3 mM ATP, 2.4 mM MgCl$_2$, 0.3 mM phenylmethanesulfonyl fluoride, 3% (v/v) polyethylene glycol, and 14% (v/v) glycerol. Processing is done at 30° for 2–3 hr, depending on the efficiency of processing of the specific RNA. Aliquots are removed at various time points, if the progress of processing is under study. When the reaction is over, the products are diluted with an equal volume of water and are made up to 0.25% (w/v) SDS. One microliter of 10 μg/μl tRNA carrier is added, and

[21] J. D. Dignam, R. M. Lebovitz, and R. G. Roeder, *Nucleic Acids Res.* **11**, 1480 (1983).
[22] A. R. Krainer, T. Maniatis, B. Ruskin, and M. R. Green, *Cell (Cambridge, Mass.)* **36**, 993 (1984).

the reaction is digested for 5 min with 25 μg/ml of proteinase K at room temperature. Afterward, the RNA is extracted from the digest by the same procedures described previously for the isolation of the initial transcript. The RNAs, precipitated from the aqueous extract by ethanol, are purified on 4 or 10% (w/v) polyacrylamide gels as described for the transcripts.[16] RNA lariats are excised from gels and are eluted as described previously for the RNA transcript.

Release of Branch Points from Lariats

P1 Nuclease Digestion. RNA (about 5–10 fmol of lariat) in 30 μl of 50 mM ammonium acetate, pH 5.3, 0.1 mM zinc acetate, and 10 μg of nuclease P1 was incubated for 3 hr at 37°.

RNase T2 Digestion. RNA (about 5–10 fmol of lariat) in 20 mM sodium acetate, pH 4.5, and 50 U RNase T2 (Sigma) is incubated for 3 hr at 37°. Branch points are usually purified from such digests on a small anion-exchange column described in a later section.

Characterization of Branch Points

Figure 2[23] outlines a set of reactions that can be used to characterize a branch point(s) purified either from a RNA population labeled *in vivo* or from a RNA lariat labeled *in vitro*. In some cases, a selected set of these reactions may be sufficient for the identification and quantitation of branch points. For example, a simplified pathway can be used to identify and quantitate branch points in a uniformly [32P]RNA population in which low levels of radioactivity limit a more complete secondary analysis. Measurement of the amount of [32P]phosphate released from putative RNase T2 branch points with a charge of −6 by nuclease P1 can indicate a branch structure. In this case, 50% of the labeled phosphate in a RNase T2 branch point will be released from uniformly labeled RNA as inorganic phosphate by the specific 3′-phosphomononucleotidase activity of this nuclease. Such an analysis is seen in Fig. 3.

Figure 2 describes pathways for the determination of branch point compositions, based on the susceptibility of the 2′- and 3′-phosphodiester bonds on the branching nucleotide to hydrolysis by snake venom (SV) phosphodiesterase. A possible dependence for this susceptibility on the base serving as the branching nucleotide is suggested by the report that a CMP at this site in a branch point, recovered from a lariat formed within the first intron of human growth hormone pre-mRNA, was resistant to SV phosphodiesterase.[18] This resistance was not seen for the AMP branch

---

[23] M. Edmonds, *BioEssays* **6,** 212 (1987).

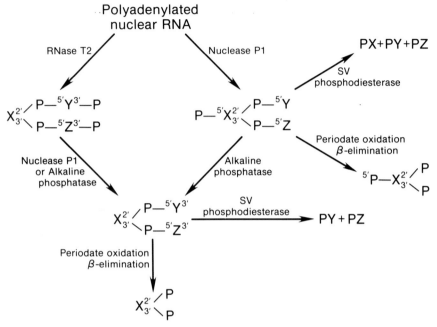

FIG. 2. Enzymatic and chemical reactions used to characterize a branch structure.[23]

points characteristic of most of the branches of HeLa cells[1] and for lariats formed *in vitro*.[17]

Figure 2 also depicts two routes that can be taken to identify the nucleotide at the branch point. Each depends on the β-elimination of the 2'- and 3'-nucleosides from periodate-oxidized branch points. In the case of purified lariats, the branching nucleotide can be obtained directly as the 2',3',5'-nucleoside triphosphate from a purified nuclease P1 branch point.[1,24] Branch point mixtures, derived from cellular RNAs, are more readily purified as RNase T2 branch points, because of their high net charge (see Fig. 1B), but require a phosphomonoesterase treatment to allow periodate oxidation of the 2'- and 3'-hydroxyl groups. A 2',3',-nucleoside diphosphate is released from this branch core structure by β-elimination.

Processing Branch Points for Secondary Analysis

The RNase T2 (Fig. 1) and nuclease P1 branch points, released from mixed RNAs, are purified on the DEAE-cellulose columns described in a previous section. Unidentified components that usually comigrated on

---

[24] R. Kierzek, D. W. Kopp, M. Edmonds, and M. H. Caruthers, *Nucleic Acids Res.* **14**, 4751 (1986).

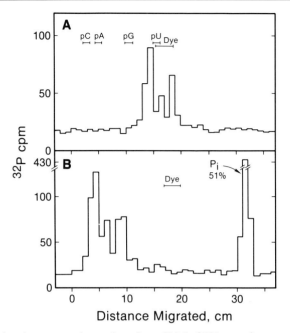

FIG. 3. High-voltage paper electrophoresis at pH 3.5 of RNase-resistant structures with a net charge of −6 (Fig. 1B). (A) Poly(A)+ nuclear RNA of HeLa cells. (B) Nuclease P1 digest of A.[1]

these columns with the nuclease P1 branch points are eliminated by pooling the ^{32}P-labeled material migrating with a charge of −4. After desalting (described later), this fraction is applied to a DBAE-cellulose column similar to that described previously. The contaminants remain unbound, while the nuclease P1 branch points, bound by their cis-glycol group to borate residues, are eluted with 0.05 M ammonium acetate.[1]

A small DEAE-cellulose column has been routinely used to free the branch points recovered on chromatography columns from salts, urea, and buffers. This same column can be used to remove from nuclease digests the mononucleotides that interfere with the secondary analysis of branch points from single RNA lariats. After appropriate dilution, the samples are passed over a DEAE-column (2 × 0.7 cm) that has been equilibrated with sterile, distilled water. The column is washed with 2 ml of 0.1 M NaCl in 25 mM Tris–HCl, pH 7.4, followed by 3 ml of 0.1 M triethylammonium bicarbonate (TEAB), pH 7.8. Branches are eluted in 2 ml of 2 M TEAB, pH 7.8, in 250-μl fractions. The eluants are assayed for Cerenkov radiation, and peak ^{32}P fractions are pooled in 12 × 75-mm siliconized borosilicate tubes. TEAB is removed in the Bio Dryer vacuum centrifuge.

Determination of Base Composition of Branch Point

Purified nuclease P1 branch points or phosphatased RNase T2 branch points from 8×10^7 HeLa cells (0.5 pmol) is incubated with 0.01 U of SV phosphodiesterase [purified as described by Laskowski[25] (Worthington, Freehold, NJ)] in 0.05 M sodium phosphate, pH 8.2, for 2 hr at 37°. Reaction products are applied directly to paper for high-voltage electrophoresis[1] or to thin-layer chromatography plates.[17] All three 5'-nucleoside monophosphates are released from a nuclease P1 branch, while the phosphatased RNase T2 branch gives only the two 5'-nucleotides attached to the 2'- and 3'-hydroxyl groups (Fig. 2). If quantitative data are available for the nucleotides recovered from each type of branch from the same uniformly labeled RNA, it is possible to calculate the base composition of the nucleotide at the branch point (X) from the following formula

$$X = 3(X + Y + Z) - 2(Y + Z)$$

where Y and Z designate the ^{32}P content of the nucleotides attached to the 2'- and 3'-hydroxyl groups of X. The first term of the equation applies to the nuclease P1, and the second term of the equation applies to the RNase T2 products. Using this equation, it was calculated that 79% of the branching nucleotides of HeLa cells are AMP,[1] which was in good agreement with the 75% value determined by direct chemical analysis.[26]

Determination of Branching Nucleotide

The method described by Sninsky *et al.*[27] was used to oxidize and β-eliminate purified nuclease P1 branches or nuclease T2 branches treated with nuclease P1. Each is suspended in 250 µl of 0.4 M cyclohexylamine–HCl/0.1 M HEPES, pH 7.5, that contains 20 mM of freshly prepared NaIO$_4$. After 30 min at 0°, an additional 50 µl of 0.1 M NaIO$_4$ is added, and incubation is continued for 30 min. Then, 100 µl of dimethyl sulfide is added, and the mixture is incubated at 0° for 30 min. The sample is evaporated to dryness at reduced pressure several times to remove dimethyl sulfide. After suspending in 250 µl of H$_2$O, the sample is incubated at 45° for 60 min. The final solution is diluted to 10 ml with water and is desalted on the DEAE-cellulose column described previously. The ^{32}P-labeled 2',3',5'-nucleoside triphosphate from a nuclease P1 branch and the 2',3',-nucleoside diphosphate from a phosphatased RNase T2 branch are analyzed either by high-voltage electrophoresis or by two-dimensional thin-layer chromatography.

[25] M. Laskowski, Sr., this series, Vol. 65, p. 276.
[26] J. C. Wallace, Ph.D. Thesis, University of Pittsburgh, Pittsburgh, Pennsylvania, 1983.
[27] J. J. Sninsky, J. A. Last, and P. T. Gilham, *Nucleic Acids Res.* **3,** 3157 (1976).

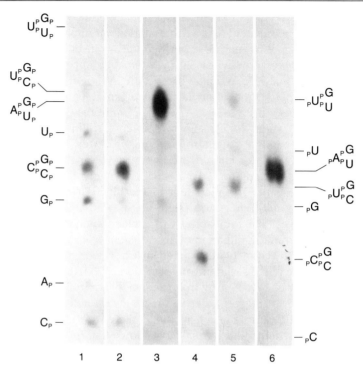

FIG. 4. High-voltage paper electrophoresis of branched oligonucleotides from RNA lariat introns. Lanes 1, 2, 4, and 5 from intron 1 (or A) of human growth hormone pre-mRNA. Lanes 3 and 6 from intron 1 that separates the first two late leader exons of adenovirus type 2. Branch points labeled with [^{32}P]CMP (lanes 1 and 4) or with [^{32}P]UMP (lanes 2, 3, 5, and 6) were from lariats digested with RNase T2 (lanes 1–3) or with nuclease P1 (lanes 4–6). Lanes 1, 2, 3, and 6 are for the excised lariat intron, while lanes 4 and 5 are from digests of the intron–exon lariat intermediate.

Separation and Quantitation of Branch Points

Our laboratory has used high-voltage paper electrophoresis[28] to separate branch points of different composition[1,24] (Fig. 4). Others[2,3,17,18] have usually used two-dimensional thin-layer chromatography in two solvent systems[29] to separate branch points from mononucleotides.

High-Voltage Paper Electrophoresis

Branch points, purified on the small DEAE columns described previously, are taken up in sterile water. Aliquots, usually 5–10 µl, containing

[28] N. P. Salzman and E. D. Sebring, *Anal. Biochem.* **8**, 126 (1964).
[29] M. Silberklang, A. M. Gillum, and U. L. RajBhandary, this series, Vol. 59, p. 58.

5–50 × 10² dpm, 10 nmol of unlabeled mononucleotide markers, and 1 or 2 μl of saturated picric acid as a visible marker, are applied about 3 cm from the end of a Whatman 3MM paper (20 × 40 cm). The paper is electrophoresed horizontally at 3000 V in the apparatus made by CAMAG, Inc. (Box 563, Wrightsville Beach, NC 28480) in a pyridine acetate buffer at pH 3.5 (60 ml of glacial acetic acid : 6 ml of pyridine : 6 g of Na_2EDTA diluted to 1.8 liters). After 60–75 min, the paper is placed in a jar of ammonia vapor for 20 min to convert pyridine acetate to free pyridine, which readily escapes from the paper. After drying in air, the paper is exposed to X-ray film covered with two Cronex Lightning Plus screens (DuPont, Wilmington, DE) at −65° overnight. Radioactive ink is used to mark the paper for alignment to the film. The mononucleotide markers viewed by ultraviolet (UV) light are used to align the autoradiogram with the paper. The radioactivity of each component is obtained by measuring the Cerenkov radiation of sections cut from the paper.

The electropherogram photographed in Fig. 4 summarizes a branch point analysis of excised lariat intron formed *in vitro* from the first intron of the pre-mRNA of human growth hormone. This pre-mRNA was transcribed from a DNA fragment, derived from the cloned gene,[18] that was subcloned into Bluescript, which included the entire intron flanked by exons 1 and 2. Hartmuth and Barta[18] have shown that three different sites are used for branch formation in this intron that has no AMP residue within 56 nucleotides of the 3′ end of the intron. The data from experiments of the type seen in Fig. 4 have allowed us to confirm their findings and to quantitate the distribution of these branch points

$$
\begin{array}{ccc}
\nearrow\text{G} & \nearrow\text{G} & \nearrow\text{G} \\
\text{C} & \text{U} & \text{and} \quad \text{U} \\
\searrow\text{C,} & \searrow\text{C,} & \searrow\text{U} \\
\text{1} & \text{2} & \text{3}
\end{array}
$$

in lariat introns and lariat intermediates (intron–exon 2). Our analyses of the excised lariat intron show CMP to be the predominant branching nucleotide (68%), while **2** was found in 24% and **3** was found in only 7% of these lariats. A striking difference in this distribution was found for the intron–exon lariat intermediate from the same experiment in that **2** was most abundant (56%), **1** was 27%, and **3** was 17% (R. F. Melhem, unpublished experiments). The reaction(s) that account for this overall difference in rates of processing of different lariats cannot of course be deduced from these data.

Figure 4 summarizes one experiment with lariat introns labeled with [³²P]UMP or [³²P]CMP that, in each case, was digested either with RNase

T2 or with nuclease P1. It is clear that these three branch points, produced by either nuclease, are separable in this high-voltage electrophoresis system. The two nuclease P1 branch points with UMP at the branch point are easily separated (lane 5) and move well ahead of the CMP branch that is not labeled by UMP, but is seen in lane 4 when the branch point is labeled by [^{32}P]CMP. All three of the RNase T2 branch points should be labeled with [^{32}P]UMP, but the low levels of the **3** branch point in this lariat intron population did not allow the branch point to be seen in this photograph, while the **2** branch point is barely detectable (lanes 1 and 2). It is necessary to correct the radioactivity measurements obtained from sections cut from these papers for the number of phosphate groups that are labeled in each branch structure to obtain the quantitative data on branch points cited above. Confidence in these analyses is enhanced by the numerical agreement in branch distribution found for the nuclease P1 and RNase T2 products in several such experiments.

Concluding Remarks

The separation, purification, and characterization of branch point oligonucleotides from a single or even a few species of RNA lariat is essentially straightforward. The techniques of nuclease digestion and oligonucleotide separations used here are based on those developed years ago for sequencing RNA by classical methods.[29] The isolation and purification of branched oligonucleotides from mixed RNA populations will usually require more elaborate purification schemes, because of the presence in nuclease digests of pre-mRNAs of modified structures, such as caps, methylated nucleotides, and other unidentified components, as well as pNps, that may copurify with branched oligonucleotides because of similarity in net charge. Such impurities are usually detected easily by secondary analysis, and some, such as caps, are easily removed. However, in some cases, positive identification of branch points may require a secondary analysis that identifies the branching nucleotide directly.[7]

Acknowledgments

We thank Drs. Klaus Hartmuth and Andrea Barta for their gift of the plasmid AAH-5, containing the gene for human growth hormone[18] (H12R1), and Drs. Brian Karger and Jim Williams for the cloned *BA*/IE fragment of adenovirus type 2. These experiments were supported by NIH Grants 5 RO1 CA 18065 and GM 3258.

[16] Enzymatic Approaches to Probing of RNA Secondary and Tertiary Structure

By GAYLE KNAPP

Both enzymatic and chemical methods are available to assist in obtaining information about the secondary and tertiary structures of RNA molecules. This chapter discusses the enzymatic methods for RNA structure probing and builds on previous reviews of RNA sequence and structure determination.[1] The chemical methods will be found in another review article.[2] Both methodologies are powerful and have particular advantages and disadvantages. Their combined use can also provide a checking system and often gives complementary information. In addition, these techniques can be expanded to the probing of RNA–protein interactions, the subject of another chapter in this volume.[3]

The choice between enzymatic and chemical methods depends largely on what and how information is to be obtained. In general, the enzymatic reactions usually allow flexibility in the choice of reaction conditions and can be closer to physiologically significant conditions. The chemical reaction conditions are often more restrictive and can require acidic pH or be altered in their specificity by changes in ionic strength. The principal objection to enzymatic structure probing is the large size of the RNases, which could contribute to probable loss of tertiary structural information due to steric hindrance. Also, nucleotide nearest-neighbor effects on the efficiency of RNase activity occasionally can cause trouble; however, confirming data, using another enzyme of similar specificity, control this minor problem. There is a wide battery of nucleases, which vary in their specificity for single-stranded or double-stranded regions or for the sequences recognized. Many of the nucleases may be used under a wide variety of conditions; although some have restrictions (e.g., S1 nuclease requires divalent cations and has a narrow, acidic pH optimum) that reduce their usefulness for physiologically valid structure probing.

A properly conducted structure-probing experiment produces data that provide information about the RNA molecule on four levels. First, the sequence is obtained for known or unknown members of a RNA

---

[1] J. N. Vournakis, J. Celentano, M. Finn, R. E. Lockard, T. Mitra, G. Pavalakis, A. Troutt, M. van den Berg, and R. M. Wurst, *in* "Gene Amplification and Analysis" (J. G. Chirikjian and T. S. Papas, eds.), vol. 2, p. 267. Elsevier/North-Holland, New York, 1981.

[2] D. A. Peattie, *in* "Methods of DNA and RNA Sequencing" (S. M. Weissman, ed.), p. 261. Praeger Scientific, New York, 1983.

[3] K. A. Parker and J. A. Steitz, this volume [31].

family [e.g., sequence variants in the proline $tRNA_{UGG}$ precursor family or leucine precursor tRNA (pre-$tRNA_{UAG}$)[4,5]] as a consequence of the alignment sequencing reactions that must always be run. Second, the assignments of double-stranded and single-stranded regions of the RNA are made by using conditions that maintain secondary structure without absolutely retaining the tertiary structure or that result in multiple cleavages by single-stranded and double-stranded-specific RNases. Third, the portions of the RNA that are exposed on the surface of the molecule are identified by selecting appropriate conditions (discussed later) that allow the maintenance of the tertiary structure and that allow only single-hit digestion. Fourth, the dynamics of the RNA structure as a function of specific changes in the environment of the RNA can be observed (e.g., $\pm Mg^{2+}$, polyamine, or binding proteins).

Rationale of Needs

Uniquely End-Labeled RNA

The most important requirement for structure probing utilizing the methods described in this chapter is a pure, uniquely end-labeled RNA. There are several methods available for both 5′ and 3′ end labeling of RNA. Some methods are general and will allow the labeling of any linear RNA; other methods take advantage of a characteristic of a particular species of RNA. This chapter presents the principal procedures for general 5′ and 3′ end labeling and one example method for each end that exploits a unique trait of the RNA under study. Coupled with the end-labeling procedure is the need for purification of the unique RNA species. RNA purification is discussed elsewhere.[6-8] In addition, the labeling of a unique terminal nucleotide or oligonucleotide should be verified using thin-layer chromatography (TLC)[1,9] and/or RNA fingerprint analysis.[10]

5′ End Labeling. The 5′ end of almost any RNA may be phosphorylated using polynucleotide kinase (PNKase) and [γ-^{32}P]ATP. PNKase preferentially labels single-stranded ends and is sterically blocked by secondary structure that recesses 5′ ends (e.g., tRNA). Thus, the RNA should be denatured prior to kinasing, if the RNA is known to have a recessed 5′

[4] M.-C. Lee and G. Knapp, *J. Biol. Chem.* **260**, 3108 (1985).
[5] R. C. Ogden, M.-C. Lee, and G. Knapp, *Nucleic Acids Res.* **12**, 9367 (1984).
[6] T. Ikemura, this volume [2].
[7] N. K. Tanner, this volume [3].
[8] G. Knapp, this volume [10].
[9] G. P. Mazzara, J. G. Seidman, W. H. McClain, H. Yesian, J. Abelson, and C. Guthrie, *J. Biol. Chem.* **252**, 8245 (1977).
[10] A. D. Branch, B. J. Benenfeld, and H. D. Robertson, this volume [12].

end or exhibits difficulty in generating efficient labeling of the full-length species. The latter can be observed in initial reactions that are conducted without prior denaturation of the RNA. No matter how carefully RNA is handled during isolation, RNA is exposed to some amount of nuclease nicking. Very low levels of nuclease contamination result in nicks in the most available sites in the single-stranded regions and thus produce more optimal substrates for PNKase. The typical pattern of labeling of a RNA with a recessed 5' end usually shows higher levels of incorporation of label into RNAs shorter than full length. If the RNA has a 5'-phosphate, dephosphorylation must precede the PNKase reaction. The reaction given below is usually adequate for dephosphorylating recessed ends. If the reaction does not work efficiently, the RNA should be denatured prior to the reaction.

One example of a special characteristic of certain RNAs that can be exploited is the presence of a 5'-triphosphate. Many RNAs undergo 5'-terminal processing, which removes the 5'-triphosphate. Any RNAs that retain the triphosphate (e.g., primary transcripts or *in vitro* synthesized RNA[11,12]) may be 5' end labeled using guanylyltransferase (GTP transferase) and [α-^{32}P]GTP in a procedure called capping.

BASIC PROCEDURE USING PNKase

1. Dephosphorylation is accomplished in a 100-μl reaction, containing 10–100 μg of RNA, 10 mM MgCl$_2$, 100 mM Tris–HCl, pH 7.4, and 0.4 U of alkaline phosphatase, calf intestine (CAP). The reaction is incubated at 37° for 30 min. To stop the reaction and remove the CAP, the reaction is extracted with an equal volume of buffer-saturated, redistilled phenol. The aqueous phase, containing the dephosphorylated RNA, is extracted two times with 2 volumes (each time) of diethyl ether. The RNA is precipitated by the addition of $\frac{1}{10}$ volume of 2.0 M sodium acetate, pH 5.0, and 3 volumes of ethanol, is stored at −70° for 30 min, and is centrifuged at 12,000–15,000 g at −5° for 15 min. The pellet is washed with 85% (v/v) cold ethanol, is dried *in vacuo,* and is dissolved in a minimal volume of distilled, diethyl pyrocarbonate (DEPC)-treated water.

2. Phosphorylation is accomplished in a 50-μl reaction, containing 10–20 pmol of a particular RNA species (on the order of 100 μg if bulk cellular RNA is used), 40 mM Tris–HCl, pH 8.0, 10 mM MgCl$_2$, 5 mM dithiothreitol (DTT), 5% (v/v) glycerol, 0.4 mM spermidine, 50–100 μCi of [γ-^{32}P]ATP (highest available specific activity), and 2–10 U of PNKase. The reaction is incubated at 37° for 30 min. The [^{32}P]RNA is then purified, most commonly by the addition of 5 μl of RNA gel dye and by electrophoretic purification in an appropriate polyacrylamide gel.

---

[11] J. K. Yisraeli and D. A. Melton, this volume [4].
[12] V. M. Reyes and J. N. Abelson, this volume [6].

3. Denaturation, if necessary, is achieved by heating the dephosphorylated RNA in the presence of 1 mM ethylenediaminetetraacetic acid (EDTA), pH 7.5, at 100° for 3 min and by plunging the tube into an ice-water bath. The appropriate amount of this RNA is immediately transferred to the PNKase reaction tube, and kinasing is initiated by the addition of PNKase. It is very important to perform the denaturation in the presence of EDTA. This prevents metallocatalyzed cleavage of the RNA by residual divalent cations.

BASIC PROCEDURE USING GTP TRANSFERASE

1. The capping of 5'-triphosphate-terminated RNAs is done in 50-μl reactions, containing RNA (0.1–1.0 μg of *in vitro* transcript or up to ~30 μg of impure isolated RNA), 50 mM Tris–HCl, pH 7.5, 1.25 mM MgCl$_2$, 6 mM KCl, 2.5 mM DTT, 50 U of RNase inhibitor (RNasin), 40 μM GTP (100–150 μCi of [α-^{32}P]GTP), and 5 U of GTP transferase. The reaction is incubated at 37° for 60 min. The reaction is terminated by the addition of 2.5 μl of 10% (w/v) sodium dodecyl sulfate (SDS), 0.5 μl of 0.5 M EDTA, pH 8.0, and 50 μl of phenol : chloroform [1 : 1 (v/v)]. After mixing, the phases are separated by centrifugation, and the RNA is precipitated in the standard manner (described previously) from the aqueous phase by the addition of 3 volumes of ethanol. The RNA pellet is dissolved in 50 μl of TE buffer. The RNA is further purified by gel filtration chromatography and/or polyacrylamide gel electrophoresis (PAGE).

2. The unincorporated [^{32}P]GTP can be removed from the RNA by chromatography using Sephadex G-50 as described elsewhere,[13] except that nonspecific binding of RNA by the resin is reduced by blocking the column. The Sephadex G-50 is equilibrated in TE buffer, is washed with 1 column volume of a solution containing 0.1% (w/v) SDS and 1 μg/ml of tRNA carrier in TE buffer, and is reequilibrated in TE buffer. The column is ready for the application of the RNA sample.

3' End Labeling. The 3' end of any linear RNA can be labeled using T4 RNA ligase and [^{32}P]pCp provided the RNA terminates with a 3'-hydroxyl. The ligase also preferentially recognizes single-stranded 3' termini and may require prior denaturation of certain RNAs. Removal of a 3'-terminal phosphate is necessary and may be accomplished using alkaline phosphatase. The reaction using alkaline phosphatase is as described previously for the 5'-phosphate. Detailed information about the ligase can be found elsewhere.[14]

Some RNAs have special posttranscriptionally added 3' termini. This fact can be exploited for end labeling, e.g., yeast pre-tRNAs.[4] Pre-tRNAs

[13] T. Maniatis, E. F. Fritsch, and J. Sambrook, "Molecular Cloning: A Laboratory Manual." Cold Spring Harbor Laboratory, Cold Spring Harbor, New York, 1982.

[14] R. I. Gumport and O. C. Uhlenbeck, in "Gene Amplification and Analysis" (J. G. Chirikjian and T. S. Papas, eds.), Vol. 2, p. 314. Elsevier/North-Holland, New York, 1981.

that have been nucleolytically processed at their termini (including *in vitro* transcripts) and tRNAs are specifically 3' end labeled using nucleotidyltransferase (NTase). The labeling is specific for tRNAs and their precursors even in bulk RNA. In this case, the 3'-CCA$_{OH}$ terminus is added by NTase using CTP and [α-^{32}P]ATP. The partially purified yeast enzyme can be used to end label tRNAs from most organisms. Successful labeling has been done using tRNAs isolated from several other yeasts (*Candida* and *Schizosaccharomyces*), bacteria (*Escherichia coli* and *Campylobacter pylori*), and a few higher eukaryotes (HeLa cells and *Xenopus*).

BASIC PROCEDURE USING T4 RNA LIGASE

1. Dephosphorylation is done as described previously. If there is a 5'-phosphate which must be maintained, the 5'-phosphate may be replaced after the ligation step using PNKase as described previously.
2. The addition of [^{32}P]pCp using ligase is done essentially as described elsewhere.[1] A 20-μl reaction, containing 50–100 pmol of RNA, equimolar (with respect to RNA) [^{32}P]pCp, 5-fold molar excess (with respect to RNA) ATP, 50 mM HEPES, pH 8.3, 10 mM MgCl$_2$, 5 mM DTT, and 2–10 U of T4 RNA ligase, is incubated at 4° for 16–96 hr. Verification of the extent of the reaction is accomplished by developing samples of the mixture on polyethyleneimine (PEI)-cellulose thin-layer chromatograms in 0.75 M KH$_2$PO$_4$, pH 3.5. The [^{32}P]RNA is purified as required using electrophoretic or chromatographic procedures described elsewhere.[6-8]

BASIC CCA ADDITION PROCEDURE

1. Partial purification of yeast NTase has been described elsewhere (Lee and Knapp[4] and references therein).
2. The NTase reaction is accomplished in a 10-μl mixture, containing 4 μg/ml of bulk RNA (or ~1–2 pmol of pure tRNA species), 10 mM Tris–HCl, pH 8.7, 10 mM MgCl$_2$, 0.06 mM CTP, 100–250 μCi of [α-^{32}P]ATP (3000 Ci/mmol), 0.4 U of RNasin, and 1 $\times$ 10^{-3} U of NTase. The reaction is incubated at 24° for 1–2 hr, and the labeled RNAs are purified by one- or two-dimensional PAGE as described elsewhere.[4,8]

Battery of Enzymes

This section will deal with the description of the range of nucleases available for RNA structure probing. The enzymes vary with respect to sequence and structural specificity and the flexibility of conditions under which each RNase retains activity. The enzymes will be divided into categories based on their structural requirements for substrates. Table I summarizes the pertinent characteristics of these enzymes.

Double-Stranded Structure Specificity. RNase V1 is the principal en-

TABLE I
ENZYMES USED FOR RNA STRUCTURE PROBING EXPERIMENTS

Endonuclease	Specificity	Special considerations
RNase V1	Double-stranded regions	Absolutely requires divalent cations
	No known sequence preferences	May also cleave in particular tertiary structures, e.g., pseudoknots
	Cleaves to leave 3'-hydroxyl	
S1 nuclease	Single-stranded regions	Requires acidic pH
	No known sequence preferences	Requires divalent cations
	Cleaves to leave 3'-hydroxyl	
Mung bean nuclease	Single-stranded regions	Requires divalent cations
	No known sequence preferences	Active at physiological pH
	Cleaves to leave 3'-hydroxyl	
RNase T2	Single-stranded regions	Optimal activity at physiological pH
	No known sequence preferences	Retains activity in EDTA
	Cleaves to leave 3'-phosphate	Active under a wide range of conditions
RNase T1	Single-stranded regions	Active at physiological pH
	Cleaves GpN bonds to leave 3'-phosphate	Retains activity in EDTA
RNase A	Single-stranded regions	Active at physiological pH
	Pyrimidine specific	Retains activity in EDTA
	Cleaves (Py)pN bonds to leave 3'-phosphate	Cleaves different sequences at varied rates (seems to have some preferences for C)
RNase U2	Single-stranded regions	Specificity for A best at acidic pH (pH 3.5)
	Cleaves ApN bonds to leave 3'-phosphate	Some cleavage after G is observed at physiological pH
Bacillus cereus RNase	Pyrimidine specific	Inhibited by urea
	Cleaves to give uniform pyrimidine ladder	Has not been used in structure-probing experiments
RNase *Phy*M	A and U specific	Has not been used in structure-probing experiments
RNase CL3	C specific (predominantly)	Has not been used in structure probing experiments

zyme used to determine which portions of a RNA are found base paired in the structure. RNase V1 can be obtained commercially from Pharmacia (Piscataway, NJ) or purified from cobra venom.[15] The standard reaction conditions are given below. The enzyme requires Mg^{2+} for activity, and

[15] S. K. Vassilenko and V. K. Rait, *Biokhimiya* **40**, 578 (1975).

addition of EDTA will quench the reaction. RNase V1 will cleave most RNA:RNA double helices, even short [4–5 base pairs (bp)] stretches, without any known, strong sequence preferences. This RNase may also recognize the helical portions of the pseudoknot structure.[16] It is assumed that failure to cleave a segment of RNA implies that enzyme access is limited due to steric hindrance by an additional structure in the RNA molecule or that the sequence is single stranded. Correlation with data collected using single-stranded-specific RNases can help confirm the latter case. The former case cannot be experimentally established due to the reaction condition requirements of RNase V1, but can be inferred if Mg^{2+} titration experiments reveal associated single-stranded regions that are buried by tertiary structure (see later).

Single-Stranded Structure Specificity. There are a number of nucleases available that cleave RNA only when RNA is single stranded (see Table I). These nucleases vary in the extent of sequence specificity and in flexibility of reaction conditions. The latter variable plays the more important role in the decision of whether an enzyme will be used in a protocol.

No SEQUENCE SPECIFICITY. Among the nucleases without strong sequence specificity are S1 nuclease, mung bean (MB) nuclease, and RNase T2. All three enzymes are commercially available. Current vendors used by the author are Pharmacia for S1 and MB nuclease and Calbiochem (San Diego, CA) or Sigma (St. Louis, MO) for RNase T2. Although S1 nuclease has been used in successful structure-probing experiments (e.g., Vournakis *et al.*[1]), use of this enzyme should be limited. The optimal conditions for S1 nuclease include acidic pH (4.5–5.0), which can have a significant effect on RNA structure. The other two enzymes have activity in more physiological ranges and will prove more useful. MB nuclease and RNase T2 are essentially equivalent with respect to their use in straight structure probing for the most accessible regions by single-hit kinetics under standard conditions. RNase T2 is more useful for structure variation experiments, because RNase T2 retains activity in EDTA, unlike MB nuclease, which requires divalent cation for activity. The standard reactions for these enzymes are given later in the section on structure-probing reactions. Another factor to be considered in selecting one of these enzymes and in interpreting the results is the cleavage mechanism. S1 and MB nuclease cleave the phosphodiester bond to generate a 5'-terminal phosphate. RNase T2 generates a 3'-phosphate. The presence or absence of a phosphate on a terminus will affect the migration rate of RNA fragments relative to the hydroxide (OH^-) and sequencing ladders.

SEQUENCE SPECIFICITY. The RNases that recognize sequence in RNA and have been most frequently used in structure-probing experiments are

---

[16] C. W. A. Pleij, K. Rietveld, and L. Bosch, *Nucleic Acids Res.* **13**, 1717 (1985).

RNases T1, U2, and A, which cleave after guanosine, adenosine, and pyrimidine nucleotides, respectively. These are also RNases that are used for RNA sequencing reactions. The author's preferred vendors for these enzymes are Sigma and Boehringer Mannheim (Indianapolis, IN). These RNases do not require divalent cations for activity and thus are very useful additions for the structure variation experiments. These nucleases are active under a very wide range of conditions. The nucleases are not inhibited by urea, temperatures up to 50–60°, the presence or absence of a variety of monovalent and divalent cations and anions, polyamines, and probably most other conditions of interest. RNase A is not inhibited by 1% (w/v) SDS. Solutions of RNases T1 and A are unaffected by freezing at −20 or−70°, but RNase U2 in dilute solutions will be inactivated by freezing. The standard conditions for sequencing and structure-probing reactions are given later. The latter will only serve as illustration, because each RNA species may have particular conditions under which its structure needs to be determined.

Solutions and Reagents

Stock solutions of RNases
   RNases T1, A, and U2 are dissolved in TE buffer in concentrations of 1 U/μl, 5 U/μl, and 5 U/μl, respectively. (Note) Solutions of RNase U2 must be made up fresh weekly. Solutions of RNases T1 and A can be stored at −20° and used indefinitely.
RNA sequencing buffer (RSB)
   20 mM Sodium citrate, pH 5.0
   1 mM EDTA
   7 M Urea
   0.025% (w/v) Xylene cyanol (XC)
   0.025% (w/v) Bromphenol blue (BPB)
Hydroxide hydrolysis buffer
   50 mM NaHCO$_3$/Na$_2$CO$_3$, pH 9.2
   1 mM EDTA
Hydroxide hydrolysis dye
   10 mM Urea
   0.05% (w/v) XC
   0.05% (w/v) BPB
2× Sequencing gel loading dye
   9 M Urea
   10% (v/v) Glycerol
   0.05% (w/v) XC
   0.05% (w/v) BPB
RNA gel dye
   20% (w/v) Sucrose

7 M Urea
0.2% (w/v) SDS
0.05% (w/v) XC
0.05% (w/v) BPB
TE buffer
    10 mM Tris–HCl, pH 7.4
    1 mM EDTA
RNase inhibitor
    RNase inhibitor (also called RNasin) is commercially available from a variety of sources; however, the level of released RNase varies from source to source. To rebind the RNase to the inhibitor, the RNasin is adjusted to 5 mM DTT (fresh) and is incubated at 37° for 15–30 min. The treated RNasin can then be restored at −20°.
Buffer-saturated, redistilled phenol
    Saturated with 100 mM Tris–HCl, pH 7.4, unless otherwise noted. Saturation is achieved when the buffer, which separates from the phenol, is at neutral pH.
DEPC-treated H$_2$O
    Distilled H$_2$O containing 0.1% DEPC is stirred at room temperature for 15–30 min, is aliquoted into acid-washed bottles, and is autoclaved 15–30 min.
tRNA carrier
    Commercially available RNA should be phenol extracted prior to use in RNA structure-probing experiments. Carrier RNA can also be prepared from commercial yeast (e.g., Budweiser pressed wet cakes or Fleischmann's lyophilized dry yeast) by phenol extraction and by batch elution from diethylaminoethyl (DEAE)-cellulose.[4,17]
"Standard" structure-probing buffer (SSPB)
    10 mM Tris–HCl, pH 7.0
    10 mM MgCl$_2$
    100 mM KCl
Mung bean nuclease buffer (MBB)
    30 mM Sodium acetate, pH 7.0
    50 mM NaCl
    1 mM ZnCl$_2$
    5% (v/v) Glycerol
Structure-probing stop dye (SP stop dye)
    9 M Urea

---

[17] R. A. Garrett and S. O. Olesen, *Biochemistry* **21**, 4820 (1982).

10% (v/v) Glycerol
0.05% (w/v) XC
0.05% (w/v): BPB

Approaches toward Enzymatic Structure Probing

A typical protocol for the enzymatic probing of a RNA structure uses a battery of enzymes with varied specificity and ideally observes the effect of variation in ionic conditions on the cleavage pattern generated by some of these enzymes. The goal of this section is to point the way for design of individually tailored protocols. Basic conditions and the standard buffer are suggested as starting points and may vary significantly for different RNAs. There are troubleshooting pointers where pertinent, which deal with potential problems and design flaws that can be commonly encountered.

The typical protocol also consists of minimally three and ideally four parts. In each, there are common elements of design. The [^{32}P]RNA is usually present in very low chemical amounts; therefore, reactions are controlled and standardized by the addition of unlabeled carrier RNA. For all reactions, the required amount of cleavage of the [^{32}P]RNA is obtained by titrating the units of RNase per microgram of total RNA (usually equivalent to the amount of carrier RNA) and the time of digestion. Different RNAs can have significant differences in sensitivity of exposed regions to RNase digestion. Therefore, these titrations must be done for every RNA species. The titrations should also be performed whenever new RNase dilutions are made, because very small volumes are involved and the measurement error can be significant.

The first step in the protocol is the generation of sequencing ladders. These serve an essential role as reference for verification of the RNA sequence and for orientation of structure-probing hits within the RNA molecule (e.g., Figs. 1–3). Although structure-probing experiments are generally conducted using previously sequenced RNAs, variant RNAs can be sequenced as well as probed. The reaction conditions have been established for several enzymes, including RNases T1 (G specific), U2 (A specific), *Bacillus cereus* and A (pyrimidine specific), *Phy*M (A and U specific), and CL3 (predominantly C specific) (see Refs. 1 and 18 and commercial sequencing kits). The choice of enzymes is dictated by how much actual sequencing (versus A, G, and/or pyrimidine ladder alignment) is needed and by potential problems encountered. The two pyrimidine-specific RNases have idiosyncrasies that affect their usefulness.

---

[18] H. Donis-Keller, A. Maxam, and W. Gilbert, *Nucleic Acids Res.* **4**, 2527 (1977).

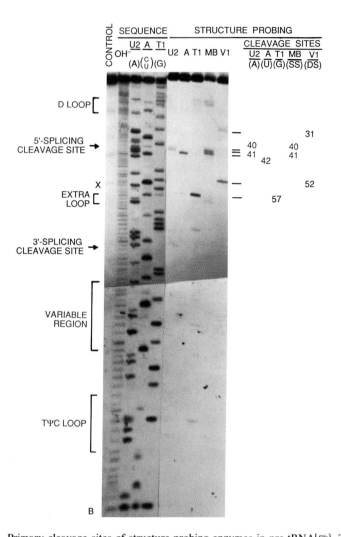

FIG. 1. Primary cleavage sites of structure-probing enzymes in pre-tRNA[Leu-3]. The enzyme–RNA ratios in the structure-probing reactions, which produced primary cleavages, were as follows: RNase U2, 1×10^{-2} U/μg of RNA; RNase A, 1×10^{-7} U/μg of RNA; RNase T1, 1×10^{-4} U/μg of RNA; and MB nuclease, 1×10^{-1} U/μg of RNA. RNase V1 was used at an appropriate dilution of the stock solution. The reactions contained approximately 10^4 cpm of 3′-end-labeled pre-tRNA[Leu-3] and were incubated at 37° for 20 min, except for RNase U2, which was incubated for 1 hr. Reaction mixtures were electrophoresed on a 85-cm-long 10% (w/v) polyacrylamide gel, which was exposed using two X-ray films to cover the whole gel length. Primary cleavage sites are indicated at the right; the numbers refer to nucleotide positions in the pre-tRNA sequence (see Fig. 4). Sequence variation is seen at nucleotide position 57 and is due to multiple genes encoding this tRNA. The portion of the gel containing the sequencing reactions was exposed for twice the time that was necessary for the structure-probing reactions to allow clear reading of the sequences. The symbols X and B indicate the positions of marker dyes, XC and BPB, respectively. The positions of features of the secondary structure with respect to the cleavage ladders are shown at the left. SS, Single stranded; DS, double stranded. (Reprinted with permission from Lee and Knapp.[4])

RNase A is essentially indestructible. RNase A also has mild preferences for which pyrimidines in a sequence RNase A will cleave—RNase A "prefers" C residues, especially those in CpA contexts. The *B. cereus* RNase generates a uniform pyrimidine ladder; however, *B. cereus* RNase is inhibited by urea. The latter characteristic can limit its use with highly structured, G- and C-rich RNAs. The reaction conditions for some of the RNases are presented in Standard Reaction Conditions and Troubleshooting. Other reaction conditions will be found in the references given previously or with the commercial kit protocols.

The second step in the protocol is the start of the actual structure-probing reactions. The standard buffer given in this chapter provides the basic conditions that maintain tRNA tertiary structure and could serve as a starting point for any other RNA. The final buffer conditions should reflect what is known about requirements for functional structure or activ-

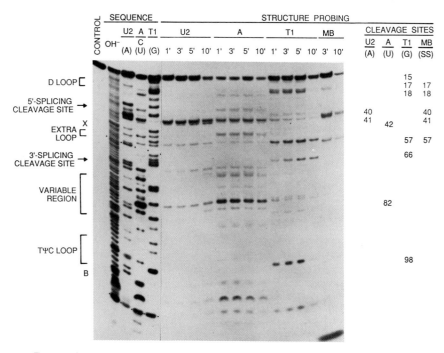

Fig. 2. Time titrations to establish secondary structure by multiple hits. The enzyme–RNA ratios for the sequencing reactions are those given in the text. The structure-probing reactions utilize the following enzyme–RNA ratios: RNase U2, 1×10^{-1} U/μg of RNA; RNase A, 1×10^{-6} U/μg of RNA; RNase T1, 1×10^{-3} U/μg of RNA; and MB nuclease, 1 U/μg of RNA. Reactions were incubated at 37° and aliquots were withdrawn at the times designated. Symbols and areas of the figure have been described in Fig. 1.

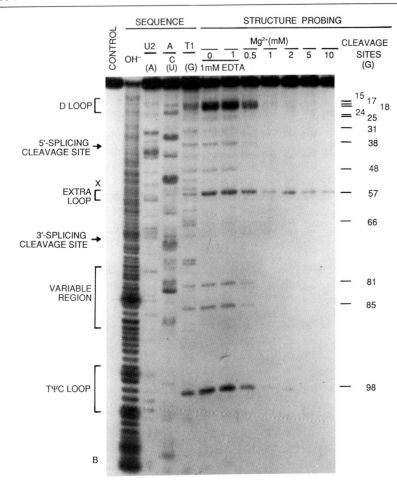

FIG. 3. Effects of Mg^{2+} on pre-tRNA[Leu-3] structure. RNase T1 at 1×10^{-4} U/μg of RNA was used to study the effect of varying $MgCl_2$ concentration on pre-tRNA[Leu-3] structure. Reaction conditions were essentially the same as those in Fig. 1, except that Mg^{2+} concentrations were varied as indicated on the figure. Cleavage sites in the absence of Mg^{2+} (1 mM EDTA) are shown at the right. Other areas of the figure and symbols have been described in Fig. 1. (Reprinted with permission from Lee and Knapp.[4])

ity of the RNA. The appropriate ratio of units of RNase per microgram of RNA and time of reaction must be titrated in the selected buffer. This ultimately results in the conditions for single-hit data. However, the titration also provides information about the secondary structure in that the multiple hits will indicate regions that are double stranded (RNase V1) or single stranded (other RNases) (e.g., Fig. 2). It is important to always include RNase V1 reactions, which confirm the major and important double-stranded portions of the RNA molecule, because this will aid in determining whether a region is not cleaved by single-stranded-specific enzymes due to base pairing or inaccessibility due to the tertiary structure. Several single-stranded-specific enzymes should also be selected to prevent misinterpretation due to artifacts inherent to one enzyme or the RNA sequence. For example (see Fig. 1), use of RNase T1 alone would misassign the most available single-stranded region, if that region lacked a guanosine. Reaction conditions and data for an example are given later for illustration.

The third step in the protocol is the accumulation of single-hit results for a variety of single-stranded- and double-stranded-specific enzymes. Various combinations of the enzymes listed in the above battery have been used in structure-probing experiments. The minimal experiment would include single hits by RNase V1 and MB nuclease or RNase T2. The ideal experiment would also include several sequence-specific RNases, which are selected on the basis of the RNA sequence. Care must be taken to obtain single hits by the RNases and in any evaluation of multiple hits. The difficulty with the multiple-hit situation is that, after the first hit, any succeeding hits are suspect. The first nick in the RNA will probably affect the integrity of the tertiary structure. If multiple hits are unavoidable in reactions using RNA labeled at one end, RNA labeled at the opposite end should be tested to see if one of the hits appears first in these alternate reactions. Even this control does not always resolve the problem. The rate of cleavage by the single-stranded-specific RNases is affected by whether the nucleotide to be hit is stacked or unstacked in the unpaired sequence. A stacked nucleotide in the most available single-stranded region and an unstacked nucleotide in a less accessible region might react at equivalent rates. Hence, the strong recommendation to use more than one RNase.

The fourth step in the protocol is the introduction of conditions that result in the variation of the structure. The effects of various ionic conditions on the structure of RNA have been investigated, and among the masses of data are observed the stabilizing effects of Mg^{2+} and, to a lesser extent, polyamines on RNA tertiary structure. For tRNA[Phe], correlation between the location of Mg^{2+} in the crystal structure and the effects of Mg^{2+} presence or absence on the appearance of tertiary hydrogen bond resonances in NMR has been made. This correlation has been expanded to the effects of Mg^{2+} withdrawal on the sites of cleavage of RNases in structure-probing experiments. An example using a yeast pre-tRNA is shown (see Fig. 3). Variation of Mg^{2+} concentration has been used with good results with other small RNAs (e.g., Garrett and Olesen[17]). Polyamines, binding proteins, sequence variants (or mutants), and any other factor that affects the function of the RNA are likely candidates to alter the tertiary or even the secondary structure of the RNA. Polyamines would be predicted to have a stronger effect on stability of the secondary rather than the tertiary structure.

Standard Reaction Conditions and Troubleshooting

Sequencing and Hydroxide Ladders

The reactions are performed as explained elsewhere[1,18] with only minor changes. The ratio of units of RNase to micrograms of RNA and reaction time must be established for each RNA species and each RNase preparation. The examples given in this section provide beginning parameters for other experiments. Sequencing reactions for only the enzymes used in the structure-probing examples will be given in this section. Full-sequencing protocols will be found elsewhere.[1,19]

Preparation

1. Make serial dilutions from RNase stock solutions into RSB. The dilutions proceed in "order-of-magnitude" steps of 1×10^{-1}–1×10^{-4} U/μl for RNases T1 and U2 and 5×10^{-2}–5×10^{-8} U/μl for RNase A. The dilutions for RNases T1 and A need only to be made once and stored between uses at $-20°$.
2. Pour and prerun sequencing gel (see later).
3. Preheat water baths at 50 and 90°.
4. Mix 2,000–10,000 cpm (Cerenkov) of [^{32}P]RNA and 10 μg of tRNA carrier per tube for the RNase sequencing reactions and two times (2×) these amounts for the OH$^-$ and control reactions. (Note) You will usually load 1× OH$^-$ and control reactions on the sides of the sequencing gel. Lyophilize.
5. Prepare ice-slurry bath.

Reactions

1. RNase sequencing reactions.
   a. Dissolve the RNA sample in 3 μl of the appropriate RNase dilution solution. Typically, the sequencing is accomplished using 1×10^{-2} U/μl of RNase T1, 1×10^{-1} U/μl of RNase U2, and 5×10^{-4} U/μl of RNase A. However, these concentrations should be verified empirically for each RNA using the serial dilutions.

---

[19] Y. Kuchino and S. Nishimura, this volume [13].

   b. Incubate at 50° for 15 min; then, immediately plunge into ice-slurry bath.
2. OH⁻ ladder reaction.
   a. Dissolve a 2× RNA sample in 5 μl of hydroxide hydrolysis buffer.
   b. Incubate at 90° for 5–7 min; then, immediately plunge into ice-slurry bath.
3. Control reaction.
   a. Dissolve the other 2× RNA sample in 6 μl of RSB.
   b. Incubate at 50° for 3 min; then, immediately plunge into ice-slurry bath.

Gels

1. The sequencing gels used for RNA sequencing and structure probing have been described elsewhere.[18,20] The dimensions of the gels were 40 × 28 × 0.2 cm for standard gels or 85 × 18 × 0.4 cm for long, resolving gels. Both 10 and 20% (w/v) polyacrylamide [acrylamide : bisacrylamide (29 : 1)] in 7 M urea and TBE are useful for resolving fragments of different sizes. The long, resolving gels are useful when a large region of the RNA needs to be seen in one run.

2. Prerunning of the gel is essential to improve resolution by heating the gel and to reduce band distortion due to ionic differences between the gel and the buffer reservoirs. The gel is prerun at 40 W for approximately 1–2 hr (until the BPB in a preloaded dye sample has electrophoresed about half the length of the gel).

3. Following thorough flushing of the wells, the reactions are loaded in the following order: control OH⁻, U2, T1, A, OH⁻, control. The gels are electrophoresed at 40–70 W for times determined by the lengths of the RNA fragments to be resolved. The BPB and XC migrate on 10% gels with oligonucleotides of ~10 and 60 nucleotides, respectively. On 20% gels, these dyes migrate ~5 and 25 nucleotides, respectively.

Troubleshooting

Inspection of Fig. 1 (left) shows the alignment of OH⁻ and sequencing ladders for pre-tRNA^{Leu-3}. The specificity problem with RNase A is very apparent, especially in the lower half of the gel. This problem is unimportant to this experiment, since the full sequence of the RNA was known. The *B. cereus* RNase should be run in parallel whenever the sequence of the RNA must be clearly obtained. In addition, variation in the intensity of certain bands in the RNase U2 lane can be seen when there are runs of two or more As. This variation is not consistent, but seems to occur when the run is at the transition from loop to base-paired stem. This variation does not significantly affect the readability of the RNase U2 lane. Under the conditions given previously, the accuracy of cleavage by this RNase is not absolute. A small degree of cleavage after Gs may be seen (e.g., TΨC loop region). Vournakis et al.[1] have improved the RNase U2 sequencing reaction by dropping the RSB pH to 3.5.

This protocol uses sequencing gels containing 7 M urea. Occasionally a G- and C-rich RNA will exhibit severe compression of bands in regions where this urea concentration is insufficient to denature all secondary structure. In this case, the urea concentration should be increased to 8.3 M.

Structure Probing Reactions

As iterated above, the conditions given in this section serve as a starting point for the design of individualized structure-probing experiments. Optimal reaction times and RNase–RNA ratios must be empirically determined for every RNA and RNase preparation.

Preparation

1. Make serial dilutions as above, except that SSPB is used. In addition, dilutions ($10–1 \times 10^{-3}$ U/μl) of MB nuclease are made in MBB, and dilutions ($1 \times 10^{-1}–1 \times 10^{-5}$) of RNase T2 are made in SSPB.
2. Pour and prerun sequencing gel (see previously).
3. Powder dry ice in an ice bucket.
4. Perform RNA sequencing reactions or have them previously run and stored at $-70°$. The reactions have been stored for at least 2 weeks with no observable loss of resolution.
5. Mix 10,000 cpm (Cerenkov) of [^{32}P]RNA and 10 μg of tRNA carrier per tube for structure-probing reactions. Lyophilize.
6. Preheat water bath at $37°$.

Reactions

1. Reactions using RNases T1, A, U2, and V1.
   a. Dissolve the RNA in 2 μl of the appropriate RNase dilution. For the pre-tRNA example shown in this section, the RNase–RNA ratios are as follows: RNase T1, 1×10^{-4} U/μg; RNase A, 1×10^{-7} U/μg; and RNase U2, 1×10^{-2} U/μg. For RNase V1, units had not been established for the preparation; thus, an appropriate dilution of the stock was determined. In other investigator's experiments, the ratio for RNase V1 is in the range of 1×10^{-2} U/μg.[17]

b. Incubate at 37° for 20 min.
c. Halt the reaction by adding 2 μl of SP stop dye and by immediately freezing in dry ice.
d. Incubations with RNase T2 can be conducted similarly.[21]
2. Reactions using MB nuclease.
a. Dissolve the RNA in 2 μl of an appropriate dilution of MB nuclease.
b. Incubate at 37° for 20 min.
c. Halt the reaction as in reaction 1c above.

Gels

The sequencing gel is run as described above in the section on RNA sequencing ladders with the exception of the loading pattern. See the examples in Figs. 1 and 2.

Example Results and Troubleshooting

1. Multiple-hit probing experiments are conducted either by varying the RNase–RNA ratio over several orders of magnitude or by taking time points with a fixed RNase–RNA ratio as shown in Fig. 2. In this experiment, 10-fold higher RNase–RNA ratios were used in time points ranging from 1 to 10 min. All major single-stranded regions of pre-tRNA^{Leu-3} are observed in the cleavage time courses for RNases U2, A, and T1. At the later time points, the effects of severe reduction in the amount of full-length substrate can be seen in the loss of some of the longer partial fragments (e.g., RNase T1 cleavages at residues 15, 17, and 18).

2. The MB nuclease concentration used illustrates another problem that can sometimes be encountered. The single-stranded 3′ terminus of the pre-tRNA is too readily cleaved, compared with sites other than the most accessible loops in the intron. Therefore, we observe only the most accessible sites as longer fragments and the 3′ nucleotide. There is no adequate pool of full-length substrate to feed into the fragments generated by the slower cleavages at less accessible sites.

3. Figure 1 (right) illustrates the near approach to single hits at the lower RNase–RNA ratio for the RNases. Other reactions, titrating RNase–RNA ratios at fixed times as given in the figure legends, had confirmed that unique hits were obtained for RNase T1 at G-57, RNase V1 at G-52, and RNase A at U-42. The cleavages at A-40 and A-41 by RNase U2 and MB nuclease were inseparable, but can be viewed as occurring at positions of equal accessibility in a loop. These cleavages indicate that the intron in this pre-tRNA (see Fig. 4) contains the most

[21] C. P. H. Vary and J. N. Vournakis, *Nucleic Acids Res.* **12**, 6763 (1984).

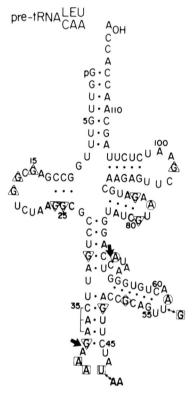

FIG. 4. Secondary structure of pre-tRNA[Leu-3]. The secondary structure of pre-tRNA[Leu-3] is shown with the nucleotides numbered starting from the mature 5' end. Sequence variations observed in the two genes that have been sequenced are shown by the double-headed arrows. The three-pronged bracket indicates the anticodon triplet. Bold arrows indicate the positions of the 5'- and 3'-splicing cleavage sites. The primary cleavage sites for the single-stranded-specific RNases are indicated by squares and are the following residues: RNase U2 (A-40 and A-41), RNase A (U-42), RNase T1 (G-57), and mung bean nuclease (A-40 and A-41). The primary cleavage site for duplex-specific RNase V1 (G-52) is indicated by the diamond. The circles indicate other nucleotides that are secondary sites of cleavage under the standard structure-probing conditions. Triangles indicate nucleotides that become very susceptible to RNase cleavage on removal of Mg^{2+}. The inverted triangles indicate nucleotides of lesser intensity of cleavage under the latter conditions.

accessible single- and double-stranded regions in the RNA molecule. The other regions are cleaved (see in Fig. 2) only later, presumably after the tertiary structure has been destabilized. These data allow the assumption of the cloverleaf for the secondary structure of this pre-tRNA.

4. The data in Fig. 1 indicate that the intron is the most accessible region of the pre-tRNA and is presumably on the outside of the molecule.

The data also indicate that the D and TΨC loops should be involved in the tertiary structure. However, the single- and multiple-hit data cannot distinguish what the tertiary structure might be. Additional experimentation is required to suggest correlations with other physical observations of tertiary structure, e.g., variation in ionic conditions (see later).

5. The unique cleavage at G-57 can be seen in Fig. 3, as well, in the lane on the right. This lane is a repeat of the previous standard structure-probing reaction. The other lanes exhibit the effect on pre-tRNA structure of variation of Mg^{2+} concentrations. At no or low Mg^{2+} concentrations, prominent cleavages are observed at residues in the D and TΨC loops. Similar results with tRNA[Phe] allow us to assume that the pre-tRNA structure is analogous to that of the tRNA.

6. In setting up experiments, it has been observed that RNAs of similar length can require RNase–RNA ratios that differ by as much as an order of magnitude. This observation predominates in the structure probing rather than in the sequencing reactions.

7. Structure-probing experiments are only as good as the care taken in their design and interpretation. The more information obtained about the functional requirements, the more attention paid to "discrepancies" in the probing results, the better will be the answers obtained from the structure-probing experiments. For example, the author's unpublished data from structure-probing pre-tRNA[Lys] showed that the pre-tRNA assumed two alternate structures in which there appeared to be RNase V1 and single-stranded-specific cleavages in the same sequence. Slow renaturation in the presence of Mg^{2+} had no affect, suggesting that we had detected structural isomers. On the other hand, tRNA[Leu-3] exhibits two forms, but one is converted to the other by renaturation in the presence of Mg^{2+}.[22] Thus, it is important to attempt to resolve conflicting data by renaturation of RNA molecules. This is readily accomplished by heating the RNA in the presence of Mg^{2+} and buffered salts to 65° for 3–5 min and by cooling slowly to room temperature.

Summary

To make strong statements about possible tertiary structure or the relative stability of regions of secondary structure, the structure-probing experiments must go further than single-hit reactions. Some elements of the environment of the RNA molecule must be altered systematically. Knowledge of the effects of ions or other interacting factors on the activ-

22 O. C. Uhlenbeck, J. G. Chirikjian, and J. R. Fresco, J. Mol. Biol. 89, 495 (1974).

ity or physical parameters (e.g., NMR and melting cooperativity) of the RNA help in experimental design. For example, the copious work on tRNA[Phe] compared the crystal and solution structures and allowed the direct correlation of Mg^{2+} stabilization of the tertiary structure of that molecule. Figure 3 demonstrates that pre-tRNA[Leu-3] responds to Mg^{2+} depletion in the same manner as detected by the appearance of highly sensitive RNase cleavage sites in the D and TΨC loops. Similar experiments titrating polyamine concentrations suggested that secondary structure was more efficiently stabilized by polyamines than by Mg^{2+}. The variation of Mg^{2+} concentrations has been used to gain additional information about other RNA structures.[17] Others have used protein–RNA interactions to approach the question of the functional structure of a RNA (for examples, see Ref. 3). Thus, the ideal parameters to choose would be those known to affect the function of the RNA. The variation of Mg^{2+} and polyamine concentrations would minimally suggest regions of greater or lesser secondary or tertiary structure stability.

[17] A Guide for Probing Native Small Nuclear RNA and Ribonucleoprotein Structures

By ALAIN KROL and PHILIPPE CARBON

Introduction

The precise mapping of secondary and tertiary interactions in RNA molecules is of fundamental importance for a detailed understanding of the role of these structural features in RNA function. Conformational modeling of RNA molecules results from the combination of three approaches. The first two approaches serve the purpose of predicting a model and consist of (1) collecting a library of computer-generated stem loops and long-range base pairings and (2) a comparative sequence analysis of homologous RNA from different organisms, so as to find the largest number of compensatory base changes in helices. Computer foldings that are not phylogenetically proved are thus discarded. (3) The third step is employed to test the validity of the model. This approach is based on an experimental investigation of the solution structure of the RNA molecule using a variety of chemical and enzymatic probes. Good agreement between prediction and experiment has been found in the case of large

METHODS IN ENZYMOLOGY, VOL. 180

ribosomal RNAs.[1-4] Other RNA molecules whose structure has been largely investigated are the small nuclear RNAs (snRNAs).[5,6] The aim of this chapter is to describe two methods that can be used for probing snRNA and small nuclear ribonucleoprotein (snRNP) structures, with special emphasis on the difficulties that may be encountered due to their content in highly structured regions. Both of the methods involve structure-specific modifications or cleavages along RNA. In the first approach, chemically modified nucleotides or phosphodiester bond cleavages are directly detected on 3'-end-labeled molecules. The other approach does not require prior end labeling of RNAs. This method relies on the reverse transcription extension of selected oligodeoxynucleotide primers to map at sequence resolution chemically modified nucleotides or phosphodiester bond cleavages. Although these two methods may be used in parallel experiments, the latter has several advantages that will be developed hereafter, one of them being the possibility of collecting structural information on native particles or on a mixture of snRNAs without prior purification.

Chemicals and Enzymes

Aniline (redistilled once) is from Fluka (Mulhouse, France) (double-distilled brand)

1-Cyclohexyl-3-(2-morpholinoethyl)carbodiimide   metho-*p*-toluene sulfonate (CMCT) is from Merck (FRG)

Dimethyl sulfate (DMS) is from Aldrich [Gold label, 99% (Milwaukee, WI)]

Diethyl pyrocarbonate (DEPC) and *N*-nitroso-*N*-ethylurea (ENU) are from Sigma (St. Louis, MO)

Hydrazine [95% (v/v) is from Eastman Kodak (Rochester, NY)

Deoxynucleoside triphosphates (dNTPs) and dideoxynucleoside triphosphates (ddNTPs) are obtained from Boehringer Mannheim (Meylan, France)

[γ-^{32}P]ATP and [5'-^{32}P]pCp are Amersham (Les Ulis, France) products

Oligodeoxyribonucleotide primers are synthesized on an Applied

[1] C. Branlant, A. Krol, M. A. Machatt, J. P. Ebel, K. Edwards, and H. Kössel, *Nucleic Acids Res.* **9**, 4303 (1981).
[2] R. Brimacombe, P. Maly, and C. Zwieb, *Prog. Nucleic Acid Res. Mol. Biol.* **28**, 1 (1983).
[3] D. Moazed, S. Stern, and H. F. Noller, *J. Mol. Biol.* **187**, 399 (1986).
[4] M. Mougel, F. Eyermann, E. Westhof, P. Romby, A. Expert-Bezançon, J. P. Ebel, B. Ehresmann, and C. Ehresmann, *J. Mol. Biol.* **198**, 91 (1987).
[5] C. Branlant, A. Krol, J. P. Ebel, H. Gallinaro, E. Lazar, and M. Jacob, *Nucleic Acids Res.* **9**, 841 (1981).
[6] A. Krol, C. Branlant, E. Lazar, H. Gallinaro, and M. Jacob, *Nucleic Acids Res.* **9**, 2699 (1981).

Biosystems synthesizer using the phosphoramidite method[7] and are purified by high-performance liquid chromatography (HPLC) Nuclease S1, RNases T1 and V1, T4 polynucleotide kinase, and T4 RNA ligase are from Pharmacia (St. Quentin en Yvelines, France) Avian myeloblastosis virus (AMV) reverse transcriptase is from Life Sciences (St. Petersburg, FL)

Buffers for Chemical Probes

Buffer I
    200 mM HEPES, pH 8, 10 mM MgCl$_2$, 50 mM KCl
Buffer II
    200 mM HEPES, pH 8, 1 mM ethylenediaminetetraacetic acid (EDTA)
Buffer III
    50 mM sodium borate, pH 8, 10 mM MgCl$_2$, 50 mM KCl
Buffer IV
    50 mM sodium borate, pH 8, 1 mM EDTA
Buffer V
    500 mM HEPES, pH 8, 25 mM MgCl$_2$, 125 mM KCl
Buffer VI
    500 mM HEPES, pH 8, 2.5 mM EDTA

Buffers for Enzymatic Probes

Buffer A (5×)
    50 mM Tris–HCl, pH 7.5, 50 mM MgCl$_2$, 250 mM KCl
Buffer B (5×)
    50 mM Tris–HCl, pH 7.5, 50 mM MgCl$_2$, 250 mM KCl, 5 mM ZnCl$_2$
Buffer C (5×)
    250 mM sodium acetate, pH 4.5, 50 mM MgCl$_2$, 250 mM KCl, 5 mM ZnCl$_2$

Methods

Procedure I: Direct Detection of Chemical Modifications or Enzymatic Cleavage on 3'-End-Labeled Molecules

General Principle of Method. This method was initially developed by Peattie and Gilbert[8] to monitor the conformational interactions of a RNA molecule in solution. A 3'-end-labeled snRNA is modified with base-

---

[7] N. D. Sinha, J. Biernat, J. MacManus, and H. Kester, *Nucleic Acids Res.* **12**, 4539 (1984).
[8] D. A. Peattie and W. Gilbert, *Proc. Natl. Acad. Sci. U.S.A.* **77**, 4679 (1980).

TABLE I
PROBES FOR STRUCTURAL ANALYSIS OF RNAs

Probes	Molecular weight	Target
Enzymes		
RNase T1	11,000	Guanines in single strands
S1 nuclease	32,000	Single strands
RNase V1	15,900	Double-stranded, stacked, or helical regions
Chemicals		
DMS	126	N-1-A and N-3-C not involved in Watson–Crick interactions
		N-7-G not involved in ligand binding, tertiary interactions, or stacking
CMCT	423	N-1-G and N-3-U not involved in Watson–Crick interactions
DEPC	174	N-7-A not involved in tertiary interactions or stacking
ENU	117	Phosphates not involved in ligand binding or hydrogen bonding

specific reagents or is enzymatically cleaved in limited conditions, allowing less than one modification or cut to occur in the RNA molecule. After chemical modifications, an aniline-induced β-elimination generates chain scission and provides polynucleotide fragments ending at the position of the modification. The distance of the strand scission from the terminal label locates the position of the attacked base or the phosphodiester bond susceptible to enzymatic cleavage. Chemical reactions can be conducted under different conditions: native conditions (low temperature and presence of Mg^{2+} and K^+), semidenaturing conditions (low temperature, absence of cations, and presence of EDTA), and totally denaturing conditions (high temperature, absence of cations, and presence of EDTA). This allows an estimation of the degree of stability of the different helical domains of the snRNA. Tertiary interactions, which are less stable than canonical Watson–Crick interactions, are expected to melt under semidenaturing conditions.

Chemical Probes. Table I summarizes the base specificities of the chemical probes that will be used, and whose detection of modification is allowed using direct 3' end labeling. The detailed mechanism of the chemical reactions is given by Kochetov and Budowski[9] and Ehresmann et al.[10]

[9] N. K. Kochetov and E. I. Budowski, "Organic Chemistry of Nucleic Acids." Plenum, New York, 1972.
[10] C. Ehresmann, F. Baudin, M. Mougel, P. Romby, J. P. Ebel, and B. Ehresmann, *Nucleic Acids Res.* **15**, 9109 (1987).

Dimethyl Sulfate (DMS). GUANINE REACTION. The reaction results in the alkylation of free N-7 positions of guanosines. A reduction of the 7,8-double bond of the alkylated ring in the presence of sodium borohydride results in a weakened glycosidic bond, providing a site for aniline-induced β-elimination. Absence of a reaction provides an indication for the involvement of N-7-G in non-Watson–Crick base pairs or in magnesium coordination.

CYTOSINE REACTION. DMS alkylates N-3 position of cytosines that are not involved in Watson–Crick pairing. Further, mild hydrazinolysis is required in order to provide a site for aniline-catalyzed β-elimination.

Diethyl Pyrocarbonate (DEPC). This reagent carbethoxylates N-7 of adenines. DEPC, therefore, monitors the sites that are not involved in tertiary interactions. Furthermore, the DEPC reaction appears to be sensitive to the stacking of adenines.

N-Ethyl-N-Nitrosourea [Ethylnitrosourea (ENU)]. ENU is an alkylating reagent that reacts with the phosphate group oxygens of nucleic acids. The resulting phosphotriesters can be hydrolyzed by mild alkaline treatment. Alkylation specifically maps phosphates nonengaged either in tertiary interactions through hydrogen bonding or in magnesium coordination.

Enzymatic Probes

RNase T1 specifically cleaves phosphodiester bonds 3′ to unpaired guanines. RNase T1 is used in mild conditions, essentially for mapping nucleotides located in loops.

Nuclease S1 is a single-stranded-specific endonuclease that senses unpaired RNA regions. Its optimal pH is around 4.5. It is generally better to carry out S1 experiments at a neutral pH to preserve the native conformation of RNAs or ribonucleoparticles. In these conditions, a high S1 concentration is required to compensate for the loss of activity.[4]

RNase V1 is an endoribonuclease extracted from *Naja oxiana* venom, which has a preference for helical regions[11] or single strands arranged in a stacked conformation.[12]

Recipes. The procedure consists of the following four steps.

1. 3′ End labeling of the snRNA with [5′-^{32}P]pCp.
2. Isolation of the ^{32}P-end-labeled snRNA on polyacrylamide gel.
3. Enzymatic cleavage or chemical modifications and subsequent chain scission at the site of chemical attack.

---

[11] S. K. Vassilenko and V. C. Ryte, *Biokhimiya* **40**, 578 (1975).
[12] H. B. Lowman and D. E. Draper, *J. Biol. Chem.* **261**, 5396 (1986).

4. Fractionation according to length on denaturing polyacrylamide gels of the nested set of radioactive fragments.

STEP 1: 3′ END LABELING OF snRNA WITH [5′-^{32}P]pCp. The unlabeled purified snRNA should be stored and aliquoted as ethanol precipitates. It would then keep for months without significant damage. Spin down the precipitate, rinse with 70% (v/v) ethanol, and air dry. Resuspend the pellet in water at a concentration of 1 mg/ml. Keep on ice. The 3′-end-labeling reaction is carried out according to Bruce and Uhlenbeck[13] in 30 μl of 50 mM HEPES, pH 7.5, 10 mM MgCl$_2$, 100 μM ATP, 10% (v/v) dimethyl sulfoxide (DMSO), 3 mM dithioerythritol (DTE), 10 μg/ml of bovine serum albumin (BSA), containing 1 μg (1 μl) of snRNA, 5 U of T4 RNA ligase, and 100 μCi of [5′-^{32}P]pCp (10 mCi/ml; specific activity, ~3000 Ci/mmol). Incubation is for 6 hr at 16° in a water bath. After the reaction is completed, precipitate the labeled snRNA with 3 volumes of ethanol in the presence of 300 mM sodium acetate, pH 5.4, and 5 μg of carrier tRNA, chill in dry ice for 5 min, and pellet at 13,000 g for 5 min.

STEP 2: ISOLATION OF ^{32}P-END-LABELED snRNA ON POLYACRYLAMIDE GEL. Rinse the pellet with 70% (v/v) ethanol, air dry, and resuspend in 5 μl of H$_2$O. Vortex and check that the pellet is properly resuspended before adding 5 μl of dye mix solution [0.025% (w/v) xylene cyanol (XC) and 0.025% (w/v) bromphenol blue (BPB) in deionized formamide]. Apply to a 12% polyacrylamide [19% (w/v) acrylamide and 1% (w/v) N,N′-methylenebisacrylamide] and 8 M urea gel (height, 40 cm; width, 20 cm; thickness, 0.35 mm). The buffer in the gel is 1× TBE (90 mM Trizma base, 90 mM boric acid, 1 mM EDTA, pH 8.3). Electrophoresis is carried out overnight at 600 V in 1× TBE electrophoresis buffer until the XC marker has moved to 10 cm from the bottom of the gel. After autoradiography, the band containing the snRNA is cut out from the gel and is transferred to a microfuge tube.

Cut the band containing the snRNA in small pieces and add 200 μl of 500 mM ammonium acetate, 10 mM magnesium acetate, 0.1 mM EDTA, and 0.1% (w/v) sodium dodecyl sulfate (SDS). Leave at room temperature for at least 10 hr. Transfer the supernatant to a fresh tube, add 10 μg of carrier tRNA and 600 μl of ethanol, vortex, and chill in dry ice for 5 min. Spin down the ethanol precipitate for 5 min at 13,000 g, wash the pellet at 4° with 70% (v/v) ethanol, and add 50 μl of H$_2$O. Take an aliquot for Cerenkov counting. The specific activity can be in the range of 7–8 × 10^6 cpm/μg, depending on the snRNA species being labeled. Remove the total counts needed for further reactions (0.3–1 × 10^5 cpm depending on

[13] A. G. Bruce and O. C. Uhlenbeck, *Nucleic Acids Res.* **4**, 2527 (1978).

the reaction to be performed). The rest is ethanol precipitated and kept at −20° for further use.

STEP 3a: ENZYMATIC CLEAVAGES AND
BASE-SPECIFIC MODIFICATIONS

1. RNase T1 or RNase V1 cleavages:
   a. In a reaction tube, combine 4 μl of 3′-end-labeled RNA (~40,000 cpm), 1 μl of tRNA (10 mg/ml) to adapt the probe–RNA ratio, and 16 μl of Buffer A (5×). Add water to an 80-μl final volume. Before the enzyme is added, remove a 20-μl aliquot for use as an undigested control. Leave at 20° and process simultaneously with the final time point.
   b. Add 0.1 U of RNase T1 or 0.2 U of RNase V1. Mix, vortex, and leave at 20°. Remove aliquots at 2-, 5-, and 10-min intervals. To each time point, add 8 μg of tRNA as carrier and 20 μl of phenol and vortex for 2 min. Spin for 30 sec to separate the phases. Transfer the upper aqueous phase to a fresh tube, precipitate the RNAs with 3 volumes of ethanol in the presence of 300 mM sodium acetate. Reprecipitate the pellet from 100 μl of 300 mM sodium acetate and rinse with 100 μl of cold 70% (v/v) ethanol. Air dry. Dissolve the RNA in 5 μl of H_2O, then add 5 μl of formamide dye mix (see Step 1).
2. S1 nuclease: Proceed as described above, except that Buffers B or C are used. With Buffer C, add 1 U/μg of S1 nuclease. Whereas if Buffer B is preferred, 25–50 U/μg are required. Then phenol extract and treat as above.

STEP 3b: CHEMICAL MODIFICATIONS

1. G reaction: To probe the snRNA under native conditions, add 3 μl of 3′-end-labeled snRNA (30,000 cpm) to 200 μl of Buffer I. Mix and spin. Then add 1 μl of DMS, mix, and centrifuge. Incubate at 30° for 5 min. To probe the RNA under semidenaturing conditions, the same procedure is followed, except that Buffer II is used. For reaction under totally denaturing conditions, the RNA is modified in 300 μl of Buffer II and is incubated at 90° for 30 sec–1 min. Terminate the reactions by chilling on ice; add 1 μl (10 μg) of carrier tRNA, 30 μl of 3 M sodium acetate, 1 ml of cold ethanol, and chill on dry ice for 5 min. Centrifuge the ethanol precipitate for 5 min at 13,000 g. Resuspend the pellet in 100 μl of 300 mM sodium acetate, then add 300 μl of cold ethanol. Repeat the precipitation step. Rinse the pellet with 100 μl of cold 70% (v/v) ethanol. Let the RNA dry at room temperature. Redissolve the RNA in 10 μl of 1 M Tris–HCl,

pH 8, and add 10 μl of fresh 200 mM NaBH$_4$. Incubate on ice, in the dark, for 30 min. Add 200 μl of cold 600 mM sodium acetate–600 mM acetic acid, pH 4.5. Mix and then add 600 μl of cold ethanol. Chill, pellet, rinse with 70% (v/v) cold ethanol, and dry. Then treat with aniline. Set up a control reaction by omitting the reagent.

2. A reaction: The procedure is identical to that described for the G reaction (use the same buffers). Add 10 μl of DEPC and incubate for 45 min at 30° (with occasional mixing during the incubation) to probe native and semidenaturing conditions. Add 1 μl of DEPC and incubate for 7 min at 90° to probe the RNA under totally denaturing conditions. Terminate the reactions by adding 30 μl of 3 M sodium acetate, 5 μg of carrier tRNA, and 600 μl of cold ethanol. Chill, pellet, reprecipitate from 100 μl of 300 mM sodium acetate, chill, pellet, rinse with cold 70% (v/v) ethanol, and dry. Then treat with aniline. Set up a control reactions by omitting the reagent.

3. C reaction: The RNA is modified with DMS and is precipitated as for the guanosine reaction. Then, instead of the reduction step with sodium borohydride, resuspend the pellet in 10 μl of an ice-cold solution of 50% (v/v) hydrazine–50% (v/v) water. Incubate at 0° for 5 min. Precipitate twice with 100 μl of 300 mM sodium acetate and treat with aniline. Set up a control reaction by omitting the reagent.

4. Aniline treatment: Dissolve the chemically modified RNAs in 20 μl of 1 M aniline–acetate buffer, pH 4.5 (10 μl of aniline, 93 μl of H$_2$O, and 6 μl of acetic acid), and incubate in the dark at 60° for 15 min. Then add 100 μl of 300 mM sodium acetate, 5 μg of carrier tRNA, and 300 μl of cold ethanol. Reprecipitate once from 100 μl of 300 mM sodium acetate and rinse the pellet with cold 70% (v/v) ethanol. Dissolve the samples in 5 μl of water. Then add 5 μl of formamide dye mix (see Step 1).

5. Phosphate alkylation:

   a. Prepare the ENU solution as follows (handle with care as ENU is a highly carcinogenic reagent). Add 100 μl of ethanol in a microfuge tube. Dissolve ENU in ethanol and keep adding powder until the solution gets saturated and yellowish. Spin and use the supernatant as the reagent.

   b. To probe the RNA in native conditions, combine 1 μl of labeled snRNA ($\sim 10^5$ cpm), 2 μl of tRNA (2 μg), 8 μl of Buffer V, 4 μl of H$_2$O, and 5 μl of ENU in ethanol. Incubate at 37° for 15–30 min. Set up a control reaction by omitting the reagent. Replace by 5 μl of ethanol. Incubate in the same conditions. To probe the RNA in denaturing conditions, use Buffer VI. Incubate at 80°for 30 sec–1 min. All other conditions remain the same. When the reaction is completed, add 10 μg of carrier tRNA, 100 μl of 300 mM sodium acetate, and 300 μl of cold ethanol. Precipitate and reprecipitate from 100 μl of 300 mM sodium acetate. Pellet, rinse, and dry.

c. Hydrolysis of the phosphotriester bond is as follows. Dissolve the pellet in 10 μl of 100 mM Tris–HCl, pH 9. Incubate all samples at 50° for 5 min. Ethanol precipitate in the presence of 300 mM sodium acetate (final concentration). Pellet, rinse, and dry. Redissolve the RNA in 5 μl of H_2O. Add 5 μl of the formamide dye mix. Sequencing ladders for mapping the cleavage positions along the sequence are done using the chemical method for RNA sequencing.[14]

STEP 4: SIZE FRACTIONATION ON POLYACRYLAMIDE GELS. Gel plates are 33 × 40 cm. Spacers are 0.35 mm thick. Run the samples on 10% polyacrylamide gels [19% (w/v) acrylamide and 1% (w/v) bisacrylamide] and 8 M urea made up in 1× TBE. To obtain information on the very first nucleotides adjacent to the 3′ end label, a 15% gel is better (BPB halfway from the bottom). Load 2–3 μl of the samples from Step 3. Gels should be prerun and run warm to avoid band compression. The migration conditions must be adapted to the length of the RNA to be analyzed, knowing that on 10% gels, XC migrates to 55 nucleotides. At the end of the run, the gel is fixed in 10% (v/v) acetic acid and is dried under vacuum. The dried gel is exposed to Kodak XAR film using an intensifying screen.

Comments on Individual Steps. STEP 1. If a mixture of snRNAs, instead of a purified species, is to be labeled, then start with 2–3 μg of RNA and use 200 μCi of the label. Other conditions remain the same. If a T7 RNA transcript is used, unincorporated NTP must be removed on Sephadex G-25; otherwise T4 RNA ligase will not work, due to the vast excess of competing ATP. Incubation times for chemical reactions have been determined for U1 RNA. The incubation times may need to be adapted for other RNA species.

STEP 3. Time points for enzymatic cleavages may be adjusted if necessary. If cleavage with RNase V1 or nuclease S1 is hard to obtain, it is generally better to increase enzyme concentration than reaction time; for long periods of incubation, spontaneous splitting of the RNA chain at the phosphodiester bond between pyrimidines and purines occurs frequently.

STEP 4. Ten percent polyacrylamide gels are generally appropriate for fractionating enzymatic cleavages. For the fractionation of chemically induced strand-scission products, 15% gels are generally best suited to minimize blurring of the bands corresponding to long 3′-end-labeled frag-

[14] D. A. Peattie, *Proc. Natl. Acad. Sci. U.S.A.* **76**, 1760 (1979).

ments. This blurring may originate from unspecific chemical cleavage. If only enzymatic cleavages are to be run on a gel, a short procedure for G ladder consists of a RNase T1 partial digest containing 1 μl (10,000 cpm) of snRNA, 1 μl (1 μg) of tRNA, 8 μl of 25 mM sodium citrate, pH 5, 10 M urea, 0.025% XC, and 0.025% BPB, and 0.01 U of RNase T1. Incubate at 55° for 10 min. Load 5 μl onto the gel.

A formamide ladder is obtained by mixing 2 μl (20,000 cpm) of snRNA to 3 μl of deionized formamide. Incubate at 90° for 25 min, then add 5 μl of the formamide dye mix. Load 5 μl onto the gel.

Nuclease S1 and ribonuclease V1 both cleave nucleic acids by leaving fragments possessing 5'-phosphates. Ribonuclease T1 and formamide ladders release 5'-OH-ending fragments. Therefore, great care must be taken in assigning bands.

Procedure II: Detection of Modifications and/or Cleavages Using Primer Extension

General Principle of Method. The structure mapping method reported in Procedure I suffers from a series of limitations inherent to the necessity of a prior end labeling of RNA before structure probing. snRNAs U1–U5 show a highly structured 3' stem loop that may lead, in some cases, to a decreased yield of labeling. Besides, the presence of a cap structure prevents the direct 5' end labeling. The obligatory use of a decapping enzyme, followed by a phosphatase treatment in order to label the 5' terminus, is not appropriate for preserving the integrity of a RNA molecule. Last, this is not the method of choice for probing snRNA structures in the form of native snRNP particles, since the method requires large snRNP quantities to avoid utilization of carrier tRNA that would otherwise interfere with subsequent labeling. The use of a reacted RNA molecule as a template for reverse transcriptase extension of labeled primers complementary to selected portions of the RNA circumvents any prior labeling. This method, initially developed by HuQu et al.,[15] allows premature termination of cDNA synthesis at the position of base modification or chain cleavage on the template. Modification at any of the Watson–Crick positions blocks the extension, whereas the effect of modification at non-Watson–Crick positions on the extension depends on the bulkiness of the modified base. Reverse transcriptase does not stop at methylated N-7-G, whereas reverse transcriptase pauses at carbethoxylated N-7-A. Therefore, aniline-induced chain scission is required in the former case prior to extension. In all cases, the resulting labeled cDNAs

[15] L. HuQu, B. Michot, and J. P. Bachellerie, *Nucleic Acids Res.* **11,** 5903 (1983).

are sized at sequence resolution in parallel reactions using chain-termina-
tor dideoxynucleotides.

Chemical Probes. This section is complementary to that in Procedure
I and will describe those base modifications that induce chain termination
without the chain-scission requirement. Therefore, base modifications
can only be detected by the primer extension method.

Dimethyl Sulfate (DMS). In addition to N-7-G methylation, DMS also
reacts with positions N-1-A and N-3-C in the order N-1-A > N-3-C (see
Table I). A stop in the elongation products at the nucleotide preceding the
modified residue provides an indication that these positions are not involved in Watson–Crick base pairing.

*1-Cyclohexyl-3-(2-morpholinoethyl)carbodiimide   Metho-p-toluene
Sulfonate (CMCT).* This reagent reacts primarily with N-3-U and N-1-G
in the order N-3-U > N-1-G (see Table I). Again, a stop of the elongation
products at the nucleotide preceding the modified residue is an indication
for unpaired uridines and guanosines.

Recipes. The procedure consists of the following four steps.

1. 5′ End labeling and gel fractionation of the primer.
2. Probing snRNAs–snRNPs.
3. Primer annealing, followed by extension with reverse transcriptase
   of the modified RNA templates.
4. Sizing the resulting cDNAs.

Step 1: 5′ End labeling and gel fractionation of primer. Dissolve the synthetic oligodeoxynucleotide in H_2O at a concentration of 100
pmol/μl. For kination, in a microfuge tube combine 1 μl of the oligodeoxynucleotide (100 pmol), 1 μl of 10× kinase buffer (100 mM Tris–HCl,
pH 7.5, 100 mM MgCl$_2$, 30 mM DTT), 2 μl of H_2O, 5 μl of [γ-^{32}P]ATP (10
mCi/ml; specific activity, 3000 Ci/mmol), and 1 μl (5 U) of T4 polynucleotide kinase. Incubate at 37° for 30 min. Then add 5 μl of formamide dye
mix and apply to a 15% polyacrylamide gel (height, 20 cm; width, 20
cm; thickness, 0.35 mm) and 8 M urea (see Step 2 in Procedure I). Electrophoresis is carried out at 15 mA for about 1 hr (12–20 mers migrate
between XC and BPB). After autoradiography, the band containing the
labeled primer is cut out, is placed in a microfuge tube, and is eluted
overnight at room temperature with 200 μl of elution buffer (see Step 2 in
Procedure I). Transfer the supernatant to a fresh tube and add 5 μg of
carrier tRNA and 600 μl of ethanol. Vortex to mix, chill in dry ice for 5
min, and spin. Rinse the pellet with 70% (v/v) cold ethanol and resuspend
the labeled primer in 100 μl of H_2O. Take an aliquot for Cerenkov counting. Adjust if necessary to roughly 10^5 cpm/μl.

STEP 2a: PROBING snRNAs–snRNPs. Conditions for probing snRNAs with DMS, DEPC, or ENU are identical to those described in Procedure I. The quantity of starting material required for probing depends on the number of primer extentions to be done, knowing that 500 ng of modified template/extension is a reasonable amount. Typically, a reaction with DMS or DEPC is achieved by mixing 4 μl of snRNA (4 μg of pure snRNA or of a mixture of all snRNAs) with 200–300 μl of Buffer I or II. Proceed as indicated in Procedure I. Set up a control reaction by omitting the reagent.

For the ENU reaction, combine 4 μl of snRNA (4 μg), 3 μl of H$_2$O, 8 μl of Buffer V or VI, and 5 μl of ENU in ethanol. Set up a control reaction by omitting the reagent. Proceed as described in Procedure I.

PROBING N-1-G AND N-3-U WITH CMCT. Prepare a CMCT solution at 42 mg/ml in H$_2$O just before use. For probing snRNAs in native conditions, add 4 μl of snRNA (4 μg), 50 μl of Buffer III, 2 μl of 1 M MgCl$_2$ (10 mM final concentration), 10 μl of 1 M KCl (50 mM final concentration), and 84 μl of H$_2$O. Then add 50 μl of CMCT solution. Vortex to mix. Incubate at 30° for 10 min. To stop the reaction, add 20 μl of 3 M sodium acetate, 10 μg of carrier tRNA, and precipitate with 600 μl of cold ethanol.

For probing in semidenaturing conditions, proceed as follows. Add 4 μl of RNA (4 μg), 50 μl of Buffer IV, 95 μl of H$_2$O, and 1 μl of 200 mM EDTA (1 mM final concentration). Then add 50 μl of CMCT solution. Vortex to mix. Incubate at 30° for 5 min. For totally denaturing conditions, incubate at 90° for 1 min. Stop the reactions as described above. Control reactions are performed for all conditions by replacing the reagent with water.

Chemical probing of snRNAs in native snRNP particles is generally possible only with DMS, due to side reactions of DEPC, CMCT, and ENU with proteins. For methylation with DMS, add 1–2 μg of snRNP (use twice as much if an aniline β-elimination is required for probing N-7-G) to 200 μl of Buffer I supplemented with 5% (v/v) glycerol. Precipitate at the end of the reaction in the presence of 10 μg of carrier tRNA. Dissolve the pellet in 50 μl of H$_2$O, add an equal volume of phenol, and vortex for 1 min. Precipitate the RNAs from the aqueous phase. Use directly for primer extension or treat with aniline (as previously described) prior to primer extension for N-7-G probing. Again, a control reaction without DMS is carried out.

ENZYMATIC CLEAVAGES. Use 4 μg (4 μl) of snRNA as the starting material and proceed as described in Procedure I. If analyzing snRNPs, start with 1–2 μg of snRNP (~300 μg/ml), which is added to 20 μl of the reaction mixture (see Procedure I) containing 5% (v/v) glycerol. Terminate the reaction by a phenol extraction. Set up the control reactions as described in Procedure I.

STEP 3: ANNEALING OF PRIMERS TO TEMPLATE AND ELONGATION WITH REVERSE TRANSCRIPTASE. Each RNA sample (both the modified templates and the control reactions) is redissolved in $N \times 2$ μl of H_2O, N being the desired number of different primers to be elongated. Add 2-μl fractions in N tubes.

For the annealing, combine 2 μl of the modified template and 1 μl of the 5'-^{32}P-labeled primer ($\sim$10^5 cpm). Heat the hybridization mixture at 90° for 1 min in a heat block. Return the tubes to room temperature and leave to anneal for 10 min.

For the extension reactions, prepare just before use the following extension mix: $N \times 1$ μl of reverse transcription buffer (5$\times$ RTB), containing 25 mM Tris–HCl, pH 8, 35 mM MgCl$_2$, 250 mM KCl, and 25 mM DTT, $N \times 1$ μl of dNTP mix, containing 850 μM of each of the four dNTPs, and $N \times 1$ U of reverse transcriptase. N stands for the total number of extension reactions to be performed (including the control reactions). Add 2 μl of the relevant extension mix just inside the rim of the tube. Spin briefly to mix and incubate at 37° for 30 min.

If samples are to be electrophoresed immediately, add 5 μl of the formamide dye mix. If the samples are not electrophoresed immediately, store without formamide at $-20°$ for no longer than 24 hr. Then add 5 μl of the formamide dye mix before continuing to the next step.

Unmodified snRNAs are used as templates for sequencing reactions in the presence of chain terminators. Prepare solutions and dN/ddN mixes as follows. Use 10 mM dNTP and ddNTP as stock solutions. Dilute samples of the stock solutions in water to give 750 μM dNTP and 150 μM ddNTP solutions. Use these to prepare the dN/ddN sequencing mixes.

Concentration		dA/ddA	dC/ddC	dG/ddG	dT/ddT
10 mM	dATP	—	2.5 μl	2.5 μl	2.5 μl
10 mM	dCTP	2.5 μl	—	2.5 μl	2.5 μl
10 mM	dGTP	2.5 μl	2.5 μl	—	2.5 μl
10 mM	dTTP	2.5 μl	2.5 μl	2.5 μl	—
750 μM	dATP	8 μl	—	—	—
750 μM	dCTP	—	8 μl	—	—
750 μM	dGTP	—	—	8 μl	—
750 μM	dTTP	—	—	—	8 μl
150 μM	ddATP	4 μl	—	—	—
150 μM	ddCTP	—	4 μl	—	—
150 μM	ddGTP	—	—	4 μl	—
150 μM	ddTTP	—	—	—	4 μl

Add H_2O to a 50-μl final volume

The hybridization mixtures consist of 2 μl (2 μg) of unmodified snRNA, 2 μl of H$_2$O, and 4 μl of 5'-^{32}P-labeled primer (~4 × 10^5 cpm). Anneal as described above. Mark four tubes A, C, G, and T (or use colored microfuge tubes) and, to each tube, add 2 μl of the annealed template–primer.

To the rim of each tube, add 1 μl of 5× RTB, 1 μl of dNTP/ddNTP sequencing mix, and 1 μl (1 U) of reverse transcriptase (use RT dilution buffer, containing 50 mM Tris–HCl, pH 8, 2 mM DTT, and 50% (v/v) glycerol to obtain the desired unit number from the enzyme stock). Spin to mix. Incubate at 37° for 30 min. Add 5 μl of the formamide dye mix or freeze at −20° without adding the formamide dye mix if not electrophoresed immediately.

STEP 4: GEL FRACTIONATION OF EXTENDED PRIMERS. Electrophoresis is in 10% polyacrylamide–8 M urea (see Procedure I). Denature the DNA by heating at 90° for 3 min before loading 3 μl of the samples on the gel. Electrophoresis conditions depend on the region to be analyzed. The XC and BPB migrate to 50–55 and 17–20 nucleotides, respectively, on this type of gel. Unextended primers (12–15), therefore, migrate slightly faster than BPB. Gels are processed as described in Procedure I.

Comments on Individual Steps. STEP 1. The length of the oligodeoxynucleotides that are routinely used vary from 12 to 15 nucleotides. This range of length confers enough specificity to the binding, even if primers must find their target in a mixture of different snRNA species and carrier tRNA. Single-stranded regions and/or loops in snRNAs are preferentially chosen as sites for complementary binding of the oligodeoxynucleotides. Binding of oligodeoxynucleotides to base-paired regions requires high energy in the form of heat (especially for melting the highly structured 3' stem loops of snRNAs). This may lead to heat-induced splitting of snRNA molecules and to unexpected data that will be further discussed.

STEP 2. Chemical modifications and enzymatic cleavages can be performed on T7 RNA transcripts without prior elimination of unincorporated NTP. All conditions (both chemical and enzymatic) that have been described so far were established for probing the structure of U1 RNA. Fine tuning may be required for probing other snRNA molecules. Chemical probing of snRNAs in the particles is essentially restricted to the use of DMS, since other reagents, such as DEPC and ENU, also react with proteins, thereby denaturing snRNPs.

STEP 3. Annealing conditions (90° for 1 min) have been selected in order to maximize the unfolding and to minimize RNA degradation. As described previously, hybridization conditions depend largely on snRNA structure and on snRNA species.

STEP 4: GEL READING. The reading of the first nucleotides extended 3' to the primer is generally obscured on the film by the scattering of β^--

particles emitted by the unextended labeled primer. Use of several primers spanning the whole molecule allows the obtention of overlapping sequences. Replacement of ^{32}P with ^{35}S helps obtain a better resolution.

All experiments described so far should be done at least in triplicate to obtain consistent data, since the yield of the chemical modifications or enzymatic cleavages observed may vary from one experiment to the other, as well as from one primer to the other, for the following various reasons:

1. Number and intensity of pyrimidine–purine spontaneous cleavages in the RNA template, which can be superimposed to specific modifications or cleavages.

2. Breathing of some RNA helices. Due to low stability of a few consecutive A–U base pairs, there may be some breathing of the RNA helix. Therefore, in a population of molecules, nucleotides that are supposed to be base paired may show chemical reactivity toward DMS or CMCT, if located in the region of a helix that breathes. The pattern of modification observed on the film may lead to the erroneous conclusion that these nucleotides are unpaired. It is worth performing chemical modifications at 0° to stabilize base pairing in those areas. Much longer exposition to modification reagents is then required to compensate for the lower activation energy driving chemical reactions performed at 0°.

3. The primer that is chosen for the extension. When complementary to structured regions, the binding of the primer is largely facilitated if an enzymatic cut occurs in the strand opposite to the complementary region. The cleavage destabilizes the base pairing, allowing the primer to bind to its target. The primer selects preferentially those molecules that are cleaved at this relevant location; premature chain termination (a few extended nucleotides) is therefore observed, leading to the wrong observation that this site of cleavage is highly susceptible to enzymes.

Concluding Remarks

Although both approaches provide detailed data at the nucleotide level and should be used in combination, we will discuss advantages and limitations of either method. Analysis of reacted templates by primer extension can be performed on minute amounts of snRNAs, on mixtures of snRNA species, or on snRNAs in the form of native snRNPs. Analysis allows the monitoring of base modifications that cannot be detected without subsequent chain scission, essentially Watson–Crick positions. If primers are conveniently chosen, one can theoretically have access to the whole snRNA molecule.

The use of reverse transcriptase is a very powerful and elegant method, but this method suffers from a few limitations, at least in its application to probing snRNAs; as all U1–U5 snRNAs end with a highly structured 3' stem loop, the 3'-terminal nucleotides can never be analyzed. In addition, U snRNAs are rather small molecules, compared to ribosomal RNAs; U snRNAs are highly structured and display a limited number of accessible single-stranded regions or loops. Therefore, the choice for different primers that can be selected in one snRNA species is limited accordingly. The presence of highly structured regions in snRNAs induces pauses in reverse transcriptase extension products (reverse transcriptase cannot melt the 3'-most stem loop of snRNAs). It is generally worth checking the integrity and the location of the pauses and stops in a RNA template by extending a primer in a sequencing reaction before starting a conformational study.

The use of 3'-end-labeled snRNAs serves to monitor the status of those nucleotides located in the vicinity of the 3' terminus. But, as discussed previously, it is essentially restricted to naked snRNAs.

It must be kept in mind that, even with the combined utilization of two detection methods, there will subsist a few uncertainties inherent to the intrinsic properties of a RNA molecule: spontaneous pyrimidine–purine cleavages, breathing, and highly ordered 3' stem loop, this latter property being characteristic for snRNAs.

Acknowledgments

This work was carried out in the laboratory of Professor Jean-Pierre Ebel. His support is gratefully acknowledged. Grants were from the Centre National de la Recherche Scientifique (CNRS) (LP 6201).

[18] Phylogenetic Comparative Analysis of RNA Secondary Structure

By Bryan D. James, Gary J. Olsen, and Norman R. Pace

Nucleotide sequences of RNAs or their genes are now readily determined. However, specific, three-dimensional foldings of the RNAs are necessary for their *in vivo* functions. Consequently, the inference of secondary structure is a crucial step in the study of a functional RNA. Secondary structure is defined by intramolecular associations (pairings) of complementary sequences at least two base pairs in length. The base pairs may be canonical (A–U and G–C) or noncanonical (e.g., G–U, A–G, etc.). A folding model must be chosen carefully, for subsequent investigations of the RNA are guided by the model.

The most incisive *a priori* approach to inferring higher order RNA structure has proved to be the use of phylogenetic comparisons. Possible helices in a RNA, as indicated by the occurrence of complementary sequences, are tested by inspecting the equivalent pairings in "homologous" RNAs in which the sequences vary. We use the term homologous in its strictest sense: homologous sequences have common ancestry and function, hence homologous sequences are expected to have similar higher order structures. The covariation of paired residues in putative helical regions, such that pairing potential is maintained, offers support for the structure. On the other hand, if equivalent complementarity is not present in homologous sequences from different organisms, the structure is unlikely to exist *in vivo*. Confirmation of sequence pairings by the phylogenetic comparative approach is formally a genetic analysis of naturally occurring mutations and corresponding second-site reversions that maintain the complementarity.

Comparative analyses have been instrumental in the elucidation of the structures of ribosomal RNAs (rRNAs),[1–3] transfer RNAs (tRNAs),[4] class I[5,6] and class II[7] introns, and small nuclear RNAs (snRNAs).[8] We describe in this chapter the application of phylogenetic comparisons to the evaluation of RNA secondary structure, using as an example the catalytic, RNA moiety of ribonuclease P (RNase P), a ribonucleoprotein, tRNA-processing endonuclease.[9] The sequences that were used for the analysis are aligned in Fig. 1. Figure 2 shows the inferred secondary structures for the RNase P RNAs of *Bacillus subtilis* and *Escherichia coli*. The reader will also find useful the discussion by Woese *et al.*[1] on the comparative analysis of 16 S rRNA secondary structure.

[1] C. R. Woese, R. Gutell, R. Gupta, and H. F. Noller, *Microbiol. Rev.* **47**, 621 (1983).
[2] G. E. Fox and C. R. Woese, *Nature (London)* **256**, 505 (1975).
[3] R. R. Gutell and G. E. Fox, *Nucleic Acids Res.* **16**, Suppl., r175 (1988).
[4] M. Sprinzl, T. Hartmann, F. Meissner, J. Moll, and T. Vorderwülbecke, *Nucleic Acids Res.* **15**, Suppl., r53 (1987).
[5] R. W. Davies, R. B. Waring, J. A. Ray, T. A. Brown, and C. Scazzochio, *Nature (London)* **300**, 719 (1982).
[6] T. R. Cech, N. K. Tanner, I. Tinoco, Jr., B. R. Weir, M. Zuker, and P. S. Perlman, *Proc. Natl. Acad. Sci. U.S.A.* **80**, 3903 (1983).
[7] F. Michel and B. Dujon, *EMBO J.* **2**, 33 (1983).
[8] P. G. Siliciano, M. H. Jones, and C. Guthrie, *Science* **237**, 1484 (1987).
[9] B. D. James, G. J. Olsen, J. Liu, and N. R. Pace, *Cell (Cambridge, Mass.)* **52**, 19 (1988).

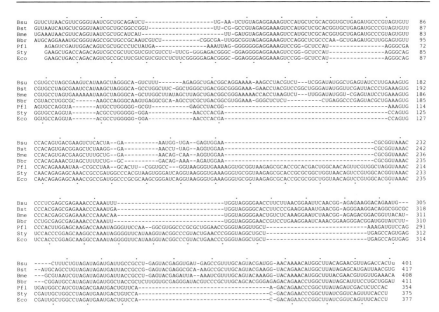

Fig. 1. Alignment of RNase P RNA sequences. The sequences of the RNase P RNAs (as inferred from the respective genes) were aligned as described in the text. *Bacillus subtilis* (*Bsu*), *Bacillus stearothermophilus* (*Bst*), *Bacillus megaterium* (*Bme*), and *Bacillus brevis* (*Bbr*) are members of the "gram-positive and relatives" group. *Escherichia coli* (*Eco*), *Salmonella typhimurium* (*Sty*), and *Pseudomonas fluorescens* (*Pfl*) are members of the "purple bacteria and relatives" group. Dashes (–) indicate alignment gaps. The nucleotide positions for each sequence are given at the end of the respective lines. Dots are placed at every tenth nucleotide of the *B. subtilis* and *E. coli* sequences. (From James *et al.*[9])

Selection of Organisms for Comparative Structure Analysis

A phylogenetic structure analysis requires comparing the sequences of homologous RNAs from diverse organisms, among which the sequence varies. However, the determination of a RNA secondary structure usually begins with the availability of a single sequence. The question then is, what additional organisms should be inspected for an efficient comparative analysis? The molecules compared must be sufficiently different to provide numerous instances of sequence variation with which to test pairing possibilities, yet, the molecules must not differ so much that homologous residues cannot be "aligned" (see Alignment of Sequences) with confidence. The choice of appropriate organisms for RNA structure comparisons requires a quantitative view of evolutionary relatedness,

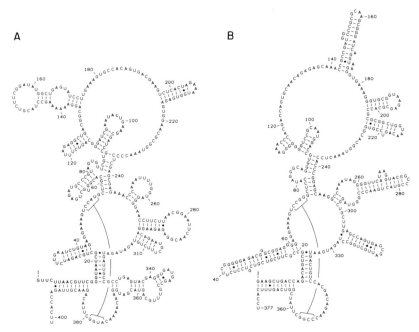

FIG. 2. Secondary structure models for the *B. subtilis* (A) and *E. coli* (B) RNase P RNAs based on phylogenetic comparisons. The complementary sequences connected by the arc form a "pseudoknot" in the structures. See the text for discussion. (From James *et al.*[9])

now available from studies of rRNA sequences.[10,11] The rRNAs are conservative molecules; homologous sequences are identifiable in the rRNAs of all organisms. These homologous sequences can be used to quantify evolutionary distances between organisms and hence to identify appropriate organisms for a comparative structure analysis of another RNA. There are several compendia of phylogenetic relationships among prokaryotes and eukaryotes, based on rRNA[10,12–15] or protein[16] sequence comparisons, that can be consulted to identify candidate organisms for

[10] C. R. Woese, *Microbiol. Rev.* **51**, 221 (1987).

[11] N. R. Pace, G. J. Olsen, and C. R. Woese, *Cell* (*Cambridge, Mass.*) **45**, 325 (1986).

[12] W. F. Walker, *BioSystems* **18**, 269 (1985).

[13] L. Hendriks, E. Huysmans, A. Vandenberghe, and R. De Wachter, *J. Mol. Evol.* **24**, 103 (1986).

[14] H. Hori and S. Osawa, *Mol. Biol. Evol.* **4**, 445 (1987).

[15] E. Dams, L. Hendriks, Y. V. de Peer, J.-M. Neefs, G. Smits, I. Vandenbempt, and R. De Wachter, *Nucleic Acids Res.* **16**, Suppl., r87 (1988).

[16] K. E. Sidman, D. G. George, W. C. Barker, and L. T. Hunt, *Nucleic Acids Res.* **16**, 1869 (1988).

study. However, because different RNAs do not evolve at the same rate, rRNA sequence similarities do not necessarily point immediately to organisms of choice. The studies with RNase P offer an example.

When we first undertook the analysis of the RNase P RNA secondary structure, the sequences of the functionally equivalent, hence presumably homologous, RNAs from *B. subtilis*,[17] *E. coli*,[18,19] and *Salmonella typhimurium*[20] were available. The latter two sequences were nearly identical (Fig. 1), offering little variation for tests of structure. In contrast, the *B. subtilis* and *E. coli* RNase P RNAs were so different that the sequences could not be aligned reliably over most of their lengths. This is graphically illustrated in Fig. 3A, a "dot plot" of all 12-base sequences that are at least 75% identical in both RNAs. Few stretches of substantial sequence similarity are indicated, consistent with the inability of the genes to cross-hybridize.[21] In contrast, the alignment of the corresponding 16 S rRNAs is evident (Fig. 3B). The RNase P RNAs of *B. subtilis* and *E. coli* have diverged more from their common ancestor than have the more slowly changing 16 S rRNAs. Therefore, the structure analysis required the examination of the RNase P RNAs from organisms of intermediate evolutionary distance from *B. subtilis* and *E. coli*.

From the standpoint of 16 S rRNA sequence comparisons, *B. subtilis* and *E. coli* are distantly related eubacteria, members of different eubacterial "phyla."[10] *Bacillus subtilis* is a member of the "gram-positive and relatives" group, and *E. coli* is one of the "purple bacteria and relatives" group. Thus, additional organisms from within each of those groups were chosen for analysis. Fairly distant (75–80% 5 S rRNA sequence similarity) intraphylum relatives were inspected to ensure significant RNase P RNA sequence divergence. In-hand RNase P RNA genes were used as hybridization probes for the isolation of the additional genes needed for the comparative study. In practice, the genes that are sufficiently similar in sequence that they can be aligned reliably are also those genes that are detectable by interspecific hybridization in Southern blot experiments. As the number of homologous sequences identified in this manner increases, some sequence stretches will likely be identical, or nearly so, in all the

[17] C. Reich, K. J. Gardiner, G. J. Olsen, B. Pace, T. L. Marsh, and N. R. Pace, *J. Biol. Chem.* **261**, 7888 (1986).

[18] R. E. Reed, M. F. Baer, C. Guerrier-Takada, H. Donis-Keller, and S. Altman, *Cell* (*Cambridge, Mass.*) **30**, 627 (1982).

[19] K. Sakamoto, N. Kimura, F. Nagawa, and Y. Shimura, *Nucleic Acids Res.* **11**, 8273 (1983).

[20] M. Baer and S. Altman, *Science* **228**, 999 (1985).

[21] C. Guerrier-Takada, K. Gardiner, T. Marsh, N. Pace, and S. Altman, *Cell* (*Cambridge, Mass.*) **35**, 849 (1983).

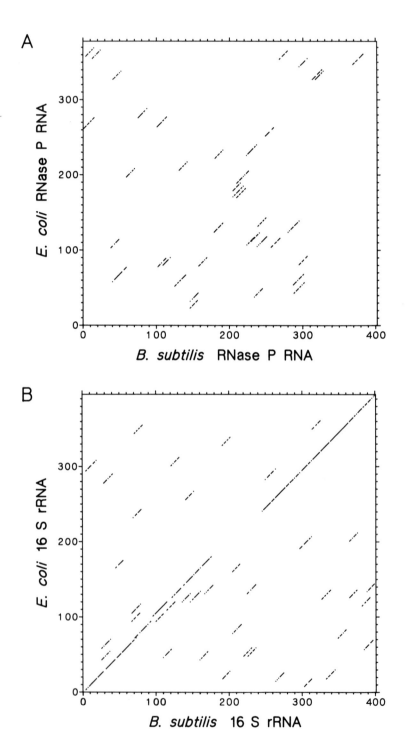

A

E. coli RNase P RNA

B. subtilis RNase P RNA

B

E. coli 16 S rRNA

B. subtilis 16 S rRNA

molecules. Synthetic oligodeoxynucleotide hybridization probes for such highly conserved sequences could, in principle, be used to seek still more distant homologs.

Alignment of Sequences

The sequences that are available for comparison must be rigorously aligned, so that evolutionarily homologous residues are juxtaposed, commonly in columns (Fig. 1). The identification of homologous residues is essential to the search for homologous secondary structure, for it is variation in homologous base pairs that defines the structure. If the compared sequences are too dissimilar, then their alignment cannot be established with confidence. Sequences that are too similar are easily aligned, but these sequences offer few changes with which to test pairing possibilities. As a rule-of-thumb, sequence similarities of 60–80% are optimal for secondary structure analysis.

It is convenient to order the aligned sequences so that the most similar ones are adjacent in the listing. This requires an initial, approximate alignment and calculation of the pairwise similarities of the sequences. These operations are conveniently executed using one of a variety of computer programs for the optimal matching of sequences. If the sequences are relatively similar, much of the computer alignment is likely to prove correct. However, alignments must be scrutinized manually and must be refined repeatedly as the secondary structure model develops. If the sequences of compared RNAs are sufficiently different to be useful for comparative analysis of the secondary structure, it is unlikely that the sequences can be unambiguously aligned over their entire lengths simply on the basis of primary structural similarity.

The alignment process begins by juxtaposing and then extending from sequence stretches that are identical, or nearly so, among the compared molecules, and hence are likely homologs. It is useful to identify and build on several, ostensibly homologous segments distributed throughout the molecules. These segments are extended as the correspondence of additional residues is established by primary or secondary structural similarity.

As the alignment of the sequences proceeds, it is usually necessary to insert "alignment gaps," shifting the register of one or more sequences in

---

FIG. 3. "Dot plot" analysis of similar sequences in the RNase P RNAs from *B. subtilis* and *E. coli*. Dots indicate the regions where the two sequences are identical in at least 9 of 12 consecutive nucleotides. (A) Comparison of the RNase P RNAs. (B) Comparison of the first 400 nucleotides of the *B. subtilis* 16 S rRNA with the corresponding region (396 nucleotides) of the *E. coli* 16 S rRNA. (Adapted from Reich *et al.*[17])

the collection in order to maximize their similarities. However, differences in sequence length due to evolutionary insertions and deletions are rare relative to nucleotide substitutions. Thus, alignment gaps should be employed sparingly, with an eye toward the overall alignment in the vicinity of the gap. If an alignment gap inserted in one sequence improves similarity over only a short stretch before a gap must be inserted in the second sequence to restore the optimal match, then it is unlikely that the adjustment in register afforded by the gaps reflects true homology. If gaps spanning only a few consecutive positions are required for the credible alignment of sequences, the gaps are likely to occur in hairpin stems of variable length.

As exemplified by the RNase P RNAs (Figs. 1 and 2), the largest length variations in functional RNAs occur through the acquisition or deletion of structural domains, commonly helical elements. The alignment of a sequence that contains such an element with one that lacks the element requires the introduction of several alignment gaps in the latter in order to reestablish the register of homologous positions. Thus, the alignment process and the search for phylogenetically consistent secondary structure naturally proceed in concert. As pairings are located, questionable alignment can be refined, because the end points of homologous secondary structural elements are expected to align.

The alignment of particularly disparate sequences can be problematic, because of the potential for length variation, as well as extensive sequence changes. It is therefore useful during alignment to rely most heavily on sequence elements that are the most conserved among all the members of the data set. This is illustrated in the RNase P RNA analysis by considering the homolog in *B. subtilis* of nucleotides 327–336 in the *E. coli* sequence, UAGAUGAAUG. The *B. subtilis* sequence UAGAUGAUUG (*Bsu* 317–326, Fig. 1) is identical in 9 out of 10 positions; however, only 6 of these residues are conserved in all of the available sequences. In contrast, the sequence UAGAUAGAUG (*Bsu* 313–322) is identical in only 8 of the 10 positions, but 7 of these are conserved in all the sequences. The latter alignment was chosen on this basis.

Search for Secondary Structure

A putative helix, indicated by the occurrence of continuous complementary sequences, is considered nominally proved when (1) complementary sequences occur in homologous segments of each of the molecules inspected and (2) two or more independent covariations occur in the complementary sequences such that their canonical pairing potential is preserved. The initial consideration of complementary sequences in a

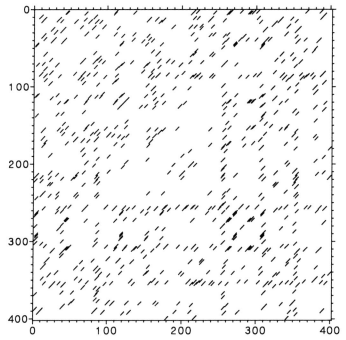

FIG. 4. Dot plot analysis of complementary sequences in the *B. subtilis* RNase P RNA. Dots indicate complementary sequences of at least five consecutive nucleotides (accepting A–U, G–C, and G–U pairs). Axes numbers correspond to nucleotide positions in the *B. subtilis* RNase P RNA sequence (Fig. 1).

particular RNA concentrates on the canonical (A–U and G–C) base pairs and on G–U. Although other base pairs (A–G, U–U, etc.) also occur in natural RNAs, these base pairs cannot be taken as initial evidence for a pairing, because there are no *a priori* rules that identify pairing partners unequivocally.

All complementary sequences within molecules in the data set must be considered as potential pairing partners. It is a general theme that continuous canonical complements in natural RNAs tend to be fairly short, say two to six base pairs, before the occurrence of noncanonical pairs or other helix irregularities. The number of randomly occurring, potential pairing partners in large RNAs is, therefore, very large. For instance, Fig. 4 shows the distribution of complements consisting of five or more base pairs (accepting G–U pairs) in the *B. subtilis* RNase P RNA sequence. Because of the large number of pairing possibilities, it is useful to focus first on local hairpins and lengthy pairing elements. As these are identi-

A

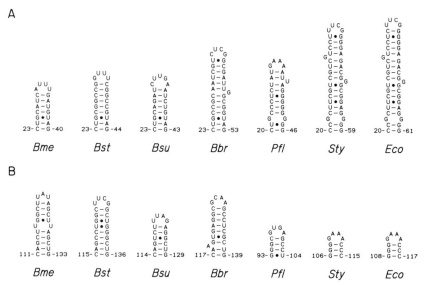

B

FIG. 5. Variations in two RNase P RNA hairpin structures. Structural variations within conserved hairpins are shown. Homologs from different phyla tend to exhibit the greatest differences. (A) The homologs of the *B. subtilis* 23–30/36–43 (*E. coli* 20–38/43–61) pairing. Unusual base pairs sometimes replace the canonical ones. For instance, the purple bacteria helices contain a G–G juxtaposition at the location occupied by the more familiar G–U pairing in the gram-positive versions of the helix. This suggests that the G–G juxtaposition does not disrupt the structure *in vivo*. (B) The homologs of the *B. subtilis* 114–119/124–129 (*E. coli* 108–110/115–117) pairing. Another unusual base pair, a U–U juxtaposition in the *Bme* helix, is observed. Such irregularities need not render helices unstable; approximately 1 out of 300 base pairs in nonmitochondrial tRNAs involves a U–U juxtaposition. Organism abbreviations are as in Fig. 1. (From James *et al.*[9])

fied, the sequence length that must be inspected for shorter pairings is reduced. Ultimately, however, all covariations must be sought, even those involving single pairs.

Two general types of complementary sequence pairs are likely to occur in large RNAs: (1) short-range pairings that create hairpins and (2) long-range pairings that order several helices into discrete structural domains. Hairpin structures are the most volatile of the pairings in sequence and length; examples of two highly variable elements in the RNase P RNAs are shown in Fig. 5. Such variability in hairpin architecture can make the identification of homologous sequences difficult or impossible. Consequently, it is sometimes necessary to rely heavily on features, such as the sequence of the helix loops and the endpoints of the pairings relative to sequences that are clearly homologous. Some of the helices shown in Fig. 5 cannot be considered as homologs simply on the basis of sequence and complementarity. Instead, we recognize them because the helices are flanked by sequences that are unequivocally homologous. That is, these hairpins are in exactly the same location in a phylogenetically conserved portion of the RNase P RNAs. The variability in the structure of these hairpins suggests that the hairpins are peripheral in the tertiary structure of the RNA.

In contrast to hairpin structures, long-range pairings (with the exception of terminal pairings) are likely to be of constant length and tend to involve more conserved sequences. Presumably, the conservative nature of the long-range pairings reflects their importance in establishing the overall superstructure of the RNA. The sequence conservation may render these pairings difficult to document, however, because instances of covariation are less common, and may be observed only when very diverse sequences are compared. For example, evidence for the long-range pairing that results in helix *Bsu* 57–61/240–244 (*Eco* 74–78/243–247) required the interphylum comparison (Fig. 1). This pairing also illustrates that the evidence for secondary structural features sometimes outweighs maximizing sequence similarity. In Fig. 1, the residue *Eco* G-74 is aligned with *Bsu* C-57, postulating that *Bsu* G-56 lacks a homolog in the purple bacteria sequences. An alignment based on the greatest primary structural similarity would not result in this conclusion. The *E. coli* and *B. subtilis* sequences would have three additional identical residues if *Bsu* 56–62 were assumed homologous with *Eco* 74–80. However, with this alternative alignment, the pairings *Bsu* 57–61/240–244 and *Eco* 74–78/243–247 would not be homologous, and the proposed structure would be contradicted by the comparative evidence.

It is convenient if long-range pairings can be identified early in the structure analysis, since the long-range pairings segregate the sequences within the loop from the remainder of the molecule. Pairings within the loop that results from a long-range association are far more likely than pairings between the loop and an element outside the loop. The latter pairings have been dubbed "pseudoknots," because the formation of a lengthy helix between a loop and a region outside the loop would pass the region through the loop, creating a knot.[22] Shorter pairings may occur between loops, however, and such pseudoknots are seen to occur in RNase P RNA (*Bsu* 52–55/376–379 and *Eco* 70–73/354–357) and other natural[23,24] and synthetic[25] RNAs. The proposal of pairings between

[22] G. M. Studnicka, G. M. Rahn, I. W. Cummings, and W. A. Salser, *Nucleic Acids Res.* **5,** 3365 (1978).

[23] C. W. A. Pleij, K. Rietveld, and L. Bosch, *Nucleic Acids Res.* **13,** 1717 (1985).

[24] D. A. Stahl, T. A. Walker, B. Meyhack, and N. R. Pace, *Cell (Cambridge, Mass.)* **18,** 1133 (1979).

[25] J. D. Puglisi, J. R. Wyatt, and I. Tinoco, Jr., *Nature (London)* **331,** 283 (1988).

loops, of course, must consider other structural features that might constrain the pairing. For instance, the potential pairing partners might be separated by a helical element sufficiently long to prevent them from reaching one another.

Because long-range pairings order larger portions of the sequence and guide further searches for structure, their identification should be considered more cautiously than seems necessary for the acceptance of local pairings. There are complementary, potential extentions of pairings in the core of the RNase P RNA that are not shown as paired in Fig. 2, because the sequences are identical in all the organisms so far inspected; in the absence of sequence variation, there can be no covariation to support the pairings. For instance, complementary residues are present such that the pseudoknot pairing *Bsu* 52–55/376–379 (*Eco* 70–73/354–357) could be extended three nucleotides if *Bsu* U-51 (*Eco* U-69) were bulged. However, because variation in these additional complementary sequences has not yet been encountered, we consider only the continuous helix *Bsu* 52–55/376–379 (*Eco* 70–73/354–357) as nominally demonstrated.

The manual scrutiny of sequence collections for covariation is a laborious task and might overlook subtle features of RNA folding, for example, noncanonical pairings or associations that involve three-base interactions. Such features are commonly seen in tRNA crystal structures and, in principle, may be revealed by covariation analysis. As the collection of sequences of a particular RNA expands to include many examples, it becomes fruitful to employ computer searches for residues that covary in any way. The application of computer-assisted covariance analyses to tRNAs[26] or rRNAs[27] successfully identifies secondary structures and even many long-range tertiary contacts in tRNA that are known only from crystallographic analyses.

Prediction of Structure from Sequence

Computer analyses that use thermodynamic estimates of helix stability to predict the "global folding" of a RNA sequence played little role in inferring the structures of RNase P RNA[9] or the rRNAs.[28] Such folding algorithms can be used to identify pairing candidates within sequences for closer inspection of conservation and covariations within the data set. However, the results of global-folding programs are unreliable. When applied to rRNAs or tRNAs, phylogenetically proved helices are pre-

[26] G. J. Olsen, Ph.D. Thesis, University of Colorado, Boulder, 1983.
[27] R. R. Gutell, H. F. Noller, and C. R. Woese, *EMBO J.* **5,** 1111 (1986).
[28] H. F. Noller and C. R. Woese, *Science* **212,** 403 (1981).

dicted only about half the time. There is a much greater tendancy for the correct prediction of hairpin structures than for long-range pairings. There are many reasons for the current unreliability of minimum energy structure prediction. A significant problem is that our knowledge of the stabilities of different sequence pairs is incomplete for the canonical pairs, and there is far less information regarding the contributions of noncanonical pairs to helix stability. Even more serious, however, is the inability to account for the consequences of tertiary structure that might promote otherwise unstable pairings or prevent the formation of associations that in another context would be favored.

Acknowledgments

The authors' research is supported by NIH grant GM34527 to Norman R. Pace and by the Office of Naval Research grants N0014-86-K-0268 to Gary J. Olsen and N14-87-K-0813 to Norman R. Pace.

[19] Mapping the Genetic Organization of RNA by Electron Microscopy

By LOUISE T. CHOW and THOMAS R. BROKER

Electron microscopic (EM) analysis of DNA–DNA heteroduplexes permits the evaluation of sequence and organizational relationships among related genomic and subgenomic molecules. Formamide is used to dissociate adventitious intrastrand pairing during EM grid preparation, facilitating the measurement of both double-stranded and single-stranded DNA segments.[1–3] This technique was extended to the study of relationships of RNAs and their template DNAs, following the realization that RNA–DNA duplexes are more stable than equivalent DNA–DNA duplexes when the solvent contains 70% (v/v) or more formamide.[4,5] There are two forms of RNA–DNA heteroduplexes. In a R-loop,[6] the DNA remains duplexed, except where RNA has hybridized and caused the displacement of its DNA counterpart, leading to the formation of a loop.

[1] A. K. Kleinschmidt, this series, Vol. 12, Part B, p. 361.
[2] B. C. Westmoreland, W. Szybalski, and H. Ris, *Science* **163**, 1343 (1969).
[3] R. W. Davis, M. Simon, and N. Davidson, this series, Vol. 21, p. 413.
[4] J. Casey and N. Davidson, *Nucleic Acids Res.* **4**, 1539 (1977).
[5] R. L. White and D. S. Hogness, *Cell (Cambridge, Mass.)* **10**, 177 (1977).
[6] M. Thomas, R. L. White, and R. W. Davis, *Proc. Natl. Acad. Sci. U.S.A.* **73**, 2294 (1976).

In the RNA–single-stranded DNA heteroduplexes (R–D heterodu-plexes),[7,8] the DNA remains unpaired, except where RNA has hybrid-ized.

Electron microscopic analyses of R-loops and R–D heteroduplexes led to the discovery of mRNA splicing.[7,8] These methods have remained instrumental in deciphering the structures of eukaryotic genes and are particularly useful in elucidating complex patterns of alternatively spliced mRNAs, for instance, those produced by human adenoviruses and papil-lomaviruses in which each early and late transcribed regions generate several partially overlapping mRNA species that share either the same 5' end, the same 3' end, or both.[8–11] This chapter is a practical guide to the design, preparation, interpretation, and analysis of R-loops and R–D het-eroduplexes.

Thermal Stability of Double-Stranded DNA

Double-stranded DNA denatures gradually over a temperature range of 5–20°, the breadth of which depends largely on the range of GC–AT base composition in different segments of the polymer. The melting tem-perature (T_m) is defined as the midpoint of this helix–coil transition in standard solvents having an ionic strength $\mu = 0.18$ (0.18 M NaCl or 0.12 M sodium phosphate, pH 6.8, or 0.15 M NaCl and 0.015 M sodium citrate, pH 7). By definition, the lowest temperature at which the duplex com-pletely denatures, the strand-separation temperature (T_{ss}), is generally about 7–10° above T_m. Both T_m and T_{ss} are complex functions of the concentrations of electrolytes and chaotropic agents, base composition, length of complementary strands or segments of strands, degree of base complementarity or mismatch, and the nature of the duplex (DNA–DNA, DNA–RNA, or RNA–RNA). To achieve optimal hybridization condi-tions, all these factors must be considered, and adjustments in one or more parameters usually made for individual experimental situations. Each element is elaborated in this chapter and can be applied to all nucleic acid hybridizations irrespective of the eventual methods of analysis.

Effects of Base Composition and Formamide Concentration

As the temperature of a DNA solution is raised, the most AT-rich regions are the first to denature, and the most GC-rich regions are the last

[7] S. M. Berget, C. Moore, and P. A. Sharp, *Proc. Natl. Acad. Sci. U.S.A.* **74**, 3171 (1977).

[8] L. T. Chow, R. E. Gelinas, T. R. Broker, and R. J. Roberts, *Cell (Cambridge, Mass.)* **12**, 1 (1977).

[9] L. T. Chow and T. R. Broker, *Cell (Cambridge, Mass.)* **15**, 497 (1978).

[10] L. T. Chow, T. R. Broker, and J. B. Lewis, *J. Mol. Biol.* **134**, 265 (1979).

[11] L. T. Chow, M. Nasseri, S. M. Wolinsky, and T. R. Broker, *J. Virol.* **61**, 2581 (1987).

TABLE I
T_m^* OF DNA: EFFECTS OF GC CONTENT
AND FORMAMIDE[a]

	$T_m^*(^\circ)$		
GC content	40%	50%	60%
70% (v/v) Formamide	47.5	53.4	59.2
80% (v/v) Formamide	41.0	47.1	53.2

[a] In 0.1 M HEPES, pH 7.9, 0.01 M ethylene-diaminetetraacetic acid (EDTA), and 0.4 M NaCl.

to denature, a consequence of three hydrogen bonds formed between G–C base pairs versus two hydrogen bonds for A–T base pairs. DNA held near its T_m is in a state of dynamic equilibrium of denaturation and renaturation, and the AT-rich regions spend a greater percentage of time denatured than do the GC-rich regions. As a result, when RNA is mixed with DNA held near its T_m, the AU-rich RNAs will preferentially anneal. The melting temperature of DNAs of various AT and GC base compositions dissolved in the standard solvents can be calculated from Eq. (1).[12]

$$T_m = 69.5^\circ + 0.41 \times (\% \text{ GC}) \tag{1}$$

T_m^* is the effective T_m under nonstandard conditions, when one or more parameters have been changed. For instance, formamide lowers the T_m of duplex polynucleotides. In the presence of formamide, T_m^* is related to T_m by Eq. (2).[4,13–16]

$$T_m^* = T_m - [0.5(\text{mole} \cdot \text{fraction GC}) + 0.75(\text{mole} \cdot \text{fraction AT})] \times (\% \text{ formamide}) \tag{2}$$

Therefore, base composition not only affects the overall T_m of the DNA, but can impose a differential response to formamide in regions of the DNA that differ in AT and GC content (Table I).

Effect of Electrolytes

By shielding the negative phosphate charges of the polynucleotide backbone, sodium ions, expressed in molar concentration, raise the T_m^*

[12] J. Marmur and P. Doty, *J. Mol. Biol.* **5**, 109 (1962).
[13] B. L. McConaughy, C. D. Laird, and B. J. McCarthy, *Biochemistry* **8**, 3289 (1969).
[14] H. B. Bluthmann, D. Bruck, L. Hubner, and A. Schoffski, *Biochem. Biophys. Res. Commun.* **50**, 91 (1973).
[15] C. Tibbetts, K. Johansson, and L. Philipson, *J. Virol.* **12**, 218 (1973).
[16] J. R. Hutton, *Nucleic Acids Res.* **4**, 3537 (1977).

according to Eqs. (3)[17] or (4).[18] The relations are valid up to about 0.4 M Na^+, whereupon increasing salt concentration has relatively less incremental effect.

$$T_m^* = T_m - 18.5° \times \log_{10}(0.18/[Na^+]) \tag{3}$$
$$T_m^* = 16.6 \log_{10}[Na^+] + 0.41(\%GC) + 81.5° \tag{4}$$

Although high-salt concentration increases the T_m, it also substantially increases the rate of reannealing.[16,19,20] For this kinetic reason, it is advantageous to hybridize RNA to DNA in 0.4 M NaCl. However, it is not advisable to anneal at a temperature in excess of 55–60°, because long polynucleotide chains suffer breaks during prolonged incubation.[4,16,21] Di- and trivalent cations stabilize duplex nucleic acids, and much more significantly raise T_m^* than do monovalent cations. They should be avoided or chelated with ethylenediaminetetraacetic acid (EDTA) during heteroduplex formation.

Effect of Duplex Length

T_m^* is related to the length of duplex, existing or anticipated, by the approximate Eq. (5).[19] This effect is trivial for DNA or RNA segments several hundred nucleotides long, but results in a significant adjustment for very short RNA segments, such as those in many spliced RNAs.

$$T_m^* = T_m - (500/\text{duplex length in base pairs}) \tag{5}$$

Effect of Partial Homology

When imperfect heteroduplexes are expected between polynucleotide strands, the T_m depression due to base-pair mismatch can be approximated by Eq. (6).[22–24]

$$T_m^* = T_m - 0.95° \times (\% \text{ mismatch}) \tag{6}$$

Principles and Basic Considerations of RNA–DNA Hybridization

Formamide

The stability of RNA–DNA heteroduplexes in aqueous solution containing 70% or more formamide has been measured to be about 10° higher

[17] W. F. Dove and N. Davidson, *J. Mol. Biol.* **5**, 467 (1962).
[18] C. Schildkraut and S. Lifson, *Biopolymers* **3**, 195 (1965).
[19] J. Wetmur and N. Davidson, *J. Mol. Biol.* **31**, 349 (1968).
[20] R. J. Britten, D. E. Graham, and B. R. Neufeld, this series, Vol. 29, p. 363.
[21] J. Bonner, G. Kung, and I. Bekho, *Biochemistry* **6**, 3650 (1967).
[22] T. I. Bonner, D. J. Brenner, B. R. Neufeld, and R. J. Britten, *J. Mol. Biol.* **81**, 123 (1973).
[23] J. R. Hutton and J. G. Wetmur, *Biochemistry* **12**, 558 (1973).
[24] T. R. Broker and L. T. Chow, *Cancer Cells* **4**, 589 (1986).

than that of the DNA–DNA counterpart.[4] RNA–RNA duplexes are even more stable. This observation becomes the basis of R-loop formation: RNA can hybridize to its DNA complement in solvents containing 70% (v/v) formamide, while the equivalent DNA strand is denatured. A second advantage of including formamide in the hybridization solution is that it lowers the T_m and therefore the necessary hybridization temperature, as discussed previously, and thus minimizes thermal degradation of nucleic acids.[16,21]

Temperature

DNA–DNA renaturation rates are maximal over a broad range of temperatures centered around $T_m^* -25°$.[19] This condition does not apply to RNA–DNA hybridization for two reasons. First, RNA tends to have rather stable intrastrand secondary structures, and the intrastrand secondary structures must be disrupted for efficient hybridization. Second, during R-loop formation, the two DNA strands must be partially or completely dissociated (near or above the T_m^* of the DNA) to be available for pairing with RNA. Accordingly, RNA–DNA hybridization is usually carried out between the T_m^* and T_{ss}^* [i.e., strand-separation temperature (T_{ss}) under nonstandard solvent conditions] of the DNA. Only when fractionated single DNA strands are used will RNA–DNA heteroduplex formation proceed without competition at temperatures below the T_m^* of the DNA.

Buffer

Formamide is subject to nucleophile-catalyzed hydrolysis, which has several undesirable consequences: decreased denaturing power of the solvent and increased ionic strength, both resulting in increased T_m^*, and acidification of the solution as ammonia derived from ammonium formate evaporates, causing nucleic acid degradation.[16] Strongly nucleophilic amine buffers, such as Tris, promote significant hydrolysis of formamide[4,6] and should be avoided. The pH of Tris buffer also has a large temperature coefficient and is not suitable for R-loop reactions at high temperatures. Sulfonate buffers are less likely to catalyze hydrolysis. To provide good buffering capacity, use 0.1 M HEPES (pK_a 7.5) adjusted with NaOH to pH 7.9.

Nucleic Acid Concentrations

RNA–DNA hybridizations are performed at DNA excess. The upper limit of DNA concentration in a reaction mixture is dictated by the sequence complexity of the DNA, as well as by the incubation temperature relative to T_m^* or T_{ss}^*. For a DNA molecule with a complexity of 15,000 bp (base pairs), 50 μg/ml is about the maximum. Very high DNA concentration could result in cross-annealing of multiple DNA molecules, when the incubation is terminated. Multistrand networks might also form during long incubations, when formamide starts to break down. Dilution with fresh and preheated incubation buffer and formamide prior to quenching (see Methods) partially alleviates this problem. At very high DNA concentrations, occasionally one RNA molecule may hybridize to two DNA molecules. The amount of RNA to use obviously depends on the relative abundance of the molecule of interest in the preparation. Up to 10–15 μg of RNA can be used in a 10-μl reaction mixture. The upper limit is set by the amount of free RNA on the grids that can be tolerated without sacrificing the clarity and contrast of the nucleic acids (2 μg/ml in the hyperphase spreading solution) and the effort the investigator is willing to spend scanning the grids, which is inversely related to the amount of DNA on the grids.

Due to the above considerations, the composition of the RNA–DNA hybridization solution is narrowly dictated by the high concentration of formamide required. The concentration of NaCl should also be as high as practicable to promote the maximal rate of annealing. Therefore NaCl, HEPES buffer, and EDTA are made up as a 7× concentrated solution to be added in a minimal volume. The RNA and DNA samples must be very concentrated or dried down in the hybridization tube to minimize the volume added, allowing the maximum (86%) concentration of formamide and thus the lowest possible temperature for the incubation. The optimal RNA–DNA annealing mixture is 70–86% (v/v) formamide, 0.4 M NaCl, 0.1 M HEPES, 0.01 M EDTA, pH 7.9, 50–150 μg/ml of DNA (adjusted proportionally to its complexity), and RNA sufficient to have any given species present in, preferably, 1–10% the molar concentration of the complementary DNA sequences.

Mapping by the R-Loop Method

Applications

R-loops should be formed and analyzed in any of the following instances.

1. To determine the general genetic organization of chromosomes known to contain several transcriptional units.[8]
2. To map common boundaries of convergent or divergent transcription.[8,25]

[25] L. T. Chow, J. M. Roberts, J. B. Lewis, and T. R. Broker, *Cell (Cambridge, Mass.)* **11**, 819 (1977).

3. To examine a rather long DNA molecule. All the relevant measurements are of duplex segments (DNA–DNA and RNA–DNA) and can be measured more accurately than when RNA is annealed to single-stranded DNA. This method also accommodates DNA molecules that contain a few single-stranded nicks; on lowering the temperature, the partially denatured strands will snap back and be largely or completely restored to duplex.
4. To map rare or unspliced RNAs. R-loops are relatively easy to see and are unambiguous even when infrequent.

Limitations

In comparison to the analysis of RNA paired with single-stranded DNA, R-loop mapping has the following disadvantages.

1. R-loops form less efficiently in GC-rich regions than in AT-rich domains of the DNA, since GC-rich regions remain paired during incubation near T_m^*.
2. RNA segments, especially if shorter than 150 nt (nucleotides), can be displaced, and small R-loops could collapse, if the sample is not quenched or stabilized swiftly.
3. R-loops do not provide information on the identity of the template strand, if branch migration obscures the identity of polyadenylate [poly(A)] at the 3′ end of the RNA.
4. Topological constraints may prevent complete hybridization of multiply spliced RNAs, particularly when exons are short or remote.

Mapping by RNA–Single-Stranded DNA Heteroduplex Method

Applications

The study of RNA–single-stranded DNA heteroduplexes is particularly informative about interrupted genes and their spliced transcripts, once a basic transcription pattern has been established by the R-loop technique. The advantages of the R–D heteroduplex methods are manyfold.

1. RNAs that are GC-rich, as well as those that are AU-rich, can be mapped.
2. All segments of spliced RNAs can and usually will hybridize to their complements. The hybridizing strands have total freedom of axial rotation to achieve complete pairing of all exons.
3. RNAs complementary to the same DNA strand can be identified.

4. The presence of a poly(A) branch helps to define the 3′ end of the R–D heteroduplex and to orient the transcript.

Limitations

There are several disadvantages or technical limitations to the R–D heteroduplex methods.

1. The instability of nucleic acids and formamide held at high temperature limits exhaustive hybridization. Therefore, only fairly abundant RNA (more than 0.1% of the mRNA population) can be readily studied. Minor RNA species may not hybridize during the short 30- to 60-min reactions possible at 60° or above.
2. Any nicks in the DNA will result in the loss of the defined termini against which map position measurements are generally made; this problem is particularly severe with long DNA molecules.
3. The distinction of single-strand/double-strand boundaries must be good to permit the visualization of the termini of the RNA. Unless the RNAs are long or are spliced and create deletion (intron) loops in the DNA, the R–D heteroduplexes are much less prominent than R-loops.
4. Measurements along single-stranded DNA are inherently less reliable than those along duplex segments and may also be subject to differential contraction of GC-rich regions.

Fortunately, the relative advantages and limitations of R-loop and R–D heteroduplex methods are complementary, and both can be employed to good advantage in the course of defining a transcription map. Moreover, they can be combined in single or multistep incubations, as described in Methods.

Materials

Glassware, Utensils, and Water Baths

Glassware, pipets, and utensils should be clean, free of grease and detergent and baked at 150° for several hours to inactivate RNases. Hybridization is performed in 6 × 50-mm culture tubes (Kimax #45048), which should be rinsed several times with ultrapure water prior to baking; these tubes are used once and discarded. Do not use siliconized glassware or plastic microcentrifuge tubes, because the transfer of small volumes of solutions is difficult and inaccurate. Several covered water baths with good temperature control are needed for empirical determination of T_m^* and for multistep hybridization procedures.

Water and Formamide

House water partially purified by distillation, deionization, or reverse osmosis is usually not adequate for the preparation of EM grids. House water should be redistilled or processed through a Millipore (Bedford, MA) MilliQ or Barnstead (Hicksville, NY) Nanopure water purification system. Formamide should be spectral quality and need not be purified by deionization, distillation, or recrystallization, as long as the stock is tightly sealed, absorption of water vapor is minimal, and no ammonia smell is detectable.

Hybridization Buffer and Glyoxal

New bottles of stock chemicals should be set aside for making solutions for the preparation of RNAs and for hybridization. The $7\times$ hybridization buffer contains 2.8 M NaCl, 0.7 M HEPES, and 0.07 M ethylenediamine tetraacetic acid adjusted to pH 7.9 with NaOH (EDTA). Pass the buffer through a sterile Millipore HAWP 0.45-μm filter that has been prerinsed to flush out surfactants. To inactivate nucleases, add several aliquots of diethyl pyrocarbonate (DEPC) (*caution:* hazardous), totaling 1/2000 total volume, and mix vigorously. Store the stock at room temperature in the dark to decrease photooxidation of the buffer. Glyoxal used for stabilizing R-loops for EM is purchased as a 40% (w/v) solution and is used as such. Precipitates in the glyoxal solution, which develop upon storage at 4°, should be removed by centrifugation.

DNA

DNA should be a homogeneous population of linear molecules, such as intact linear chromosomes or those resulting from restriction endonuclease digestion, and must be free of protein. The DNA should have few, if any, single-stranded nicks. Therefore, plasmid DNA should be form I supercoils prior to restriction digestion. The quality of the DNA can be determined by denaturation followed by renaturation to 50% completion. The single-stranded DNA should be of uniform length by visual inspection. Renatured DNA should have few, if any single-stranded ends or branches. Deletion, insertion, and substitution loops should also be rare or absent in the self-annealed DNA. A substantial percentage of DNA molecules with such major mutations or physical damage can severely complicate interpretation of structures and assignment of heteroduplex map positions.

To orient RNA on a circular chromosome or recombinant plasmid, the DNA should be linearized with a restriction enzyme outside the transcribed region. The DNA insertion should be located asymmetrically in the linearized molecule. In either case, the cleavage site should be at least 200–300 bp away from the region of interest. A shorter end would have a lower T_m^*, according to Eq. (5), than the overall T_m^* of the DNA and will tend to denature. Proximity of the RNA–DNA heteroduplex to such a single-stranded DNA end would make accurate measurement difficult. To map the RNA unequivocally, it may be necessary to use a second preparation of linear DNA generated by digestion with a different restriction enzyme. Similarly, restriction endonuclease digestions can also be used to orient a R-loop on a linear chromosome.

To increase the efficiency of hybridization, it is sometimes advantageous to excise the cloned genomic DNA from the vector. For instance, when the DNA is cloned in phage λ vectors, the excluded volume effect of random coils makes the internal sequences less accessible for hybridization than sequences near the ends.[19] Removal of the flanking λ arms reduces this problem. It is not necessary to purify the cloned DNA away from the λ DNA fragments by gel electrophoresis, because the vector DNA will not interfere with the hybridization.

The DNA sample should be completely free of proteins, such as restriction enzymes, used in its preparation or covalently bound proteins, such as those on the termini of certain viral DNAs. Proteins cause aggregation during hybridization reactions. Do not vortex samples during extraction of the DNA with phenol or chloroform; shear forces can break long duplex DNA. The DNA stock should be 100 μg/ml or higher, dissolved in water or in low concentrations of Tris–EDTA buffer.

RNA

RNA should be prepared by one of several standard methods, which generate molecules with as little degradation as possible. As with DNA, RNA preparations should be free of protein contaminants. The RNA solution is generally prepared at concentrations up to several milligrams per milliliter in water. One of the great powers of R-loop formation and analysis by electron microscopy is that the RNA species of interest usually need not be isolated, although enrichment will reduce the background, the hybridization time, and the EM survey time. A species that is 0.1–0.5% of the poly(A^+) RNA can easily be mapped using total cytoplasmic RNA. Those as low as 0.005% of the poly(A) RNA can be found with effort. Poly(A) or hybridization selection should be used to enrich rare mRNAs of interest. Hybridization-selected RNA should be put through subsequent poly(A) selection to remove DNA molecules that might be also eluted from the membrane. Low-molecular-weight RNAs, such as tRNAs, impair contrast in the EM sample and should not be used as carrier for the recovery of the selected mRNAs.

RNA preparations must be of high quality, with a low percentage of broken molecules, because analysis depends on statistical definition of populations with reproducible branch points and ends. If total or cytoplasmic RNA is used, the quality of mRNA can be inferred by the sizes of the ribosomal RNAs (which constitute over 95% of the mass), as visualized by spreading in the presence of formamide. The collapsed rRNA bushes should be big (see Figs. 2 and 3). If not, degradation has occurred, and a new preparation should be made.

Electron Microscopic Reagents

Stock solutions of 2 M Tris base plus Tris–HCl, pH 8.5 (174.4 g/liter plus 88.4 g/liter), and 0.2 M Na$_3$EDTA, adjusted to pH 8.5, are each filtered through Nalgene 0.45-μm nitrocellulose filters. The two solutions are then mixed 1 : 1 (v/v) and are used as a working stock at room temperature. Uranyl acetate (0.05 M) (*caution:* a toxic and radioactive heavy metal) dissolved in 0.05 N HCl is filtered and then stored in the dark at 4°; the solution is freshly diluted 500× into 90% (v/v) ethanol (10 μl into 5 ml) for staining and fixation of EM sample grids at room temperature. Equine heart cytochrome c is prepared as a 1 mg/ml of stock solution in 0.1 M Tris, 0.01 M EDTA, pH 8.5, and is kept at 4°. The quality of cytochrome c varies a great deal with the manufacturer and batch. The most common problems are the formation of DNA "flower" aggregates and excessive graininess of the protein monolayer films. These problems can sometimes be alleviated by filtration through nitrocellulose membranes. The concentration of cytochrome c in the hyperphase should be about 50 μg/ml to achieve the best results.

Methods

Empirical Determination of T_m^* in R-Loop Buffer–Formamide

To a first approximation, all the equations for calculating T_m^* of DNA can be applied additively to estimate the conditions for hybridization. If the GC content of the gene of interest cloned in a plasmid or λ vector is not known, the following simple procedure can be used to determine the T_m^*. Digest a few micrograms of the plasmid or λ DNA with restriction enzyme(s) to separate the insert (or as much of the insert as possible) from the vector. Purify the digests by phenol and ether extractions and by ethanol precipitation. Dissolve the digest in the 1× R-loop buffer–formamide cocktail and aliquot the mixture into 6 × 50-mm tubes (5–10 μl each), then seal with Parafilm. Incubate the mixture in water baths with temperatures set at about 3–5° apart, centered around the T_m^* calculated from the overall GC content of the organism. After incubation for approximately 10 min, stabilize the reaction with glyoxal (see Method 1: R-Loop Formation). Glyoxal reacts with amino groups of guanine, adenine, and cytosine bases in nucleic acids and prevents DNA renaturation. Keep over ice until all samples are ready for agarose gel electrophoresis. An aliquot of unheated sample is used as a standard. When the incubation temperature exceeds T_{ss}^*, the band of double-stranded DNA insert will disappear. There should be an obvious transition. This experiment also reveals the T_{ss}^* of the vector relative to the gene of interest. If the former is higher than the latter, then one can use the vector sequence as a "GC-rich clamp" to hold the strands together during incubation at the T_{ss}^* of the insert.

It is likely that different regions in a long piece of genomic DNA will be heterogeneous in their GC contents. If there is no additional information to localize the segment of interest to a smaller restriction fragment, then hybridizations should be set up about 2–3° apart near and above the anticipated T_{ss}^* or T_m^* (Eqs. (1)–(6); Table I). The optimal condition where R-loops or R–D heteroduplexes are found most frequently is determined by EM examination.

Hybridization

Deliver the concentrated RNA and DNA solutions in small volumes to the bottom of the 6 × 50-mm culture tube with sterile glass micropipets. Dry them down in a Speed Vac Concentrator (Savant, Boston, MA) while centrifuging, except when a short incubation time is planned during which DNA may not completely redissolve. Dissolve the sample in a total volume of 5–10 μl of buffer plus formamide cocktail. Measurement and transfer must be carried out with great accuracy. Avoid carrying in or retaining liquid on the outside surface of the pipets. Tightly seal the tube with several layers of Parafilm and spin the tube briefly at low speed (in the Speed Vac without vacuum) to ensure that all solution is at the bottom. Incubation is performed in covered water baths. In most respects, including the composition of the hybridization solution, the method of R–D heteroduplex mapping is closely analogous to that with R-loops. The differences are in the temperature of incubation and the methods of quenching the reaction (see the following three sections).

Method 1: R-Loop Formation

Generally, it is not advisable to carry out the incubation at T_m^* longer than 24 hr, because the nucleic acids and the formamide begin to break down, and the T_m^* increases as the formamide is hydrolyzed, as discussed previously. Consequently, the DNA strands reassociate, and hybridization of RNA to DNA slows down or ceases. To terminate R-loop formation, the reaction tube is chilled quickly in a water bath set at 7–12°. A $\frac{1}{7}$ volume of a 40% (w/v) glyoxal solution (4°) is quickly added to the mixture, and the mixture is incubated for 10–15 min at 7–12°. The mixture is then quenched with 10 volumes of ice-cold water. This solution remains stable for many days for the preparation of EM grids. RNA displacement by renaturation of the homologous DNA strand is prevented by the fixation of the displaced single-stranded DNA with glyoxal.[26]

Method 2: RNA–Single-Stranded DNA Heteroduplex Formation

To denature DNA, the hybridization mixture containing duplex DNA and RNA is heated to a temperature about 4–5° above the T_{ss}^* for 5–10 min. The solution is then transferred to the desired RNA–DNA annealing temperature at or above T_m^*. Hybridization should not exceed 30–60 min at temperatures above 60°; otherwise, nicking of DNA about 25 kb or longer and of RNA would become significant. With shorter DNA molecules or lower temperatures, the incubation can be proportionally lengthened up to several hours. When predenatured, the degree of DNA–DNA reannealing at T_m^* will not be very significant during the short duration of the hybridization, provided the DNA concentration is moderate. The reaction is quenched by addition of 10 volumes of ice-cold water, and the sample may be stored for many days without glyoxal fixation.

To allow pairing of short or AU-rich RNA exons in spliced RNAs that might not have annealed to their DNA template at T_m^* or T_{ss}^*, the incubation temperature can be lowered by a few degrees for 5–10 min or longer prior to quenching. To prevent DNA–DNA renaturation from occurring during this secondary incubation, dilute the sample severalfold into fresh hybridization cocktail, which has been preheated to the same temperature. Since renaturation of separated DNA strands is a second-order reaction, dilution will decrease DNA–DNA hybridization considerably.

Method 3: Combination Hybridization Procedures

To visualize rare RNA species containing multiple exons, a combination of the R-loop method and the R–D heteroduplex method can be used. RNA is first driven into R-loops by prolonged hybridization with double-stranded DNA, then the DNA strands are fully separated by a brief incubation (7–10 min) at T_{ss}^*.[27] Before this secondary incubation, it is advisable

[26] D. B. Kaback, L. M. Angerer, and N. Davidson, *Nucleic Acids Res.* **6**, 2499 (1979).
[27] B. W. Stillman, J. B. Lewis, L. T. Chow, M. B. Mathews, and J. E. Smart, *Cell (Cambridge, Mass.)* **23**, 497 (1981).

to dilute the R-loop mixture (3-fold) into fresh cocktail to compensate for any loss of formamide during the long incubation for R-loop formation. Again, the temperature can be lowered for a few minutes to allow short exons to hybridize, before terminating the reaction by quenching with 5–10 volumes of ice-cold water. Once the DNA strands separate, each exon can quickly anneal to its complement by intramolecular pairing with pseudo-first-order kinetics.

When the vector has higher GC content and, therefore, higher T_m^* than the insert of interest, the vector can serve as a clamp during the formation of R-loops or R–D heteroduplexes (see Fig. 2b). The incubation is carried out at the T_{ss}^* of the insert, but below the T_{ss}^* of the vector. The result is, of course, the best of both the R-loop method and the RNA–single-stranded DNA heteroduplex method, providing the advantage of mostly double-stranded segments for measurement and the clear distinction of all exons separated by single-stranded DNA loops corresponding to introns. When dealing with a large chromosome with a wide range of GC content in different intervals, careful selection of incubation temperature can achieve similar results. Stabilization with glyoxal is necessary to prevent the displaced, single-stranded DNA from reassociating with the DNA deletion loops, which can cause tangles and make interpretation difficult.

Electron Microscopic Procedures

Mounting Samples on Grids

Spreading of nucleic acid preparations in cytochrome c monolayers[3] should be carried out with a 50-μl hyperphase composed, by sequential addition, of water, 10× TE buffer (final concentration of 0.1 M Tris and 0.01 M EDTA, pH 8.5), cytochrome c (final concentration of about 50 μg/ml), formamide [final concentration of 40% (v/v)], single-stranded (ss) DNA length standard (ΦX174 DNA, 5386 nt) and double-stranded (ds) DNA length standard (open circles of ΦX174 RF or other plasmids of known lengths) at 0.01 μg/ml, and about 0.1–0.5 μg/ml of sample DNA (1–2 μl of 10 times diluted reaction mixture). Linear dsDNA (no more than half the length of phage λ DNA) without R-loops present in the sample are ideal standards. Use less sample, if the background RNA exceeds 2 μg/ml. The hypophase (90 ml) consists of 0.01 M Tris, 0.001 M EDTA, pH 8.5, and 10% (v/v) formamide, the latter added just prior to pouring the solution into a plastic bacteriological petri dish, containing an acid-cleaned glass microscope slide (1 inch × 3 inch) resting against a Teflon bar (15 × 1 × 0.6 cm) serving as a ramp. A second bar is placed about $1\frac{3}{8}$ inch in front of the first to define the area in which the cytochrome *c* film is allowed to form in the midsection of the petri dish. Formamide and the nucleic acid sample should be added to the hyperphase within seconds prior to spreading. If other spreading conditions are preferred, it is advisable to balance the formamide concentrations of the hyperphase and hypophase solutions to be isodenaturing [e.g., hyperphases containing 45, 50, or 60% (v/v) formamide onto hypophases of 15, 20, and 30% (v/v) formamide, respectively].[28] Spreading samples across water or very low-salt hypophases can cause anomalous stretching of single strands or may disrupt duplex segments shorter than about 150 bp. The cytochrome *c* film, containing nucleic acid molecules, is picked up on a Parlodion-covered 200-mesh EM grid, stained immediately for 30 sec with uranyl acetate in 90% (v/v) ethanol, and is rinsed for 5 sec in 90% (v/v) ethanol. Three grids can be prepared from each spread. The grids are rotary shadowed in a vacuum evaporator at a 7° angle with a platinum–palladium (80 : 20 alloy) wire, wrapped around a tungsten filament. The vacuum should be better than 2.5×10^{-5} Torr. Pumping time should be kept less than 15 min to minimize contamination by backstreaming diffusion pump oil. Use liquid nitrogen in the evaporator trap to shorten pumping time if necessary.

Evaluation of Specimens

The dsDNA and DNA–RNA heteroduplexes should appear as sinuous curves without particular orientation. The single strands are thinner, with some kinks. The overall contrast of the DNA and the distinction between single-stranded and double-stranded polynucleotides should be good. The concentration of RNA in the background must not impose on the free layout of the DNA. It is far easier and takes less scanning time to see unencumbered DNA at low concentration, than DNA at higher concentration crowded by free RNAs, which also reduce the contrast and confuse the identification of heteroduplexes. The length : nucleotide ratio of ds and ssDNA should be within 5–10% of one another, especially when using the RNA–ssDNA heteroduplex method. A bigger difference is symptomatic of anomalous spreading, and measurements may not be accurate. Hybridization of more than one R-loop per DNA molecule can assist in determining orientation and map coordinates relative to one another, but too many RNAs per DNA molecule may reduce the information obtainable (see Interpretation of Structures). If R-loops are not reasonably reproducible, consider the possibility that the RNA is partially degraded; it may be necessary to prepare a new RNA sample. Because

[28] R. W. Davis and R. W. Hyman, *J. Mol. Biol.* **62,** 287 (1971).

the reaction is always carried out at DNA molar excess, most DNA molecules will not have RNA hybridized. Not all R-loops or R–D heteroduplexes will be well laid out. Considerable patience while examining a grid is mandatory.

Photography and Tracing

Electron micrographs should be exposed at a magnification of ×8000 to ×10,000. Molecules should be traced with an electronic planimeter or digitizer directly from a projected image of the negative, for example, using a photographic enlarger. To achieve high accuracy, images should be enlarged so that 120 nt measure about 3 mm (i.e., ΦX174 DNA measures about 13 cm). Since there is substantial directional stretching of photographic paper during processing and drying, tracing should not be performed with prints.

With R-loop molecules, all relevant portions of the structure are duplexed. Segments are most accurately represented as percentages of the entire molecule which, in turn, is calibrated against other length standards in the same micrograph to be sure of its integrity. RNA–DNA is assumed to have the same length : base ratio as DNA–DNA duplex within ±4%, without introducing a systematic error larger than other measuring and interpretive uncertainties. With RNA–ssDNA heteroduplexes, the single-stranded and double-stranded segments can be evaluated with respect to single-stranded and double-stranded length standards, respectively, that are present in the same micrograph. Generally, six or more standard molecules of each type should be traced from each micrograph and averaged.

Interpretation of Structures

R-Loops

RNA annealed to dsDNA can create many types of R-loops, depending on the structure of the RNA and the number of RNA molecules annealed side by side. Figures 1 through 4 illustrate structures one might encounter.

1. RNA transcripts that are unspliced will create simple R-loops in which a single RNA is completely paired with the DNA, and the homologous DNA sequences are displaced as a single-stranded loop (Figs. 1a and 2a). If branch migration occurs, sequences at one or both ends of the RNA may be displaced to variable extents by partial reannealing of the homologous DNA strand, and RNA branches at the forks of the loop may be visible. However, the branches at the ends of R-loops may have other

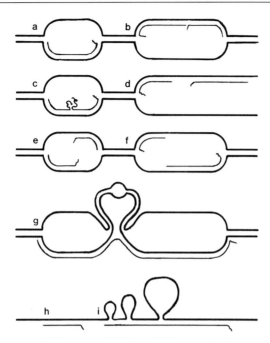

FIG. 1. Structures of R-loops and RNA–ssDNA heteroduplexes. (a) A R-loop of a collinear RNA. (b) An expanded R-loop formed with two adjacent and collinear mRNAs. (c) A R-loop formed between two identical or overlapping mRNAs and one DNA template. (d) An expanded R-loop formed between two mRNAs that are close, but not directly adjacent, to each other. The single-stranded region between the mRNAs fails to renature when glyoxal fixation is carried out swiftly. This situation can also happen to the end of DNA, when the T_m^* of the segment is lower than the incubation temperature, due to its short length or low GC content, as illustrated at the right end of the expanded R-loop. (e) Convergent R-loops. (f) Divergent R-loops. (g) A R-loop formed with a spliced mRNA. The DNA deletion loop can contain partially unpaired regions (underwound) as shown here. When one of the exons is very short, the exon may not be visible, but the DNA loop will remain, indicative of the presence of such a short exon. (h) A RNA–ssDNA heteroduplex of a collinear mRNA. (i) A RNA–ssDNA heteroduplex of a spliced mRNA with three introns. Spliced mRNA can also form all the structures in b–f. The poly(A) tail, represented by the short, unhybridized end, reveals the orientation of each mRNA on the DNA template.

origins: at the 3′ ends, poly(A) tracts will appear as short branches (see Figs. 1–4). If the RNA contains sequences that are derived from remote coding segments of the genome, but fail to hybridize, branches may also be seen at the 5′ end of the R-loop.[25]

2. Two identical RNAs, or pieces of identical RNAs, can coanneal, creating a full-size R-loop for that species, with branches (bushes) of unpaired RNA at a random position in which the two molecules meet and

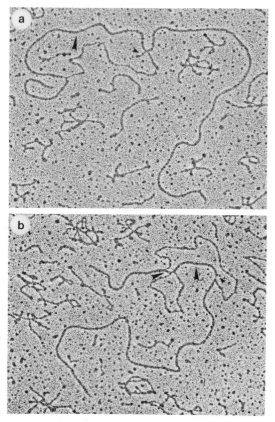

FIG. 2. (a) An electron micrograph of a R-loop formed between cloned human papillo-mavirus (HPV) type 11 DNA and a collinear mRNA isolated from a genital wart. (b) An electron micrograph of a spliced HPV-11 mRNA hybridized to a cloned viral DNA using the RNA–ssDNA method at a temperature above T_{ss}^* of the viral DNA, but below that of the vector. The intron remains single stranded. The 5′ exon is about 120 nt long, and the second exon is about 1060 nt long. The cloned DNA is linearized in the vector close to the cloning site. The asymmetric disposition of the viral sequence in the linear DNA allows unequivocal orientation of the mRNA in the viral genome. The collapsed RNA bushes in the background are mostly rRNA; the large sizes of rRNA indicate that the RNA preparation experienced little degradation. Large arrowheads point to the RNA–DNA strands, which are thicker and have smoother contours than the displaced ssDNA. A small arrowhead points to the 3′ poly(A) tail.

overlap (Fig. 1c). A similar structure can result if two different transcripts share common sequences.

3. Two different, but adjacent, transcripts from the same strand can coanneal and form an expanded R-loop (Fig. 1b); expanded loops can be recognized because the individual loops will be smaller and will have new,

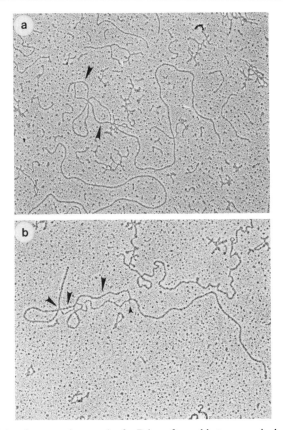

FIG. 3. (a) An electron micrograph of a R-loop formed between a singly spliced HPV-6 RNA from a genital wart and a linearized cloned DNA in which both exons are long and visible. (b) An electron micrograph of a R-loop of a twice-spliced HPV-1 RNA from a plantar wart hybridized to a linearized cloned viral DNA. The short 5' R-loop collapsed and is invisible. Nonetheless, the dsDNA loop, corresponding to the intron, remained in place. The stem regions of the second DNA deletion loop are denatured. A small arrowhead points to the 3' poly(A) tail. Large arrowheads point to the RNA–DNA strands.

reproducible termini, when the hybridization is repeated at a lower RNA concentration.

4. Two convergent or divergent transcripts complementary to opposite strands can coanneal, forming a R-loop in which only part of one DNA strand of the loop is paired with one RNA, and the remaining part of the other strand forming the loop is paired with the second RNA (Fig. 1e and f).[8,25] The single-strand/double-strand junctions on the two arms of the loop indicate the strand switch point. The position of the poly(A) tail

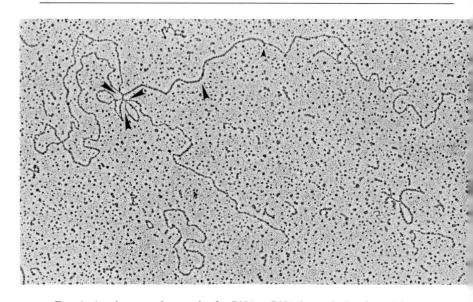

FIG. 4. An electron micrograph of a RNA–ssDNA heteroduplex formed between an adenovirus type 2 late mRNA and a viral ssDNA. The mRNA consists of a tripartite leader and a main coding exon, causing the DNA to form three loops, corresponding to the three introns. The RNA–DNA heteroduplex of the first exon (41 nt long by cDNA sequencing) cannot be distinguished from the flanking ssDNA. The existence of this short exon is inferred by the formation of the DNA loop, which is not present in the DNA strand when there is no hybridizing mRNA. The second exon (72 nt) and third exon (90 nt) are clearly demarcated by the DNA loops. Large arrowheads point to the RNA–DNA strands. A small arrowhead points to the 3' poly(A) tail.

reveals the orientation of transcription. In some genetic regions, opposite strand transcripts may actually overlap.

5. A RNA annealed to a DNA molecule, truncated by a restriction enzyme, will extend past the restriction site and create a R-fork rather than a loop closed by DNA–DNA duplexes at both ends.[25]

6. If the R-loop hybridization reaction is quenched or fixed quickly, occasionally the reassociation of denatured, complementary DNA segments that potentially should occur does not for kinetic reasons. This failure of DNA reassociation could happen in short internal segments flanked by RNA–DNA heteroduplex regions (Fig. 1d). Such expanded R-loops should be evident by careful examination for single-strand/double-strand junctions in the heteroduplex. Failure of DNA–DNA renaturation at the ends of the molecule may result in single-stranded forks (Fig. 1d) (L. T. Chow and T. R. Broker, unpublished results).

7. If the RNA is spliced, a composite R-loop can form, with one or more segments of dsDNA equivalent to the intervening sequences constrained into DNA deletion loops between the R-loops (Figs. 1g and 3a). When one segment of a spliced RNA, such as a 5' leader, is very short, it may pair without creating a visible R-loop. But the intervening dsDNA loop will still be evident (Fig. 3b). The dsDNA loop may exhibit superhelical twists, because it is a topologically closed circle that may change helical pitch when the solvent is altered between the hybridization and spreading steps. Depending on the temperature and the speed of R-loop stabilization, such DNA deletion loops may remain partially single stranded (Figs. 1g and 3b).

RNA–Single-Stranded DNA Heteroduplexes

The sample preparation should be of sufficient quality that the double-strand/single-strand junctions at the boundaries of the hybridized RNA are discernible. A poly(A) tail is often long enough to appear as a branch at the 3' end of a RNA–DNA heteroduplex. In addition, the pattern of RNA splicing can suggest polarity: very short RNA segments tend to be at the 5' end of the RNA, and longer exons tend to be toward the 3' end. The following types of structures can be observed.

1. A fully collinear transcript will make a simple RNA–DNA heteroduplex flanked by ssDNA (Fig. 1h).
2. Spliced RNA–DNA heteroduplexes contain ssDNA deletion loops corresponding to the intervening sequences (Figs. 1i and 4). The lengths of internal exon segments can be accurately estimated because the segments are clearly demarcated by flanking deletion loops; the short terminal RNA–DNA duplex (usually at the 5' end) may not be measurable, although its presence is evidenced by the DNA deletion loop (Fig. 4). If, however, the terminal exon is very short and shear forces are strong, it may not hybridize or may be pulled apart, but remains close to the homologous DNA sequence on the grid in some or all of the RNA–DNA heteroduplexes. This observation also applies to R-loops. As described earlier, R–D heteroduplexes can also be prepared in a mostly duplexed molecule, when a more GC-rich region of the DNA is used as a clamp (Fig. 2b).

Some of the same "artifacts," as described previously for R-loops, can also be found when the RNA concentration is too high. It is therefore advisable to perform the annealing at different RNA concentrations: at low RNA concentration, individual RNA species can be distinguished; at somewhat higher RNA concentration, RNAs originating from the same template strand can be identified, provided that the RNAs do not overlap.

Data Analysis

Statistically, measurements of about 20 examples of a homogeneous RNA species are sufficient to achieve reasonable accuracy. In practice, more molecules are usually measured to ascertain whether there are multiple overlapping species. Obviously, truncated molecules must be eliminated from statistical analysis. Other broken molecules are recognized only after initial analysis and should be also excluded from final compilation. Accuracies in mapping RNA termini and splice sites have been within 20–100 nt, as confirmed when cDNA sequences have become available.

Apparent accuracy can be compromised by systematic errors that affect each individual measurement in the same way. For instance, short gaps between RNA–DNA duplex segments tend to stretch slightly, because of the concentration of shear forces in the intervening single-stranded segment. Very short deletion loops tend to collapse; as a result, small deletion loops and ends of ssDNA seem to measure about 50–80 nt shorter (less than 2 mm at the recommended magnification for image tracing). Deletion loops shorter than 100–150 nt are not reproducibly detected by electron microscopy. However, the existence of such a short intron can be recognized because the deletion loop frequently causes a sharp bend in the DNA contour at the location of the intron. Measurements of R-loops are subject to errors caused by branch migration due to DNA–DNA renaturation, which displaces the ends of the RNA in the R-loop, masking heterogeneity in the termini. This effect can be minimized by rapid fixation with glyoxal or can be circumvented by using the R–D heteroduplex method.

Errors in map coordinates increase slightly when multiple segments are measured to arrive at the desired coordinate; such a situation occurs with the series of deletion loops and duplexes formed in spliced RNA–DNA heteroduplexes. Errors also increase in proportion to the length of the measurement, especially with ssDNA.[3] Accumulated errors are reflected in larger standard deviations of measurements of RNA coordinates near the center of a large chromosome than near its ends. To some extent, both types of errors can be reduced by making measurements to a closer reference point, for instance, to a second RNA–DNA heteroduplex, to a secondary structure in ssDNA or to a restriction site in a truncated DNA molecule (which is, of course, only as accurate as the restriction map).[25]

The complex sensitivities of hybridization rates to temperature, GC composition of the different regions of the DNA, and excluded volume effects, when dealing with a long chromosome, can affect accuracy in assessing the relative abundances of different RNA species. The R–D heteroduplex method is more reliable than the R-loop method, in which hybridization of GC-rich transcripts is at a considerable disadvantage. Nonetheless, it is possible to use the R-loop method to compare the relative abundances of several RNA species present in different RNA preparations (represented as a percentage of the total hybridized RNA). For instance, RNAs may be prepared at different times after infection, with different mutants or in the presence of different metabolic inhibitors. In these cases, base composition and position effect cancel out, and the only variables, beside actual differences, are the reproducibility of temperature and solvent.

Conclusions

Electron microscopy of RNA–DNA heteroduplexes has been a major technique for the structural analysis of eukaryotic RNA transcripts. The methodology is based on the observation that RNA–DNA heteroduplex stability exceeds that of DNA–DNA duplexes in the presence of high concentrations of formamide. The total characterization of a transcription system is most efficient when both the R-loop method and the RNA–ssDNA method are used whenever practicable. Mapping accuracy is usually within 50–100 nt, which is generally sufficient to guide correlations with genomic DNA sequences.[11–29] Exons as short as 41 nt can easily be recognized, since they cause diagnostic DNA deletion loops. In contrast to methods involving bulk RNA analysis, multiple-spliced RNAs and families of related RNAs can be deciphered readily with EM heteroduplex analysis, and rare messenger RNA species can be mapped, with effort, in a background of 20,000 mRNAs. Very small amounts of RNA and DNA are needed for a complete project. Furthermore, the RNA need not be pure; heterologous RNA will not hybridize to the DNA and simply appears in the background. Neither the RNA nor the DNA need be radioactively labeled, and mapping can be done without detailed restriction endonuclease cleavage maps of the DNA. Together with alternative, biochemical methods described in this volume, EM heteroduplex mapping can generate comprehensive structural descriptions of genes and their transcripts.

Acknowledgments

    The authors are supported by research grant CA36200 from the United States Public Health Service. We thank Shirley Thomas for assistance in the preparation of the manuscript.

[29] M. O. Rotenberg, L. T. Chow, and T. R. Broker, *Virology*, in press, (1989).

[20] Computer Prediction of RNA Structure

By MICHAEL ZUKER*

Introduction

The biological activity of a RNA molecule is determined by its three-dimensional conformation. This structure is achieved by the molecule bending back on itself and forming helical regions stabilized by hydrogen bonds between complementary bases. Base pairing can be of three types: G with C, A with U, and the weaker G with U. This chapter does not deal with three-dimensional structure or with the complications introduced by interactions with other biomolecules. Instead, this chapter deals with computer methods to predict which hydrogen bonds will occur. It is the collection of hydrogen bonds that comprise the secondary structure, or folding, of the RNA molecule.

RNA structure is essential to the functioning of transfer RNA (tRNA) and to the assembly and functioning of ribosomal RNA (rRNA). The structure of yeast tRNA[Phe] has been determined by crystallographic means,[1] and several hundred other tRNAs, which have been sequenced, possess the same cloverleaf folding potential. Common foldings have been determined for many 5 S RNA molecules, and closely related structures have been computed for the small and large subunits of rRNA. In both cases, the models are well supported by phylogeny. Much of the interest in RNA secondary structure prediction comes from the desire to fold messenger RNA (mRNA). The structure of mRNA controls translation and splicing of introns in eukaryotes and transcriptional regulation in bacterial systems. Thus, mRNA structure controls whether some proteins are expressed and the level of expression as well. It has been proposed that mRNA structure might even affect the structure of the expressed protein.

The usual criterion for computing a RNA secondary structure is to minimize the free energy of the folded molecule. This model is a great simplification of reality. Three-dimensional effects are ignored, and the energy rules used to assign free energies to structures are derived from melting data on small oligonucleotides. In addition, the model itself, as a mathematical entity, has a disturbing property. In general, many different foldings are possible close to the minimum energy. If energy minimization

* National Research Council of Canada.
[1] S. H. Kim, F. L. Suddath, G. J. Quigley, A. McPherson, J. L. Sussman, A. H. J. Wang, N. C. Seeman, and A. Rich, *Science* **185**, 435 (1974).

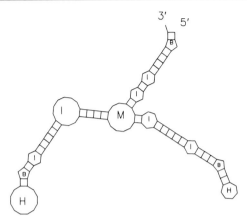

FIG. 1. Minimum energy folding of a 126-base sample RNA. The minimum free energy using Turner's rules is −35.5 kcal/mol. All the different kinds of loops occur and are designated by letters: B, Bulge loop; I, interior loop; H, hairpin loop; M, multiloop or multibranched loop. The unmarked areas are stacking regions between consecutive hydrogen bonds.

is the only folding criterion, then one must accept multiple solutions and find ways to compute them. Methods to determine the reliability of foldings become important. If additional information is available, this must be incorporated into the folding algorithm to narrow the range of solutions.

Definitions and Nomenclature

The ribonucleotides of a RNA molecule of length N will be denoted by $r_1, r_2, \ldots, r_N$. A base pair between r_i and r_j will be written as $r_i \cdot r_j$. The topology of RNA secondary structure has been well discussed.[2-4] A region of consecutive hydrogen bonds is called a helix or a stack. Single-stranded regions of the molecule comprise various types of loops, as shown in Fig. 1. A collection of helices interrupted only by bulge or interior loops is called a stem. A stem which ends in a hairpin loop is called a hairpin.

Only a few rules are imposed. The molecule cannot bend back too sharply on itself, so that at least three bases are required in hairpin loops.

[2] G. M. Studnicka, G. M. Rahn, I. W. Cummings, and W. A. Salser, *Nucleic Acids Res.* **5**(9), 3365 (1978).

[3] D. Sankoff, J. B. Kruskal, S. Mainville, and R. J. Cedergren, *in* "Time Warps, String Edits, and Macromolecules: The Theory and Practice of Sequence Comparison" (D. Sankoff and J. B. Kruskal, eds.), p. 93. Addison-Wesley, Reading, Massachusetts, 1983.

[4] M. Zuker and D. Sankoff, *Bull. Math. Biol.* **46**(4), 591 (1984).

A base can pair with at most one other base. Two base pairs $r_i \cdot r_j$ and $r_{i'} \cdot r_{j'}$ which satisfy

$$i < i' < j < j'$$

are said to form a pseudoknot. Although structures with pseudoknots have been shown to occur,[5] are one of two universal foldings for 5 S RNA molecules,[6] and have been invoked in intron splicing models,[7] structures with pseudoknots have been excluded by all RNA folding algorithms. Some algorithms would break down altogether if pseudoknots were permitted. Other algorithms simply become more complicated and slower. Another reason for their exclusion is that the pseudoknots would create complicated loops for which energy assignment would be very difficult. Complicated rules would have to be developed to predict which pseudoknots were possible and which were not.

Energy Rules

RNA foldings are usually computed to minimize free energy. At a finite temperature, the molecule is in some sort of equilibrium, and a minimum energy solution must be regarded as a most probable folding. Energy is assigned not simply to hydrogen bonds, but to the stacking of one hydrogen bond over another. Various loops are given destabilizing energies. A principle of additivity is assumed, so that the overall free energy of a folding is the sum of the energies of the stacked base pairs and the loops. This principle is certainly not correct, but the additivity assumption is essential to some prediction algorithms, and better experimental data are not yet available to produce more sophisticated rules.

Free-energy data come from melting studies performed on small oligonucleotides.[8–14] This information was summarized by Salser.[15] Standard base pairs of the form G–C and A–U are allowed. In addition, the non-

[5] J. D. Puglisi, J. R. Wyatt, and I. Tinoco, Jr., *Nature (London)* **331**, 283 (1988).

[6] E. N. Trifonov and G. Bolshoi, *J. Mol. Biol.* **169**, 1 (1983).

[7] R. W. Davies, R. B. Waring, J. A. Ray, T. A. Brown, and C. Scazzocchio, *Nature (London)* **300**, 719 (1982).

[8] I. Tinoco, Jr., O. C. Uhlenbeck, and M. D. Levine, *Nature (London)* **230**, 362 (1971).

[9] T. R. Fink and D. M. Crothers, *J. Mol. Biol.* **66**, 1 (1972).

[10] O. C. Uhlenbeck, P. N. Borer, B. Dengler, and I. Tinoco, Jr., *J. Mol. Biol.* **73**, 483 (1973).

[11] J. Gralla and D. M. Crothers, *J. Mol. Biol.* **73**, 497 (1973).

[12] J. Gralla and D. M. Crothers, *J. Mol. Biol.* **78**, 301 (1973).

[13] I. Tinoco, Jr., P. N. Borer, B. Dengler, M. D. Levine, O. C. Uhlenbeck, D. M. Crothers, and J. Gralla, *Nature (London) New Biol.* **246**, 40 (1973).

[14] P. N. Borer, B. Dengler, I. Tinoco, Jr., and O. C. Uhlenbeck, *J. Mol. Biol.* **86**, 843 (1974).

[15] W. Salser, *Cold Spring Harbor Symp. Quant. Biol.* **42**, 985 (1977).

```
BASE PAIRING ENERGIES IN TENTHS OF A KCAL/MOL
SALSER'S DATA
STACKING ENERGIES (UG = GU)
       |  GU  |  AU  |  UA  |  CG  |  GC  |
  GU     -3     -3     -3    -13    -13
  AU     -3    -12    -18    -21    -21
  UA     -3    -18    -12    -21    -21
  CG    -13    -21    -21    -48    -43
  GC    -13    -21    -21    -30    -48
BULGE LOOP DESTABILIZING ENERGIES BY SIZE OF LOOP
            1 |  2 |  3 |  4 |  5 |  6 |  7 |  8 |  9 | 10| 12| 14| 16| 18| 20| 25| 30|
           28   39   45   50   52   53   55   56   57   58   59   61   62   63   64   65   67
HAIRPIN LOOP DESTABILIZING ENERGIES BY SIZE OF LOOP
            1 |  2 |  3 |  4 |  5 |  6 |  7 |  8 |  9 | 10| 12| 14| 16| 18| 20| 25| 30|
CG CLOSING 999  999   84   59   41   43   45   46   48   49   50   52   53   54   55   57   59
AU CLOSING 999  999   80   75   69   64   66   68   69   70   71   73   74   75   76   77   79
INTERIOR LOOP DESTABILIZING ENERGIES BY SIZE OF LOOP
CLOSED BY   1 |  2 |  3 |  4 |  5 |  6 |  7 |  8 |  9 | 10| 12| 14| 16| 18| 20| 25| 30|
  CG-CG    999    1    9   16   21   25   26   27   28   29   31   32   33   34   35   37   39
  CG-AU    999   10   18   25   30   34   35   36   37   38   39   40   41   42   43   45   47
  AU-AU    999   18   26   33   38   42   43   44   45   46   48   49   50   51   52   54   56
```

FIG. 2. Reproduction of the Salser energy input file for the author's RNAFOLD program discussed in the text. Energies are in tenths of a kilocalorie per mole so that integer arithmetic can be used. For the stacking energies, the column base pairs are 5′–3′, while the row base pairs are 3′–5′. The "999" energies are simply large numbers used to prevent hairpin loops which are too small, or impossible interior loops.

standard G–U base pair is allowed, although the nonstandard G–U base pair has sometimes not been allowed to occur at either end of a helix.[2,16] Loop destabilizing energies depend on the size (number of single-stranded bases) of the loop and on the nature of the base pairs which close them. There are no measured energies for multiloops. Multiloops can be treated as interior loops of the same size[16] or in a slightly more complicated way.[17] These so-called Salser's rules are given in Fig. 2. The folding temperature implicit in these rules is 25°.

Various alternatives and modifications to the above rules have appeared. Tinoco altered the rules slightly in 1982,[18] eliminating the dependence on closing base pairs in hairpin and interior loop destabilizing energies. Ninio adjusted the energy rules so that the phylogenetically determined structure would also be a minimum energy structure as frequently as possible. This was done first for tRNAs[19] and later extended to 5 S RNAs.[17] The set of nonstandard pairs is expanded to cover G–G,

[16] M. Zuker and P. Stiegler, *Nucleic Acids Res.* **9**(1), 133 (1981).

[17] C. Papanicolaou, M. Gouy, and J. Ninio, *Nucleic Acids Res.* **12**(1), 31 (1984).

[18] T. R. Cech, N. K. Tanner, I. Tinoco, Jr., B. R. Weir, M. Zuker, and P. S. Perlman, *Proc. Natl. Acad. Sci. U.S.A.* **80**, 3903 (1983).

[19] J. Ninio, *Biochimie* **61**, 1133 (1979).

```
Base pairing energies in tenths of a kcal/mol
Ninio/Turner loops + Soo NN at 37°
Stacking Energies
      |  GU  |  AU  |  UA  |  CG  |  GC  |
 GU     -5     -5     -7    -15    -13
 AU     -5     -9    -11    -18    -23
 UA     -7     -9     -9    -17    -21
 CG    -19    -21    -23    -29    -34
 GC    -15    -17    -18    -20    -29
Bulge loop destabilizing energies by size of loop
             1 | 2 | 3 | 4 | 5 | 6 | 7 | 8 | 9 | 10| 12| 14| 16| 18| 20| 25| 30|
            32  52  60  67  74  82  91 100 105 110 118 125 130 136 140 150 158
Hairpin loop destabilizing energies by size of loop
             1 | 2 | 3 | 4 | 5 | 6 | 7 | 8 | 9 | 10| 12| 14| 16| 18| 20| 25| 30|
CG CLOSING  999 999  74  59  44  43  41  41  42  43  49  56  61  67  71  81  89
AU CLOSING  999 999  74  59  44  43  41  41  42  43  49  56  61  67  71  81  89
Interior loop destabilizing energies by size of loop
closed by    1 | 2 | 3 | 4 | 5 | 6 | 7 | 8 | 9 | 10| 12| 14| 16| 18| 20| 25| 30|
   CG-CG    999   8  13  17  21  25  26  28  31  36  44  51  56  62  66  76  84
   CG-AU    999   8  13  17  21  25  26  28  31  36  44  51  56  62  66  76  84
   AU-AU    999   8  13  17  21  25  26  28  31  36  44  51  56  62  66  76  84
```

FIG. 3. Turner's energy rules at 37° in the same format as Fig. 2. A computer program exists for creating these energy files for folding at arbitrary temperatures.

U–U, C–C, C–A, A–A, A–G, and U–C base pairs. A variety of special rules are introduced, which make the overall energy assignment nonadditive. In addition, these Ninio rules distinguish between (for example)

$$5'\text{-UGC-}3' \quad \text{and} \quad 5'\text{-UUC-}3'$$
$$3'\text{-AUG-}5' \qquad\qquad 3'\text{-AGG-}5'$$

when considering stacking involving nonstandard base pairs, while the Salser rules do not.

More recently, new experimental data on RNA duplex stability, made possible by breakthroughs in oligoribonucleotide synthesis, have resulted in energy rules which supercede Salser's rules.[20] Fig. 3 contains some of these new data. Note that 37° is the folding temperature. The only nonstandard base pair is G–U, but energy does depend on whether G or U is the 5' base in a stack. As with the Ninio rules, single-stranded, terminal nucleotides, or dangling ends, are given free-energy increments. A new rule adjusts the energies of hairpin and interior loops depending on the nature of the terminal mismatched pair(s). These rules will be referred to as Turner's rules. Work continues to refine them.

Thermodynamic calculations allow one to alter energy rules for folding at different temperatures. This was done by Steger *et al.*,[21] starting

[20] S. M. Freier, R. Kierzek, J. A. Jaeger, N. Sugimoto, M. H. Caruthers, T. Neilson, and D. H. Turner, *Proc. Natl. Acad. Sci. U.S.A.* **83,** 9373 (1986).

[21] G. Steger, H. Hofmann, J. Förtsch, H. J. Gross, J. W. Randles, H. L. Sänger, and D. Riesner, *J. Biomol. Struct. Dyn.* **2**(3), 543 (1984).

TABLE I

ADJUSTABLE PARAMETERS OF SEQL PROGRAM: IDEAS SEQUENCE ANALYSIS SYSTEM
FOR RNA LOCAL SECONDARY STRUCTURE

SEQL parameters	Meaning (default value)
GMAX	Hairpins with lower free energies are printed (−10.0)
MODE	0, Do not print sequence; 1, print sequence; 2, print and annotate sequence (0)
LWID	Linewidth for terminal or output file (80)
LHMAX	Maximum size of a hairpin loop (20)
LIMAX	Maximum size of an interior loop (10)
LBMAX	Maximum size of a bulge loop (5)
LEN	Maximum size of any hairpin structure (100)

from most of the original data used by Salser. Because of differences in theoretical treatments, the Steger energy rules do not agree with Salser's rules even at 25°. A computer program written by Jaeger[22] adjusts the Turner rules for folding at arbitrary temperatures. Folding algorithms usually read in stacking and loop energies from external files. These files are easily changed for folding at alternate temperatures.

Types of Folding Programs

Four types of RNA folding programs are discussed. No single method is sufficiently precise to yield a definitive answer, and so different approaches are necessary to complement one another.

The first type of folding program will be called basic. This includes methods that predict hairpin structures or simply helices. Energy considerations may or may not enter. Programs which compute hairpins fall short of full secondary structure prediction by excluding the prediction of multibranched loops and by limiting the distance along the sequence between base pairs. These programs are good for finding local folding motifs. A good example of such a program is SEQL,[23] now part of the IDEAS analysis package of the Laboratory of Mathematical Biology at the National Cancer Institute of the National Institutes of Health (Bethesda, MD). This FORTRAN program has been written for VAX/VMS. The user can adjust seven parameters, which are explained in Table I. A sample output is shown in Fig. 4. The output includes hairpins, which may be incompatible with one another in a single global folding either

---

[22] J. A. Jaeger, personal communication (1987).
[23] M. I. Kanehisa and W. B. Goad, *Nucleic Acids Res.* **10**(1), 265 (1982).

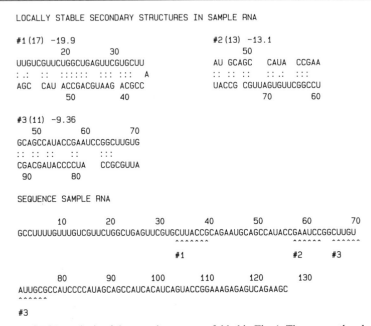

FIG. 4. SEQL analysis of the sample sequence folded in Fig. 1. The energy threshold for display is −5.0 kcal/mol. LHMAX, LIMAX, LBMAX, and LEN are 10, 8, 3, and 50, respectively. The annotated sequence shows the hairpin loops (including the closing base pairs) that were found.

because they involve hydrogen bonding of the same bases or because pseudoknots are created. The energies are according to Salser.[15]

Another kind of secondary structure program can be called combinatorial. A list of all possible helices is generated. These helices are then pieced together in all possible ways to form secondary structures. The best such program in use today was developed by Ninio, Dumas, Papanicolaou, and Gouy.[17,19,24,25] Called CRUSOE, it is written in FORTRAN 77 and is available from Gouy.[26] The computer memory required by the program is proportional to the number of helices used, and this grows as the square of the sequence size. It is computation time, not memory, which limits this algorithm. Although the program uses a clever tree search procedure to limit the amount of searching, computation time increases exponentially with sequence size, and the practical limit for a thorough

[24] J.-P. Dumas and J. Ninio, *Nucleic Acids Res.* **10**(1), 197 (1982).
[25] M. Gouy, P. Marliere, C. Papanicolaou, and J. Ninio, *Biochimie* **67**, 523 (1985).
[26] M. Gouy, in "Nucleic Acid and Protein Sequence Analysis: A Practical Approach" (M. J. Bishop and C. J. Rawlings, eds.), p. 259. IRL Press, Washington, D.C., 1987.

analysis is about 150 nucleotides. Larger molecules can be folded by increasing the minimum helix size from two and by eliminating the nonstandard base pairs. Both of these options are available. Pseudoknots are not allowed. This program has several advantages. Strict additivity of locally assigned energy is not necessary. Multiple foldings are easily predicted.

Recursive programs build optimal foldings one base at a time. They comprise two distinct parts. The first part, called the *fill,* computes and stores minimum folding energies for all fragments based on minimum folding energies of smaller fragments, starting with pentanucleotides. The final part, called the *traceback,* computes a structure by searching through the matrix of folding energies. The fill algorithm requires the bulk of computing time, while the time for a single traceback is negligible. If multiloops are treated in a simple manner, such algorithms can be much faster than the combinatorial type, executing in time proportional to the cube of sequence length. Computer storage requirements grow as the square of the sequence length and it is usually memory, not time requirements, which limit this kind of algorithm. Pseudoknots cannot be handled, and the additivity assumption for free-energy contributions is necessary. A recursive algorithm was first used by Nussinov *et al.*[27] to maximize base pairing and later extended to free-energy minimization.[28] Zuker and Stiegler developed a recursive algorithm independently, emphasizing the need for imposing constraints.[16] The algorithm was reprogrammed in 1983 by the author and has been adapted for a variety of different computers. Recursive algorithms of this kind are also called dynamic programming algorithms. By their nature, recursive algorithms predict the optimal folding of all subfragments. This means, for example, that optimal foldings of a growing RNA sequence can be simulated without any additional cost, once the entire sequence is folded. Recursive algorithms are designed to yield a single solution, but with some effort, the traceback algorithm can be extended to yield multiple solutions. This has been achieved by Williams and Tinoco.[29]

Another kind of folding algorithm can be called dynamic, because this algorithm simulates the folding of a RNA molecule in time. A good example of such an algorithm is the MONTECARLO program of Martinez, written in C.[30,31] The program first compiles a list of stems that can form.

[27] R. Nussinov, G. Pieczenik, J. R. Griggs, and D. J. Kleitman, *SIAM J. Appl. Math.* **35,** 68 (1978).
[28] R. Nussinov and A. B. Jacobson, *Proc. Natl. Acad. Sci. U.S.A.* **77,** 6309 (1980).
[29] A. L. Williams, Jr. and I. Tinoco, Jr., *Nucleic Acids Res.* **14**(1), 299 (1986).
[30] H. M. Martinez, *Nucleic Acids Res.* **12**(1), 323 (1984).
[31] H. M. Martinez, *Nucleic Acids Res.* **16**(5), 1789 (1988).

These stems may contain small bulge or interior loops. The free energies of these stems are then computed, and the stems are then given weights, or Boltzmann probabilities, in proportion to $\exp(-\Delta G/RT)$, where ΔG is the stem energy, R is the gas constant per mole, and T is the temperature in Kelvin. One stem is then chosen at random, using the assigned probabilities. After $(i - 1)$ stems have been chosen, all stems in the list incompatible with the ones already chosen are removed, and the ith stem is added at random. The process terminates when no additional stems can be added. Pseudoknots are not permitted, although pseudoknots could easily be allowed. The only problems are how to assign energies and how to draw the resulting structures. The algorithm requires a modest amount of storage, similar to combinatorial algorithms, and is also very fast. Large RNA molecules can be folded quickly, but there is no guarantee of achieving a minimum energy folding. The folding procedure often terminates far from the global energy minimum as computed by a recursive algorithm. The strategy in MONTECARLO is to refold the molecule over and over again and to compute statistics on which stems tend to occur repeatedly. Thus, multiple foldings are very much part of the algorithm.

Zuker–Stiegler Algorithm and Associated Programs

The original algorithm, described by Zuker and Stiegler,[16] was written in FORTRAN 66 and ran in batch mode on an IBM 3032 processor under an earlier operating system (TSS), which is no longer in use. Several versions of the program were created for special purposes: one version forced known or desired base pairs to occur, while another version did not allow multibranched loops. These were all combined into a single, new program in the spring of 1983, which was designed to be interactive and to incorporate all the special features of the earlier programs. The programming language and computer system were the same.

This "generic" folding program will be called RNAFOLD for the purposes of this chapter. This program was adapted to the CRAY supercomputer by Michael Ess, a programmer at Cray Research, Inc. (Mendota Heights, MN). This program resides in the CRAY program library as RNAFOLD, version OCT83, but will be called CRAYFOLD in this chapter. The author adapted the program for use on the BIONET[32] DEC 2060 computer, using the TOPS 20 operating system, in which the program is called BIOFLD. When the IBM AT and associated clones appeared, the program was adapted for these microcomputers. This version, known as

[32] D. Roode, R. Liebschutz, S. Maulik, T. Friedemann, D. Benton, and D. Kristofferson, *Nucleic Acids Res.* **16**(5), 1857 (1988).

PCFOLD, also runs on the XT microcomputer and is completely menu-driven, with help files available on-line. A version was included in the University of Wisconsin Genetics Computer Group (UWGCG) (Madison, WI) package,[33] but this version would fold only a given fragment without allowing constraints or the repeat folding of subfragments. Called FOLD internally, this implementation will be called GCGFOLD in this chapter. Version 5 of the UWGCG package allows more options in FOLD. Other adaptations to the Cyber, to Unix machines, and to IBM VM/CMS have appeared. The author's personal versions now reside on a VAX 11/750 running VMS. The program closest to the original is called FOLD_VAX. The most up-to-date version is called FOLD_VAX_WISC, because this version is the basis of FOLD in version 5 of the UWGCG package. Both of these are distributed by the author along with notes, energy files, and sample runs.

Although computation time grows as the cube of sequence length, the real limits to RNAFOLD are space requirements. The program requires $3N^2/2$ half-integers (16-bit integers) plus assorted other fixed and variable arrays, which grow linearly with N. Many computers have a virtual memory capacity, which makes the working memory larger by using disk space temporarily. BIOFLD is limited to folding 280 bases, because half-integers are not available in FORTRAN on the DEC 2060, and there is no virtual memory. PCFOLD is limited to 425 bases. In fact, PCFOLD only does this well, because PCFOLD packs three integers into the space of two, utilizing only N^2 storage. PCFOLD will fail when as few as 14 base pairs are forced, because the extra energies used to force the folding must be stored in limited space. For this reason only, an "unpacked" version of PCFOLD, called PCFOLD2, has been retained. PCFOLD2 will not fail with a modest number of forced base pairs, but PCFOLD2 can fold only 345 bases. CRAYFOLD could handle about 2500 bases on the CRAY XMP. A more recent adaptation on the CRAY 2, using the vector processing capability of the machine, folded the entire human immunodeficiency virus (HIV) (9718 bases) in about 7 hr.[34] The present VAX version can easily fold 2000 bases on an 11/750. The program has been written to minimize the exchange of information between the central processing unit (CPU) and the disk. The "Wisconsin" version uses only as much space as is required for folding the given molecule. The size limit in GCGFOLD has been arbitrarily set at 1200, although the size limit could easily be increased.

The first prompt by the program is for an energy file name. The usual

---

[33] J. Devereux, P. Haeberli, and O. Smithies, *Nucleic Acids Res.* **12**(1), 387 (1984).
[34] J. V. Maizel, Jr., personal communication (1988).

energy file, currently called FOLD_VAX.ENE, uses Salser's energies. Tinoco's modifications[18] to Salser's rules are also available in a file named FOLD_VAX.NEWEN. A blank energy file called FOLD_VAX.NOEN is also distributed, which gives a zero energy to all stackings and loops. This blank energy file allows the user to set constant loop and stack energies within the program. Most recently, a new energy file using Turner's rules has been created. The fact that these energy rules distinguish between such stacks as

$$5'\text{-AGG-}3' \quad \text{and} \quad 5'\text{-AUG-}3'$$
$$3'\text{-UUC-}5' \qquad\qquad 3'\text{-UGC-}5'$$

meant that the program had to be slightly altered. This modification has been done in FOLD_VAX_WISC and in GCGFOLD. The other versions of RNAFOLD remain unchanged. Nevertheless, the new energy file can still be used with the unaltered versions, since the energy discrepancies are slight. Single-base stacking (dangling ends) have not been incorporated into the algorithm.

The next prompt is for a SAVE, CONTINUATION, or regular run. The SAVE option creates a large file containing the energies of the time-consuming *fill* algorithm. This option can be done in batch mode. A subsequent CONTINUATION run allows the user to compute an optimal folding of the chosen fragment or of subfragments. This feature must be chosen as a command line option with GCGFOLD and is not available in PCFOLD. The S or C for SAVE or CONTINUATION must be entered in uppercase, as must all other commands.

The sequence file format has changed over the years. Sequences must be in uppercase, and T is treated the same as U. This could be a problem in using GenBank data, which is in lowercase. The original sequence format originated in the author's group. This format allowed many sequences in a single file. The sequences could only be accessed sequentially. The VAX versions use a subroutine called FORMID, which examines a sequence file and decides which format is being used. The earlier format is allowed, as well as the IntelliGenetics format used on BIONET,[32] the Protein Identification Resource format of the National Biomedical Research Foundation,[35] the GenBank[36] format, and the EMBL[37] format. The program reads and displays all sequence titles in the file. The user may select any sequence by number or name. GCGFOLD requires a single sequence in a file in UWGCG format, as do other pro-

[35] K. E. Sidman, D. G. George, W. C. Barker, and L. T. Hunt, *Nucleic Acids Res.* **16**(5), 1869 (1988).

[36] H. S. Bilofsky and C. Burks, *Nucleic Acids Res.* **16**(5), 1861 (1988).

[37] G. N. Cameron, *Nucleic Acids Res.* **16**(5), 1865 (1988).

grams in this package. PCFOLD will read a file in IntelliGenetics format or in the format used by the Pustell programs licensed by International Biotechnologies, Inc. (IBI, New Haven, CT).[38] Only one sequence per file is allowed.

After a sequence has been selected, the program queries the user to choose between terminal or file output and whether to create a CT (coordinate table) file for later use in plotting. The next step is the selection of a fragment for folding. The sequence itself is numbered from 1 to N. There is no way to have the numbering begin at a negative integer, so that, for example, the -35 and -10 regions of a mRNA leader sequence show up as such. However, the fragment, which is selected, retains the "historical numbering" of the full sequence from which the fragment was selected. At this point, the user may enter the command F to begin folding, T to terminate execution, or a variety of other commands to constrain or otherwise alter the folding. These commands and their syntax will be discussed later. The F (fold) command is always used to begin folding. After the folding, a refolding of the entire fragment or of any subfragment may be obtained by entering the 5′ and 3′ ends in historical numbering. This refolding is very quick. If the number one is added after the 5′ and 3′ subfragment ends, then the refolding forces the ends to pair with one another if possible. This refolding procedure, which can also be used to compute the stability of local regions, is demonstrated in Fig. 5. At this stage, various output parameters can be altered. Program execution may then be terminated, or another fragment may be selected from the same or another sequence.

Output and Display

There are three criteria for judging the output of an RNA folding program. The output should be quickly and easily produced without requiring special devices. The results should be visually appealing. Finally, the display should be designed so that alternative foldings of the same sequence can be visually inspected for similarity. In practice, these criteria often conflict with one another.

All the programs mentioned in this chapter produce a line printer output, which either lists helices or else draws a picture of the secondary structure including all single- and double-stranded regions. This is the "quick and easy" output, which is not suitable for publication quality representations of secondary structures or for comparative studies.

Better quality displays are produced by graphics programs, which read base-pair lists created by folding programs and which output files for

---

[38] J. Pustell and F. C. Kafatos, *Nucleic Acids Res.* **14**(1), 479 (1986).

```
FOLDING BASES    1 TO   126 OF sample RNA
ENERGY  =       -35.5

          10              20        30
 - C     U  G  --      UC      A    G   UU
   GC UUUUG UU UC      GU  UGGCUG GUUC UGC
   CG AAGAC GA AG      CA  ACCGAC UAAG ACG  A
 C  -     U  G  AA     -U      G    -   CC
   120                 50        40


          60              70            80
        CGAA      ----CU       U  -  CAUC
        UCCGG         UGUGAU GC GC     C
        AGGCC         ACACUA CG CG     C
        ----      AUGACU       C  A  AUAC
        110           100          90


ENTER: T TERMINATE,  NS NEW SEQUENCE,  NF NEW FRAGMENT,
O OUTPUT PARAMETER DEFINITION,  OR THE ENDPOINTS OF A
SUBFRAGMENT BETWEEN      1 AND    126.
END WITH 1 TO FORCE ENDS TO BASEPAIR.
14 113 1 ←
FOLDING BASES   14 TO   113 OF sample
ENERGY  =       -28.7

          20          30
 --     UC      A    G   UU
 C     GU  UGGCUG GUUC UGC
 G     CA  ACCGAC UAAG ACG  A
 AA     -U      G    -   CC
        50        40


          60              70          80
        CGAA      ----CU       U  -  CAUC
        UCCGG         UGUGAU GC GC     C
        AGGCC         ACACUA CG CG     C
        ----      AUGACU       C  A  AUAC
        110           100        90
```

FIG. 5. Example of repeat foldings and the line printer output of RNAFOLD. The first structure is the minimum energy folding plotted in Figs. 1, 6, and 7. The second folding of bases 14–113 forces the C-14–G-113 base pair, which would not occur in an optimal folding of this segment. The energy difference between the two foldings gives the stability of the stem from G-1–C-125 to C-14–G-113 (−6.8 kcal/mol). The arrow points to input from the user.

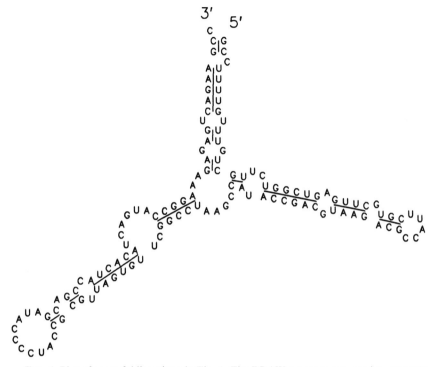

FIG. 6. Plot of same folding given in Fig. 1. The DRAW program was used to generate both. The automatic untangling feature was used, and the base letter display option was chosen.

use on standard plotting devices. One such program, originally developed by Osterburg and Sommer,[39] has been adapted for the UWGCG package and is called SQUIGGLES. Shapiro et al.[40] wrote the DRAW program in the SAIL language for use on the VAX. The DRAW program was soon translated into Pascal. Figures 1 and 6 were drawn using a version of this program adapted to the IBM VM/CMS computing environment. The program is interactive. The user can use the cross-hair feature of Tektronix-type terminals to designate pivot points for rotating portions of the structure to eliminate overlaps. This program produces a very pleasing output, although some effort is usually required. There is an automatic untangling feature, which eliminates most of the need for user intervention, but the resulting output is often not as visually satisfying. However, by placing

[39] G. Osterburg and R. Sommer, *Comput. Programs Biomed.* **13**, 101 (1981).
[40] B. A. Shapiro, J. V. Maizel, Jr., L. E. Lipkin, K. Currey, and C. Whitney, *Nucleic Acids Res.* **12**(1), 75 (1984).

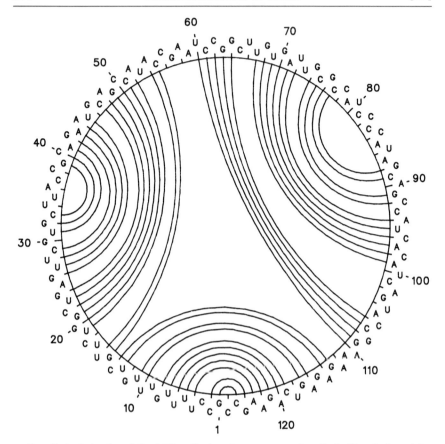

FIG. 7. A circle plot of the folding displayed more conventionally in Figs. 1, 5, and 6.

stems at prescribed angles to one another, this option is useful for the visual comparison of repeated foldings of the same molecule. Bruccoleri and Heinrich report an improvement in this algorithm, programmed in C for VAX/VMS systems.[41] Perhaps the best way to draw structures for comparative purposes is the circle representation first introduced by Nussinov et al.[27] The sequence is mapped along the perimeter of a circle, and hydrogen bonds are represented by circular arcs, which cut the circle at right angles. This presentation is one of the most abstract ways to display RNA foldings. An example is shown in Fig. 7. J. R. Thompson has created a graphics program called MOLECULE written in Pascal for use on

[41] R. E. Bruccoleri and G. Heinrich, *CABIOS* **4**(1), 167 (1988).

```
-------------- REGION table --------------        ----------- CT file ------------

(    1)      1    125    2    -3.4       126 ENERGY =    -35.5      sample RNA
(    2)      4    123    5    -3.7         1 G       0    2   125    1
(    3)     10    117    2    -0.5         2 C       1    3   124    2
(    4)     13    114    2    -2.3         3 C       2    4     0    3
(    5)     15     55    2    -2.1         4 U       3    5   123    4
(    6)     19     52    6   -11.6         5 U       4    6   122    5
(    7)     26     45    4    -3.9         6 U       5    7   121    6
(    8)     31     41    3    -5.2         7 U       6    8   120    7
(    9)     60    110    5   -10.1         8 G       7    9   119    8
(   10)     67     99    6    -8.9          .  .   and so on .   ...     .
(   11)     74     92    2    -3.4       125 C      124  126     1  125
(   12)     76     89    2    -3.4       126 C      125    0     0  126
```

FIG. 8. REGION table and part of the CT file for the optimal folding of the sample RNA sequence displayed in Figs. 1, 5, 6, and 7. The REGION table contains no sequence information. This information must be supplied in another file. Each row describes a single helix. The first column is the number of the helix. The next two columns give the exterior 5'- and 3'-closing bases of the helix. The fourth column is the length of the helix. The final column is the helix energy. The first row of the CT file contains the total number of bases in the folded sequence, the folding energy, and the sequence name. Subsequent lines contain the base number (in the fragment), the base letter, the number of the 5'-connecting base, the number of the 3'-connecting base, the number of the hydrogen-bonded base (0 if the base is single stranded), and the historical number of the base in the original sequence. The REGION table is far more compact.

the IBM AT and XT. MOLECULE is based on an original program by Lapalme *et al.*[42] The output is similar to that of DRAW, although the program is not interactive. MOLECULE is distributed along with PC-FOLD on a single diskette by the author.

Two formats have been widely used for base-pair output from folding programs. The first is the CT file introduced by Feldmann.[43] All the UWGCG graphics programs for RNA structure display accept this format, as does the MOLECULE program. The REGION table was introduced by Shapiro *et al.*[40] for use with DRAW. REGION does not contain any sequence information. Examples of these files are given in Fig. 8.

The author's RNAFOLD program will produce a line printer output, as well as CT files and REGION tables. The program has six output parameters, three of which can be set by the user to control output. These parameters can be set before folding begins or before a repeat folding is requested. The command syntax before folding is

$$PO \ i_1, j_1 \qquad i_2, j_2 \ \ldots$$

[42] G. Lapalme, R. J. Cedergren, and D. Sankoff, *Nucleic Acids Res.* **10**(24), 8351 (1982).
[43] R. J. Feldmann, "Manual for Programs NUCSHO and NUCGEN of Nucleic Acid Structure Synthesis and Display." Division of Computer Research and Technology, National Institutes of Health, Bethesda, Maryland, 1981.

TABLE II
THREE DIFFERENT KINDS OF OUTPUT FROM RNAFOLD
PROGRAM CONTROLLED BY SINGLE PARAMETER

Parameter value	Output parameter number 2 from RNAFOLD		
	Line printer output	REGION table output	CT file output
1	Yes	No	No
2	No	Yes	No
3	No	No	Yes
4	Yes	Yes	No
5	Yes	No	Yes
6	No	Yes	Yes
7	Yes	Yes	Yes

and has the effect of setting output parameter i_1 equal to j_1, parameter i_2 equal to j_2, and so on. All three variable parameters may be set in a single command. Before refolding, the command syntax is

$$O\ i_1, j_1 \qquad i_2, j_2 \ldots$$

Although PCFOLD is completely menu-driven, the above command is still valid at the refolding stage and makes the program a bit more flexible. Parameter number 2 controls which kinds of output will be produced by the program. Table II shows the possibilities. This parameter is 1 by default and is automatically set to 5 if a CT file is requested. Setting this parameter to 4, 6, or 7 is the only way of getting a REGION table, which will show up as file FOR025.DAT or FILE.REG on the VAX, depending on which version of RNAFOLD is used. With PCFOLD, the user is prompted by the operating system for a file name. Parameter number 5 controls the number of columns in the line printer output. Parameter number 5 is 132 by default on the VAX and 79 in PCFOLD. Parameter number 6 is not normally used. Parameter number 6 is the FORTRAN unit number used for the line printer output. This parameter is 6 (terminal output) normally and is set to 24 when file output is selected. If a file is named for the output, the user will not see the output on the screen. After folding, this parameter can be set to 6, and a refolding can be generated, which will be sent to the terminal. Setting this parameter back to 24 will send further output to the file. Setting the parameter to 7, for example, would create the new file FOR007.DAT on the VAX (user-defined name with PCFOLD), which would collect all further line printer output. These features are not available in GCGFOLD, in which the line printer output and CT file are created by default.

Experimental Data and Constrained Folding

Because free-energy minimization is not sufficient to determine a folding with confidence, it is essential to utilize whatever other information is available. This information is usually about specific single- or double-stranded regions, which are believed to occur. It is best to fold a RNA molecule first without constraints and see just how much of the folding conflicts with data on single- or double-stranded regions. This is called a free folding. In general, the energy of the refolded molecule (using constraints) will be greater than the energy of the free folding. If this energy difference is of the order of 10%, there is no problem. When the energy difference rises to 50%, this is an indication that something is wrong. Although the examples given refer almost entirely to the author's program, the principles are general and could be applied to other methods.

Various kinds of data indicate single-stranded regions. Some chemically modified bases lose their ability to base pair. A ribosome or protein, which binds to a RNA molecule, may prevent many bases in a row from pairing. Some enzymes cleave a RNA molecule preferentially in single-stranded regions. Analysis of digestion fragments leads to predictions that some bases are single stranded. One way to prevent these bases from pairing in computer foldings is to assign large destabilizing energies to loops or stacks which involve paired bases that should be single stranded. This is what is done in RNAFOLD, in which single-stranded regions can be designated in two different ways. The first way is to modify the sequence. An X can be used to mark a base which cannot pair, or, better still, lowercase a, c, g, and u can be used. This will not work with GCGFOLD. There is also a command within the program that can be used. Before folding begins, the command

$$AP\ i,\ 0,\ k$$

will force r_i, r_{i+1}, ..., r_{i+k-1} to be single stranded. In PCFOLD, this can be accomplished using the menu system. In GCGFOLD, the option

$$/PREVENT = i,\ 0,\ k$$

on the command line will have the same effect, although this is not stated in the program manual. The CRUSOE program allows single-stranded regions to be designated by using an X. This has the effect of reducing execution time.

It may be desirable to prevent certain base pairs or entire helices from forming. This can be especially useful when folding regulatory RNA.[44] When one set of base pairs is disrupted, another interesting folding may appear. As with forced single-stranded regions, penalty energies can be used to prevent unwanted base pairs. In RNAFOLD, the command

$$AP \ i, j, k$$

issued before folding begins will prevent the formation of the base pairs $r_i \cdot r_j$, $r_{i+1} \cdot r_{j-1}$, ..., $r_{i+k-1} \cdot r_{j-k+1}$. This is accomplished in GCGFOLD by the option

$$/PREVENT = i, j, k$$

on the command line and through the menu system in PCFOLD.

Double-stranded-specific enzymes, such as cobra venom RNase, may suggest that certain bases are in helices. A nonenzymatic double-stranded-specific probe has been reported.[45] Such data must be used carefully, since the data really indicate probable double-stranded regions. If a number of consecutive bases are thought to be double stranded, it is safer to force only a few of them to be paired. In RNAFOLD, the command

$$AF \ i, 0, \ k$$

issued before folding begins will force r_i, r_{i+1}, ..., r_{i+k-1} to be double stranded if possible. PCFOLD uses a menu system, while in GCGFOLD the option

$$/FORCE = i, 0, \ k$$

on the command line will have the same effect, although this is not reported in the notes. The forced pairing is accomplished by giving a bonus energy to stacks or loops closed by a designated base. The value of the bonus energy is -50.0 kcal/mol by default (-20.0 in PCFOLD). This quantity is totally arbitrary; all that is important is that the bonus be sufficient to cause the desired base to be double stranded. The bonus energy number is actually an adjustable parameter and can be set before folding commences by the command

$$PE \ 9, \ e$$

where e is the new bonus energy. The bonus energy is stored in tenths of a kilocalorie per mole, so that -50.0 is stored as the integer number -500. This energy cannot be altered in PCFOLD or GCGFOLD, except by altering the value of EPARAM(9) in the source code. A total of 66 forced base pairs will cause a computer memory overflow, because half-integer

[44] C. Yanofsky, *Nature (London)* **289,** 751 (1981).
[45] C. P. H. Vary and J. N. Vournakis, *Proc. Natl. Acad. Sci. U.S.A.* **81,** 6978 (1984).

variables are used and the largest (absolute) energy that can be stored is −3276 kcal/mol. In such cases, fewer bases should be forced to pair or else the bonus energy could be set to −30.0 or −20.0 kcal/mol, as long as the base pairing still occurs. The bonus energies are subtracted by the traceback algorithm so that the output contains the correct energy. PC-FOLD will overflow with as few as 14 forced base pairs. One solution is to force fewer base pairs. Another is to use PCFOLD2, which can tolerate larger energies, although PCFOLD2 is limited to smaller molecules.

Cross-linking experiments can be valuable in indicating the existence of specific base pairs. Phylogenetic comparison between homologous RNA sequences may also point to conserved helices, which are thought to occur in a RNA folding. Enzymatic digestion data may also suggest the existence of certain base pairs. In this case, specific base pairs can be forced to occur. In RNAFOLD, the command

$$AF\ i, j, k$$

issues before folding begins will force the base pairs $r_i \cdot r_j$, $r_{i+1} \cdot r_{j-1}$, ..., $r_{i+k-1} \cdot r_{j-k+1}$. This is handled by the menu in PCFOLD and by the command line option

$$/FORCE = i, j, k$$

in GCGFOLD. Forced base pairs between noncomplementary bases will not occur. Bonus energies are used to achieve these base pairs, and the same comments given above hold. In case of overflow, there is one extra trick that can be used. The command

$$AF\ i, j, k$$

which forces k base pairs causes k bonus energies to be added to the overall energy. The commands

$$AF\ i, j, 1$$
$$AF\ i + k - 1, j - k + 1, 1$$

force just the first and last base pairs of the helix. This adds just two bonus energies, and the intervening base pairs will form automatically. The CRUSOE program allows for forced helices, and this option greatly reduces execution time. In RNAFOLD, execution time does not improve.

It is desirable to be able to excise fragments from a sequence and fold the remaining sequence. In RNAFOLD, the command

$$AX\ i, j$$

excises r_i through to r_j and covalently links r_{i-1} with r_{j+1}. This command may be repeated to cut out other segments. The most obvious application is to intron splicing. The program retains the historical numbering of nucleotides in the processed sequence, making it easier to compare foldings of a mRNA with or without introns. Another feature, which seems to be unique to RNAFOLD, is what the author calls the closed excision. The command

$$AE \ i, j$$

excises r_{i+1} through to r_{j-1} and links r_i and r_j together by a hydrogen bond, not by a covalent bond. The base pair $r_i \cdot r_j$ must be possible. Multiple, closed excisions are allowed. If closed excisions are made at i_1, j_1; i_2, j_2; $\ldots, i_k, j_k$; then the condition $i_1 < j_1 < i_2 < j_2, \ldots, < i_k < j_k$ must hold. The folding of the "processed" sequence and the foldings of all the excised fragments with their 5' and 3' ends forced to base pair yield a composite folding of the entire sequence with base pairs $r_{i_1} \cdot r_{j_1}$ to $r_{i_k} \cdot r_{j_k}$ forced. This composite folding is much faster than using the AF i, j, k option above, and because the folding is split into smaller parts, this procedure might turn a very large folding problem into a tractable one. The negative side to this procedure is that the output is fragmented into $k + 1$ separate files. If REGION table output is produced, these files could all be appended into a single file for immediate use by DRAW or another plotting program. PCFOLD handles excisions with the menu system, while GCGFOLD offers only the regular excision with the option

$$/REMOVE = i, j$$

on the command line.

RNAFOLD has five parameters which are called folding parameters. The correct syntax for setting the folding parameters is

$$PF \ i_1, j_1; \ i_2, j_2 \ \cdots$$

issued before folding commences. One up to all five parameters can be set in a single command. The effect of this command is to set parameter i_1 equal to j_1 and so on. These parameters are set using the menu feature of PCFOLD and are not available in GCGFOLD. The use of the first three parameters can greatly restrict the range of allowable base pairs, so that a large increase from the free folding energy is to be expected in general.

The first folding parameter is the minimum size of a hairpin loop and equals three by default. This parameter can be set larger to force base pairs between distant regions of the molecule. The second parameter is the maximum value of $j - i$ allowed in a base pair $r_i \cdot r_j$ and is equivalent to the LEN parameter in SEQL. Normally set to infinity so that there is no constraint, this parameter can be set to 100, 50, or even 30 to force foldings with only short-range base pairs. When this parameter is small, folding is faster. The third parameter is either 0 or 1. When 0, normal folding occurs. When set to 1, multiloops are not allowed, and the output is just a sequence of hairpin structures, called an open folding. This parameter is often used in conjunction with the second parameter. The program runs much faster, with execution time growing as the square of the fragment size, rather than as the cube with the regular algorithm. The fourth parameter takes effect only when an open folding is selected using parameter 3. This parameter is normally 1. In this case, repeat foldings of subfragments will yield correct results only when the 5′ end is included. If the 5′ end is not included, only a single optimal hairpin structure will be produced. When this parameter is set to 2, repeat foldings may omit the 5′ end, but repeat foldings must include the 3′ end to produce an optimal open folding. If this parameter is set to 0, then arbitrary subfragments may be chosen for repeat foldings, but the price paid is that improvements in program speed are lost.

The fifth folding parameter offers another way to deal with single-stranded-specific information. RNase T1, for example, cuts the covalent linkage of a G nucleotide with its 3′ neighbor. The crudest way to handle such a situation is to prevent the G from pairing. A more subtle approach is to allow G or its 3′ neighbor to pair, but not both. In RNAFOLD, bases are called accessible (to RNase attack) if their 3′-phosphodiester bond is cleaved by a RNase. The accessible bases are designated using the letters B, Z, H, and V for A, C, G, and U, respectively. The default value of this fifth parameter is 0. In this case, B, Z, H, and V are treated as bases that cannot pair. When the parameter equals 1, B, Z, H, and V are treated just like A, C, G, and U, respectively. When the parameter equals 2, accessible bases are allowed to pair only if their 3′ neighbor is not paired. This is a minimum constraint for nuclease accessibility. This constraint is achieved using large penalty energies. When single-stranded-specific probes are used, initial cuts may allow the molecule to unfold somewhat, exposing further sites to attack. For this reason, such data must be used conservatively, with only the very best sites designated as accessible.

Another feature of RNAFOLD is that G–U base pairs at the ends of helices are either allowed or not allowed. The parameter that controls this is set within the program, so the user has no choice. However, the line

$$EPARAM(1) = 1$$

in the main program, which prevents these external G–U base pairs, can be changed to

$$EPARAM(1) = 0$$

to allow the external G–U base pairs. The earlier versions of RNAFOLD, including PCFOLD, do not allow the external G–U base pair. The newer versions, including GCGFOLD, do allow the external G–U base pairs. The author's latest VAX version of RNAFOLD offers the user a choice. There no longer seems to be any reason to exclude the external G–U base pairs.

Confidence and Reliability

When a free folding and a folding using constraints give the same answer, the user's confidence is increased. The same is true when phylogenetic data support a folding. However, even when data are available to constrain a folding, it is desirable to know how robust the computer prediction is. How can one determine which parts of a structure are well determined and which are not? One solution is to predict a number of alternative foldings close to the minimum energy and to observe what motifs, if any, are shared by these foldings. Another method is to compare the minimum folding energy of a molecule with what would be expected from folding a random sequence with the same A, C, G, and U content.

By their nature, both CRUSOE and MONTECARLO produce a variety of foldings. With MONTECARLO, there is no guarantee that any of the solutions will be close to the minimum energy, but this is not the case with CRUSOE. Unfortunately, this program can only fold small RNA molecules, and so its usefulness is limited. Recursive programs, such as RNAFOLD, do not naturally yield multiple solutions. Williams and Tinoco[29] extended the traceback algorithm in RNAFOLD, so that multiple solutions not far from the energy minimum are predicted. One cannot expect to predict all foldings within, for example, 10% of the energy minimum, because there may be an astronomical number of foldings, and in any case, there may be many foldings that are similar to one another. The author has extended the RNAFOLD program so that all base pairs that can occur in foldings close to the energy minimum are predicted.[46–48] Individual suboptimal foldings can be predicted, but it is the collection of all possible base pairs that shows some motifs to be well defined and other motifs to be poorly defined.

Virologists, folding very large viruses using RNAFOLD, have been forced to break up the problem into the folding of overlapping segments. Structural agreement in overlapping areas increases confidence in the overall folding. This technique is in general use.

[46] A. B. Jacobson, M. Zuker, and A. Hirashima, in "Molecular Biology of RNA: New Perspectives" (M. Inouye and B. S. Dudock, eds.), p. 331. Academic Press, Orlando, Florida, 1987.
[47] M. Zuker, in "Mathematical Methods for DNA Sequences" (M. S. Waterman, ed.), p. 154. CRC Press, Boca Raton, Florida, 1989.
[48] M. Zuker, Science 244, 48 (1989).

Because recursive methods compute optimal foldings on subfragments, the folding of a RNA sequence as the sequence grows from the 5' end can be simulated. Motifs appear and disappear as the simulated molecule grows into its final configuration. Substructures which form and are not destroyed as the molecule continues to grow may be regarded as significant. Modelevsky and Akers[49] monitor the percentage of bases in fixed windows (continuous segments of RNA) that are double stranded in a growing RNA sequence. The simulation uses RNAFOLD. Modelevsky and Akers correlate this statistic with the level of gene expression.

Le et al.[50] fold overlapping segments of RNA and then randomly permute the bases in that segment a number of times, refolding at each stage. This Monte Carlo method gives the mean and standard deviation of the folding energy which could be expected in that segment by chance alone. The segments used are not long; 100 bases is a typical length. This technique gives a basis for deciding which local folding motifs are significant. The size of these segments are varied, so that significant local structures of optimal size can be predicted. The method does not assess the significance of long-range interactions.

Another way to produce alternative structures is to perturb the folding rules and to repeat secondary structure prediction. This method is unnecessary with combinatorial or Monte Carlo folding programs for reasons already discussed, but this method is useful with traditional recursive programs, such as RNAFOLD, which have not been adapted to produce multiple solutions. The usual perturbation is to change the energy rules slightly and to refold the molecule a number of times. This method is as costly as the Monte Carlo assessment of folding significance developed by Le et al.[50] The reason is that the time-consuming fill algorithm must be performed over and over again. RNAFOLD has six "energy" parameters, which allow some energy assignments or some other folding rule changes to be made by the user without altering any input file. The energy parameters are set before folding commences with the command

$$\text{PE } i_1, j_1; i_2, j_2 \ldots$$

which has the effect of setting parameter i_1 equal to j_1, parameter i_2 equal to j_2, and so on. One or all parameters may be set with a single command. PCFOLD allows these parameters to be changed with the menu system, and GCGFOLD does not allow these parameters to be changed. Parameter 1 controls the existence of G–U base pairs at the ends of helices and has already been discussed. Parameters 2 through 5 inclusive are the extra energies, in tenths of a kilocalorie per mole, which are to be added to all

[49] J. L. Modelevsky and T. G. Akers, *CABIOS* **4**(1), 161 (1988).

[50] S.-Y. Le, J.-H. Chen, K. M. Currey, and J. V. Maizel, Jr., *CABIOS* **4**(1), 153 (1988).

base-pair stackings, bulge or interior loops, hairpin loops, or multiloops, respectively. These parameters are 0 by default. Parameter 6 is the maximum size allowed for bulge or interior loops. Parameter 6 is 30 by default. Although RNAFOLD has an automatic feature which causes the program to give up looking for arbitrarily large bulge or interior loops, this parameter does speed up program execution as the value is lowered. Thirty is a safe value for folding at 37°, but this value should be increased for folding at higher temperatures. Parameter 7 is the maximum value for |SIZE1–SIZE2|, where SIZE1 and SIZE2 are the number of single-stranded bases on the two sides of an interior or bulge loop. Parameter 7 can be called the maximum lopsidedness of an interior loop. Parameter 7 is also the maximum size of a bulge loop. Equal to 30 by default, this parameter takes effect only when this parameter is smaller than parameter 6. It is worthwhile to note that Ninio's CRUSOE program has a gradually increasing penalty for lopsided interior loops. This is a good idea and has already been incorporated into a trial version of RNAFOLD.

Altering the energy parameters does not produce random folding perturbations. The parameter changes bias the folding in predictable ways. Adding an extra 1–2 kcal/mol to all bulge or interior loops will usually produce a refolding with fewer bulge and interior loops. The addition of energy will tend to eliminate weakly stable helices and to consolidate single-stranded regions into fewer but larger loops. Adding a few extra kilocalories per mole to hairpin loops, say ε kcal/mol, will often result in a folding with fewer hairpin loops. If it does not, then the conclusion is that there are no alternative foldings with fewer hairpin loops within ε kcal/mol from the minimum energy. The same sort of analysis applies to multiloops. Unlike the situation with bonus energies used to force base pairs, these extra energies are included in the overall folding energy. Decreasing parameters 6 or 7 by one or two from the maximum size found in a first folding is another good way to perturb the original folding. Setting parameter 6 to 0 when folding a tRNA is an excellent way to force the desired cloverleaf structure, which is often not the minimum energy structure in the folding model. This brief description is meant to give the reader the idea of "playing" with the rules while refolding a sequence of interest. An example is given in Fig. 9.

Concluding Remarks

Three-dimensional prediction of RNA structure will not be possible for the foreseeable future. The problem is still unsolved for proteins, and in that case, there is at least a database of over 400 solved proteins, so that predictions can be compared with structures deduced from X-ray diffrac-

```
ENTER:  T TERMINATE, NS NEW SEQUENCE, NF NEW FRAGMENT,
O OUTPUT PARAMETER DEFINITION, OR THE END POINTS OF A
SUBFRAGMENT BETWEEN     1 AND    126.
END WITH 1 TO FORCE ENDS TO BASE PAIR.
NF  ←
ENTER END POINTS OF FRAGMENT TO BE FOLDED. (DEFAULT  =  1,   126 )
1 126  ←
ENTER:  F BEGIN FOLDING,  T TERMINATE,  A AUXILIARY INFORMATION,
P PARAMETER DEFINITION
PE 3, 10  ←
F  ←
    10    20    30    40    50    60    70    80    90    100   110   120


FOLDING BASES    1 TO   126 OF sample RNA
ENERGY  =      -27.9


                   10                20          30
--------    -    GUUUG     UUC      A     G   UU
        GC CUUUU          UCG    UGGCUG GUUC UGC
        UG GAGAA          AGC    ACCGAC UAAG ACG   A
CCGAAGAC  A      -----    CAU      G    -   CC
   120                     50                40


                  60                70             80
                  A     ----CU      U   -  CAUC
                  UCCGG        UGUGAU GC GC    C
                  AGGCC        ACACUA CG CG    C
                  -     AUGACU      C  A  AUAC
                  110      100           90
```

FIG. 9. Perturbation folding of the sample RNA shown folded in Figs. 1, 5, 6, and 7. A penalty of 1 kcal/mol is given to all bulge and interior loops using the command PE 3, 10. The result is an alternate folding, which differs in energy from the optimal folding by only 0.6 kcal/mol after the energy correction is made (7 kcal/mol). User input is indicated by the arrows.

tion experiments. Nevertheless, some three-dimensional aspects of RNA structure should be considered. The most important of these is the prediction of pseudoknots. In the absence of energy rules to deal with the complicated loops created by pseudoknots, and without ways of knowing whether hypothetical structures are stereochemically possible, the most sensible course of action is to use a two-stage procedure. First, predict an ordinary secondary structure and then look for tertiary interactions, which might induce pseudoknots at a final stage in the folding process. These tertiary interactions would be potential helices between single-stranded regions leftover after the molecule is folded. The single-stranded regions could be enlarged by opening up short or weak helices in the secondary structure.

The treatment of enzymes and other chemicals, which indicate single- or double-stranded regions, has been crude to date. The method of forcing a helix is valid when there is phylogenetic evidence or cross-linking data. Some chemically modified bases cannot pair with certainty. However, very often the available data are saying that some bases are probably single stranded or probably double stranded. Forcing the folding one way or another, or even using the nucleotide accessibility option in RNA-FOLD, is too heavy handed. What is needed is a multiple folding procedure in which bases likely to be single stranded would be so most of the time, or else a certain fraction of the time corresponding to a given probability. The same would hold for bases likely to be double stranded.

It is often desirable to predict a common folding of the leader sequences or coding regions of homologous mRNAs. The existence of a common folding greatly increases confidence in the predicted structure and can be used to explain the similar regulation of the related genes. Although Sankoff[51] reports an algorithm for simultaneously aligning and folding homologous RNA sequences, the method is not practical. Even if the related sequences are already aligned, there is no reliable and automatic way of predicting a common folding. The usual procedure is to use a program, such as SEQL, to find lists of potential helices in each sequence. These lists can be scanned for common helices. At this point, each sequence can be folded separately using an ordinary folding program, with the common helices forced. Depending on how many helices are forced, how similar the sequences are, and chance, the results will vary from a common folding of all sequences to dissimilar structures with some shared motifs.

Acknowledgment

M. Zuker is a Fellow of the Canadian Institute for Advanced Research.

[51] D. Sankoff, *SIAM J. Appl. Math.* **45**, 810 (1985).

[21] RNA Pseudoknots: Structure, Detection, and Prediction

By C. W. A. PLEIJ and L. BOSCH

RNA pseudoknots are structural elements of the RNA architecture. Although detected first in tRNA-like structures at the 3' ends of various plant viral RNAs,[1-4] pseudoknot occurrence is much more widespread,[5-8] and the pseudoknots may be looked on as expressing one of the basic principles of the folding of RNA chain segments in space. It should be emphasized that our knowledge of the spatial folding of RNA in general is limited. Crystal structures of only a few tRNA species have been elucidated,[9,10] and, although these analyses have revealed a wealth of new intramolecular interactions, it has not become clear whether these structural features have a more general significance and whether these structural features have been used more frequently for realizing the intricate folding of RNA. It is conceivable, however, that a few more general building principles will become apparent when the three-dimensional structure of a larger number of naturally occurring RNA species becomes available. Clearly, the determination of the higher order structures of RNA is a great challenge and probably will prove difficult, because these molecules remain refractory to crystallization.

A basic feature of the secondary structure of RNA is the classical stem–loop structure. Pseudoknots arise on base pairing of the single-stranded loops enclosed by these stems, with complementary unpaired regions elsewhere in the RNA chain. In this chapter, we describe the principle underlying this folding of the secondary structure into the higher order pseudoknot. Some of the steric problems encountered by the fold-

[1] K. Rietveld, K. Linschooten, C. W. A. Pleij, and L. Bosch, *EMBO J.* **3**, 2613 (1984).

[2] A. Van Belkum, J. P. Abrahams, C. W. A. Pleij, and L. Bosch, *Nucleic Acids Res.* **13**, 7673 (1985).

[3] C. W. A. Pleij, K. Rietveld, and L. Bosch, *Nucleic Acids Res.* **13**, 1717 (1985).

[4] C. W. A. Pleij, J. P. Abrahams, A. van Belkum, K. Rietveld, and L. Bosch, *UCLA Symp. Mol. Cell. Biol.* [N.S.] **54**, 299 (1986).

[5] H. F. Noller, *Annu. Rev. Biochem.* **53**, 119 (1984).

[6] I. C. Deckman and D. E. Draper, *J. Mol. Biol.* **221**, 235 (1987).

[7] R. W. Davies, R. B. Waring, J. A. Ray, T. A. Brown, and C. Scazzocchio, *Nature (London)* **300**, 719 (1982).

[8] D. S. McPheeters, G. D. Stormo, and L. Gold, *J. Mol. Biol.* **201**, 517 (1988).

[9] J. P. Robertus, J. E. Ladner, J. T. Finch, D. Rhodes, R. S. Brown, B. F. C. Clark, and A. Klug, *Nature (London)* **250**, 546 (1974).

[10] S. H. Kim, F. L. Suddath, G. J. Quigley, A. McPherson, J. L. Sussman, A. H. J. Wang, N. C. Seeman, and A. Rich, *Science* **185**, 435 (1974).

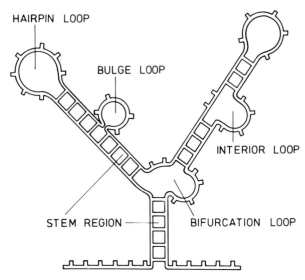

FIG. 1. Elementary secondary structure elements of RNA.

ing will be addressed, and some of the characteristic properties of pseudoknots will be discussed. Methods for the detection and prediction of pseudoknots will also be presented and illustrated by a number of well-documented examples.

Principle of Pseudoknotting

Standard secondary RNA structures, such as the well-known clover-leaf of tRNA, are characterized by a small number of typical features as depicted in Fig. 1. Beside double-helical or stem regions and hairpins, which result from intramolecular base pairing, interior and bulge loops, bifurcations, and bifurcation loops can be recognized. All current computer programs predict these features.[11]

However, as first realized by Ninio[12] and later by others,[13] it is conceivable that the single-stranded regions of the various loops in turn can participate in base pairing with complementary single-stranded regions elsewhere in the chain. The result of such a base pairing was initially called a knotted structure, because of the potential formation of real

[11] M. Zuker and P. Stiegler, *Nucleic Acids Res.* **9,** 133 (1981).
[12] J. Ninio, *Biochimie* **53,** 458 (1971).
[13] G. M. Studnicka, G. M. Rahn, I. W. Cummings, and W. A. Salser, *Nucleic Acids Res.* **5,** 3365 (1978).

knots, in particular when stretches of 10 or more nucleotides are involved.[14] Presently, the term pseudoknot, originally proposed by Studnicka *et al.,*[13] is used.

A general definition of a pseudoknot may be given as follows: a structural element of RNA formed on base pairing of nucleotides within one of the four loops in an orthodox secondary structure (see Fig. 1) with nucleotides outside that loop. According to this definition, base pairing is not restricted to interactions between one of the loops and the free single-stranded stretches at either the 3′ or 5′ terminus of the RNA chain as depicted in Fig. 1. Also loop–loop interactions satisfy the definition, for instance, that of the hairpin loop with either side of the interior loop. Such an interaction may be difficult to realize physically, in some cases, for steric reasons. Clearly, steric limitations have to be considered when searching for pseudoknots of any kind (see Searching for Pseudoknots in RNA).

In principle, the location and the number of nucleotides within the various loops involved in pseudoknotting are not subject to limitations, but for reasons outlined below, we will restrict ourselves to interactions involving at least three base pairs or more. This means that the well-known interaction between the D- and T-loop of tRNA, involving, among others, the conserved $G \cdot C$ and $G \cdot \psi$ base pairs, will not be considered in this chapter. We further assume that the double-helical segments in pseudoknots are of the RNA–A type and that the interaction of the complementary strands occurs in an antiparallel fashion, such as that in standard secondary structures.

Before discussing methods to find and predict pseudoknots in RNA, the characteristics of a relatively simple type of pseudoknot will be described, because of its widespread occurrence and since its geometric properties are reasonably well understood.

Representative Example of Pseudoknot

Figure 2 shows one of the simplest examples of a pseudoknot, displaying all the characteristics of this structural element. A stretch of three or more nucleotides of a hairpin loop base pairs with a complementary region outside the hairpin. The loop sequence, participating in this tertiary interaction, borders directly to the stem region of the hairpin. This particular type of pseudoknot will be called the H pseudoknot (H standing for hairpin loop). The H pseudoknot was detected first as an essential part of

[14] C. R. Cantor, *in* "Ribosomes" (G. Chambliss, G. Craven, J. Davies, K. Davis, L. Kahan, and M. Nomura, eds.), p. 23. University Park Press, Baltimore, Maryland, 1980.

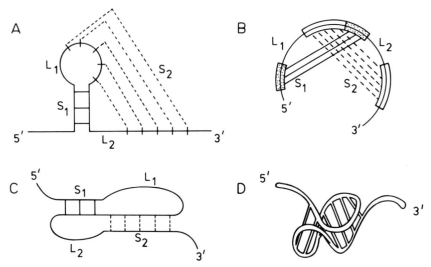

FIG. 2. The formation of a pseudoknot of type H. S_1 and S_2 represent stem regions obtained by Watson–Crick base pairing, and L_1 and L_2 represent the single-stranded connecting loops. (A) Conventional secondary structure presentation. (B) Schematic presentation in which the RNA sequence is given in a circular form. (C) A "two and one-half-dimensional" presentation. (D) Artistic view of the three-dimensional folding, illustrating the coaxial stacking of the eight base pairs.

the aminoacyl acceptor arm of the tRNA-like structure at the 3' terminus of turnip yellow mosaic virus RNA (TYMV RNA).[15] This plant viral RNA is able to specifically accept valine, analogous to its standard tRNA counterpart. This suggests that the 3'-terminal structure of TYMV RNA displays a rather close resemblance to the structure of tRNA^{Val}. Probing end-labeled 3'-terminal fragments of TYMV RNA for their secondary structure did not reveal a cloverleaf structure, however.[15,16] The techniques used in studies of this type include chemical modification with diethyl pyrocarbonate (DEPC) and dimethyl sulfate (DMS) according to Peattie and Gilbert[17] and enzymatic digestion with S1 nuclease, RNase T1, and the double-stranded-specific nuclease from the venom of the cobra *Naja naja oxiana*. These and a few other probes consist of the standard tools for mapping RNA structures and will not be described in

[15] K. Rietveld, R. van Poelgeest, C. W. A. Pleij, J. H. van Boom, and L. Bosch, *Nucleic Acids Res.* **10**, 1929 (1982).
[16] C. Florentz, J. P. Briand, P. Romby, L. Hirth, J. P. Ebel, and R. Giege, *EMBO J.* **1**, 269 (1982).
[17] D. Peattie and W. Gilbert, *Proc. Natl. Acad. Sci. U.S.A.* **77**, 4769 (1980).

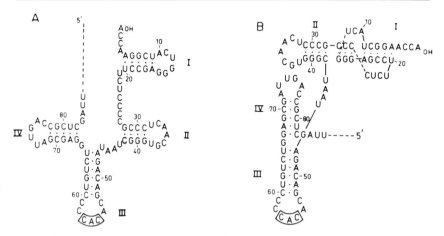

FIG. 3. Secondary and tertiary structure of the 3' terminus of TYMV RNA. (A) Secondary structure. Hairpins are indicated with roman numerals (I–IV). Numbering of the nucleotides is from the 3' end. (B) L arrangement of the tRNA-like structure of TYMV RNA. The aminoacyl acceptor arm contains the RNA pseudoknot.

detail in this chapter. The reader is referred to Ehresmann *et al.*[18] for a recent review, to the chapter by Knapp[19] in this volume.[20] In the case of the 3' end of TYMV RNA, the secondary structure illustrated in Fig. 3A was found. This structure consists of four regular hairpins, but lacks the conventional aminoacyl acceptor stem of tRNA generated by base pairing of the 3' and 5' ends.[15,16]

Indications that the secondary structure of Fig. 3A can be folded into a tertiary structure resembling the familiar L shape of canonical elongator tRNA were provided by the following data.

1. The ACCA end is immediately followed by a stem region (stem I), reminiscent of the aminoacyl acceptor arm of tRNA.
2. The double-stranded-specific cobra venom nuclease cuts in the triple G sequence (G-13–G-15) of the loop of hairpin I and somewhat more weakly in the triple C sequence adjacent to stem II (C-25–C-27), suggesting base-pairing interactions.
3. The triple C sequence is protected from chemical modification with DMS under so-called native conditions (10 mM Mg^{2+}), but not

[18] C. F. Ehresmann, F. Baudin, M. Mougel, P. Romby, J. P. Ebel, and B. Ehresmann, *Nucleic Acids Res.* **15**, 9109 (1987).
[19] G. Knapp, this volume [16].
[20] C. W. A. Pleij and L. Bosch, this volume [21].

under semidenaturing conditions [1 mM ethylenediaminetetraacetic acid (EDTA)].

4. The complementarity of the possible triple G–triple C sequence interaction is conserved among related viral RNAs.

Base pairing of the triple G and triple C sequences enables building of a model in which stem I, the (G-13–G-15) : (C-25–C-27) segment, and stem II are stacked coaxially on top of each other. This model closely resembles the aminoacyl acceptor arm of canonical tRNA (Fig. 3B) (compare also Refs. 15 and 21). Note that the model contains a pseudoknot of the same type as the one shown schematically in Fig. 2. An additional feature is that the pseudoknot itself is stacked on top of stem II.

It may be realized that the discovery of the pseudoknot at the 3' end of TYMV RNA, as outlined above, was strongly guided by the aim of finding a resemblance with the classical three-dimensional structure of tRNA. It is doubtful whether the outcome would have been the same without the knowledge of the tRNA data.

Pseudoknots of the type illustrated in Figs. 2 and 3 are relatively widespread in the noncoding sequences of viral RNAs, as became apparent when the 3' termini of various plant viral RNAs were studied.[4,22] These phylogenetic comparisons also permit derivation of some general rules concerning the steric requirements for their formation.[3] The rules are underscored and extended by model building using computer graphics.[21]

Properties of H Pseudoknot

The main properties of the H pseudoknot are briefly reviewed in this section, since these properties are relevant in searching for pseudoknots in other RNAs. The properties illustrate the importance of the double-helix geometry of RNA in considerations of pseudoknots.

The H pseudoknot of Fig. 2 always consists of two stem regions, S_1 and S_2, and two single-stranded loops, L_1 and L_2. S_1 and S_2 are thought to be coaxially stacked, so that a quasi-continuous double helix is formed. It should be emphasized that this coaxial stacking, though strongly suggested by the models of the tRNA-like structures, is an assumption for which direct proof remains unavailable. The minimal number of base pairs in each stem is found presently to be three base pairs. When three base pairs are found in one stem, then five to seven base pairs are found in the

[21] P. Dumas, D. Moras, C. Florentz, R. Giege, P. Verlaan, A. van Belkum, and C. W. A. Pleij, *J. Biomol. Struct. Dyn.* **4**, 707 (1987).

[22] C. W. A. Pleij, unpublished results (1988).

other stem.[3,4] To date, we have not been able to derive any rule for the nature or sequence of the base pairs in either S_1 or S_2, except that a short run of three base pairs usually contains two or three $G \cdot C$ pairs. It is crucial to realize that, due to the polarity of the RNA chain and the geometry of the RNA–A double helix, the connecting loops L_1 and L_2 are not equivalent. L_1 always crosses the deep groove, and L_2 crosses the shallow groove of the quasi-continuous helix formed by S1 and S_2. Analysis of the RNA–A double helix shows that surprisingly few nucleotides are needed for L_1 to span the deep groove. No steric problems are encountered when L_1, containing two nucleotides, has to span five to seven base pairs, but even a L_1, with one single nucleotide, is capable of doing so.[3,21] In the latter case, for reasons unknown to date, this single nucleotide is often a G residue. An example of a H pseudoknot with a single A residue in L_1 is found in Ref. 8. The distance for bridging the shallow groove is larger (reason that L_2 has to consist of two or more nucleotides). In principle, there is no upper limit for the size of L_1 and L_2. These single-stranded loops can have lengths of hundreds of residues and can possess an elaborate secondary structure of their own.[7,23] As we will see below, the very steric properties of these connecting loops are responsible for the anomalous behavior of the constitutive base residues in structure mapping experiments.

Searching for Pseudoknots in RNA

In some cases, pseudoknots can be traced by the following simple approach. Starting from an orthodox secondary structure, one inspects visually the loop regions for potential base pairing with other single-stranded regions. Though this method appears time-consuming and unscholarly, the approach certainly is rewarding, when the size of the RNA is relatively small or the secondary structure is well established (e.g., rRNAs) or when pseudoknots can be expected (e.g., noncoding sequences of viral RNAs). This approach has revealed the existence of various consecutive pseudoknotted structures in a number of plant viral RNAs.[4,22] Finding pseudoknots in this way is not a rare phenomenon. Statistically, one can expect that roughly 1 out of every 10 hairpin loops can form a pseudoknot of the H type, provided that the connecting loops L_1 and L_2 do not exceed some 10 nucleotides. Support for the existence of pseudoknots found in this way should come from other techniques, such as structure mapping or sequence comparisons (see Detecting Pseudoknots and Phylogenetic Sequence Comparisons). The search for H

[23] K. Rietveld, C. W. A. Pleij, and L. Bosch, *EMBO J.* **2,** 1079 (1983).

pseudoknots can be expedited by using the program of Zuker and Stiegler[11] in the mode that yields a so-called open structure (Zuker-2). If one also varies the maximal length of the hairpins to be folded, one obtains a collection of hairpin structures that can be rapidly screened for possible pseudoknots. A more sophisticated method is to develop a program or a subroutine that enables the search for extra base pairing in an otherwise fixed secondary structure. This was performed by Salser, who enumerated these tertiary interactions in the rabbit β-globin mRNA.[24] Finally, we provide some additional suggestions for finding pseudoknots of the H type. Good candidates for forming pseudoknots are regular hairpins having 5–10 nucleotides in the loop and 5–8 base pairs in the stem. Bulge loops are seldom present in S_1 and S_2 (Fig. 2), the only documented examples being a pseudoknot in TMV RNA[2] and the rather complicated pseudoknot in *Escherichia coli* α mRNA reported by Deckman and Draper.[6] In some cases, a disruption of the upper base pair of a hairpin stem by increasing the hairpin loop size by two nucleotides can be helpful. Similarly, a temporary disruption of other structural elements can sometimes reveal an interesting alternative structure of the pseudoknot type.

On screening hairpin loops, one should not infer from Fig. 2A that only the 3' side of a loop can be involved in base pairing. If the hairpin with the other stem segment (S_2 in Fig. 2) would have been drawn (or found) first, then the 5' side of the loop would have been part of the base-pairing interaction. The final result, however, remains the same. Thus, one should examine both the 3' and 5' side of any hairpin loop under consideration. Stacking of the loop nucleotides on top of the S_1 double helix occurs toward the 3' end, whereas stacking of the loop nucleotides on top of the S_2 double helix occurs toward the 5' end. Finally, there is also the question of which connecting loop crosses the deep groove. In the H-type pseudoknots, the first connecting loop encountered following the RNA chain from the 5' end always crosses the deep groove.

Detecting Pseudoknots

Currently available chemical and enzymatic probes can provide information specific for pseudoknots. The tertiary interactions of these structural elements can be utilized. Lowering the salt concentration or elevating the temperature leads characteristically to a relatively rapid unfolding of these tertiary interactions and thus of pseudoknots. On pseudoknotting, the RNA chain adopts a more compact conformation, which requires screening of the negative phosphate groups (e.g., by Mg^{2+}). Structural

---

[24] W. Salser, *Cold Spring Harbor Symp. Quant. Biol.* **42**, 985 (1977).

mappings, carried out under varying ionic conditions or at varying temperatures, may thus reveal which RNA sequences are involved in tertiary interactions and in pseudoknotting. Specifically, the reader is referred to a study by Peattie and Gilbert of tRNA[Phe] from yeast.[17] Probing with DMS or DEPC can be performed at 37° in the presence of 10 mM Mg^{2+} (native conditions) or 1 mM EDTA (semidenaturing conditions). Chemical probing is also possible under so-called denaturing conditions at 90°. A good example of this experimental approach is the study of the tRNA-like structure of brome mosaic virus RNA,[23] revealing the disappearance of the pseudoknot under semidenaturing conditions with concomitant transition of the involved RNA chain segment into a new secondary structure.

Chemical probing in the presence and absence of Mg^{2+} is therefore recommended strongly. Probing of the structure at Mg^{2+} concentrations varying over a whole range may present a clearer picture and moreover may provide some insight into the dynamic behavior of the RNA molecule. Similar information may be derived from melting transitions observed on chemical modifications and nuclease digestions at various temperatures. Studies of this type were first reported by de Bruyn and Klug, who were able to propose a three-dimensional model for a mammalian mitochondrial tRNA, mainly on the basis of base modifications with DEPC and DMS at temperatures from 0 to 90°.[25] Similar experiments were carried out for ribosomal 5 S RNA and for the so-called 3′-terminal cloacin fragment of 16 S rRNA from *E. coli*.[26,27] The method can be extended to other chemical probes, such as sodium bisulfite, which converts unpaired cytosine residues into uridines. A drawback of bisulfite is that the modification reaction has to be carried out at high Na$^+$ concentrations, hampering comparisons of the results with those results obtained by other means.[28] Information concerning the proper reaction conditions for each of the chemical probes mentioned above may be found in Refs. 25 and 28.

Unfolding of the RNA secondary and tertiary structure generally renders the base residues more accessible to modifying agents. An interesting exception has been found for the connecting loop L$_2$, spanning the shallow groove of the pseudoknot H type (compare Fig. 2). When this loop is short, such as that of the pseudoknot at the 3′ end of TYMV RNA (three nucleotides), its base residues are forced to point outward into the solvent, rendering the base residues very accessible to probes, such as

[25] M. H. L. de Bruyn and A. Klug, *EMBO J.* **2**, 1309 (1983).
[26] T. Pieler, M. Digweed, and V. A. Erdman, *J. Biomol. Struct. Dyn.* **3**, 495 (1985).
[27] H. A. Heus and P. H. van Knippenberg, *J. Biomol. Struct. Dyn.* **5**, 951 (1988).
[28] A. van Belkum, P. Verlaan, J. Bing Kun, C. W. A. Pleij, and L. Bosch, *Nucleic Acids Res.* **16**, 1931 (1988).

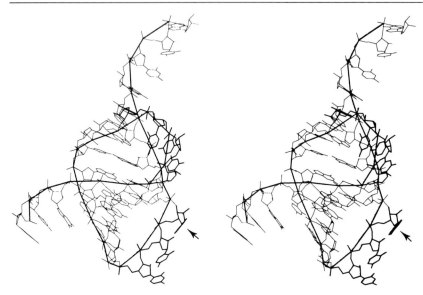

FIG. 4. Stereoscopic view of the pseudoknot in the aminoacyl acceptor arm of the tRNA-like structure of TMV RNA, as modeled with computer graphics. The model represents residues A-1–C-30 (see Fig. 3). The two connecting loops [(A-10–U-12) and (U-21–C-24)] are indicated by thick lines. The arrows indicate residue A-10.

DMS or DEPC. This rather unusual conformation of unstacked bases, not taken up in a more hydrophobic environment, is clearly visualized by model building.[21] The A residue at position 10 (compare Figs. 3 and 4) appears to be highly reactive with DEPC at 10 mM Mg^{2+} even at 0°. Similarly, C residues have been found to be exposed, when probed with DMS or sodium bisulfite.[15,28] On unfolding, however, the adenine becomes much more shielded against DEPC, probably as the result of its incorporation into the stem of a new hairpin. A similar behavior has been found for an adenine residue in TMV RNA[2] (see Fig. 5) and for an adenine in an identical position in a pseudoknot at the 3′ end of tobacco rattle virus RNA.[29] Such an anomalous behavior of a single base can therefore form an indication for the presence of a pseudoknot.

Enzymes, such as nuclease S1 and RNase T1, can also be used for temperature-dependent probing.[26,28] RNase T1, though cutting only after single-stranded G residues, appears to be superior to nuclease S1 in that RNase T1 is active up to 80° and in the presence or absence of Mg^{2+},

[29] A. Van Belkum, B. Cornelissen, H. Linthorst, J. Bol, C. W. A. Pleij, and L. Bosch, *Nucleic Acids Res.* **15**, 2837 (1987).

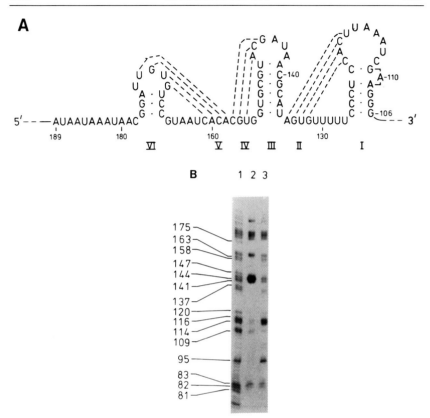

FIG. 5. Pseudoknots in the 3'-terminal noncoding region of TMV RNA, located up-stream of the tRNA-like structure. (A) Conventional secondary structure presentation. The dashed lines represent the pseudoknotting. Roman numerals (I–VI) indicate the six stem segments that can be stacked into one quasi-continuous double helix. Numbering is from the 3' terminus of TMV RNA. (B) Probing of adenosine residues with DEPC under denaturing (lane 1), native (lane 2), and semidenaturing (lane 3) conditions. The numbering of adenosine residues corresponds to that given in A. Note the strong accessibility of A-144 in lane 2.

whereas nuclease S1 is inactive above 60°, needs Zn^{2+}, and has a pH optimum in the acidic range.

It is noteworthy that the two connecting loops, L_1 and L_2, in the H-type pseudoknot, in contrast to what one expects, usually are not very sensitive to the single-stranded-specific nuclease S1 and RNase T1. This is more remarkable for the loop spanning the shallow groove. Possibly the protruding base residues and the strong steric constraints make this loop less easy to handle by the nucleases. As such, this property may serve as another characteristic to distinguish loops in pseudoknots from other more orthodox loops.

Phylogenetic Sequence Comparisons

Probably the most powerful approach for establishing the secondary structure of RNA is the approach based on phylogenetic sequence comparisons.[30] Double-stranded regions are considered proved, when an equivalent pairing is maintained in related homologous RNAs despite differences in nucleotide sequence. To date, the most reliable secondary structures are largely based on this principle. A detailed discussion of this method is given elsewhere in this volume.[31]

The demonstration of so-called covariations or compensatory base changes is also of utmost importance for proposing pseudoknotted structures. Pseudoknots differ essentially, however, from other structural elements in the sense that pseudoknots can be considered proved only when covariations are found in at least two stem regions (S_1 and S_2 of Fig. 2). The criteria for proof or disproof of a helical segment in a pseudoknot do not differ from those of any other putative helix. The rule adopted to date is that at least two independent covariations are needed per stem region.[31] This means that four or more covariations have to be found in a pseudoknotted structure.

As outlined previously, sequence comparisons in combination with computer-aided predictions of secondary structure can be quite successful in detecting pseudoknots and can be sufficient, provided that enough related sequences are available. A number of pseudoknots were thus revealed in the noncoding regions of plant viral RNAs.[4,22] The evidence for a proposed pseudoknot can be reinforced when the complementarity is not preserved only among various strains of a particular viral RNA, but also among duplications of a pseudoknot in one RNA chain. This was found to be the case in the 5′-noncoding region of foot and mouth disease virus (FMDV) RNA.[32]

Predicting Pseudoknots

One of the major aims in molecular biology is to predict a functionally meaningful three-dimensional structure of a protein or nucleic acid on the basis of sequence data alone. Especially for RNA, the fulfillment of this

[30] G. E. Fox and C. R. Woese, *Nature (London)* **256**, 505 (1975).
[31] B. D. James, G. J. Olsen, and N. R. Pace, this volume [18].
[32] B. E. Clarke, A. L. Brown, K. M. Currey, S. E. Newton, D. J. Rowlands, and A. R. Carroll, *Nucleic Acids Res.* **15**, 7067 (1987).

aim is yet to be realized. Pseudoknots have been ignored or not allowed thus far in all attempts to deduce a RNA structure from a sequence, for reasons of inherent complexity of the algorithms to be developed or because naturally occurring examples were not known at the time[11,13] (see elsewhere in this volume[33]). Based on our findings on some plant viral RNAs, an attempt was made by Abrahams et al.[34] to develop a computer program that considers pseudoknotting as a possible step during the folding of a RNA chain. A summary of the principles and the merits of this program follows.

The algorithm, developed by Abrahams et al., turns out to be basically the same as the one reported by Martinez[35] in that the algorithm simulates the RNA folding process itself. It is postulated that this folding is equivalent to adding one stem at a time to a growing structure. Abrahams and co-workers assumed that each next stem to be added is the one which contributes most to lowering the free energy and which is compatible with the intermediate structure. For the calculation of the ΔG of the stem regions to be added to the intermediate structure, stacking interactions with the structure are taken into account. This procedure leads to a series of intermediate structures that are well defined, stable, and not subject to rearrangements.

An essential feature of this algorithm is that the formation of a pseudoknotted structure is allowed on the addition of the next double-helical stem region. In a pseudoknot of type H (see Fig. 2), the two segments are stacked, and this energy gain is taken into account by the program. The structure thus obtained does not necessarily have the global energy minimum as is found by the algorithm developed by Zuker and Stiegler.[11,33]

More specifically the following three steps can be discerned in the algorithm.

1. List all possible stems in a given sequence.
2. Incorporate that stem in the structure which adds most to the stability of the structure.
3. Update the list of remaining stems, dependent on the new intermediate structure formed, by trimming incompatible stems and by calculating the new ΔG values.

These steps are repeated until no further energy gain is possible. The calculation of the free energies associated with the stem regions uses currently available thermodynamic parameters for base-pair stacking and

[33] M. Zuker, this volume [20].
[34] J. P. Abrahams, E. van Batenburg, and C. W. A. Pleij, to be published.
[35] H. M. Martinez, Nucleic Acids Res. 12, 323 (1984).

loop formation. An important exception was made for large loops. For instance, energy values for hairpin loops containing more than six nucleotides were not derived from the Jacobson–Stockmayer equation (see Ref. 24), but rather were derived from a logarithmic extrapolation of the experimental values for small loop sizes. The consequence of this calculation is an extra penalty for long-range interactions. A fundamental problem in such an algorithm is to introduce energy parameters for each structural motif that is created by pseudoknotting. Until now, no experimental values have been published whatsoever. A first attempt to collect data for the pseudoknot of the H type was published recently.[36] Based on the experimental work on the pseudoknotted structures in the noncoding regions of some plant viral RNAs, such as TMV RNA,[2] it is estimated that the loop energies for both L_1 and L_2 in the H-type pseudoknot (see Fig. 2) will be of the order of 3–5 kcal/mol. The current value used is set to 4.1 kcal/mol. This seems to be reasonable, when these so-called connecting loops are compared to the unstacked regions in normal hairpin loops. Apart from possible modifications, the seven-membered anticodon loop of tRNA should have a ΔG of 4.5 kcal/mol.[24] In this loop, five of the seven bases are stacked, and the two unstacked ones are more or less equivalent to L_1 in the H-type pseudoknot (Fig. 2). Also, the findings about the minimal size of L_1 and L_2, as outlined previously, were introduced in the program. Some types of more complicated pseudoknotted structural elements were ruled out by assigning them high, positive ΔG values. The RNA structure-predicting program of Abrahams and co-workers, written in APL, was adapted for running on an IBM or IBM-compatible personal computer and is able to fold a sequence of 1000 nucleotides overnight.

This program was tested on a large number of different RNAs, some of which contained pseudoknots. A better score was obtained for the prediction of the cloverleaf of tRNAs, as compared to the widely used program of Zuker and Stiegler.[11] The current version of the program of Abrahams and co-workers is able to predict the pseudoknot in the tRNA-like structure of TYMV RNA (Fig. 3) and finds five out of the six stems of the three consecutive pseudoknots in the 3'-noncoding region of TMV RNA (Fig. 5a). The program is therefore reasonably successful in predicting pseudoknots of type H. The program also revealed some other new pseudoknot structures both in plant viral and animal viral RNAs, which could be confirmed by sequence comparisons.[22,32] However, occasionally loop–loop interactions are predicted which appear sterically less likely. It is clear that the current program needs further improvement on this and

[36] J. D. Puglisi, J. R. Wyatt, and I. Tinoco, Jr., *Nature (London)* **331**, 283 (1988).

other points, which, in turn, is dependent on new experimental data or on new pseudoknots found in natural RNAs.

Concluding Remarks

There is little doubt that pseudoknots are important structural elements in RNA that can play a decisive role in the higher order folding. Pseudoknots occur in a variety of natural RNAs, such as ribosomal RNA,[5] mRNAs,[6,8] ribozymes,[37] and plant viral RNAs.[4] The main techniques, which give evidence for their existence, are chemical modification, enzymatic digestion, computer-aided prediction, and phylogenetic sequence comparisons. Of these techniques, the search for covariations in the double-helical stem regions of the pseudoknot is probably the most powerful method to date. Reliable proposals can usually be made only when a combination of these methods is used, as is also true for most other elements in a RNA (secondary) structure.

A serious drawback in the analysis of RNA pseudoknots is the almost complete absence of structural data from biophysical techniques, such as nuclear magnetic resonance (NMR) or X-ray diffraction. These methods not only may prove pseudoknotting itself, but may also answer the important questions about the stacking of the constitutive double-helical segments or about the conformation of the connecting loops. These techniques also may yield the thermodynamic parameters that are required for a successful prediction of pseudoknots in a given RNA sequence.

Acknowledgments

We thank Krijn Rietveld, Alex van Belkum, and Paul Verlaan of our laboratory for their valuable contributions to the study of RNA pseudoknots. The pseudoknot-predicting program was developed in collaboration with Eke van Batenburg of the Department of Theoretical Biology, Leiden University, and Jan Pieter Abrahams of the Department of Biochemistry, Leiden University. We gratefully acknowledge the collaboration with Richard Giege and Dino Moras of the Institut de Biologie Moleculaire et Cellulaire du Centre National de la Recherche Scientifique (CNRS), Strasbourg, France.

[37] B. D. James, G. J. Olsen, J. Liu, and N. R. Pace, *Cell* (*Cambridge, Mass.*) **52,** 19 (1988).

[22] Absorbance Melting Curves of RNA

By Joseph D. Puglisi and Ignacio Tinoco, Jr.

The transition between an ordered, native structure and a disordered, denatured state in a nucleic acid can be conveniently monitored using ultraviolet (UV) absorbance. As the ordered regions of stacked base pairs are disrupted, the UV absorbance increases. The increase in absorbance is called hyperchromicity; the absorbance in the disordered state approaches the sum of the absorbances of the constituent nucleotides. Thus, the absorbance of a native nucleic acid is hypochromic relative to its nucleotides; the amount of hypochromicity is a measure of the base pairing and stacking of the secondary structure. The easiest way to denature a nucleic acid is by heating. The resulting profile of absorbance versus temperature is called a melting curve, due to its similarity in appearance to a phase transition. From the absorbance data, a curve of the fraction of one component versus temperature can be constructed; the midpoint of the transition is defined as the melting temperature, T_m. Qualitative and quantitative information about conformations of RNA molecules can be obtained from measurement of UV absorbance melting curves. The percentage of hyperchromicity on melting at a chosen wavelength depends on the number and type of base pairs broken. The dependence on RNA strand concentration of the T_m of a melting transition yields information on the molecularity of a transition (unimolecular hairpin to coil, bimolecular duplex to single strand, etc.). This information can be obtained from a qualitative analysis of the melting curves. Further analysis can yield quantitative thermodynamic data for the melting transition. This chapter will deal with the experimental methods needed to acquire a melting curve and the analysis and interpretation of the data.

Experimental Methods

Any standard commercial UV spectrophotometer can be equipped to measure melting curves. A useful instrument is a single-beam Gilford (Oberlin, OH) spectrophotometer (Model 2530) with an automated reference compensator that allows melting curves to be obtained on three separate samples simultaneously. Four cuvettes are placed in a thermistor-controlled thermoelectric cell holder, which is connected to a thermoprogrammer that controls the heating rate; three cuvettes are for samples, the fourth cuvette contains the reference solvent. The temperature also

METHODS IN ENZYMOLOGY, VOL. 180

can be controlled using a circulating water–ethylene glycol bath with equal effectiveness. Data can be collected on any microcomputer, which can be interfaced to the spectrometer through a RS-232 port. Absorbance data can be collected over the temperature range from ~0 to 90° at a heating rate of 1–0.25°/min. Absorbance data should be acquired about every 0.2°. The sampling rate, controlled by the Gilford spectrophotometer, should be set to measure at least 1 sec at each temperature point. Each data point can be the average of several absorbance readings and temperature readings per point; this gives smoother data at lower absorbance.

A heating rate should be chosen that is consistent with the rates of the processes being observed. The Gilford thermoprogrammer can heat at rates of 1, 0.5, and 0.25°/min. For the normal melting of short duplexes to single strands, the rates of the forward and reverse reactions in the transition region are fast enough to justify a heating rate of 1°/min. Since a correct melting curve is an equilibrium measurement, this assumption should be tested by measuring the melting curve at a lower heating rate; the two curves should be identical. Some processes have very slow kinetics. Sequences that form hairpins can exist in either a hairpin (monomer) form or in a duplex with an internal loop (dimer). At higher strand concentrations, the dimer may be favored at low temperature, with the dimer first melting to hairpins and then to single strands. The equilibrium melting curve for these processes should be measured at a very low heating rate, since the dimer to hairpin transition can be very slow. Melting curves measured at too high a heating rate will give erroneous results. A decade divider circuit can be added to the Gilford thermoprogrammer, which allows heating rates as low as 0.025°/min to be used. Once the heating rate has been set, the data acquisition rate should be adjusted to acquire data every 0.1–0.4°. Normally, a melting curve data set consists of 200–400 data points. After a melting experiment, the sample should be cooled to the starting temperature, and the final absorbance should be compared to the initial absorbance. Any evaporation or hydrolysis of the sample will result in a rise in absorbance. Differences of ~1% or less are acceptable.

A wide range of cell path lengths are available for UV melting experiments. The standard 1-cm path-length cell is usually the longest path length used; 5-, 2-, and 1-mm cells are commercially available, and path lengths of 0.1 mm are obtainable with spacers. This means a concentration range of about a factor of 1000 can be measured by using an absorbance of 0.2 in a 1-cm cell and an absorbance of 2 in a 0.1-mm cell (absorbance $A = -\log(I/I_0) = \varepsilon cl$, where c is the strand concentration, l is the path length, and ε is the extinction coefficient per strand). Of course, one must realize that an absorbance of 2 means only 1% of the light is transmitted, and thus a poor signal-to-noise ratio results. Although the absorbance is usually measured relative to a reference cell containing only solvent, the absolute absorbance of all sample cells relative to air should not be above 2.

Cells that are 0.5 cm wide, rather than the normal 1-cm-wide cells, are preferred, since only 250–300 μl of sample are required; these cells also allow faster and more even heating of the sample. Shorter path-length cells should be used in conjunction with aluminum adapters, such that the cells are in thermal contact with the sample holder. The 10- and 5-mm path-length cells are sealed with Teflon stoppers. Since 1- or 2-mm path-length cells are usually stopperless, they are sealed with a small amount (10 μl) of Dow silicone oil (Corning, 200-fluid 20-centipoise viscosity). This oil completely prevents evaporation. Sealing shorter path-length cells requires careful attention. For a 0.1-mm path, about 7 μl of sample is placed at the bottom of a 2-mm cuvette, and the spacer is carefully slid in. Any bubbles in the sample should be removed before adding the oil. This can be done by briefly (30 sec) centrifuging the cuvette placed in an Eppendorf tube. A small amount of oil is then added; if excess oil is added, the oil can creep down the sides of the cell during an experiment and cause spurious results. If the sample is to be recovered, the sample must be separated from the silicone oil. This can be accomplished by pipetting the sample–oil mixture onto a Teflon dish. The oil will stick to the Teflon, as the bead of sample is rolled around the dish.

One major advantage of using UV spectroscopy is the high sensitivity of the method. Normally, the absorbance of the sample used should be between 0.2 and 2.0. The value of the extinction coefficient for RNA molecules at their absorbance maximum is typically 10^5–10^6, so for an absorbance of 0.2, concentrations as low as 0.5 μM in RNA strands can be studied. A typical volume for a sample is 0.3 ml, so that nanomoles of RNA are needed.

Sample preparation for UV melting studies is straightforward. The RNA stock solution is prepared by dialysis against the desired buffer, and different concentrations are made by dilution. Samples should then be degassed in preparation for a melting experiment. Oxygen dissolved in the sample will form bubbles at higher temperatures, which will scatter light and affect absorbance measurements. Simple degassing procedures are to either bubble N_2 or argon through the sample for about 10 min to saturate the sample with these gases or to subject the sample to a vacuum for about 5 min. Care should be taken by moderating the vacuum to avoid vigorous bumping of the sample. Degassing is especially important for short path-length (2 mm or less) samples; bubbles seem to form easier in these cells. Each sample should be heated above its melting point (to 80–

90°) and allowed to equilibrate at the starting temperature (normally 0–5°) for a minimum of 15 min before the melting curve is determined. After heating, check for air bubbles. For equilibria with slow kinetics, such as hairpin–duplex transitions, longer equilibration times are necessary. Equilibration can be monitored by the change of absorbance as a function of time. For measurements below 20°, the sample compartment should be purged with N_2 to prevent moisture condensation on the cells.

Since the melting behavior will depend on the solution conditions, choice of solvent is very important. Most work on model compounds[1,2] was done in 1 M NaCl and 10 mM sodium cacodylate or phosphate, pH 7.0 [with 0.1 or 1 mM ethylenediaminetetraacetic acid (EDTA); note that 1 mM EDTA has a high A_{260}]. The high salt concentration was chosen to minimize electrostatic repulsion between strands and to avoid divalent ions, which catalyze hydrolysis of RNA and favor triple-strand formation. This solvent provides a standard condition for measuring melting curves and for comparing results with previously published data. However, other salt conditions may be needed depending on the sequence and structure being studied. For example, the T_m of a structure may be too high (>80°) in 1 M NaCl to allow analysis of the melting curve (see Fig. 2b).

The formation of more complex secondary and tertiary structures often requires the addition of divalent ions, which bind very specifically to stabilize the structure (a good example is tRNA).[3] Lower monovalent ion concentrations are also used in these situations to avoid competition with divalent ion binding. The buffer of choice is usually phosphate, which has a very low temperature coefficient. Cacodylate is sometimes used because of its lower binding constant to divalent ions. Traditional biochemical buffers, such as Tris and HEPES, should not be used because of their high temperature variability ($\Delta p K_a/° = -0.031$ and -0.014, respectively).[4] Ideally, RNA samples should be dialyzed into the desired buffer, especially for melting curves in low salt (<100 mM). A flow dialysis apparatus (Microdialysis System, BRL, Gaithersburg, MD) is ideal for the dialysis and recovery of small volumes (approximately 300 μl). The use of 1000 molecular-weight cutoff dialysis tubing (Spectrum Medical Industries, Los Angeles, CA) allows the safe dialysis of oligonucleotides as small as 8 nucleotides. A typical dialysis sequence for the preparation of a sample in 50 mM NaCl, 10 mM sodium phosphate, and 0.1 mM

[1] S. M. Freier, R. Kierzek, J. A. Jaeger, N. Sugimoto, M. H. Caruthers, and D. H. Turner, *Proc. Natl. Acad. Sci. U.S.A.* **83**, 9373 (1986).

[2] I. Tinoco, Jr., P. N. Borer, B. Dengler, M. D. Levine, O. C. Uhlenbeck, D. M. Crothers, and J. Gralla, *Nature (London) New Biol.* **246**, 40 (1973).

[3] A. Stein and D. M. Crothers, *Biochemistry* **15**, 157 (1976).

[4] Calbiochem, "Buffers" Behring Diagnostic, La Jolla, California, 1985.

EDTA, pH 7, is as follows: dialyze once against 50 mM NaCl, 10 mM sodium phosphate, and 10 mM EDTA, pH 7 (12 hr) and once against final buffer. The dialysis against a high EDTA concentration, in the presence of NaCl, is crucial for the removal of any divalent impurities.

In general, salt will bind preferentially to the state with the greatest charge density. For nucleic acids, the double-stranded state has a greater charge density than the single strands. We can write the equilibrium as follows:

$$M_n \cdot D \rightleftharpoons M_{n'} \cdot S_a + M_{n''} \cdot S_b + \Delta n \, M$$

where M is an ion or any small molecule that binds differentially to the states and Δn is the net number of salt molecules released or bound per mole of double strands (D) melted to single strands (S) ($\Delta n = n - n' - n''$). The equilibrium constant for this dissociation is

$$K = [(M_{n'} \cdot S_a)(M_{n''} \cdot S_b)(M)^{\Delta n}]/(M_n \cdot D) \tag{1}$$

and $d \ln K/d \ln(M) = \Delta n$, which is the change in the number of ions bound. Usually we want to know how the T_m changes with salt concentration. Using the van't Hoff equation in the form

$$d \ln K/dT = \Delta H^\circ/RT^2 \tag{2}$$

one obtains

$$dT_m/d \ln(M) = -\Delta n \, RT_m^2/\Delta H^\circ \tag{3}$$

where ΔH° is the standard enthalpy change of melting the duplex; ΔH° is a positive quantity. Since, Δn is positive (ions are released on melting), the T_m will increase with the salt concentration. Experimentally, up to about 1 M salt concentration, the T_m increases with increasing salt. For higher salt concentrations the T_m may decrease; salts, such as $NaClO_4$ at 6 M or above can lower the T_m by 30–40°. Below 0.1 M salt, T_m is linear in the logarithm of the salt concentration for DNA, RNA, and hybrid polynucleotide helices. For each factor of 10 increase in NaCl, the T_m increases 17–19°, but depends on base composition; the increase reaches a plateau by 1 M salt.[5] The increase for oligonucleotides is smaller, but the increase is still substantial. The melting of triple strands will be more dependent on salt concentration than double strands; the increase of T_m with 10-fold salt increase is about 30° for triple-stranded polynucleotides.

The wavelength of UV light that is most useful for a melting curve varies between 240 and 280 nm. The absorbance maximum for most RNAs is 260 nm, and this is the wavelength of maximum hyperchro-

[5] M. T. Record, Jr., *Biopolymers* 5, 975 (1967).

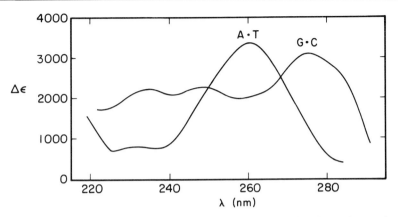

FIG. 1. Change in extinction coefficient ($\Delta\varepsilon$) for melting of G–C and A–T base pairs in DNA as a function of wavelength. (Data from Felsenfeld and Hirchman.[6])

micity. This is the wavelength normally monitored in a melting experiment. However, RNA structures with a preponderance of A · U or G · C base pairs should be monitored at different wavelengths. As shown in Fig. 1, A · T base pairs in DNA show a maximum hypochromicity at 240 nm, while G · C base pairs have a maximum hypochromicity at 280 nm[6]; similar effects occur for RNA. A comparison of melting curves (the percentage of hypochromicity) at these different wavelengths can give information about the base composition of the double-stranded structures that are melting (A · U rich versus G · C rich).

Data Analysis

Typical absorbance versus temperature profiles are shown in Fig. 2a and b. The following sections will describe how to analyze and interpret these data. The first important parameter is the total strand concentration. This can be determined spectrophotometrically using $A = \varepsilon c l$. The extinction coefficient, ε, for any native structure can be determined experimentally by hydrolyzing the RNA to nucleotides and by measuring the A_{260} of the resulting mixture. This gives the molar concentration of the nucleotides (the base composition of the RNA must be known), and thus the extinction coefficient of the RNA can be determined. The extinction coefficient for a single-stranded molecule can be estimated, if one assumes only nearest-neighbor interactions among the bases in the sequence. For

[6] G. Felsenfeld and S. Z. Hirschman, *J. Mol. Biol.* **13,** 407 (1965).

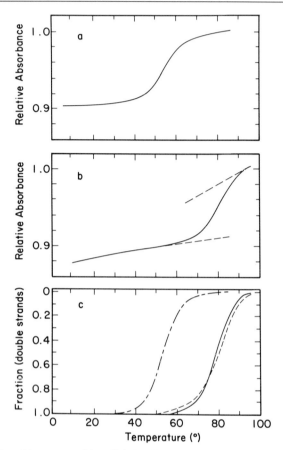

FIG. 2. (a) A melting curve with well-defined base lines; 5′-GCGAUUUCUGACCGC-3′ in 50 mM NaCl, 10 mM sodium phosphate, and 0.1 mM EDTA, pH 6.5. Normalized absorbance at 260 nm is plotted versus temperature. (b) A melting curve with undefined upper base line. 5′GGGAGUUUCCGCUCCC-3′ in 1 M NaCl, 10 mM sodium phosphate, and 0.1 mM EDTA, pH 7.0. Normalized absorbance (——) at 260 nm is plotted versus temperature with the approximate upper and lower base lines (-----) used for data analysis. (c) Fraction double strands (f) versus temperature (T) curves corresponding to a (——) and b. The two f versus T profiles for melt (b) correspond to unsubtracted base lines (-----) and subtracted base lines (——).

example, the calculation of the extinction coefficient for 5′-ApCpGpUp … is

$$\varepsilon(\text{ApCpGpU} \ldots) = 2[\varepsilon(\text{ApC}) + \varepsilon(\text{CpG}) + \varepsilon(\text{GpU}) = \ldots]$$
$$- [\varepsilon(\text{Cp}) + \varepsilon(\text{Gp}) + \varepsilon(\text{Up}) + \ldots]$$

TABLE I
EXTINCTION COEFFICIENTS FOR NUCLEOTIDES AND DINUCLEOSIDE PHOSPHATES TO
CALCULATE EXTINCTION COEFFICIENTS OF SINGLE STRANDS[a]

Nucleotide or dinucleoside phosphate	$\varepsilon(260)\ M^{-1}\ cm^{-1} \times 10^{-3}$ RNA	DNA	Nucleotide or dinucleoside phosphate	$\varepsilon(260)\ M^{-1}\ cm^{-1} \times 10^{-3}$ RNA	DNA
Ap	15.34	15.34	CpG	9.39	9.39
Cp	7.60	7.60	CpU(CpT)	8.37	7.66
Gp	12.16	12.16	GPA	12.92	12.92
Up(Tp)	10.21	8.70	GpC	9.19	9.19
ApA	13.65	13.65	GpG	11.43	11.43
APC	10.67	10.67	GpU(GpT)	10.96	10.22
ApG	12.79	12.79	UpA(TpA)	12.52	11.78
ApU(ApT)	12.14	11.42	UpC(TpC)	8.90	8.15
CpA	10.67	10.67	UpG(TpG)	10.40	9.70
CpC	7.52	7.52	UpU(TpT)	10.11	8.61

[a] At 260 nm, 25°, 0.1 ionic strength, pH 7. Data calculated by K. H. Johnson and D. M. Gray, Program in Molecular Biology, The University of Texas at Dallas, based on nucleotide extinction coefficients from the Ph.D Thesis of M. Alexis, University of London, 1978, provided by E. G. Richards, and hypochromicity data from the Ph.D. Thesis of M. M. Warshaw, University of California, Berkeley, 1966. The extinction coefficients are estimated to be valid to ±0.10.

where $\varepsilon(ApC)$, etc. are the extinction coefficients for the component dinucleoside phosphates per mole of nucleotides (this is the reason for the factor of two), and $\varepsilon(Cp)$ . . . are the extinction coefficients for the nucleotides (note that the two end nucleotides are not subtracted). Shown in Table I are the extinction coefficients for the mononucleotides and dinucleoside phosphates needed to calculate the single-strand extinction coefficient at 25°. These calculations are good to within 10% of the true values. As can be seen in Fig. 2a, the extinction coefficients for the double-stranded and single-stranded states depend on temperature. This dependence is usually approximately linear, so that the single-strand absorbance can be extrapolated to 25° for use with the calculated single-strand extinction coefficient.

Melting curves can be analyzed either semiquantitatively to determine hypochromicity and T_m or quantitatively to determine thermodynamic parameters. First, one must determine the type of transition represented by the melting curve; melting curves can be normalized to the same absorbance to allow comparison of different curves. Table II gives the relevant equations for the common transitions that can be studied by

TABLE II
TWO-STATE ANALYSIS OF NUCLEIC ACID TRANSITIONS

Reaction type	Equilibrium constants	ΔH° from slope of f versus T	Concentration dependence of T_m
Monomolecular $S = H$	$K = \dfrac{[H]}{[S]} = \dfrac{(f)}{(1-f)}$	$\Delta H^\circ = 4RT_m^2 \left(\dfrac{df}{dT}\right)_{T=T_m}$	
Bimolecular (self-complementary) $2S = D$	$K = \dfrac{[D]}{[S]^2} = \dfrac{f}{2(1-f)^2 c_t}$	$\Delta H^\circ = 6RT_m^2 \left(\dfrac{df}{dT}\right)_{T=T_m}$	$\dfrac{1}{T_m} = \dfrac{R}{\Delta H^\circ} \ln c_t + \dfrac{(\Delta S^\circ - R \ln 4)}{\Delta H^\circ}$
Bimolecular (non-self-complementary) $S_A + S_B = D$	$K = \dfrac{[D]}{[S_A][S_B]} = \dfrac{2f}{(1-f)^2 c_t}$	$\Delta H^\circ = 6RT_m^2 \left(\dfrac{df}{dT}\right)_{T=T_m}$	$\dfrac{1}{T_m} = \dfrac{R}{\Delta H^\circ} \ln c_t + \dfrac{(\Delta S^\circ - R \ln 4/3)}{\Delta H^\circ}$
Trimolecular (identical strands) $3S = T$	$K = \dfrac{[T]}{[S]^3} = \dfrac{f}{3c_t^2(1-f)^3}$	$\Delta H^\circ = 8RT_m^2 \left(\dfrac{df}{dT}\right)_{T=T_m}$	$\dfrac{1}{T_m} = \dfrac{2R}{\Delta H^\circ} \ln c_t + \dfrac{[\Delta S^\circ - R \ln 4/3]}{\Delta H^\circ}$
Trimolecular (nonidentical strands) $S_A + S_B + S_C = T$	$K = \dfrac{[T]}{[S_A][S_B][S_C]} = \dfrac{9f}{c_t^2(1-f)^3}$	$\Delta H^\circ = 8RT_m^2 \left(\dfrac{df}{dT}\right)_{T=T_m}$	$\dfrac{1}{T_m} = \dfrac{2R}{\Delta H^\circ} \ln c_t + \dfrac{(\Delta S^\circ - 2R \ln 6)}{\Delta H^\circ}$
Tetramolecular (identical strands) $4S = Q$	$K = \dfrac{[Q]}{[S]^4} = \dfrac{f}{4c_t^3(1-f)^4}$	$\Delta H^\circ = 10RT_m^2 \left(\dfrac{df}{dT}\right)_{T=T_m}$	$\dfrac{1}{T_m} = \dfrac{3R}{\Delta H^\circ} \ln c_t + \dfrac{(\Delta S^\circ\ R \ln 2)}{\Delta H^\circ}$
Tetramolecular (nonidentical strands) $S_A + S_B + S_C + S_D = Q$	$K = \dfrac{[Q]}{[S_A][S_B][S_C][S_D]} = \dfrac{64f}{c_t^3(1-f)^4}$	$\Delta H^\circ = 10RT_m^2 \left(\dfrac{df}{dT}\right)_{T=T_m}$	$\dfrac{1}{T_m} = \dfrac{3R}{\Delta H^\circ} \ln c_t + \dfrac{(\Delta S^\circ - 3R \ln 8)}{\Delta H^\circ}$

melting experiments. Unimolecular (hairpin to coil) and multimolecular transitions (duplex or triple strand to coil) can be distinguished by varying the concentration of RNA strands. The T_m and the shape of the melting curve [fraction of species versus temperature (T)] should be independent of strand concentration for a unimolecular transition. For multimolecular transitions, T_m and the shape of the melting curve will depend on concentration. The concentration should be varied over a factor of about 100 to test the concentration dependence; a smaller concentration variation may miss the change in T_m. It should be noted that the melting of long RNA duplexes, such as poly[(rA)·(rU)], is concentration independent, although bimolecular; the concentration-dependent initiation event is negligible for a long enough polynucleotide.

It is difficult to distinguish between types of multimeric transitions, since the T_m values for all the transitions will depend on strand concentration. Varying the stoichiometry of the reacting strands[7] or determining the size of the constituent species by electrophoresis[8] should be done to determine the molecularity of a structure. One should be especially careful when working with sequences that are partly self-complementary, since these sequences can exist as either duplex or hairpin structures. However, once the type of transition has been identified, the melting curve can be analyzed using the theory of helix-to-coil transitions.

Standard helix-to-coil theory[9] describes transitions between a native structure and a melted structure (single strands). A helix-to-coil transition can be induced by changing any of a number of thermodynamic variables—pressure, concentration of reactants, concentration of salts, and most commonly, temperature. The transition can be monitored as a function of temperature by any physical property that is dependent on the number of base pairs formed. For example,

$$A = fA_D + (1 - f)A_S \qquad (4)$$

where f is the fraction of bases paired and A_D and A_S are the values of the property for the single-stranded (S) and double-stranded (D) species, respectively. The physical property A is usually UV absorbance, but can also be circular dichroism, NMR chemical shift, etc. The origin of hypochromicity is the electronic interactions between neighboring stacked bases. Theoretical calculations[10] and experimental data[11] show that per-

[7] C. Stevens and G. Felsenfeld, *Biopolymers* **2**, 293 (1964).
[8] N. R. Kallenbach, R.-I. Ma, and N. Seeman, *Nature (London)* **305**, 829 (1983).
[9] D. Poland and H. A. Scheraga, "Theory of Helix-Coil Transitions in Biopolymers." Academic Press, New York, 1970.
[10] I. Tinoco, Jr., *J. Am. Chem. Soc.* **82**, 4785 (1960).
[11] A. Rich, and I. Tinoco, Jr., *J. Am. Chem. Soc.* **82**, 6409 (1960).

centage of hypochromicity

$$\text{Hypochromicity } (\%) = (A_S - A_D)/A_S$$

is approximately a linear function of the number of stacked bases. UV absorbance therefore monitors the fraction of bases that are stacked as a nucleic acid molecule is melted.

A melting curve can be analyzed using Eq. (4); this curve relates the absorbance (or other property) to a profile of fraction of bases paired (f) versus temperature. The T_m is the temperature where $f = 0.5$. So far, we have assumed no explicit model for how the native double-stranded state is in equilibrium with single strands. In order to derive thermodynamic parameters for the transition ($\Delta H°$, $\Delta S°$, and $\Delta G°$), the absorbance melting curve must be translated into the concentrations of the single-stranded and double-stranded states. This is done most simply by assuming a two-state (all-or-none) model.[12,13] This model assumes that single strands are in equilibrium with only one base-paired native structure; there are no partially base-paired intermediates in the melting process. This approximation is most appropriate for short (<12 bp) duplexes.[14] If the two-state model is accurate, f is the fraction of fully base-paired strands. Expressions are given in Table II for equilibrium constants as a function of f and c_t, where c_t is the total concentration of RNA strands (equimolar amounts of complementary strands are assumed). Expressions are also given for obtaining $\Delta H°$ from the slope of f versus T at the T_m and from the concentration dependence of T_m. Equations can easily be derived for reactions of molecularity greater than three (for example, four-stranded cruciform structures).[15] If we assume a two-state model, a melting curve can be converted to a fraction native structure versus temperature profile, which in turn gives an equilibrium constant at each temperature. These data can then be treated using the van't Hoff relation

$$d \ln K/d(1/T) = -\Delta H°/R \qquad (5)$$

and standard thermodynamic equations

$$\Delta G° = -RT \ln K \qquad (6)$$
$$\Delta S° = (\Delta H° - \Delta G°)/T \qquad (7)$$

to obtain the standard enthalpies, entropies, and free energies per mole of the reaction: $\Delta H°$, $\Delta S°$, and $\Delta G°$ (see Thermodynamic Parameters for details).

[12] J. Applequist and V. Damle, *J. Am. Chem. Soc.* **87**, 450 (1965).
[13] C. R. Cantor and P. R. Schimmel, "Biophysical Chemistry," Vol. 3. Freeman, San Francisco, California, 1980.
[14] K. J. Breslauer, J. M. Sturtevant, and I. Tinoco, Jr., *J. Mol. Biol.* **99**, 549 (1975).
[15] L. A. Marky, N. R. Kallenbach, K. A. McDonough, N. C. Seeman, and K. J. Breslauer, *Biopolymers* **26**, 1621 (1987).

For longer base-paired sequences (>12 bp), the helix-to-coil transition is usually not two state.[14] RNA molecules with complex structures often melt in stages,[16] with separate regions of structure melting independently (see Complex RNA Molecules). This type of non-two-state transition may produce a shoulder in the melting curve. However, even melting curves that appear two state can involve intermediates in the melting process.[14] The analysis of transitions, where the double-stranded (native) and single-stranded states are in equilibrium with a significant number of partially base-paired intermediates, usually requires the use of statistical mechanics.[12]

Thermodynamic Parameters

Once the type of transition (unimolecular, bimolecular, etc.) is known, thermodynamic quantities can be derived using several different methods; Table II gives the relevant equations. The methods are (1) plot ln K versus $1/T$ at a single concentration, (2) plot (df/dT) versus T for a single concentration, (3) fit an absorbance versus T profile at a single concentration, and (4) plot ln c_t versus $1/T_m$. Clearly, the last method cannot be used for unimolecular transitions. The advantages and disadvantages of each method are evaluated below; all methods, except for the third method above, require that absorbance curves be converted to f versus T curves.

The absorbance versus temperature profile is converted into the fraction of molecules base paired (f) versus T, using Eq. (4). The temperature dependence of the extinction coefficients of the double strands and single strands must be taken into account. The sloping upper and lower base lines in absorbance versus T plots (the so-called base-line problem) is caused by the single-stranded and double-stranded states changing with temperature. The temperature dependence is usually approximated by assuming a linear dependence of the extinction coefficients on temperature where m and b are the slope and the intercept, respectively.

$$\varepsilon_D = m_D T + b_D \tag{8}$$
$$\varepsilon_S = m_S T + b_S \tag{9}$$

If the base lines are well defined (10–15° of linear absorbance versus T) as in Fig. 2a, ε_D and ε_S can be determined using a linear least-squares fit of the absorbance data in the base-line region. However, many times either one or both base lines are not well defined. If the T_m is too high (>80°) or too low (<20°), the upper or lower base lines may not be sufficiently defined for a good least-squares fit. In these situations, the data should be analyzed using different approximations for the base lines. Figure 2c

---

[16] P. E. Cole, S. K. Yang, and D. M. Crothers, *Biochemistry* **11**, 4358 (1972).

shows the fraction versus temperature profiles for data from Fig. 2b, generated using two different choices for the base lines. Recognizing which base line is best will be discussed below. A poor choice of base line can be the largest source of error in determining thermodynamic parameters from UV melting data.

Method 1: ln K versus 1/T. Values of K are calculated at each temperature from f, using the appropriate equation from Table II. Normally, only points with $0.15 \leq f \leq 0.85$ are used in the van't Hoff plot, because this is the region where K values are most precise. From Eqs. (5–7), one sees that a plot of ln K versus $1/T$ yields $(-\Delta H^\circ/R)$ as the slope and $(\Delta S^\circ/R)$ as the intercept; this is a van't Hoff plot. If ΔH° is independent of temperature, the plot should be linear. A nonlinear van't Hoff plot can result from several factors: temperature dependence of ΔH°, poor choice of base lines, or a non-two-state transition.[17] The most likely candidate is a poor choice of base lines.[14] In this case, base lines should be adjusted, and the data reanalyzed. It is important that the van't Hoff data actually be plotted, so that a poor least-squares fit due to errors in the data can be distinguished from a nonlinear plot.

Method 2: df/dT. This is a variation of the van't Hoff analysis that involves numerical differentiation of the f versus T data. The differentiation can be done using a nonlinear least-squares fit to a quadratic equation at each data point.[18] ΔH° can be obtained from df/dT at the T_m (at $f = 0.5$), as shown in Table II; note that in general, the T_m is not the maximum in the derivative (df/dT) versus T plot. Expressions relating df/dT to ΔH° at any value of f have been derived by Gralla and Crothers[19] and discussed by Marky and Breslauer.[20] Thus, ΔH° can be determined from any portion of the derivative curve, such as the full-width, or half-width, of the derivative curve at half-height. For example, ΔH° can be determined from only the upper half of a melting curve (i.e., for a molecule with very low T_m). The advantage of this method is that the results are less sensitive to the choice of base lines than the first method. A good test of the choice of base lines is comparison between the ΔH° determined using the van't Hoff plots and the derivative method. If the results using both methods do not agree to within ~5%, then the base lines should be readjusted, and the data reanalyzed. For longer duplexes, the transitions are sharper and the base lines are better defined; in these cases, disagreement between the two methods is more likely due to a non-two-state transition.

[17] M. Petersheim and D. H. Turner, *Biochemistry* **22**, 256 (1983).
[18] P. R. Bevington, "Data Reduction and Error Analysis for the Physical Sciences." McGraw-Hill, New York, 1969.
[19] J. Gralla and D. M. Crothers, *J. Mol. Biol.* **73**, 497 (1973).
[20] L. A. Marky and K. J. Breslauer, *Biopolymers* **26**, 1601 (1987).

Method 3: A versus T. The experimental absorbance versus temperature curve is fit directly to six parameters: ΔH°, ΔS°, and the four parameters that specify the slopes and intercepts of the upper and lower base lines. The raw melting curve is fit[17] by the Marquardt nonlinear least-squares method[18] to the following equations

$$
\begin{aligned}
A &= fA_D + (1 - f)A_S \\
A_D &= \varepsilon_D c_t l = (m_D T + b_D)c_t l \\
A_S &= \varepsilon_S c_t l = (m_S T + b_S)c_t l \\
K &= \exp(-\Delta H^\circ / R + \Delta S^\circ / RT)
\end{aligned}
\tag{10}
$$

Petersheim and Turner[17] report differences of only 0.5% between the raw data and the calculated curve. This method is similar to the $\ln K$ versus $1/T$ method, except that the whole curve is used in the fit, not just the central part of the transition curve, and the base line parameters are fit simultaneously with ΔH° and ΔS°. One must be careful in using this method, when either base line is not well defined or has an anomalous shape. The six parameters fit the experimental curve well enough, but the thermodynamic parameters are not meaningful.

Method 4: $\ln c_t$ versus $1/T_m$. For all of the above methods, thermodynamic data should be calculated from replicate experiments at more than one concentration. A very easy and effective method to obtain thermodynamic data is from the concentration dependence of the T_m. The relevant equations relating T_m and total strand concentration c_t are listed in Table II; a van't Hoff plot of $\ln c_t$ versus $1/T_m$ yields ΔH° and ΔS°. All that needs to be determined precisely for each experiment is the strand concentration and T_m. Normally, the melting curves are measured at a minimum of 10 different strand concentrations over 2–3 orders of magnitude (micromolar–millimolar). In principle, as c_t is raised, only the T_m should change, with no change in the percentage of hypochromicity. Experimentally, melts at higher concentrations often show a greater percentage of hypochromicity than at lower concentrations; this is usually ascribed to end-to-end aggregation of the RNA double strands.[21] A nonlinear plot of $\ln c_t$ versus $1/T_m$ indicates possible non-two-state behavior, of which aggregation is a specific example. Nevertheless, this method of measuring thermodynamic parameters has advantages over the other averaging methods, since ΔH° and ΔS° just depend on measuring c_t and T_m, which are less sensitive to the choice of base lines.

Table III summarizes the enthalpies for the melting of d(GC)$_3$, determined by the different methods outlined above.[22] The ΔH° values deter-

[21] J. W. Nelson, F. H . Martin, and I. Tinoco, Jr., *Biopolymers* **20**, 2509 (1981).

[22] D. D. Albergo, L. A. Marky, K. J. Breslauer, and D. H. Turner, *Biochemistry* **20**, 1409 (1981).

TABLE III
ENTHALPIES FOR COIL-TO-HELIX TRANSITION OF d(GC)$_3$[a]

Parameter	Method 1 (ln K versus $1/T$)	Method 2 (slope at T_m)	Method 3 (curve fit)	Method 4 (ln c_t versus $1/T_m$)
No base lines subtracted	−39.0	−42.1	−33.0	−58.1
Upper base line subtracted	−37.0	−41.4	—	−57.6
Both base lines subtracted	−58.9	−56.8	−56.6	−57.4

[a] Data in kilocalories per mole from Albergo et al.[22]

mined using all four methods agree with each other only if base lines are properly subtracted. All these values agree with the enthalpy determined calorimetrically, which indicates that the two-state model is valid for this molecule. Only the ln c_t versus $1/T_m$ is in agreement with the calorimetric enthalpy if the base lines are not subtracted; this demonstrates the insensitivity of this method to base lines.

One should keep in mind the implicit and explicit assumptions made in deriving thermodynamic parameters from melting curves. Activities are replaced by molar concentrations in the equilibrium constants; this determines the standard state. The standard values of $\Delta H°$, $\Delta S°$, and $\Delta G°$ refer to a reaction at 1 M concentration of each species, but with each species having the properties corresponding to an infinitely dilute solution (an ideal solution). Although we are interested in the thermodynamic properties of the reaction extrapolated from infinite dilution, we must specify a concentration (1 M), because ΔS and ΔG depend on concentration even for ideal solutions. ΔH is independent of concentration for ideal solutions. The standard thermodynamic values obtained are valid only for the solvents used in the experiments. Other concentrations of salt, or other kinds of salt, may produce different values. Second, as discussed earlier, the absorbance is assumed to be linearly related to the fraction of bases paired (an assumption based on the theory of hypochromicity). Third, the native to single-strands transition is assumed to proceed in a two-state (all-or-none) manner. The validity of thermodynamic data, derived from melting curves, depends on the validity of this model. A good test is agreement between data derived using the different methods outlined above. In general, any assumption of a two-state model can be experimentally tested by observing the melting by a different technique [calorimetry, nuclear magnetic resonance (NMR), circular dichroism, etc.]. If the transition is indeed two state, the thermodynamic data should be independent of the measuring technique. Fourth, the $\Delta H°$ and $\Delta S°$ values, derived from this analysis, are assumed to be independent of temperature. Since

$$d\ \Delta H/dT = \Delta C_p \tag{11}$$

where ΔC_p is the difference of heat capacities (at constant pressure) of the native and melted states, this assumption can be checked by plotting $\Delta H°$ values obtained using methods 1, 2, or 3 versus T_m; the slope is ΔC_p. The errors introduced by this assumption are small, but Petersheim and Turner[17] have introduced a simple correction for $\Delta H°$ based on Eq. (11).

Prediction of Melting Behavior

The previous sections have described how to determine experimentally the thermodynamic stability and melting behavior of a RNA molecule. One can also calculate the approximate $\Delta G°$ and T_m for melting a RNA secondary structure. In order to predict the melting behavior of a complex RNA molecule, one needs to know the free-energy contributions of regions of duplexes, bulges, hairpin loops, and internal loops relative to the single strands. These structural elements are shown schematically in Fig. 3. Contributions from $G \cdot U$ base pairs and from dangling single strands must also be included. Prediction of secondary structure has been reviewed recently.[23] It is assumed that the free energy for a secondary structure is the sum of the free energies for the separate regions.[2] The further approximation is made that the free energy of a duplex region is the sum of nearest-neighbor interactions. According to the nearest-neighbor model, the free energy for duplex formation consists of an initiation free energy for formation of the first base pair plus a sum of propagation free energies for formation of the subsequent base pairs. There is also a small symmetry term, which arises for double-strand formation of self-complementary sequences. For example, the free energy of formation for the following non-self-complementary duplex

5'   3'

GCGUGAU

CGCACUA

3'   5'

---

[23] D. H. Turner, N. Sugimoto, and S. M. Freier, *Annu. Rev. Biophys. Biophys. Chem.* **17,** 167 (1988).

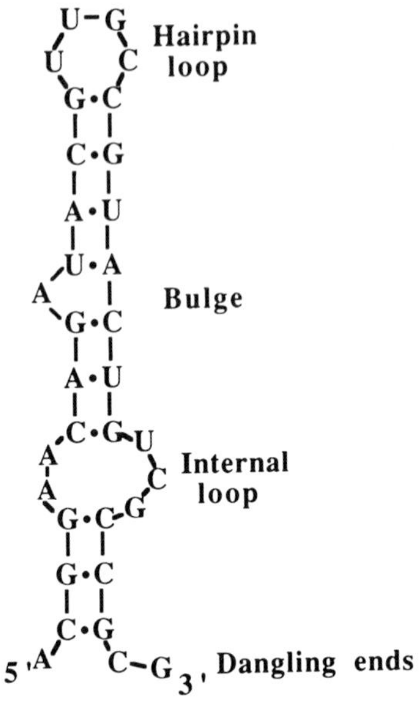

FIG. 3. A representation of the various types of secondary structures that a RNA molecule can adopt.

is calculated as the sum of nearest-neighbor contributions with zero-symmetry correction

$$\Delta G^{\circ}_{\text{total}} = \Delta G^{\circ}\,\overrightarrow{\underset{\underset{\leftarrow}{\text{CG}}}{\text{GC}}} + \Delta G^{\circ}\,\overrightarrow{\underset{\underset{\leftarrow}{\text{GC}}}{\text{CG}}} + \Delta G^{\circ}\,\overrightarrow{\underset{\underset{\leftarrow}{\text{CA}}}{\text{GU}}} + \Delta G^{\circ}\,\overrightarrow{\underset{\underset{\leftarrow}{\text{AC}}}{\text{UG}}}$$

$$+ \Delta G^{\circ}\,\overrightarrow{\underset{\underset{\leftarrow}{\text{CU}}}{\text{GA}}} + \Delta G^{\circ}\,\overrightarrow{\underset{\underset{\leftarrow}{\text{UA}}}{\text{AU}}} + \Delta G^{\circ}_{\text{sym}}\,(=0) + \Delta G^{\circ}_{\text{init}}$$

ΔS° and ΔH° values can be calculated in an analogous manner. Standard free energies, entropies, and enthalpies for all nearest neighbors at 37° in 1 M NaCl have been determined by Turner and co-workers.[1] The oligonucleotides used for the determination of these parameters were chosen such that their T_m values were near 37°; thus, the parameters required little extrapolation. The free energies depend on temperature [see Eq. (7)]; the enthalpies and entropies are assumed temperature independent, but are most useful near 37°. The T_m of any oligonucleotide can be predicted using the expressions in Table II. As noted by Freier et al.,[1] the difference between measured and predicted parameters are due not only to the inaccuracy of the data set, but also to the imperfection in the nearest-neighbor model, i.e., contributions due to next nearest-neighbor interactions.

Since RNA secondary structure consists of duplexes, loops, bulges, and single-stranded regions (Fig. 3), one needs to know the thermodynamic contributions of loops and bulges, as well as duplex regions, in order to calculate the free energy. The free-energy contributions of non-bonded loops and bulges are simply added to the free energy for the duplex region. Duplex initiation free energies are contained in the ΔG° for a loop. Bulge free energies are just the destabilization of the perfect duplex. Published free-energy parameters for loops and bulges are very limited[1]; these data assume that ΔG° of a loop only depends on its size, with a maximum stability at a loop size of six or seven. However, recent data show that loop free energy depends both on size and sequence.[24]

In view of the limited data on these structures, for precise comparison of RNA sequences, model compounds should be studied. A good example of this approach are the studies by Tuerk et al. of hairpin loop sequences, which occur with very high frequency in intercistronic regions in RNA produced by T4 bacteriophage infection.[25] These investigators showed that certain four-base loops can have stabilities much greater than predicted by the present parameters. Similarly, Groebe and Uhlenbeck studied a number of variants of the R17 coat protein binding site sequence.[26] This sequence contains a bulged A in the stem and a four-base hairpin loop. In this study, the loop sequence was kept constant, while the bulged nucleotide or flanking base pairs were varied. These variants were compared to the hairpin with a perfect duplex stem.

In order to evaluate, or to predict, a folded structure for a RNA molecule, one must consider other factors in addition to the secondary structural elements discussed previously. Specific ion binding may stabilize a particular secondary or tertiary structure. Tertiary interactions may favor, or may prevent, the formation of certain secondary structures. The free-energy contributions from tertiary structural elements, such as pseudoknots and base triplets, are not known. Thus, well-designed model compounds must be studied to provide the missing free-energy values.

[24] D. R. Groebe and O. C. Uhlenbeck, Nucleic Acids Res. **16**, 11725 (1988).
[25] C. Tuerk, P. Gauss, C. Thermes, D. R. Groebe, N. Guild, G. Stormo, M. Gayle, Y. d'Auberton-Carafa, O. C. Uhlenbeck, I. Tinoco, Jr., E. N. Brody, and L. Gold, Proc. Natl. Acad. Sci. U.S.A. **85**, 1364 (1988).
[26] D. R. Groebe and O. C. Uhlenbeck, Biochemistry **28**, 742 (1989).

Other Techniques

Temperature-Jump Methods. Derivatives of melting curves can be obtained directly by using a temperature-jump apparatus.[19] The temperature is rapidly (microseconds) increased by a few degrees, and the corresponding absorbance change (ΔA) is measured. There are two major advantages of this method over normal UV melting studies. First, the base-line problem is eliminated. The molecular processes, which give rise to sloping base lines (most likely, unstacking of single or double strands), are very fast compared to the actual melting processes. Thus, the ΔA due to these very fast steps can be resolved from the ΔA due to the slower melting. This advantage can also be used to resolve overlapping transitions, which occur at different rates. Crothers *et al.*[27] and Riesner *et al.*[28] have used this method to study tRNA melting and the stability of hairpin and internal loops. Differential melting curves can also be measured directly using a double-beam spectrometer with two identical RNA samples at slightly different (0.1°) temperature.[29]

Nuclear Magnetic Resonance (NMR). Melting transitions can be followed using the chemical shift of the nonexchangeable protons, usually aromatic protons. The chemical shift versus temperature profile can be analyzed using a two-state model

$$\delta_{obs} = f\delta_{native} + (1 - f)\delta_{coil} \tag{12}$$

where δ_{obs} is the observed chemical shift of a given proton, δ_{native} and δ_{coil} are the chemical shifts of that proton in the native and coil forms, respectively, and f is the fraction of the native form. From the calculated fraction versus temperature profile, thermodynamic parameters can be derived. The advantage of using NMR to study melting is that, in principle, each residue in the sequence can be monitored once the NMR spectrum has been assigned. This allows the validity of the two-state model to be tested: in a two-state transition, melting curves for different residues should yield the same thermodynamic data. Also, for complex molecules like tRNA, which have multiphasic melting profiles, NMR can assist in determining what is melting in each transition. This was done in tRNA by monitoring the exchangeable imino protons for each stem.[27] However, NMR has major disadvantages. The assumption of a two-state model in analyzing a chemical shift versus temperature curve is questionable. This is because kinetically fast exchange between the two states (duplex and

[27] D. M. Crothers, P. E. Cole, C. W. Hilbers, and R. G. Shulman, *J. Mol. Biol.* **87**, 63 (1974).

[28] D. Riesner, G. Maass, R. Thiebe, P. Philippsen, and H. G. Zachau, *Eur. J. Biochem.* **36**, 76 (1973).

[29] A. Wada, S. Yabuki, and Y. Husimi, *CRC Crit. Rev. Biochem.* **9**, 87 (1980).

single strand) is assumed for Eq. (12) to be valid

$$k_{ex} \gg 2\pi \, \Delta\nu \tag{13}$$

where k_{ex} is the rate constant for exchange between the two states and $\Delta\nu$ is the difference in resonant frequencies for a proton in the two states. This often is not the case; if the rate constant is too slow, intermediate exchange ($k_{ex} \cong 2\pi \, \Delta\nu$), resulting in line broadening, or slow exchange ($k_{ex} \ll 2\pi \, \Delta\nu$), resulting in two lines for the two states, can occur. Another obvious difficulty with NMR melting experiments is the high concentration required for NMR in general (millimolar strand concentrations). NMR melting curves should be compared to optical melting curves taken on the same sample at the high NMR concentration and at the lower optical concentrations to ensure that the same phenomenon is being studied.

Calorimetry. Unlike the other methods presented in this section, calorimetry allows the determination of ΔH directly; the method of differential scanning calorimetry (DSC)[30,31] is most often used. In DSC, the excess heat capacity of the transition (ΔC_p) is measured as the temperature is varied. Since $\Delta H = \int \Delta C_p \, dT$, the integrated area under a DSC versus T curve is the transition enthalpy. The major advantage of DCS over optical melting is that no assumptions about the transition need to be made; DSC values of ΔH are model independent.[20] However, it is important to remember that the ΔH measured calorimetrically is the enthalpy of whatever process occurs between the two states used to define the base lines. Unlike the van't Hoff standard enthalpy, $\Delta H^{\circ}{}_{VH}$, which is obtained from the temperature dependence of the equilibrium and refers to the standard-state conditions (1 M strand concentrations, but with the properties of infinitely dilute solutions), the calorimetric ΔH refers to the conditions of the experimental measurement (strand concentration measured in millimolar). These values will be equal only (for the same solvents) if the enthalpy of the reaction is independent of the concentration of oligonucleotides over the concentration range studied optically and calorimetrically.

A schematic DSC transition curve is shown in Fig. 4. In order to determine the area under the transition curve, upper and lower base lines (states of approximately constant heat capacity) must be defined, and initial and final temperatures for integration must be chosen. If the base lines are not colinear, the initial and final states have different C_p. Incomplete transitions, or very broad transitions, present the same problems as

---

[30] L.-H. Chang, S.-J. Li, T. L. Ricca, and A. G. Marshall, *Anal. Chem.* **56**, 1502 (1984).
[31] P. L. Privalov, *FEBS Lett.* **40**, S140 (1974).

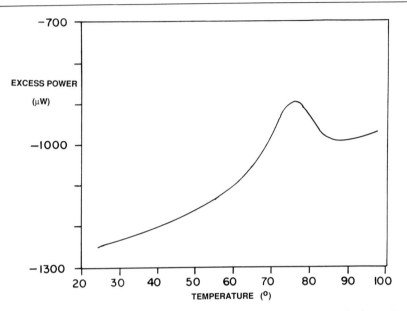

Fig. 4. A sample DSC trace for d(CGCGAATTCGCG)$_2$ in 1 M NaCl obtained on a Hart Scientific DSC. The vertical axis is the excess power required to maintain the same temperature in the sample and reference cells, which is proportional to the excess heat capacity due to the nucleic acid, and the horizontal axis is temperature. The data represent raw data before base-line correction. [Data courtesy of Dr. D. Kallick (Department of Chemistry, University of California, Berkeley).]

in the van't Hoff methods. However, comparison of ΔH from an optical melting experiment to ΔH from calorimetry can be very helpful. For example, if $\Delta H_{VH} < \Delta H_{CAL}$, this indicates the presence of intermediates in the melting process. The major drawback of DSC is the large amount (0.5 ml at millimolar concentrations) of material needed. The small concentrations and small amount of material needed make optical melting the current method of choice for studies of RNA secondary structure.

Complex RNA Molecules

We will outline briefly in this section some of the information that can be obtained from melting studies of complex RNA molecules. Examples of such molecules studied to date include tRNA, 5 S RNA, viroids, and viral RNA. The melting transition of a complex RNA will not be two state; the melting curve is often multiphasic. Comparison of hypochromicities at 260 and 280 nm will allow an estimate of A · U and G · C base pairs broken for each transition in the melting curve; this will aid in assigning transitions. The T_m values of the transitions usually are affected differently by addition of Mg^{2+} or Na^+. Cole *et al.* constructed a phase diagram of tRNA^{Phe} as a function of salt.[16] Each transition may be interpreted as an independent melting of a portion of the molecule. The temperature-jump method is especially useful for resolving slow ($t >$ milliseconds) and fast ($t <$ milliseconds) transitions. For example, the melting of the tertiary structure to an extended, base-paired structure is slow, while the melting of individual hairpin regions is fast. The ΔH^o for each transition can be determined from the differentiated melting curve, assuming the transition is independent of others.

Acknowledgments

We are very greatful to Professor Donald Gray, The University of Texas at Dallas for providing the data in Table I and for making helpful comments. Dr. Susan Freier (Molecular Biosystems, Inc., La Jolla, CA) carefully read the manuscript and made valuable suggestions; her help is greatly appreciated. The work was supported by grant GM 10840 from the National Institutes of Health and grant DE-FG03-86ER60406 from the Department of Energy.

[23] Nuclease Digestion: A Method for Mapping Introns

By SHELBY L. BERGER

Genomic exons are characterized by means of nuclease protection assays. Cloned, single-stranded genomic DNA probes are hybridized with cellular RNA; the hybrids are treated with a single-strand-specific nuclease to remove unreacted probe together with irrelevant RNA and unprotected stretches of DNA; and finally, the protected hybrids are characterized by means of gel electrophoresis. The number of surviving fragments equals the number of exons. The mobility of each fragment is a measure of the size of that exon.

The nuclease digestion method presented in this chapter is the inverse of nuclease protection; exons are degraded, while introns are preserved. The method is as follows. Synthetic, genomic sense-strand RNA is first hybridized to an excess of single-stranded antisense cDNA. Then, the purified hybrids are treated with ribonuclease H (RNase H), an enzyme that degrades only RNA in DNA–RNA hybrids. As a consequence, the exons in the RNA moiety are degraded; the RNA fragments that survive are the introns. These fragments can be analyzed electrophoretically,

A   B

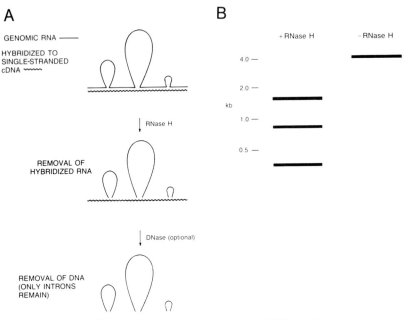

FIG. 1. Isolation of introns from a hypothetical gene. (A) Top: Genomic sense-strand RNA is hybridized with antisense cDNA. The looped out segments are the introns. Middle: Digestion with RNase H removes exonic RNA sequences, but leaves introns and cDNA intact. Bottom: Digestion with DNase removes the cDNA, leaving only intronic RNA sequences. (B) A diagrammatic representation of the results from gel electrophoretic analysis. Bands representing introns are shown in the +RNase H lane. In the absence of the enzyme [−RNase H (control lane)], only the genomic RNA transcript is visualized. In an experiment performed in the absence of cDNA, but in the presence of RNase H, the minus DNA control (not shown in the figure), the pattern should closely resemble the −RNase H (control lane). In practice, the addition of the enzyme usually causes some degradation of the synthetic, genomic RNA transcript.

and, as shown in Fig. 1, the number of introns and their sizes can be inferred.

The two techniques are complementary. In principle, if the number of exons is known from a nuclease protection assay and is represented by n, the number of introns must be $n - 1$. If the lengths of the gene and the mRNA have been ascertained, the composite length of intronic material can be deduced, but not the sizes of the individual introns. By performing a nuclease digestion assay, the number of introns can be measured directly, and the size of each intron can be determined. Since the techniques are entirely independent, the investigator has the option of counting and characterizing introns first, by the nuclease digestion technique, and confirming the results, in part, with the nuclease protection approach. The nuclease digestion technique detailed in this chapter is an extension of a method for mapping specific mRNA molecules using cloned cDNA fragments or oligomers and bulk, uncharacterized mRNA as the starting materials.[1] In both techniques, partly characterized DNA sequences are used to direct RNase H to cleave RNA molecules at preselected locations in order to generate fragments of RNA for subsequent analysis.

Materials

The starting materials for a straightforward nuclease digestion experiment are as follows: a vector, containing both the gene of interest and one or more promoters for RNA polymerase, and a cDNA clone, representing sequences in the mRNA transcribed from that gene. The cDNA need not be full length, but for ease of interpretation, the cDNA should contain at least 60 bases of the first and last exons and complete information from all internal exons.

Single-stranded RNA and DNA are highly recommended, particularly for initial nuclease digestion assays. Using the appropriate promoter, a genomic sense-strand RNA transcript must be generated. As shown below, the RNA strand can be labeled or unlabeled, depending on the method of analysis. Since all unhybridized RNA will contribute to the results, 5'- and 3'-flanking regions of the gene must not be transcribed into RNA or, failing that, must be distinguished from the introns by further analysis at the end of the experiment.

Single-stranded cDNA can be obtained directly as an insert within the plus strand of M13 phage or within the rescued strand of a plasmid vector with an origin for single-stranded replication [Bluescript vector (Stratagene, La Jolla, CA) or pGEM-3Zf(+) and pGEM-5Zf(+) (Promega, Madison, WI)]. The cDNA can be circular or linear; vector sequences within circular molecules or at either or both ends of linear molecules will not interfere. Clearly, the cDNA must be free of ribonuclease.

In the following protocol, the use of sense genomic RNA and antisense cDNA is described, although antisense genomic RNA and sense cDNA will work as well. Regardless of which strand is sense and which antisense, it should be apparent from Fig. 1 and the preceding discussion that the gene must be represented by RNA and that the mature, processed

[1] S. L. Berger, *Anal. Biochem.* **161,** 272 (1987). See also A. Leone, M. S. Krug, R. E. Manrow, and S. L. Berger, *Anal. Biochem.,* in press (1989).

transcript of that gene (mRNA sequences) must be represented by complementary DNA.

Nucleic Acid Requirements

Genomic RNA. Prepare a synthetic RNA transcript of the gene of interest. (See this volume [4] or elsewhere in this series.[2–4]) This means that the gene must be cloned into a vector containing a strong promoter for a commercially available RNA polymerase. In this laboratory, use of the T7 RNA polymerase generally results in higher yields of longer transcripts. RNA may be labeled in the presence of 10–50 μCi of the chosen [α-^{32}P]ribonucleoside triphosphate in a 50-μl reaction, containing 50 μM of the same unlabeled triphosphate. Alternatively, the RNA may be made without a tracer in the presence of 0.5 mM of each ribonucleoside triphosphate. At the completion of RNA synthesis, add 5 μg of tRNA. Then digest DNA with ribonuclease-free DNase and extract the mixture with an equal volume of chloroform : isoamyl alcohol [94 : 6 (v/v)]. Retain the aqueous phase and extract the organic phase with an equal volume of a RNA extraction buffer [10 mM sodium acetate, pH 5.1, 20 mM NaCl, and 3 mM disodium salt ethylenediaminetetraacetic acid (Na$_2$EDTA) is standard in this laboratory]. Mix the aqueous phases. The RNA should be drop dialyzed against H$_2$O to remove DNase I digestion products.[5]

The quality of the synthetic RNA should be ascertained by gel electrophoresis, preferably in denaturing agarose gels. It is usually possible to characterize introns unambiguously with RNA transcripts that are composed predominantly of full-length molecules with a smear of smaller molecules. If the RNA is heavily contaminated with truncated transcripts, however, it will be necessary to purify full-length material by excising it from the gel.

For the determination of introns with unlabeled RNA as described in this chapter, approximately 0.5 pmol of genomic RNA is convenient for one determination, including all necessary controls. For the determination with labeled RNA, considerably smaller amounts will suffice.

cDNA. Prepare single-stranded DNA using any convenient vector. Measure the concentration. For small amounts of DNA, the concentration can be estimated electrophoretically in gels using commercial plus strand of ΦX174 as a standard. The DNA strand must be complementary to the genomic RNA strand specified previously.

---

[2] D. A. Melton, this series, Vol. 152, p. 292.
[3] G. M. Wahl, J. L. Meinkoth, and A. R. Kimmel, this series, Vol. 152, p. 578.
[4] G. A. Evans and G. M. Wahl, this series, Vol. 152, p. 607.
[5] D. M. Wallace, this series, Vol. 152, p. 41.

Reagents

All reagents must be free of ribonuclease. [See this volume [1] for methods for inactivating ribonucleases in solution with diethyl pyrocarbonate (DEPC) or for removing ribonucleases by filtration through nitrocellulose.]

> 5× Hybridization buffer: 0.2 M HEPES adjusted to pH 7 with NaOH, 2 M NaCl, and 5 mM ethylenediaminetetraacetic acid (EDTA)
> 10× RNase H buffer 1: 0.5 M Tris–HCl at pH 8.3, 0.5 M NaCl, and 0.1 M MgCl$_2$
> 10× RNase H buffer 2: 1 M Tris–HCL at pH 8.3, and 1 mM MgCl$_2$
> Ribonuclease H from *Escherichia coli* (BRL)
> Deionized formamide[6]
> Carrier tRNA at 25 μg/μl, DEPC treated
> Other useful solutions include 2 M potassium acetate for ethanol precipitation and chloroform : isoamyl alcohol [94:6 (v/v)]

Methods

1. Mix ~0.5 pmol of genomic sense-strand RNA with 0.5–2 pmol of single-stranded antisense cDNA and bring to dryness (tube 1).

2. Bring 0.25 pmol of genomic sense-strand RNA to dryness (tube 2). This material will serve as the minus DNA control.

3. Redissolve the nucleic acids in tube 1 in 3 μl of H$_2$O. Add 2 μl of 5× hybridization buffer and 5 μl of deionized formamide. For the RNA in tube 2, use 1.5 μl of H$_2$O, 1 μl of 5× hybridization buffer, and 2.5 μl of formamide.

4. Heat both samples in a water bath at 85° for 3 min. Recover the drop by brief centrifugation in a microfuge.

5. Transfer both samples to a 50° water bath. Hybridize for 15 min. Note that the large amount of material of known sequence complexity renders hybridization for extended periods unnecessary. These conditions are satisfactory for exons of 70 base pairs (bp) and larger.

6. Precipitate both samples in ethanol as follows. Add to tube 1, 100 μl of H$_2$O, 1 μl of DEPC-treated carrier tRNA at 25 μg/μl, 11 μl of 2 M potassium acetate, and 305 μl (2.5 volumes) of ethanol. Use half the amount of each addition for tube 2. Chill both samples at −55° in a dry ice–ethanol bath for 15 min and recover the pellets by centrifugation at 4° for 15 min in a microfuge at top speed.

---

[6] D. D. Blumberg, this series, Vol. 152, p. 22.

7. Dissolve the contents of tube 1 in 13.0 μl of H_2O and divide into two aliquots called tubes 1A and 1B, respectively. Dissolve the RNA in tube 2 in 6.5 μl of H_2O.

8. To each of the three tubes add 1 μl of $10\times$ RNase H buffer. (Separate reactions with both recommended buffers should be run.) Add 0.02–0.2 U of RNase H to tubes 1A and 2. Do not add enzyme to tube 1B. Bring all samples to 10 μl with H_2O.[7] Note that the amount of RNase H and the digestion time must be determined for each experiment.

9. Incubate at 37°. Remove four aliquots from tube 1A at intervals of 2–20 min.

10. Remove an aliquot from each of the controls at 20 min.

11. Transfer each aliquot to a fresh tube containing 10 μl of H_2O and 1 U of RNase-free DNase. Incubate at 37° for 20 min.

12. Add 50 μl of a RNA extraction buffer (see Nucleic Acid Requirements) and extract each sample with 100 μl of chloroform : isoamyl alcohol [94 : 6 (v/v)]. Recover the aqueous phase by centrifugation.

13. Precipitate the RNA in each sample with 10.5 μl of 2 M potassium acetate, 300 μl of ethanol, and 0.3 μl of carrier tRNA at −55° for 15 min. Recover the RNA pellets by centrifugation for 15 min in a microfuge and dissolve each pellet in 4 μl of H_2O.

14. Analyze 2 μl of each sample in a 2% (w/v) agarose formaldehyde gel.[8] Save the remainder for a second gel if needed. RNA ladders (BRL) can be used as markers.

15. Expose either the wet gel or the gel after drying to X-ray film, if the introns are radioactive. For experiments without a tracer, prepare a Northern blot from the gel, hybridize the blot with the desired probes (see Fig. 2, for examples), and expose the blot to X-ray film. To visualize the markers, in either case, stain only the relevant lanes with ethidium bromide (EtBr). Details of these techniques can be found elsewhere.[4,9–11]

Analysis of Results

Experiments with Labeled RNA

The labeled fragments that remain after treatment with RNase H are the introns. The number of fragments found on electrophoretic fractiona-

---

[7] The use of RNasin (placental ribonuclease inhibitor) is not recommended. The inhibitor contains bound ribonuclease, which can be released in active form during preparation of the sample for electrophoresis or in the absence of a high concentration of mercaptoethanol or DTT.

[8] H. Lehrach, D. Diamond, J. M. Wozney, and H. Boedtker, *Biochemistry* **16,** 4743 (1977).

[9] G. M. Wahl and S. L. Berger, this series, Vol. 152, p. 419.

tion in a gel represents the number of introns; the size of each fragment reflects intron size. Experiments with labeled RNA cannot be used to deduce the order of the introns within the gene.

Experiments with Unlabeled RNA

The use of unlabeled genomic RNA makes possible a determination of the order of the introns, as well as the total number of introns and their sizes. If the gene has been partially mapped, gene fragments representing known locations can be oligolabeled or nick translated and used to probe the blot containing the introns. The position of an intron within the gene is given by the map location of the probe fragment. If the gene is not mapped, but the cDNA is well characterized, the opposite promoter (either SP6 or T3, depending on the vector) of the same genomic clone already in hand can be used to generate labeled antisense RNA transcripts. By cleaving the gene at exonic restriction sites that are absent from the introns lying between the chosen promoter and the chosen exonic restriction site, a series of labeled RNA probes can be synthesized and used to order the blotted introns as shown in Fig. 2. In this example, three exonic restriction sites have been selected, designated 1, 2, and 3. Cleavage at 1, for example, gives rise to an antisense RNA transcript (called probe 1 in Fig. 2). This transcript will hybridize to all introns. In contrast, probe 2 will hybridize to two introns, while probe 3, made by cleaving the cloned DNA at restriction site 3, will hybridize to nothing at all, because the probe contains exonic sequences exclusively. The results of the experiment in Fig. 2 show that the intermediate-sized intron lies nearest the 5' end of the gene and that all introns lie 5' of restriction site 3.

RNA transcripts of varying lengths can also be generated by varying the time of RNA synthesis on an uncleaved template.[4] Hybridization to a transcript slightly longer than probe 3 (Fig. 2) positions the intron near the 3' end of the gene, whereas hybridization to a long transcript similar to probe 1 (Fig. 2), but not to a short transcript, directs attention to the 5' end of the gene.

Those who prefer DNA probes can synthesize single-stranded material by primer extension using a SP6 or T3 promoter-specific oligonucleotide and Klenow fragment.

If the flanking region of the gene were included in the original RNA sense genomic transcript used in the digestion experiment, five RNA fragments would have been produced as a result of RNase H treatment.

---

[10] W. M. Bonner, this series, Vol. 152, p. 57.
[11] R. C. Ogden and D. A. Adams, this series, Vol. 152, p. 70.

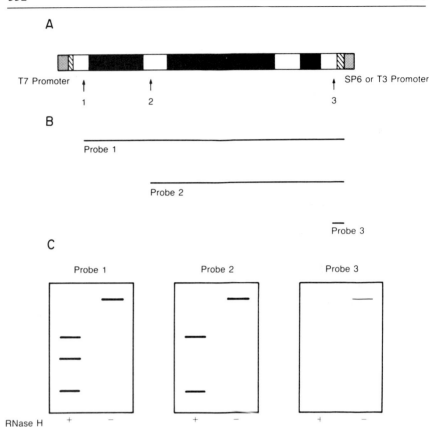

FIG. 2. Analysis of unlabeled intronic RNA sequences. (A) A hypothetical cloned gene consisting of four exons (□) and three introns (■) inserted into a vector containing promoters for RNA polymerases (▨) and polylinker sequences (▧). The inserted gene contains three exonic sites, indicated by arrows, that are recognized by restriction enzymes 1, 2, and 3, respectively. (B) Antisense probes 1, 2, and 3 generated by cleavage of the cloned genomic DNA with restriction enzymes 1, 2, and 3, respectively. The probes can be RNA, in which case SP6 or T3 RNA polymerase is used for synthesis, or DNA, in which case a SP6 or T3 promoter-specific oligomer is hybridized to denatured DNA and extended with Klenow fragment. (C) Northern blots of the intronic RNA sequences hybridized with probes 1, 2, and 3. Note that probe 1 recognizes all introns, whereas probe 2, by hybridizing to only two introns, identifies the missing intron as the one lying furthest from the SP6 or T3 promoter. To order the remaining two introns, an additional probe (not shown) would be required. The blot hybridized with probe 3 is included to emphasize the fact that the probes, composed only of exonic sequences, will not hybridize to the RNA fragments that survive treatment with RNase H, i.e., introns.

Intron ordering experiments, as described in this section, could then be used to distinguish true introns from flanking material.

Interpretation of Results: Troubleshooting

The intron mapping technique depends on generating a recognizable single-stranded intronic RNA and retaining it in the presence of RNase H and DNA. The RNase H solution is the cause of two problems: loss of large introns and production of spurious bands. In the absence of DNA, contaminant ribonucleases can catalyze the cleavage of RNA into discrete fragments. In the presence of DNA, very short hybrids between genuine intronic RNA and exonic DNA, vector DNA, or fragments of the template used for genomic RNA synthesis, can form substrates for RNase H. Cleavage of such transient minihybrids results in underrepresentation of the larger introns. To solve the problem, two RNase buffers are recommended, one which stabilizes hybrids (buffer 1) and one which destabilizes them (buffer 2). Use of both buffers in sequential experiments results in the production of the same intronic RNAs and a different collection of artifacts; thus, the intronic RNAs should be clearly revealed. Further discrimination between introns and artifacts can usually be achieved by changing the time of the reaction, the concentration of enzyme, and the ratio of DNA to RNA. For example, the presence of the genomic RNA transcript is indicative of insufficient enzyme or time of digestion.

Difficulties can also arise from the presence of truncated RNA genomic transcripts in nuclease digestion experiments. In general, the more 3' the location of the intron, the more underrepresented the intron will be among molecules in the initial T7 polymerase-catalyzed transcript of the gene. If the last intron is also very large, it will be difficult to recognize; on electrophoretic fractionation and transfer to a solid support, the last intron might be a faint band with a smear below the intron or simply a smear. In some cases, a legitimate intronic band or smear can be recognized by the absence of material of the same size in the lanes containing the minus RNase H control reaction and the minus DNA control reaction (not shown in Fig. 1B). In other cases, it will be necessary to purify full-length genomic RNA transcripts before attempting the procedure. If the yield of suitable genomic RNA transcripts is too low, gene fragments can be subcloned, transcribed into full-length RNA copies of the subclone, and analyzed separately.

Nuclease Digestion Experiments and Polymerase Chain Reaction

Nuclease digestion experiments, as described in this section, require a cloned gene, as well as a cDNA clone. In principle, the cloned cDNA can be replaced by bulk cDNA synthesized with reverse transcriptase,[12] using unfractionated cellular poly(A)+ RNA as a template. If the desired cDNA is sufficiently abundant, so as to be in excess of the sense genomic transcript with which the cDNA must hybridize, the experiment can be done. Failing that, the polymerase chain reaction can be used to ampify the specific cDNA.[13] This cDNA will necessarily be composed partly or wholly of double-stranded molecules. Theoretically, double-stranded cDNA can substitute for the single-stranded cDNA, as suggested in Methods, but, in practice, the protocol has not been worked out. However, if double-stranded cDNA is available, one might as well clone cDNA, characterize cDNA, and produce homogeneous single-stranded cDNA at will.

Similarly, in the absence of a genomic clone, unprocessed RNA transcripts of the gene of interest, if present in detectable quantity among bulk nuclear RNA, can be used a source of genomic sense-strand RNA. With this approach, the problem of recognizing the protected introns becomes severe; recall that the cDNA clone, consisting of exons, will not hybridize to the fragments that survive RNase H treatment. Once again, the polymerase chain, with specific primers that reflect exonic sequences and genomic DNA as a template, can generate probes containing intronic sequences. However, if the gene of interest is amplified successfully, the smart move is to clone the gene first and to search for introns later.

---

[12] M. S. Krug and S. L. Berger, this series, Vol. 152, p. 321.
[13] K. B. Mullis and F. A. Faloona, this series, Vol. 155, p. 335.

[24] Characterization of RNA Molecules by S1 Nuclease Analysis

By ARNOLD J. BERK

Strategy

S1 nuclease analysis[1] can be used to map specific RNA molecules on well-characterized DNA templates. The method can also be used to quantitate specific RNAs accurately in complex mixtures of RNAs of different sequence. S1 nuclease analysis depends on the availability of pure preparations of specific DNA sequences (cloned DNAs, viral DNA restriction

---

[1] A. J. Berk and P. A. Sharp, Cell (Cambridge, Mass.) 12, 721 (1977).

fragments, or synthetic DNA) and the exquisite specificity of S1 nuclease. S1 nuclease isolated from *Aspergillus oryzae*[2] is an endodeoxyribonuclease and an endoribonuclease, which specifically hydrolyzes single-stranded nucleic acids to mono- and oligonucleotides with 5′-phosphate and 3′-hydroxyl groups. The phosphodiester bonds of base-paired DNA and RNA strands are digested at rates many thousands of times slower than for single-stranded nucleic acids. Under appropriate conditions of salt and temperature to maintain duplex DNA structure, S1 nuclease purified by the method of Vogt[2] does not nick an 8-kb (kilobase) fragment of adenovirus type 2 DNA under conditions in which the same amount of enzyme digests 250 times the amount of the heat-denatured DNA fragment to >95% acid solubility (A. J. Berk, unpublished).

The strategy for the analysis is outlined in Fig. 1 for a RNA molecule encoded in three exons (Fig. 1A). A complex RNA sample, such as total cellular RNA, is hybridized to a radiolabeled, purified DNA fragment, which includes all (Fig. 1B) or a portion (Fig. 1C) of the coding sequence. Because of the specificity of hybridization, the DNA "probe" hybridizes only to complementary RNA in the complex mixture. Even when the RNA being analyzed is at very low concentration, all the complementary RNA can be driven into RNA–DNA hybrids by using an appropriately high concentration (see later) of DNA probe to drive the bimolecular hybridization reaction. Following hybridization, the reaction products are digested with nuclease S1 under conditions of salt and temperature, which stabilize the hybrid structures (see later). Sufficient S1 is used to digest all the unhybridized excess probe to small oligonucleotides and to digest all the single-stranded regions of the hybrid molecules. Since S1 is a ribonuclease as well as a deoxyribonuclease, the unhybridized RNA is also digested. The S1-resistant RNA-labeled DNA hybrids are then recovered by ethanol precipitation.

In the simplest application of the method, the size of the S1-resistant, labeled DNA is determined by denaturing the S1-resistant RNA–DNA hybrids, by subjecting the denatured material to gel electrophoresis with single-stranded DNA standands of known size, and by detecting the S1-resistant DNA in the gel by autoradiography. In the analysis with a DNA probe containing the entire coding region, the S1-resistant DNA is equal in length to the exons of the RNA (Fig. 1B). When the probe contains a portion of the sequence encoding the RNA (or exon), only the length of the probe complementary to RNA is protected (Fig. 1C). By using a combination of different probes with alternative known end points (usually restriction sites) through the coding region, the sizes of the S1-pro-

[2] V. M. Vogt, this series, Vol. 65, p. 248.

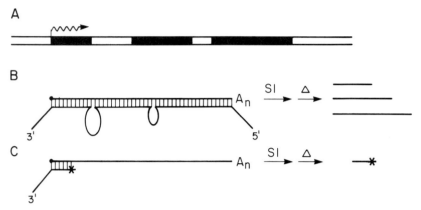

FIG. 1. The strategy of S1 nuclease mapping. (A) This diagram represents a simple transcription unit encoding three exons, represented by the black bars. The arrow indicates the direction of transcription. (B) The diagram at the left represents a RNA–DNA hybrid between the fully processed RNA from the gene depicted in A and the full length of the coding strand depicted in A. The top strand represents the RNA molecule with a small circle at the 5' end, signifying the cap structure, and A_n, indicating the poly(A) tail. The bottom strand represents the coding strand of the gene, with the 5' and 3' ends indicated. Loops of single-stranded DNA occur at the positions of the two introns, and both the 5' and 3' ends are single-stranded DNA in the hybrid structure, sensitive to digestion by S1 nuclease. Base pairing is represented by the short vertical lines between the complementary regions of RNA and DNA. S1 represents digestion by S1 nuclease as described in the text. Δ represents denaturation of the S1 digestion products. The lines at the right represent the labeled DNA fragments produced by S1 digestion of the RNA–DNA hybrid. The labeled DNA fragments are nearly precise complements of each of the exons. (C) The diagram at the left represents the hybrid formed between the same RNA as in B, hybridized to a 5'-end-labeled restriction fragment. The bottom line represents the coding strand of a restriction fragment from the 5' end of the gene extending from within the first exon to upstream of the gene. The asterisk represents the 5'-labeled phosphate, and the 3' end of the DNA strand is indicated. Other symbols are as in B. The asterisk and line at the right symbolize the fragment of labeled DNA resulting from the procedure. This S1-protected fragment is nearly equal in length to the distance from the restriction enzyme cut site to the 5' end of the first exon.

tected DNA fragments determine a unique map of the exons along the DNA sequence.

For quantitation of specific RNAs in a complex mixture, the hybridization is performed with DNA probe in molar excess over the complementary RNA. If the complementary RNA is hybridized to completion, the quantity of the S1-protected fragment is equal to the amount of RNA in the hybridization reaction. If the specific activity of the DNA is known, the molar amount of S1-protected DNA can be calculated from the counts per minute of S1-protected DNA, determined by cutting and counting the gel used to resolve the S1-protected fragment. Alternatively, the counts per minute of S1-protected DNA can be estimated from comparison of the intensity of the band of the S1-protected DNA on the autoradiogram to standards of known counts per minute of DNA of similar size applied to the same gel. More frequently, molar amounts of specific RNAs are not calculated, but rather relative amounts of the RNA in different samples are determined by densitometry of the autoradiogram. The method can be made extremely sensitive by using high-specific-activity DNA as the probe (see later) and can be applied to multiple samples analyzed on the same gel and autoradiogram. The method is frequently used, for example, to determine the relative amounts of a specific RNA transcribed from transfected DNAs with wild-type and mutagenized transcription control regions in separate, transient transfection assays.

The probe DNA can be uniformly labeled, labeled either at the 5' end with a 5'-phosphate (using $[\gamma\text{-}^{32}P]ATP$ and polynucleotide kinase[3]) or at the 3' end [by incorporation of an $[\alpha\text{-}^{32}P]dNTPs$ (deoxynucleotide triphosphate) using DNA polymerase I[3]]. When using an end-labeled probe, the labeled nucleotide must base pair with complementary RNA, or the nucleotide will be hydrolyzed by S1. When a RNA is mapped using an end-labeled probe, the polarity, as well as the map position of the RNA (or exon), is determined[4] (Fig. 1C).

If the labeled end of the DNA restriction fragment is within ~300 bases of the end of the RNA (or exon), high-resolution mapping can be obtained by subjecting the denatured S1-resistant RNA–DNA hybrid to electrophoresis on a DNA sequencing gel in parallel with the products of Maxam–Gilbert sequencing reactions prepared from the same end-labeled DNA fragment.[4] For 5'-end-labeled probes, one can also use the products of dideoxy-sequencing reactions[5] prepared using a primer labeled at the same nucleotide as the S1 probe. One must keep in mind that Maxam–Gilbert chemistry cleaves the DNA strand at the position of the modified nucleotide.[3] Consequently, bands on a sequencing gel resulting from the G-specific reaction (for example) are actually one nucleotide shorter than molecules ending in that G residue.

At the resolution of sequencing gels, end-labeled S1-resistant fragments often resolve into two to six bands, each separated by a single nucleotide in length. This results from heterogeneity in the cleavage of the probe at the transition from RNA–DNA duplex to single-stranded DNA

---

[3] A. M. Maxam and W. Gilbert, this series, Vol. 65, p. 499.
[4] R. F. Weaver and C. Weissman, *Nucleic Acids Res.* **7**, 1175 (1979).
[5] A. J. H. Smith, this series, Vol. 65, p. 560.

in the RNA–DNA hybrid. This is particularly apparent using 5'-end-labeled probes to map the capped 5' end of eukaryotic mRNAs. Weaver and Weissman[4] found that a S1-resistant fragment, produced following hybridization of a 5'-end-labeled probe to globin mRNA, was 2–5 nt (nucleotides) longer than the product predicted by the globin mRNA sequence determined by direct analysis of labeled globin mRNA. S1 does not digest the 5'–5' phosphotriester bonds of the cap structure.[6] Weaver and Weissman[4] suggested that the 5' cap structure on the mRNA molecule may sterically hinder S1 digestion at the phosphodiester bond immediately adjacent to the first base-paired nucleotide of the probe. Consequently, microheterogeneity in S1-protected products *cannot* be taken as evidence for microheterogeneity in the 5' end of a capped RNA. Also, in general, S1 mapping of the 5' end of a capped RNA can only be performed to a resolution of $\pm \sim 5$ nt. Proof of microheterogeneity at the 5' end of a RNA and precise mapping to the resolution of a single nucleotide must come from direct analysis of the RNA. This is often not practical for RNAs of low abundance. Often, the most intense S1-protected fragment is assigned a +1 position for a transcript. But it must be kept in mind that the actual capped nucleotide may be at position $+1 \pm \sim 5$.

Assignment of the 5' end(s) of a RNA should be confirmed by primer extension using reverse transcriptase.[7] If the primer is 5' labeled at the same base, which is used to prepare a 5'-labeled S1 probe, then the S1-protected fragment and the fully extended primer should comigrate on a sequencing gel $\pm \sim 5$ nt. If primer extension is not used, a short 5' exon can easily be overlooked in a S1 analysis with a 5'-end-labeled probe (Fig. 2).

Most mRNAs of higher eukaryotes are spliced RNA molecules. RNAs composed of two or three exons can be mapped by further enzymatic and electrophoretic analyses of RNA–labeled DNA hybrids.[8,9] The first step is to map exons along the length of a genomic DNA as described previously. Once exons are mapped on the DNA, the manner in which the exons are spliced together can be analyzed as follows. The rate of digestion of single-stranded DNA is 7-fold faster than digestion of single-stranded RNA.[2] When the single-stranded DNA tails and the single-stranded DNA loop (intron) in a hybrid between a genomic DNA fragment and a spliced RNA (as in Fig. 1B) are trimmed by nuclease S1 digestion, a significant fraction of the products recovered by ethanol precipitation contain an

---

[6] G. Haegeman, D. Iserentant, and W. Fiers, *Nucleic Acids Res.* **7,** 1799 (1979).
[7] W. R. Boorstein and E. A. Craig, this volume [25].
[8] A. J. Berk and P. A. Sharp, *Proc. Natl. Acad. Sci. U.S.A.* **75,** 1274 (1978).
[9] A. J. Berk and P. A. Sharp, *Cell (Cambridge, Mass.)* **14,** 695 (1978).

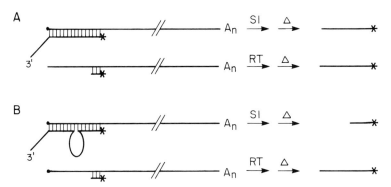

FIG. 2. Mapping the 5' end of a RNA by S1 nuclease mapping and primer extension. Symbols are as in Fig. 1. RT, Reverse transcriptase. (A) When a S1 probe with a 5' end in the first exon of a RNA is used in S1 mapping, the S1-protected fragment is approximately equal in length to the primer extension product of a primer labeled at the same site as the S1 probe. (B) When the labeled 5' end of the S1 probe is in the second exon, the S1-protected fragment is shorter than the product of primer extension. In this situation, the S1 fragment maps the 5' end of the second exon.

intact RNA molecule (with the poly(A) tail removed) base paired to the S1-resistant DNA exon sequences. S1 digestion at room temperature or lower temperature produces a greater fraction of S1-digested products with an intact RNA molecule than does digestion at 37° or higher.[10] This RNA–DNA hybrid migrates through neutral agarose gels with a mobility very close to that of a duplex DNA molecule equal to the length of the two exons. Thus, the total length of the spliced RNA can be accurately estimated relative to DNA restriction fragment markers. Digestion of the RNA-labeled genomic DNA hybrid with the single-stranded DNA-specific exonuclease, Exo VII, yields an Exo VII-resistant fragment equal in length to the exons plus the intervening introns present in the probe DNA.[8,9]

While useful for mapping spliced RNAs of only two or three exons, interpretation of the data becomes increasingly more complex for spliced RNAs composed of larger numbers of exons. Consequently, cDNA cloning and comparison of the sequences of (nearly) full-length cDNA clones to the sequence of the corresponding genomic DNA is probably the method of choice for defining the exon–intron structure of transcription units containing multiple introns. However, S1 mapping is useful for confirming that the deduced exon structure is found in RNA isolated from the original source. Also, for transcription units which give rise to alterna-

[10] J. Favaloro, R. Treisman, and R. Kamen, this series, Vol. 65, p. 718.

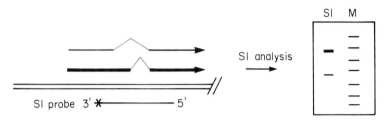

FIG. 3. Quantitating the relative abundance of alternatively spliced RNA molecules. The figure represents an analysis of the early mRNAs from region E1A of adenovirus type 2. The left end of the viral genome is represented by the double line in the diagram to the left. The exons of spliced RNAs are represented above, the exons being connected by caret symbols. The more abundant E1A mRNA represented by the thick lines has a first exon, which extends further in the 3' direction than the first exon of the less abundant E1A early mRNA. The line marked S1 probe represents a 3'-end-labeled fragment of the coding strand extending from within the 5' exon of the shorter mRNA to a region in the common 3' exon. The final autoradiogram of the procedure is represented to the right. S1 indicates the lane of S1-protected fragments. M indicates markers. The longer E1A mRNA protects a longer labeled S1-protected fragment than the shorter E1A mRNA. The relative intensities of the two bands indicates the relative concentrations of the two mRNAs.

tively spliced RNA molecules, the relative abundance of different alternatively spliced RNAs can be measured by S1 analysis with the appropriate probe. An example is diagrammed in Fig. 3. Similarly, the relative abundance of RNAs with alternative 5' ends or RNAs with alternative 3' ends can be quantitated by S1 analysis using the appropriate probes.

Hybridization

Using Single-Stranded DNA Probes. Hybridization of specific RNA to the coding strand of DNA is straightforward in hybridization reactions with single-stranded DNA. Single-stranded DNA probes can be obtained in several ways. With the advent of current automated DNA synthesis methods, it is possible to synthesize single-stranded DNA of lengths in the 60- to 100-nt range, which can be use as probes in S1 analysis. Such probes are convenient to use for the quantitation of well-characterized RNAs. Synthetic DNA is usually labeled at the 5' end with T4 polynucleotide kinase and [γ-^{32}P]ATP.[3] The probe should include 10 or more nt at the 3' end, which are not complementary to the RNA being quantitated. This is because a small fraction of the full-length probe (usually <1%) often escapes S1 digestion and is observed on the final autoradiogram of the procedure. Since only a small fraction of the probe is hybridized to specific RNA when analyzing a low-abundance transcript, this "background" of undigested probe can be of similar intensity to S1-protected DNA fragments. If a probe is used, which is protected in its entire length

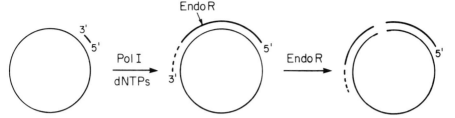

FIG. 4. Preparation of single-stranded DNA probe. A synthetic oligonucleotide of ~20 bases is used to prime synthesis of labeled DNA. The template is a M13 clone of the sense strand of a gene of interest. Following DNA synthesis, the duplex DNA is digested with a restriction enzyme, which will define the 3' end of the probe. As a result, the labeled probe DNA strand is much shorter than the complementary template strand. After denaturation, the probe is readily purified by electrophoresis on a polyacrylamide gel in TBE plus 7 M urea.

by hybridization to RNA, it is not straightforward to distinguish background undigested full-length probe from DNA protected from S1 digestion by hybridization to specific RNA. If the probe is designed so that the S1-protected fragment is shorter than the full-length probe (as in Fig. 1B and C), then the true signal of a S1-protected fragment is separated from the background of full-length undigested probe during gel electrophoresis and can be readily distinguished.

Another method for preparing single-stranded DNA as probe for a S1 analysis is to separate restriction fragment strands by electrophoresis of the denatured DNA through neutral agarose[11] or polyacrylamide gels.[3] However, the yield of separated strands is often low because of renaturation of the complementary strands during the procedure. A method we have found useful for preparing labeled single-stranded DNA is depicted in Fig. 4. A M13 clone of the sense strand of the RNA to be analyzed is used as the template for *in vitro* synthesis of labeled DNA complementary to the RNA. A 5'-end-labeled probe can be prepared by labeling a synthetic oligodeoxyribonucleotide of 15–20 nt with T4 polynucleotide kinase and [γ-^{32}P]ATP. Following hybridization to the M13 clone, the primer is extended by polymerization using *Escherichia coli* DNA polymerase I and dNTPs. Alternatively, high-specific-activity, uniformly labeled DNA can be prepared by extending a primer using DNA polymerase I and [α-^{32}P]dNTPs. In this case, the "universal primer" can be used to prime synthesis on the mp phage vectors,[12] since the 5' end of the probe need not be complementary to the RNA when using a uniformly labeled probe. Following polymerization, the DNA is cut with a restric-

[11] G. S. Hayward, *Virology* **49**, 342 (1972).
[12] J. Messing, this series, Vol. 101, p. 20.

tion enzyme to generate the 3' end of the probe. Using this strategy, the template strand is much longer than the labeled DNA strand synthesized *in vitro* and can be separated from the labeled DNA with good yield by denaturation and electrophoresis through a polyacrylamide gel made 7 *M* in urea to inhibit renaturation.[3] Synthetic DNA distal to the restriction site is also well separated from the probe DNA, which can be visualized by autoradiography of the wet gel. The band of labeled probe DNA is cut from the gel and is recovered for use as a S1 probe.

For preparing 5'-end-labeled probe, 2 pmol of M13 clone DNA (~5 μg) is hybridized to 3–5 pmol of 5'-phosphate-labeled oligonucleotide primer in 50 μl of 50 m*M* Tris–HCl, pH 7.2, and 10 m*M* MgSO$_4$ by incubation at 95° for 5 min, followed by 50° for 30 min. The solution is cooled and 10 μl of a mixture of 5 m*M* of the four dNTPs is added, plus 10 μl of 2 m*M* dithiothreitol (DTT), 25 μl of water, and 5 μl (25 U) of the Klenow fragment of *E. coli* DNA polymerase I, and the solution is incubated at 25° for 30 min. The solution is then diluted 2-fold into a buffer appropriate for the restriction enzyme of choice, and sufficient restriction enzyme is added to cut the DNA to completion.

For the preparation of uniformly labeled probe of high specific activity, unlabeled primer of oligonucleotide is hybridized to the template M13 clone as described previously. Polymerization of the DNA is performed as described previously, except that 0.1 m*M* high-specific-activity [α-^{32}P]dNTPs are used in an initial polymerization reaction for 30 min. Then unlabeled dNTPs are added to 0.5 m*M*, and polymerization is continued for 30 min. Following synthesis of the labeled DNA strand, the reaction mixture is diluted into restriction endonuclease buffer and is digested to completion with the appropriate restriction enzyme.

Following synthesis of 5'-end-labeled or uniformly labeled DNA, the products are recovered by ethanol precipitation. The precipitated material is dissolved in 10 μl of 2 : 1 (v/v) 98% formamide plus tracking dyes : 0.1 *M* NaOH, is incubated at 95–100° for 1 min, and is subjected to electrophoresis in a polyacrylamide gel prepared in TBE plus 7 *M* urea.[3] The unique length probe DNA is visualized by autoradiography of the wet gel, and the band is excised and eluted. High-specific-activity proves must be used within 1 week, since the probes are subject to autoradiolysis. End-labeled probes can usually be used for up to 2 months, subject to the decreasing specific activity resulting from ^{32}P decay.

When hybridization is performed with single-stranded complementary DNA, the kinetics of hybridization are similar to those of DNA–DNA renaturation, which have been thoroughly described.[13,14] The rate is maxi-

[13] J. G. Wetmur and N. Davidson, *J. Mol. Biol.* **31,** 349 (1968).
[14] R. J. Britten and D. E. Kohne, *Science* **161,** 529 (1968).

mal in high salt (1 M NaCl) at a temperature equal to T_m -25–$30°$, but the hybridization rate remains high over a broad range of temperature. Some workers prefer to perform the hybridization reaction in 50% (v/v) formamide and lower salt concentrations to lower the T_m and, consequently, the hybridization temperature. This reduces thermal degradation of RNA, which may be important in generating RNA–DNA hybrids of >300 base pairs. The melting temperature of a DNA fragment can be approximated by[15]

$$T_m = 81.5 + 0.5 \,(\% \text{ G} + \text{C}) + 16.6 \log [\text{Na}^+] - 0.6 \,(\% \text{ formamide})$$

We typically perform the hybridization in 50 μl of 10 mM Tris–HCl, pH 7.2, 1 mM ethylenediaminetetraacetic acid (EDTA), and 1 M NaCl at 55–68° (depending on the percentage of G + C) at a probe DNA concentration of 0.2 nM ($\sim$1–10 ng of probe DNA). At this DNA concentration, hybridization of probe to RNA of equal or lower concentration is complete in 30 min. For RNA–DNA hybrids with duplex regions of <100 bp (base pairs), we extend the hybridization time to 2 hr, since the rate of hybridization falls for short hybrids.[13]

Using Double-Stranded DNA Probes. Because of the simplicity of preparing end-labeled duplex DNA restriction fragments, hybridization is often performed with double-stranded DNA probes. In this case, the hybridization temperature is critical. Casey and Davidson[16] found that in 70–80% (v/v) formamide, the T_m of DNA–DNA duplex is 5–10° lower than the T_m of RNA–DNA duplex. Consequently, it is possible to hybridize RNA directly to the coding strand of DNA using a double-stranded DNA restriction fragment in the hybridization reaction. This is done by performing the hybridization at a temperature higher than the melting temperature of the DNA–DNA restriction fragment, but lower than the melting temperature of the RNA–DNA hybrid.[16] Casey and Davidson found that 80% (v/v) formamide, 0.4 M NaCl, 0.04 M PIPES, pH 6.4, and 1 mM EDTA was a particularly useful buffer, becaue the buffer maintained constant pH and ionic strength at temperatures of 40–60°, the temperature range most frequently used for RNA–DNA hybridization. The kinetics of hybridization under these conditions are approximately 12-fold slower than the kinetics of DNA–DNA renaturation in 0.4 M NaCl at T_m $-25°$.[16] The slower rate of hybridization results from the viscosity of formamide and the fact that the hybridization is performed at only a few degrees below the melting temperature of the RNA–DNA hybrid. Using a DNA concentration of 0.4 nM, hybridization is nearly complete in 3 hr.[17] For RNA–DNA hybrids of <300 bp, the rate is somewhat slower. For

[15] M. Thomas, R. L. White, and R. W. Davis, *Proc. Natl. Acad. Sci. U.S.A.* **73**, 2294 (1976).

[16] J. Casey and N. Davidson, *Nucleic Acids Res.* **4**, 1539 (1977).

[17] P. A. Sharp, A. J. Berk, and S. M. Berget, this series, Vol. 65, p. 750.

convenience, we typically perform hybridizations overnight (12–16 hr) at a probe DNA concentration of 0.1–1.0 nM for duplex regions of <300 bp.

As mentioned above, the temperature is critical for hybridization to double-stranded DNA probes at high formamide concentration. Hybridization should be at 2–4° above the DNA–DNA melting temperature. We determine the optimum hybridization temperature empirically, beginning with an estimate of the T_m for the duplex DNA based on the formula above. However, the melting temperatures of short duplex regions are usually several degrees below this estimate. Generally, we do trial hybridizations at the estimated T_m, and at this temperature −6, −3, +3, and +6°, and determine which temperature gives the strongest signal on the final autoradiogram. If the temperature optimum was not clearly distinguished, we repeat a series of hybridizations through either a higher or lower temperature range, depending on the initial results. Once the optimum temperature is determined, the conditions can be used repeatedly to quantitatively assay a specific RNA using a probe, which can be readily prepared by end-labeling a restriction fragment.[3]

To simplify dissolving high concentrations of RNA in the high-formamide hybridization buffer, we set up a hybridization as follows. Up to 300 μg of a RNA preparation (e.g., total cell RNA) in 10 mM Tris–HCl, pH 7.0, 1 mM EDTA, and 0.1% (w/v) sodium dodecyl sulfate (SDS) is mixed with 5–50 fmol of labeled probe DNA in a 1.5-ml polypropylene microcentrifuge tube. One-tenth volume of 3 M sodium acetate, pH 6, is added, and the nucleic acids are precipitated by the addition of 2.5 volumes of −20° ethanol. The precipitate is collected by centrifugation, and the pellet is washed by carefully adding 1 ml of 95% (v/v) ethanol at −20°, briefly centrifuging, and carefully decanting the supernatant. The pellet is drained dry briefly, to remove the ethanol on the walls of the tube, but the pellet is not dried. Bringing the pellet to dryness makes dissolving the RNA much more difficult. The pellet is then dissolved in 40 μl of 98% (v/v) formamide by incubating at room temperature for 20–30 min, followed by repipetting with a micropipet until the pellet is completely dissolved. Ten microliters of 2 M NaCl, 0.2 M PIPES, pH 6.4, and 5 mM EDTA is added, and the solution is thoroughly mixed by repipetting. The capped tubes are incubated at 68° for 10 min to completely denature the double-stranded DNA probe and then are submerged in a water bath at the appropriate hybridization temperature. We use a floating Styrofoam rack on a covered water bath with a constant temperature circulator.

S1 Nuclease Digestion

At the completion of the hybridization reaction (either aqueous reactions using single-stranded DNA probes or high-formamide hybridizations using double-stranded DNA probes), 9 volumes of S1 digestion buffer (0.25 M NaCl, 0.03 M sodium acetate, pH 4.5, and 1 mM ZnCl$_2$) are added. For high-formamide hybridizations with double-stranded DNA probes, S1 buffer at 0° should be added rapidly to each tube as the tube is removed from the water bath used for the hybridization reaction, and the solution should be placed on ice to prevent reannealing of the complementary DNA strands. This solution is sufficiently high in salt to stabilize most duplex regions against digestion by S1 nuclease at room temperature. S1 nuclease requires Zn^{2+} as a cofactor, and the pH optimum is 4.5 for S1 activity.[2] We typically perform hybridization reactions in 50 μl and do the S1 digestions in 0.5 ml. For a typical analysis using 50–300 μg of total RNA and 2–20 ng of probe DNA, we add 200–500 U[2] of S1 and digest at room temperature for 30 min.

Since the RNA is also a substrate for S1, lower concentrations of S1 are required to digest samples with lower concentrations of RNA. The correct amount of S1 can be determined empirically by performing separate S1 digestions with different concentrations of S1. The limit digest should be obtained through a 10-fold S1 concentration range. If the amount of S1 is too low, a high background is observed in the autoradiogram of the gel lane. At S1 concentrations which are too high, the signal observed with end-labeled probes is diminished, and the signal observed with uniformly labeled probes is slightly reduced in size due to digestion of the probe at the ends of the RNA–DNA hybrid region. A sharp band (or group of approximately two to six bands on a sequencing gel) should be observed on the final autoradiogram. S1-protected fragments of the same length ±5 nt should be obtained through a 10-fold range in S1 concentration (as in Ref. 4, for example).

The products of S1 digestion are recovered by ethanol precipitation after addition of 2 volumes of 95% (v/v) ethanol at −20°. If the hybridization reactions are in 50 μl in a 1.5-ml microcentrifuge tube, all of the steps can be performed in the same tube, simplifying the analysis of multiple RNA samples. To remove residual salt before electrophoresis, we dissolve the ethanol precipitate in 0.2 ml of 0.3 M sodium acetate, pH 6.0, and reprecipitate nucleic acids by addition of 0.5 ml of 95% (v/v) ethanol at −20°. The pellet is collected by centrifugation and is rinsed with 1 ml of chilled 95% (v/v) ethanol. The tube is drained dry, and the pellet is dissolved in a small volume (3–20 μl) of 2:1 (v/v) 98% formamide plus tracking dyes: 0.1 M NaOH. After incubation at 95–100° for 1 min, the samples are subjected to electrophoresis in a polyacrylamide gel prepared in TBE plus 7 M urea.[3] Appropriate size markers are analyzed on the same gel.

S1-protected DNA is detected by autoradiography. Drying the gel before exposure increases sensitivity, especially for gels of >0.5 mm thickness. Gels of 0.5 mm thickness or less can be directly dried onto filter paper before autoradiography. For thicker gels, the gel can be soaked in cold 5% (w/v) trichloroacetic acid (TCA) for 20 min with occasional rocking to remove the urea. After rinsing with water, the gel can be dried onto filter paper prior to autoradiography. For quantitation of the S1-protected DNA by densitometry of the final autoradiogram, autoradiography should be direct (i.e., without intensifying screens) or done with intensifying screens using preflashed film.[18] ^{32}P autoradiography with intensifying screens without preflashed film does not result in a linear relationship between intensity and disintegrations per minute.[18]

Pitfalls and Qualifications of the Method. Some sequences in RNA–DNA hybrids are unusually sensitive to S1 digestion. This probably results from instability of the hybrid duplex structure. In particular, runs of rU–dA form a surprisingly unstable duplex structure,[19] which may be digested by S1 at room temperature. Minor S1-digested products terminating at what would be (rU–dA)- and/or (rA–dT)-rich regions of a RNA–DNA hybrid are suspect. RNA–DNA hybrids with such (rU–dA)- and (rA–dT)-rich regions can be digested at temperatures below room temperature, including digestion at 0° on ice. The rate of S1 digestion falls a factor of ~2 for every 10° drop in temperature, and the time of S1 digestion or the amount of S1 used in the digestion should be adjusted accordingly.

A small (<1%) fraction of full-length probe DNA often escapes S1 digestion. Perhaps the small fraction is sterically blocked from interacting with the enzyme because of association with impurities in the probe DNA preparation. Because of this, a S1-protected band equal to the full length of the probe cannot be interpreted to infer a particular RNA structure. Only S1-protected fragments shorter than the full-length probe can be properly interpreted. Similarly, in designing a probe for use in quantitating a RNA species, the probe should be designed so that the S1-protected fragment is shorter than the full-length probe.

A useful control is obtained by subjecting an aliquot of the probe DNA to the same procedures used for RNA mapping or quantitation, but without including complementary RNA in the hybridization reaction. Either no RNA can be included or, better yet, an amount of an unrelated RNA (e.g., yeast tRNA) can be included equal to the mass of RNA in the sample with the specific RNA under analysis. This negative control should produce no bands on the final autoradiogram in the region of the specific S1-protected fragments. Any bands that are observed in this neg-

[18] R. A. Laskey, this series, Vol. 65, p. 363.
[19] F. H. Martin and I. Tinoco, *Nucleic Acids Res.* **8**, 2295 (1980).

ative control should be considered background. S1-protected fragments occurring at this position in the analysis of RNA complementary to the probe cannot be interpreted.

It is essential that the DNA probe be in molar excess over the RNA being analyzed, both for mapping and for quantitating specific RNAs. If alternatively processed RNAs are being mapped and the RNA is in excess over the DNA probe, more than one RNA molecule can hybridize to the same DNA molecule. Since the DNA is protected from S1 digestion throughout the RNA–DNA hybrid region, only the longer of the alternative exons is detected.[10] In quantitating RNAs, excess RNA will protect all the probe DNA if hybridization is taken to completion, and the quantity of specific RNA will be underestimated. To control for excess probe DNA in mapping experiments, one can test if the same relative intensities of S1-digested products are observed as the RNA concentration in the hybridization reaction is decreased through a serial dilution series. To control for excess DNA when quantitating a specific RNA, one can perform the hybridization reaction with twice the standard amount of RNA to test that the S1-protected band on the final autoradiogram has two times the intensity of the signal from the standard reaction. If a specific RNA is being analyzed in a series of RNA samples, this control need only be done for the sample with the highest concentration of the RNA being analyzed.

As mentioned above, heterogeneity on the order of ~2–6 nt in the size of a S1-digested product cannot be interpreted to indicate heterogeneity in the 5' end of a capped RNA. Also, the resolution of mapping the 5' end of a capped RNA is limited to approximately ±5 nt. The map position of the 5' end of an RNA should be confirmed by primer extension; otherwise, a short 5' exon can be easily overlooked (Fig. 2).

[25] Primer Extension Analysis of RNA

By WILLIAM R. BOORSTEIN and ELIZABETH A. CRAIG

Introduction

Primer extension analysis is used to map the 5' termini of RNA transcripts, to quantify RNA levels, and to detect low-abundance RNA species. The analysis of RNA by this technique involves four basic steps: (1) selection and preparation of a radioactively labeled primer that is complementary to the transcripts of interest, (2) hybridization of this primer to the complementary region of the RNAs, (3) extension of the primer catalyzed by reverse transcriptase using the RNA as a template, and (4) analysis of the extended DNA products utilizing polyacrylamide gel electrophoresis (PAGE) followed by autoradiography. The number of nucleotides from the 5' end of the primer to the end of the template can be determined from the rate of migration of the extended DNAs on a gel. The primer is supplied in excess of complementary RNA to allow accurate quantification of the abundance of these RNA transcripts. The method is schematically illustrated in Fig. 1.

The primer extension technique is complementary to 5' end mapping methods involving the protection of radiolabeled probes from single-stranded-specific nucleases by hybridization to complementary RNA.[1,2] Nuclease protection mapping allows determination of the length of the transcripts that are colinear with the probe. RNA transcripts that are spliced in the region upstream from the labeled end of the primer or genomic probe result in discrepancies between mapping data from the two types of methods. In addition, the two types of methods are subject to different types of artifacts. For these reasons, it is important to use both primer extension and nuclease protection experiments, when mapping the 5' ends of previously uncharacterized transcripts.

Primer extension is the simplest and most rapid technique for repeated analysis of 5' end positions and transcript abundance; only the hybridization step requires optimization. The use of oligonucleotide primers (described in detail in Oligonucleotide Primers) greatly facilitates application of the technique, overcoming previous limitations both in primer selection and in discrimination between related transcripts. The sensitivity and precision of this method result from the ability to use large amounts of RNA and to accurately determine the size of short, extended fragments. Variations of the technique allow determination of the exact identity of the extended products and distinction between 5' end heterogeneity and differential internal transcript processing (e.g., splicing). This versatility allows primer extension to be used to address fundamental quantitative and qualitative questions about the expression of individual genes.

Materials

Selected materials and reagents used in the methods described are included below along with the names of the suppliers and order numbers.

[1] A. J. Berk and P. A. Sharp, Cell (Cambridge, Mass.) 12, 721 (1977).
[2] A. J. Berk, this volume [24].

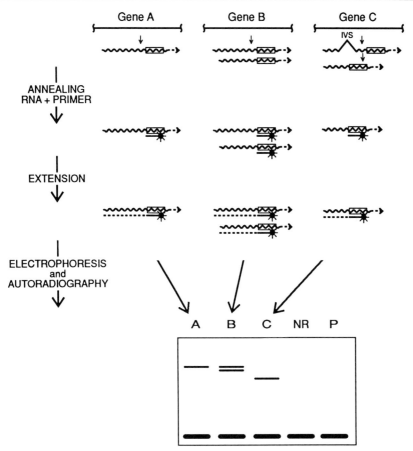

Fig. 1. A schematic diagram of primer extension analysis of transcripts from three genes. Only a portion of the 5' ends of the genes and transcripts are depicted. Gene A encodes a single transcript. Gene B encodes two transcripts with different initiation sites. Gene C encodes a transcript that is spliced near the 5' end. DNA is represented by solid lines and RNA by wavy lines. The short solid lines with stars represent 5'-end-labeled primers. The adjacent broken lines depict the extended DNAs. The boxes indicate regions of the transcripts that are complementary to the primer. (Autoradiograph) Bands corresponding to each of the diagrammed extended products are shown. The distance from the initiation site to the primer binding site is the same in gene A and gene C. However, the extension products from the C transcripts are shorter due to splicing. The dark bands at the bottom of lanes A, B, and C, which align with those in the "no RNA" (NR) and "primer only" (P) control lanes, represent unextended primer. IVS, intervening sequence.

This does not imply that these products are superior to those from other companies.

Reagents

0.45-μm Cellulose acetate syringe filters [(190-2045) Nalge Co., Rochester, NY]

[α-^{32}P]dNTPs (deoxynucleoside triphosphates) [3000–6000 Ci/mmol (Amersham, Arlington Heights, IL)]

[γ-^{32}P]ATP [3000–5000 Ci/mmol (Dupont/New England Nuclear Research Products, Boston, MA)]

Acetonitrile suitable for use in ultraviolet (UV) spectrophotometry [Photorex (9392-1) (J. T. Baker, Phillipsburg, NJ)]

Acrylamide [99.9% (161 0101) (Bio-Rad, Richmond, CA)]

Actinomycin D [(A-7890) (Sigma, St. Louis, MO)]

Analytical-grade mixed-bed resin [AG 501-X8 (D) (142-6425) (Bio-Rad)]

C$_{18}$ columns [Sep-Pak (51910) (Waters Associates, Milford, MA)]

Deoxynucleotide triphosphates [ultrapure (27-2050-01 series) (Pharmacia, Piscataway, NJ)]

Dimethyl sulfoxide (DMSO) [SpectrAR (5507) (Mallinckrodt)]

Enzymes (restriction, DNA modifying, and polymerases) (Boehringer Mannheim, Indianapolis, IN)

Formamide [Spectro grade (120 3066) (Eastman Kodak, Rochester, NY)]

Glycogen [molecular biology grade (901 393) (Boehringer Mannheim)]

N,N'-Methylenebisacrylamide [Ultra Pure (5516UA) (Bethesda Research Laboratories, Bethesda, MD)]

Polyethyleneimine (PEI) plates [Polygram Cell 300 PEI (801053) (Brinkmann Instruments, Westbury, NY)]

Phenol [AR (0028) (Mallinckrodt, Paris, KY)]

Ribonuclease inhibitor (RNasin) [(P2111) (Fisher/Promega Biotec, Madison, WI)]

Urea [Ultra Pure (5505UA) (Bethesda Research Laboratories)]

Buffers

Buffers and solutions used for hybridization and primer extension reactions should be RNase free. RNases can be inactivated by the addition of diethyl pyrocarbonate (DEPC) to a final concentration of 0.1% (v/v). Shake vigorously, wait 15 min, then autoclave for 20 min or heat to 70° for 1 hr to decompose the DEPC. DEPC cannot be used to treat buffers containing Tris; these solutions should be made with DEPC-treated H$_2$O. DEPC is a suspected carcinogen and should be treated with caution; wear gloves and work in a fume hood. Glassware and pipets used to dispense RNase-free solutions should be baked at 250° for 4 hr.

TE buffer: 10 mM Tris–hydrochloride (Tris–HCl), pH 7.5, and 1 mM ethylenediaminetetraacetic acid (EDTA), pH 8.0

Calf intestinal alkaline phosphatase (CIAP) storage buffer: 30 mM Triethanolamine, pH 7.6, 3 M NaCl, 1 mM MgCl$_2$, and 0.1 mM ZnCl$_2$

10× CIAP reaction buffer: 0.5 M Tris–HCl, pH 8.0, and 1 mM EDTA, pH 8.0

1× Klenow buffer: 130 mM Potassium phosphate, pH 7.4, 6.5 mM MgCl$_2$, 1 mM dithiothreitol (DTT)

10× TBE: 0.9 M Tris base, 0.9 M boric acid, and 20 mM EDTA

Soaking elution buffer (SEB): 0.5 M NaCl, 0.1 M Tris–HCl, pH 8.5, and 5 mM EDTA, pH 8.0

5× Single-stranded (SS) hybridization buffer: 1.5 M NaCl, 50 mM Tris–HCl, pH 7.5, and 10 mM EDTA, pH 8.0

EDTA–formamide solution: 100 : 1 (v/v) Mixture of deionized formamide and 0.25 M EDTA, pH 8.0

Double-stranded (DS) hybridization buffer: 4 M NaCl and 0.1 M piperazine-N,N'-bis(2-ethanesulfonic acid) (PIPES), pH 6.8

Phenol–CHCl$_3$: 1 : 1 (v/v) Mixture[3]

1.25× Reverse transcriptase buffer: 1.25 mM dATP, 1.25 mM dTTP, 1.25 mM dCTP, 1.25 mM dGTP, 12.5 mM DTT, 12.5 mM Tris–HCl, pH 8.4, and 75 μg/ml of actinomycin D, 7.5 mM MgCl$_2$. Dissolve actinomycin D in 95% (v/v) ethanol, then dilute 1 : 1 (v/v) with H$_2$O to make a working stock

Standard 5× loading buffer: 25% (w/v) Sucrose, 25 mM EDTA, 0.5% (w/v) sarkosyl, 0.2% (w/v) bromphenol blue (BPB), and 0.2% (w/v) xylene cyanol (XC)

Formamide loading buffer: 0.03% (w/v) BPBlue, 0.03% (w/v) XC, and 20 mM EDTA in deionized formamide. Store at room temperature for up to 1 month or at −20° for at least 1 year

10× Kinase buffer: 0.1 M MgCl$_2$, 0.6 M Tris–HCl, pH 8.0, and 50 mM DTT

Gels

Standard polyacrylamide gel: Acrylamide : N,N'-methylenebisacrylamide [19 : 1 (w/w)] and 0.5× TBE (percentage of acrylamide is

---

[3] Preparation of phenol is described in T. Maniatis, E. F. Fritsch, and J. Sambrook, "Molecular Cloning: A Laboratory Manual." Cold Spring Harbor Laboratory, Cold Spring Harbor, New York, 1982.

determined by the size of the fragments to be separated[4]). Gels are 20 cm long and 1.5 mm thick, unless otherwise specified Denaturing polyacrylamide gel: Same as the standard polyacrylamide gel, except with 7 M urea. Denaturing gels should be prerun for at least 30 min prior to the application of the samples. These gels should be warm (~50°), not hot, during electrophoresis

Primer Selection and Preparation

Mapping resolution is maximized with small, extended products; therefore, the ideal primer is a short single-stranded DNA fragment that is complementary to a region within the 5' end of the transcript to be analyzed. Extension products shorter than 150 bases allow optimal resolution. When mapping the position of a 5' end without any prior information regarding the initiation site, it is best to choose a primer with a 3' terminus just within the amino-terminal end of the coding region. This allows the detection of transcripts, regardless of the size of the untranslated leader. If some or all of the transcripts have unusually long leaders, or if the transcripts are spliced near their initiation sites, it may be necessary to repeat the analysis with a different primer that will yield smaller extended products. These criteria for the selection of a DNA primer are most easily achieved using synthetic oligonucleotides, although DNA fragments generated by restriction enzymes may also be used.

Oligonucleotide Primers

Oligonucleotides have several advantages over restriction fragments as primers. The use of oligonucleotides avoids the limitations of primer selection imposed by the chance occurrence of restriction enzyme recognition sites. Furthermore, by choosing an oligonucleotide primer that is a unique sequence, cross-hybridization to related genes can be completely eliminated. Oligonucleotides offer several technical advantages as well. Oligonucleotides can be efficiently radiolabeled to a higher specific activity than restriction fragments, because oligonucleotides have free 5'-hydroxyl groups. As single-stranded polynucleotides, oligonucleotides hybridize to RNA without competition from a complementary DNA strand. Due to their short size, self-priming (utilization of primer as template) is rarely a problem. Twenty nucleotides is a practical length for a deoxyribooligonucleotide primer, affording a high degree of specificity.

---

[4] Tables relating effective separation ranges to polyacrylamide percentages for denaturing and native gels can be found in Ref. 3.

Oligonucleotide purification. Oligonucleotides synthesized by the solid-phase method can be obtained commercially. Purification of a primer that has been cleaved from the support, uncapped, and detritylated is described in this section. The simplest and most reliable purification method involves PAGE, followed by reversed-phase chromatography (RPC). Gel purification is important to remove aberrantly short oligonucleotides formed during synthesis. These aberrantly short oligonucleotides are usually present in very small amounts, but their reduced hybridization specificity may result in background signals.

1. Dry the crude oligomer preparation to remove residual NH_4OH and to concentrate the sample.
2. Resuspend the sample in an appropriate volume of TE to yield a final concentration of approximately 10 mg/ml, based on the estimated synthesis yield. Dilute a small aliquot of this solution and determine the actual concentration from the absorbance at 260 nm (1 OD $\approx$ 33 μg/ml, assuming equal amounts of the bases G, A, T, and C).
3. Prepare a 20% (w/v) denaturing polyacrylamide gel 40 cm in length. Load ~130 μg of oligonucleotide (in $\geq$50% (v/v), formamide loading buffer)/2-cm well. Run the gel at ~17 V/cm (warm to the touch) until the BPB reaches the bottom. BPB and XC comigrate with ~11 and 57 base oligonucleotides, respectively.
4. Separate glass plates and transfer the gel to Saran wrap. Place the gel on a fluorescent surface, such as a PEI-cellulose plate, and visualize the DNA by the shadow from a hand-held shortwave (254 nm) UV light. The desired oligonucleotide should be the most abundant and slowest migrating DNA on the gel.[5] Excise this UV-absorbent band with a razor blade, being careful not to take any of the gel below the band.
5. Place the gel slice in a 50-ml polypropylene tube. Crush the piece of acrylamide with a glass rod, then add 5 ml of SEB. Elute the DNA by shaking overnight at 37°.
6. Pellet acrylamide by centrifugation at 2000 g for 5 min. Transfer the liquid to a new tube with a Pasteur pipet, leaving the acrylamide behind. Rinse the acrylamide and the elution tube with an additional 2 ml of SEB. Spin again and pool supernatants.
7. Residual acrylamide is removed by passing the eluent through a 0.45-μm cellulose acetate syringe filter.

---

[5] DNA size standards with terminal phosphate groups will migrate differently than non-phosphorylated oligonucleotides.

The oligonucleotide solution is desalted by RPC on a minicolumn. Attach the barrel of a 10-ml syringe to the long end of a Sep-Pak C_{18} cartridge to make a reservoir. The plunger can be used to apply pressure to the column.

1. Prewet the column with 5 ml of acetonitrile.
2. Wash with 15 ml of H_2O.
3. Apply the sample to the column.
4. Wash with 10 ml of H_2O.
5. Elute with 2 ml of 30% (v/v) acetonitrile in H_2O.
6. Evaporate to dryness.
7. Resuspend at approximately 200 $\mu g/ml$.

The binding and elution of the sample to the column can be monitored with absorbance readings at 260 nm. Determine the yield spectrophotometrically and make a working dilution of 25 ng/μl. $^{32}PO_4$ is transferred to free 5'-hydroxyl groups of the purified oligonucleotides, using phage T4 polynucleotide kinase as described in the kinase section for primers generated by restriction enzyme digestion.

Restriction Fragment Primers

There are many strategies for the preparation and isolation of primers from cloned DNA (see Fig. 2 for examples). Single-stranded primers are preferable for the increased ease of hybridization and resulting increased sensitivity. Double-stranded restriction fragments with a single labeled end are, however, often easier to generate and can be routinely used with satisfactory results (these double-stranded restriction fragments will be referred to as double-stranded primers). Primers can be labeled at their 5' or 3' ends. 5'-End-labeled primers are advantageous, as any exonucleolytic cleavages removing sequences from the 5' end yield unlabeled products, which will not interfere with autoradiographic identification of full-length extension products. 3'-End-labeling involves a simpler procedure, does not require DNA free of RNA, and can be used to make primers of severalfold higher specific activity. Strategies to generate primers described in this section involve end labeling of a restriction fragment, followed by the isolation of a single-labeled DNA primer (i.e., separation of the primer away from its complementary end-labeled strand and from other labeled fragments). The same general criteria described for the selection of oligonucleotide primers apply to restriction fragment primers.

Construction of a primer that is a single species and is labeled to high specific activity requires unnicked DNA. DNA that has been prepared over two CsCl gradients is ideal for construction of 5'-end-labeled

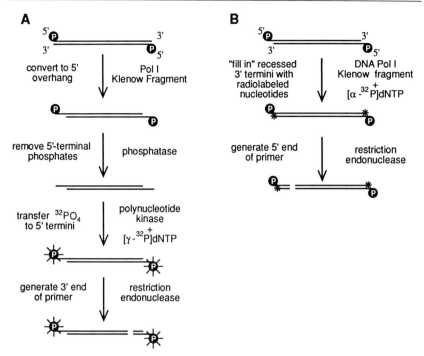

FIG. 2. Examples of two strategies for constructing ^{32}P-labeled primers. (A) A 5'-end-labeled primer is made from a restriction fragment with blunt ends. (B) A 3'-end-labeled primer is generated from a restriction fragment with recessed 3'-termini. Singly labeled single-stranded or double-stranded primers are isolated from polyacrylamide gels, following the steps diagrammed on the figure. Black circles labeled "P" represent 5'-terminal phosphate groups, those black circles with rays are radiolabeled; the asterisks represent [32P]phosphate in phosphodiester bonds.

primers, as the DNA will be free of RNA. Gel-purified restriction fragments can also be efficiently labeled. Because 5' overhangs are most efficiently labeled, it is preferable to choose a restriction enzyme that will cleave the DNA yielding this type of terminus. Blunt ends and recessed 5' termini can be labeled to a high specific activity, following their conversion into 5' overhangs, as described in Generating 5' Overhangs from Blunt or Protruding 3' Termini.

Digest 5 pmol of cloned DNA [~20 μg of a 6-kb (kilobase) plasmid] with a restriction enzyme to generate the 5' end of the primer. Significantly less DNA is sufficient to prepare primer for many extension reactions; however, this large excess is required for some primer preparation steps, such as visualization of very small unlabeled restriction fragments.

If too many ends are generated to efficiently label or if subsequent digestion will result in fragments comigrating with the primer, it may be necessary to gel purify a fragment.

Generating 5' overhangs from blunt or protruding 3' termini. If a restriction enzyme that creates blunt ends or 3' overhangs is used to generate the termini of the primers to be labeled, the following procedure can be used to remove a small number of nucleotides from the 3' ends. The resulting protruding 5' and recessed 3' termini will be more favorable substrates for 5' end labeling with polynucleotide kinase or for 3' end-labeling with the large fragment of *Escherichia coli* DNA polymerase I (Klenow fragment).[6]

Adjust the NaCl concentration of the restriction enzyme digestion to 50 mM. Add Klenow fragment, such that there is a 10-fold molar excess of DNA ends over enzyme (usually 0.5–5 U) and incubate for 30 min at room temperature.[7] If the sequence of the restriction fragment is known, the exonucleolytic digestion can be halted at a specific site. This is done by adding a single, appropriate dNTP to the reaction at a final concentration of 200 μM. The exonuclease activity will predominate over the polymerase activity only until the nucleotide complementary to that present in the reaction mixture is exposed on the opposite strand. In the example below, a 3' overhang, generated by *Sac*I, is converted to a 3-base 5' overhang with the Klenow fragment in the presence of dATP.

```
..... GGACTGAGCTC ...3'    SacI      ..... GGACTGAGCT3'   Klenow    ..... GGA3'
..... CCTGACTCGAG ...5'   ------>    ..... CCTGAC5'       ------>    ..... CCTGAC5'
           SacI           digest                           dATP
```

5' End Labeling. Phosphatase. It is necessary to dephosphorylate the 5' termini generated by the restriction enzyme digestion, leaving free-hydroxyl groups for the subsequent efficient incorporation of [^{32}P]phosphates at these positions. CIAP free of DNases and RNases and from which ammonium sulfate has been removed should be used.[8] CIAP (0.1 U) is sufficient to dephosphorylate 50 pmol of termini. Digestion of 20 μg of a 6-kb DNA plasmid with an enzyme that cleaves each molecule only once will generate ~10 pmol of ends. To facilitate complete inactivation and removal of the enzyme after this reaction, it is best to dilute the enzyme and add only the required amount. Dilute CIAP in CIAP storage buffer immediately prior to use. Since CIAP is active in most restriction

---

[6] Alternatively, T4 DNA polymerase or terminal transferase can be used to label protruding 3' ends directly.

[7] B. D. Erikson, Z. F. Burton, K. K. Watanabe, and R. R. Burgess, *Gene* **40**, 67 (1985).

[8] G. Chaconas and J. H. Van De Sande, this series, Vol. 65, p. 75.

enzyme buffers, dephosphorylation is most simply performed by adding CIAP directly to the restriction enzyme digestion. Alternatively, the digested DNA can be extracted with phenol, ethanol precipitated, and resuspended in 200 μl of 1× CIAP reaction buffer. Incubate at 37° for 30 min.

Extensive steps are required to ensure that no CIAP activity remains. Inactivate the phosphatase by the addition of $\frac{1}{10}$ volume of 0.5 M ethylene glycol bis(β-aminoethyl ether)-N,N'-tetraacetic acid (EGTA), pH 8.0, followed by heating to 65° for 45 min. If carrier is needed for efficient precipitation, add 30 μg of glycogen. Extract with phenol–CHCl$_3$ at least two times to remove the CIAP. Ethanol precipitate with sodium acetate, wash with 70% (v/v) ethanol, and dry the pellet.

Kinase. T4 polynucleotide kinase is used to catalyze the transfer of radiolabeled phosphates from [γ-^{32}P]ATP to the protruding 5'-hydroxyl groups of the primer DNAs. Ammonium sulfate, which is present in some alkaline phosphatase preparations, inhibits kinase and must be removed prior to this reaction. Nucleic acid carrier, for precipitating DNA to be labeled, must also be avoided, as nucleic acid carrier will actively compete with the primer in the kinase reaction. Glycogen is an effective carrier that will not interfere with these reactions.

Label the phosphatase-treated fragment or 8 pmol of oligonucleotide (2 μl of the 25-ng/μl stock of 20-mer described previously) as follows.

1. Combine the DNA to be labeled, a 4-fold molar excess of [γ-^{32}P]ATP (3000–5000 Ci/mmol) over 5'-hydroxyl groups (final concentration must be at least 0.7 μM[9]), 2 μl of 10× kinase buffer, and H$_2$O to a final volume of 19 μl.
2. Begin the reaction with the addition of 5 U of T4 polynucleotide kinase.
3. Incubate 30 min at 37°.
4. Stop the reaction by thermal inactivation of the kinase at 65° for 5 min.
5. Add 30 μg of glycogen carrier if necessary (for oligonucleotides and DNA at low concentrations). Remove unincorporated [γ-^{32}P]dATP with ethanol precipitations as follows. Add $\frac{2}{3}$ volumes of 5 M ammonium acetate, mix, then add 2 volumes (sample + ammonium acetate) of cold 95–100% (v/v) ethanol. Pellet DNA in a microfuge, remove supernatant, suspend pellet in 200 μl of TE buffer, and precipitate a second time. Rinse one time with 70% (v/v) ethanol and dry the pellet.

[9] J. R. Lillehaug, R. K. Kleppe, and K. Kleppe, *Biochemistry* **15**, 1858 (1976).

(Notes) Equimolar amounts of [γ-^{32}P]ATP and primer DNA can be used with a small decrease in specific activity of the resulting primer, eliminating the need for step 5 above. The efficiency of incorporation can be determined from the proportion of the radioactivity that can be precipitated with trichloroacetic acid or that will absorb to DE-81 paper.[10]

Labeled oligonucleotide primers are now ready to use. Restriction enzyme-generated primers must be separated from other fragments with labeled ends that could potentially act as primers. To isolate the desired primer with a single-labeled end, a second restriction digest and/or electrophoretic strand separation may be employed as described in Isolation of Labeled Primer.

3′ End Labeling. The large fragment of DNA polymerase I can be used to label the 3′ end of primers by "filling in" recessed 3′ termini with radiolabeled dNTPs.

1. Add a 2-fold molar excess of the appropriate [α-^{32}P]dNTP(s) (3000–6000 Ci/mmol) directly to the restriction digest (e.g., [α-^{32}P]dATP to label a 3′ terminus generated by digestion with *Hin*dIII). More than one labeled dNTP can be incorporated to generate primers of higher specific activity. This reaction is efficiently catalyzed in common restriction enzyme buffers and in the presence of active restriction enzymes. Alternatively, digested DNA can be extracted with phenol–CHCl$_3$, can be ethanol precipitated, and can be resuspended in 1× Klenow buffer prior to the addition of dNTPs.
2. Start the reaction with the addition of 5 U of the Klenow fragment of DNA polymerase I. Incubate at room temperature for 30 min.
3. Add 30 μg of glycogen carrier if necessary (for oligonucleotides and DNA at low concentrations). Extract the sample with phenol–CHCl$_3$, ethanol precipitate with ammonium acetate two times to remove unincorporated [^{32}P]dNTPs (see Kinase protocol), rinse with 70% (v/v) ethanol, and dry the pellet.

Isolation of Labeled Primer

Removal of Second Labeled End. The unwanted labeled ends can be removed simply by restriction digestion and by subsequent gel purification, yielding a double-stranded primer that is labeled at only one end.

[10] For protocols, see Ref. 3.

1. Digest the DNA with a restriction enzyme that will generate the unlabeled end of the desired primer.
2. Ethanol precipitate, rinse the pellet with 70% (v/v) ethanol, and dry.
3. Resuspend the DNA in 16 μl of TE buffer. Add 4 μl of standard 5× loading buffer. Load the sample and radiolabeled DNA size standards on a standard polyacrylamide gel of appropriate percentage. Apply up to 16 V/cm to the gel.
4. Remove one glass plate from the gel. Cover the gel with Saran wrap. Place tape with radioactive ink on the Saran wrap and expose to X-ray film for 5–15 min.
5. The desired band can be identified by comparison with DNA size markers. Using a razor blade, cut the desired exposed band out of the autoradiogram. Align the film on the gel with the ink marks, and use the aligned film as a template to excise the region of the gel containing the radioactive primer.
6. Elute the primer from the gel slice. The crush and soak method, described for Oligonucleotide Purification, works well for fragments up to 250 bp (base pairs). Larger fragments may be electroeluted from polyacrylamide or isolated from agarose gels. Monitor recovery of the primer with a hand-held radiation monitor. Add 30 μg of glycogen as carrier if necessary, ethanol precipitate, rinse the pellet with 70% (v/v) ethanol, and dry.

Strand Separation. A single-stranded primer can be isolated by one of the two following techniques.

Strand separation by denaturing gel electrophoresis. Small single-stranded fragments can often be isolated by strand separation on denaturing gels by virtue of the size differences created by restriction enzyme-generated overhangs. For example, if the 5′ end of the primer is generated by a restriction enzyme, such as *Eco*RI, which generates a 4-nt (nucleotide) 5′ overhang, and a second cut is made with *Pst*I, generating a 4-nt 3′ overhang, the two strands will differ in length by 8 nt, a substantial proportion of the total length of a small restriction fragment. Thus, the two strands will be easily separable on a denaturing polyacrylamide gel. A size difference sufficient to allow separation of small fragments can sometimes be generated by filling in one recessed 3′ end with the Klenow fragment of DNA polymerase I.

1. Prepare a 1.5-mm-thick denaturing polyacrylamide gel.
2. Resuspend the DNA in formamide loading buffer.
3. Heat the sample at 90° for 2.5 min and place immediately in ice.

4. Load on the gel and electrophorese at approximately 22 V/cm, keeping the gel warm. Up to 5 μg can be loaded/3-cm well, depending on the size of the fragment to be isolated.
5. Identify, excise, and elute the primer band as described for double-stranded primer fragments in Removal of Second Labeled End.

Strand separation by nondenaturing gel electrophoresis. Larger fragments or fragments with strands of equal length can often be successfully separated by denaturation, followed by electrophoresis on a nondenaturing, weakly cross-linked gel.[11] The molecular basis for the separation is unclear. Presumably, separated complementary strands form unique secondary structures with different migration rates on entering the gel. The success of this technique is sequence dependent and unpredictable.[12]

1. Prepare an 8% polyacrylamide gel [60 : 1 (w/w) acrylamide : N,N'-methylenebisacrylamide] in 0.5× TBE buffer (20–40 cm long, 1.5 mm thick). Use 6% (w/v) polyacrylamide for primers larger than 200 nt. Allow 2 hr for polymerization with this low percentage of cross-linker.
2. Resuspend the sample in 20 μl of 30% (v/v) DMSO, 1 mM EDTA, 0.03% (w/v) XC, and 0.03% (w/v) BPB.
3. Heat to 90° in a sealed 1.5-ml microfuge tube for 2.5 min. Remove from heat and place immediately on ice.
4. Load the samples and run the gel at a maximum of 8 V/cm. Do not allow the gel to warm; it is preferable to run the gel at 4°. The rate of migration of a given single-stranded fragment relative to marker dyes is *usually* similar to that of a double-stranded fragment of the same size on standard polyacrylamide gels of the same percentage. An aliquot of DNA that has not been denatured can be included on the gel as a marker for the renatured, double-stranded fragments. It is important not to overload strand-separation gels in order to obtain sufficient resolution.
5. Identify, excise and elute the fragments as described previously.

If the primer fragment was cut after labeling, the extent of strand separation can be detected by staining with ethidium bromide (EtBr). The correct band can be identified as the only radiolabeled band of the appropriate size. If both strands are labeled, the separated bands can be excised, and their identities can be determined by the chemical cleavage

[11] A. M. Maxam and W. Gilbert, this series, Vol. 65, p. 499.
[12] A. A. Szalay, K. Grohnamm, and R. L. Sinsheimer, *Nucleic Acids Res.* **4**, 1569 (1977).

sequencing method[11] or by their ability to hybridize to a single-stranded (e.g., M13 or F1) clone including the primer region. Labeled primers should be stored at $-20°$ and can be used for at least 1 week.

Hybridization

The success of primer extension analysis depends on efficient and specific hybridization of the primer to the transcripts of interest. For accurate quantitative analysis and mapping of the position of 5′ ends, it is imperative to include control reactions demonstrating primer excess and hybridization specificity as described in the following section.

Theory and Controls

An empirical approach (as described in the following protocols) toward optimizing hybridization conditions is recommended. A number of formulas exist equating T_m, the melting temperature or the temperature at which half of the duplexes are dissociated, with sequence composition, but these formulas are useful only for determining initial trial hybridization conditions for a given primer sequence. The formulas often assume conditions very different from those commonly used for hybridization, and the formulas do not include complex, but important, components, such as nucleotide distribution. The formula below was derived from a large number of studies relating hybridization conditions to T_m of DNA fragments greater than 50 nt in length. T_m is expressed in terms of the molar monovalent cation concentration (M), the G + C content, the length (L) in nucleotides of the hybrids, and the formamide concentration[13,14]

DNA–RNA
$$T_m = 79.8° + 18.5 \log M + 58.4 \text{ (mole fraction G + C)}$$
$$+ 11.8 \text{ (mole fraction G + C)}^2 - (820/L) - 0.5(\% \text{ formamide})$$
$$(1)$$

Generally, the maximum stringency, or conditions which maximize the requirement for the primer to hybridize to complementary sequences with no mismatches, is obtained at 5° below T_m. However, one should keep in mind that the greatest hybridization rate occurs significantly below the temperature at which the highest stringency is obtained. For example, hybridization of DNA fragments to RNA in the presence of 50% (v/v) formamide is most rapid at 16–20° below the T_m.[15]

[13] G. M. Wahl, S. L. Berger, and A. R. Kimmel, this series, Vol. 152, p. 399.
[14] J. Casey and N. Davidson, *Nucleic Acids Res.* **4,** 1539 (1977).
[15] M. L. Birnstiel, B. H. Sells, and I. F. Purdom, *J. Mol. Biol.* **63,** 21 (1972).

Hybridizations must be carried out under conditions of differing stringency to control for the specificity of hybridization of the primer to the RNA. If there are several products, which result from precise primer–transcript pairing, repeated analysis with increased stringency to conditions near the T_m will result in a *concomitant* decrease in the amount of each extension product. A *differential* loss of extension products indicates that the primer is annealing to related but nonidentical transcripts.

The specificity of hybridization can also be addressed by varying the RNA : primer ratio. A 5- to 10-fold excess of primer over complementary RNA is optimal. However, the fraction of total or poly(A)-selected RNA that is complementary to the primer is often unknown. Fifty micrograms of total RNA and 100 fmol (7 ng of a 100-bp double-stranded restriction fragment) of primer are reasonable amounts for initial hybridizations with previously uncharacterized transcripts. Many messages can be readily detected with substantially lower amounts of total RNA. More than 500 μg of RNA/reaction or poly(A)-selected RNA can be used to analyze transcripts present at low abundance. By altering both the stringency and the primer concentration and by utilizing different primers, the extension products from precise primer annealing can be unambiguously identified.

Control hybridization–extensions must be included in quantitative primer extension analysis to demonstrate that all transcripts complementary to the primer are utilized. If all RNA sites complementary to the primer are saturated, the resulting signal should be proportionate to the RNA concentration and independent of primer concentration. This can be tested by performing the hybridization and extension reactions with a 2-fold change in the amount of primer or RNA. If the primer is not in excess by these criteria, repeat the reactions with a greater primer : RNA ratio or increase the time of the hybridization reaction.

It is also important to include controls to identify contaminating labeled fragments, self-priming, or utilization of carrier or contaminating nucleic acids as templates. These background signals can be recognized by performing a hybridization–extension reaction with primer and carrier (if added), but without the transcript of interest. If possible, RNA isolated from the same species under conditions, or from tissues, in which the gene of interest is not expressed should be used. Alternatively, the RNA can be omitted.

Hybridization reactions can be carried out under completely aqueous conditions or in the presence of formamide. The aqueous hybridization protocol below is recommended for use with single-stranded primers. The hybridization protocol with formamide must be used with double-stranded primers to minimize reannealing of the primer strands. When discriminating between closely related RNA species, it may be necessary to perform hybridizations at temperatures close to the T_m (increasing the hybridization times due to the reduced rate of annealing); because degradation of RNA occurs at high temperatures in the presence of base or divalent cations, the formamide hybridization protocol should be used (with both single- and double-stranded primers) to lower the effective T_m and thus allow the use of lower hybridization temperatures. The high rate of hybridization of oligonucleotide primers makes it difficult to adjust annealing and extension reaction conditions to maintain stringency near the T_m of the desired hybrid. Therefore, reduced primer concentrations, alternative oligonucleotides, or restriction fragment primers may be required.

Aqueous Hybridization

The following hybridization procedure is recommended for use with synthetic oligonucleotides and other single-stranded primers.

1. Combine the RNA and primer and bring the volume up to 8 μl with H_2O. (It is convenient to store RNA in ethanol. The desired amount of primer and RNA can then be coprecipitated and resuspended in 10 μl of 1× SS hybridization buffer. Then proceed to step 3). Fifty micrograms of RNA and 100 fmol of primer are reasonable amounts for initial characterization of transcripts as discussed previously in Theory and Controls. Include appropriate controls (see Theory and Controls).
2. Add 2 μl of 5× SS hybridization buffer.
3. Denature the primer and RNA at 80° for 4 min (longer exposure to high temperature may result in hydrolysis of RNA).
4. Anneal at 50° for 2–3 hr. (Specificity of hybridization for primers >60 nt in length can be increased by annealing at 65°; however, the formamide protocol in Hybridization with Formamide should be used to maximize stringency.)
5. The hybridized primer + transcripts can be diluted in the appropriate buffer and extended with reverse transcriptase without any purification.

Hybridization with Formamide

It is possible to perform the hybridization under conditions in which RNA–DNA hybrids are strongly favored over DNA–DNA hybrids, allowing the efficient use of double-stranded restriction fragment primers, which have not been strand separated. Comparison of Eq. (2) for the T_m of

DNA–DNA hybrids, with Eq. (1) indicates that primer–RNA hybrids are more stable than primer–primer hybrids in the presence of high formamide concentrations.[14,16]

DNA–DNA

$$T_m = 81.5° + 16.6 \log M + 41(\text{mole fraction G} + \text{C})$$
$$- (500/L) - 0.63(\% \text{ formamide})   (2)$$

As indicated, this difference is enhanced with longer primers and primers of higher G + C content.

The following protocol is recommended for favoring DNA–RNA hybridization when using double-stranded primers or for discriminating between closely related transcripts with single-stranded primers.

1. Prepare three annealing reactions (to include a range of hybridization stringencies) plus appropriate control hybridizations (see Theory and Controls). Combine the primer and RNA in 4.5 μl of H_2O. Fifty micrograms of RNA and 100 fmol of primer are reasonable amounts for initial characterization of transcripts, as discussed previously in Theory and Controls.
2. Add 40 μl of EDTA–formamide solution.
3. Denature the primer and RNA at 80° for 4 min (longer exposure to high temperature may result in hydrolysis of RNA).
4. Quickly add 5.5 μl of DS hybridization buffer and immediately incubate the reaction at the appropriate temperature. The hybridization temperature is determined empirically for each fragment. Forty-eight and forty-two degrees are good starting points for high and medium stringency. Also include a low-stringency hybridization in which the temperature is allowed to cool over several hours from 46 to 35°.
5. Hybridize for 2–3 hr (longer times may be required).
6. Ethanol precipitate the hybridization reactions prior to the reverse transcriptase reaction. Rinse the pellets with 70% (v/v) ethanol and dry.

Extension

Primers that have been hybridized to complementary RNA sequences are elongated in a reaction catalyzed by avian myeloblastosis virus (AMV) RNA-dependent DNA polymerase, utilizing the transcripts as

[16] R. W. Davis, D. Botstein, and J. R. Roth, "A Manual for Genetic Engineering; Advanced Bacterial Genetics." Cold Spring Harbor Laboratory, Cold Spring Harbor, New York, 1980.

templates. Reverse transcriptase is known to utilize different templates with unequal efficiencies. This is at least partly due to the interference of the polymerase by secondary structure in the RNA. Generally, this is not a problem considering the short lengths of the desired products and the high nucleotide concentrations most commonly used in this type of application. The temperature of the extension reaction can be increased to 50° to reduce secondary structure in the template. Because reverse transcriptase can efficiently use DNA as a template, actinomycin D, which preferentially inhibits the polymerase activity utilizing DNA versus RNA templates, is included in this reaction.[17]

1. (A) Following aqueous hybridization, dilute annealing reactions with 40 μl of prewarmed 1.25× reverse transcriptase buffer.
   (B) Following hybridization with formamide, resuspend the precipitated pellet in 1× reverse transcriptase buffer.
2. Start the extension with the addition of 12 U (1 μl) of AMV reverse transcriptase. RNasin (0.5 μl) can be added to inhibit RNases.
3. Incubate at 42° for 30 min. Higher temperatures (e.g., 50°) may be used to reduce artifactual bands, resulting from both RNA secondary structure and from hybridization of the primer to similar, but nonidentical, sequences.
4. Stop the reactions with the addition of 1 μl of 0.5 M EDTA.
5. If more than 30 μg of RNA are used, the RNA should be hydrolyzed prior to electrophoresis. Add 6 μl of 1 N NaOH and incubate at 50–60° for 1 hr. If smaller amounts of RNA are used, steps 5 and 6 can be omitted.
6. Neutralize with 6 μl of 1 M HCl.
7. Ethanol precipitate with sodium acetate, rinse with 70% (v/v) ethanol, and dry.
8. Suspend the samples in formamide loading buffer. Denature at 90° for 2.5 min. Place the tubes immediately on ice. Load on a 0.4-mm-thick denaturing polyacrylamide gel of appropriate percentage to resolve the bands of interest and apply approximately 40–45 V/cm, maintaining the gel plates at about 50°. It is important to include a gel loading in which nonextended primer is retained on the gel to be certain that all short extension products are detected. A DNA sequencing ladder is a good size marker, allowing mapping of extension product lengths to within 2–3 bases.
9. Visualize the extended products by autoradiography of the polyacrylamide gel.

[17] W. E. G. Müller and R. K. Zahan, *Nature (London), New Biol.* **232,** 143 (1971).

Interpretation. The positions of the electrophoretically separated primer extension products relative to standards indicate the number of nucleotides in the RNA transcripts between the 3′ end of the primer binding sites and the 5′ termini. For precise mapping, it is important to note that sequencing ladders generated by the chemical cleavage method migrate as if the sequencing ladders are 1.5 nt shorter than the corresponding extension products.[18] This is because the base modified in sequencing is removed by the subsequent cleavage, leaving a 3′-phosphate group, while complete extension products contain the terminal nucleotide with a 3′-terminal hydroxyl group.

To map the initiation sites to the genomic sequence, it is necessary to demonstrate that the transcripts are colinear with the genes by a method such as S1 nuclease protection, as discussed previously. It is ideal to run primer extension and nuclease protection samples in adjacent lanes of a polyacrylamide gel, having used a primer and probe that share a common 5′ end. Comigration of DNAs from the two techniques indicates the position of the 5′ end of unspliced transcripts. Labeled DNAs resulting from *only one* of these techniques may result from transcript processing or from artifacts. The use of alternate primers, probes, and reaction conditions must then be employed to determine the true position of the 5′ end of the transcripts.

The abundance of the transcripts can be quantified by scintillation counting of DNAs excised from the gel. Alternatively, scanning densitometric ratios of autoradiographic band intensities can be used to determine the relative abundance of transcripts. The most accurate quantitative results are obtained by including a constant standard in each lane. This standard is generated by extension from a control primer that is added to all reactions. The control primer should be specific for a transcript that is present at constant levels in all of the conditions to be studied, and the control primer should give rise to extended products of different sizes from those of the experimental primer.

Primer extension is a very sensitive technique: 1.5 amol or 2 pg of a 2-kb transcript can be readily detected. This corresponds to approximately one RNA molecule per cell, depending on the species and tissue source. These estimates are based on the assumptions that (1) 10 dpm (disintegrations per minute) can be visualized by autoradiography overnight with an intensifying screen, (2) primers are labeled to a specific activity of 3 μCi/pmol (routine for oligonucleotides), (3) 2 pg of total RNA/cell are obtained, (4) 50 μg of total RNA/reaction are used, and (5) a rare transcript will comprise approximately 10^{-7} of the total cellular RNA.

[18] B. Sollner-Webb and R. H. Reeder, *Cell (Cambridge, Mass.)* **18**, 485 (1979).

Artifacts. Occasionally, bands are observed that do not result from extension of primers to the 5' end of RNAs containing primer–complementary sequences. Identification and reduction of artifacts, resulting from hybridization of primer to partially complementary sequences (within RNAs, primer, carrier, or contaminating nucleic acids), are discussed in the Hybridization section. For repeated analysis of members of multigene families, this problem can be overcome by the use of an oligonucleotide probe complementary to a less-conserved region of the transcript, usually within the 5'-noncoding sequences.

Aberrantly short extension products from complementary templates can result from degraded RNA, as well as pausing and premature termination by reverse transcriptase. The effect of secondary structure cannot be predicted from the primary sequence. The estimated free energy of formation of hairpins does not correlate well with the ability of the polymerase to transcribe through inverted repeats. The enzyme can efficiently read through some regions that would be expected to form very stable hairpin structures, while other short, inverted repeats, such as those flanking the loop sequence UUCG, cause termination.[19] Reverse transcriptase also pauses or terminates prematurely directly opposite or up to 2 nt preceding some modified template bases. m_2^6A and 3-[3-amino-3-carboxypropyl]-1-methylpseudouridine (amψ)-modified bases of rRNA and tRNA, which are predicted to interfere with base pairing, are known to result in efficient termination of reverse transcription.[20,21] The methylated bases adjacent to the cap structure of some eukaryotic mRNAs may result in a fraction of the extension products terminating 1 or 2 nt before the end of the template.

Variations

Increased Sensitivity with Continuous Labeling. The extension reaction can be performed with radiolabeled nucleotides to increase sensitivity. However, it is important to maintain a high concentration of nucleotides to minimize premature termination (at least 100 μM dATP).[22] All products of radiolysis from labeling to high specific activities, as well as degradation from contaminating nucleases, will be radioactive and thus

[19] C. Tuerk, P. Gauss, C. Thermes, D. R. Groebe, M. Gayle, N. Guild, G. Stormo, Y. D'Aubenton-Carafa, O. C. Uhlenbeck, I. Tinoco, Jr., E. N. Brody, and L. Gold, *Proc. Natl. Acad. Sci. U.S.A.* **85,** 1364 (1988).

[20] O. Hagenbüchle, M. Santer, J. A. Steitz, and R. J. Mans, *Cell (Cambridge, Mass.)* **13,** 551 (1978).

[21] D. C. Youvan and J. E. Hearst, *Nucleic Acids Res.* **9,** 1723 (1981).

[22] G. F. Gerard, *Focus (Bethesda, Md.)* **10,** 12 (1988).

will interfere with signals from intact extension products, a problem not encountered with 5′-end-labeled probes. For these reasons, a relatively high concentration of nucleotides and a moderate specific activity should be used, and the samples should be electrophoresed as soon as possible following the extension reaction. Furthermore, contaminating nucleic acids, which can act as primers, will give rise to labeled products in the presence of radioactive nucleotides. While these factors can be minimized with appropriate precautions, it is preferable to use end-labeled primers if possible. Uniform labeling is not recommended for quantitative comparison of transcripts with different 5′ ends, as the intensity of the signal is proportional to the number of incorporated labeled radionucleotides, and thus the length of the DNA.

Sequence Determination of Extension Products. The identity and size of individual extended products can be unambiguously determined by chemical cleavage or dideoxy sequence analysis. Single 5′-end-labeled extension products from a large-scale reaction can be excised from a preparative denaturing polyacrylamide gel and then can be subjected to chemical cleavage sequence analysis.[11] Alternatively, extension reactions can be carried out with end-labeled primers and the appropriate ratio of dNTPs–ddNTPs (dideoxynucleoside triphosphates) for chain-termination sequence determination.[23,24] These techniques allow precise identification of the 5′ terminus, splice junctions, and the positions of internal transcript modifications that result in premature termination, such as methylation.

Characterization of Multiple Primer Extension Products. Extension products of different lengths can arise from both heterogeneity in initiation and in splicing. The heterogeneity resulting from multiple initiation sites or alternate processing at the extreme 5′ ends can be eliminated by cleaving the extended DNAs at a single site.[25] This is accomplished by hybridizing a synthetic oligonucleotide to an appropriate region of the extension products that contains a restriction endonuclease recognition site and then by cutting with the restriction enzyme. The cleavage will be specific and unique to this short double-stranded region.

1. Suspend the DNA from step 7 of the extension reaction (see Extension) in 2 μl of 5× SS hybridization buffer. Add 3 pmol of the oligonucleotide (~20 ng of a 20-base oligonucleotide) and H_2O to a final volume of 10 μl.
2. Denature at 80° for 5 min.

[23] T. Inoue and T. R. Cech, *Proc. Natl. Acad. Sci. U.S.A.* **82**, 648 (1985).
[24] F. Sanger, S. Nicklen, and A. R. Coulson, *Proc. Natl. Acad. Sci. U.S.A.* **74**, 5463 (1977).
[25] P. J. Good, R. C. Welch, W.-S. Ryu, and J. E. Mertz, *J. Virol.* **62**, 563 (1988).

3. Anneal at 60° for 30 min.
4. Adjust the salt and buffer solution for the restriction enzyme to be used. Add the restriction enzyme and incubate under the recommended conditions.
5. Ethanol precipitate and analyze as described in steps 7 and 8 of the extension protocol (see Extension). Load the extension products, which have been treated with restriction enzyme, and the untreated extension products in adjacent PAGE lanes.

[26] Modification Interference Analysis of Reactions Using RNA Substrates

By LAURA CONWAY and MARVIN WICKENS

The modification interference method described in this chapter yields, in a single experiment, information equivalent to that obtained from the analysis of a large collection of point mutations.[1,2] The method is analogous to that developed to analyze protein–DNA interactions by Siebenlist and Gilbert.[3] Modification interference can identify those nucleotides in a RNA that are essential for any reaction of interest, provided that synthetic RNA can serve as a substrate in the reaction and that reaction substrates and products are separable.

General Method

We will describe in detail the use of two chemicals as modifying reagents: diethyl pyrocarbonate (DEPC), which carboxyethylates purine bases, and hydrazine, which removes pyrimidine bases (Fig. 1). Modification of RNA with either reagent renders the phosphodiester backbone susceptible to cleavage with aniline at the site of modification.[4] We describe only DEPC and hydrazine modifications, since these modifications enable one to examine every nucleotide in a RNA; however, other modifications are also useful (see Other Modifications).

For modification interference (Fig. 2), an end-labeled RNA substrate is prepared in vitro. The end-labeled RNA substrate is modified using

[1] L. Conway and M. Wickens, EMBO J. 6, 4177 (1987).
[2] B. Rymond and M. Rosbash, Genes Dev. 2, 428 (1988).
[3] U. Siebenlist and W. Gilbert, Proc. Natl. Acad. Sci. U.S.A. 77, 122 (1980).
[4] D. A. Peattie, Proc. Natl. Acad. Sci. U.S.A. 76, 1760 (1979).

METHODS IN ENZYMOLOGY, VOL. 180

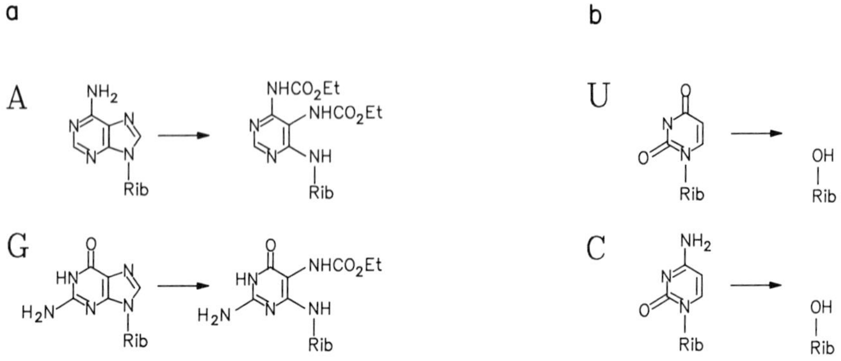

a   b

FIG. 1. Structure of bases following treatment with either (a) DEPC (purine modifications) or (b) hydrazine (pyrimidine modifications).

either DEPC or hydrazine, such that, on average, each molecule is modified only once. The modified transcripts are then used as substrate in the reaction of interest. RNAs that have successfully undergone the reaction are purified. The substrate and product RNAs are then cleaved with aniline and are analyzed by gel electrophoresis. Gaps in the sequencing "ladder" of the products correspond to sites which, when modified, prevent the reaction of interest.

Like point mutagenesis, the modification interference approach may not detect every nucleotide that is essential. Neither method will detect critical bases if more than one base must be changed to prevent the reaction; similarly, neither method will identify nucleotides that participate through their ribose moieties.

Materials

Enzymes

T4 RNA ligase was purchased from U.S. Biochemical (Cleveland, OH)

T4 polynucleotide kinase and SP6 polymerase were from New England Biolabs (Beverly, MA)

Calf intestinal alkaline phosphatase (molecular biology grade) was from Boehringer Mannheim (Indianapolis, IN).

Buffers

Buffer A: 50 mM Sodium acetate, pH 4.5, and 1 mM ethylenediaminetetraacetic acid (EDTA), pH adjusted with acetic acid

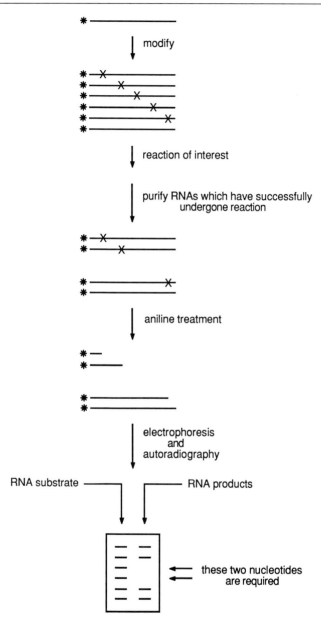

FIG. 2. Modification interference strategy. Modified nucleotides are indicated with X's. The asterisk (∗) at the end of the RNA indicates that the RNA is end labeled. See the text for details.

Buffer B: 8 M Urea, 20 mM Tris–HCl, pH 7.9, 1 mM EDTA, 0.05% (w/v) xylene cyanol (XC), and 0.05% (w/v) bromphenol blue (BPB)

Buffer C: 20 mM Tris–HCl, pH 7.6, 1 mM EDTA, and 20 mM NaCl

Buffer D: 0.5 M Ammonium acetate, pH 6.5, 1 mM EDTA, and 0.1% (w/v) sodium dodecyl sulfate (SDS)

10× T4 polynucleotide kinase buffer: 0.5 M Tris–HCl, pH 7.6, 0.1 M MgCl$_2$, 50 mM dithiothreitol (DTT), 1 mM spermidine, and 1 mM EDTA

10× T4 RNA ligase buffer: 0.5 M Tris–HCl, pH 7.9, 0.15 M MgCl$_2$, and 33 mM DTT

Reagents

1 M Aniline, pH 3.8: Dilute pure aniline (11 M) with 0.3 M sodium acetate, pH 3.8. The aniline should be distilled once under nitrogen before use and stored at −20° in the dark

Phenol–chloroform: An equal volume of each, equilibrated with 0.1 M Tris–HCl, pH 7.9

Yeast RNA, DEPC, and dimethyl sulfoxide (DMSO) were from Sigma (St. Louis, MO)

Bovine serum albumin (BSA) and glycogen (both molecular biology grade) were from Boehringer Mannheim

Radionucleotides were from Amersham (Arlington Heights, IL)

Aniline and hydrazine were both Baker brands from VWR (Chicago, IL)

Sephadex G-25 (fine) beads were from Pharmacia (Piscataway, NJ)

All solutions were made with water that had been treated with DEPC.[5] DEPC should be used within 1 week after DEPC is opened.

End-Labeling Methods

RNA is synthesized *in vitro* using an appropriate polymerase (see this volume, chapters [4] and [5]). The desired RNA species is then purified either by gel electrophoresis or by gel filtration chromatography prior to end labeling (see Purification of RNA).

5′ End-Labeling[6]

Step 1. Synthesize transcripts *in vitro* without a cap and purify.

Step 2. After the second ethanol precipitation, redissolve 5–10 pmol

---

[5] T. Maniatis, E. F. Fritsch, and J. Sambrook, "Molecular Cloning: A Laboratory Manual," p. 401. Cold Spring Harbor Laboratory, Cold Spring Harbor, New York, 1982.

[6] T. Maniatis, E. F. Fritsch, and J. Sambrook, "Molecular Cloning: A Laboratory Manual," p. 122. Cold Spring Harbor Laboratory, Cold Spring Harbor, New York, 1982.

of RNA in 94 μl of water. To remove the 5′-terminal phosphates, add 5 μl of 1 M Tris–HCl, pH 9.0, and 20 U (1 μl) of alkaline phosphatase. Incubate for 1 hr at room temperature.

Step 3. Remove the alkaline phosphatase with one phenol–chloroform extraction.

Step 4. Precipitate the RNA with ethanol. Rinse the pellet and redissolve in 7 μl of water. Add 2 μl of 10× T4 polynucleotide kinase buffer, 1 μl (10 U) of T4 polynucleotide kinase, and 10 μl (100 μCi) of [γ-^{32}P]ATP. Incubate for 1 hr at 37°. Purify the RNA by polyacrylamide gel electrophoresis (PAGE).

3′ End-Labeling[7,8]

3′-End-labeling RNA is discussed in detail in a chapter in a previous volume of this series.[8]

Step 1. Synthesize transcripts *in vitro* with or without a cap and purify.

Step 2. After the second ethanol precipitation, redissolve the RNA (at least 5 pmol) in 2 μl of 250 μM ATP, 2 μl of 0.1 mg/ml BSA, 2 μl of 10× T4 RNA ligase buffer, 2 μl of DMSO, and 10 μl (100 μCi) of [5′-^{32}P]pCp. Cool to 4° before adding 8–20 U (2 μl) of T4 RNA ligase. Incubate at 4° for at least 10 hr. Purify the labeled RNA by gel electrophoresis.

(Note) RNA that ends in a U labels 10-fold less efficiently than RNAs that end in A, C, or G.[7] We therefore avoid using RNAs with a 3′-terminal U.

Chemical Modification of RNA[4]

Amount of RNA Required

The end-labeling protocols described in End-Labeling Methods yield RNA at a specific activity of approximately 10^6 cpm (counts per minute)/pmol, corresponding to the labeling of 15% of the available termini. The recovery of RNA after the labeling, chemical modification, and aniline cleavage steps is about 80% of the starting material. To sequence a 200-nt (nucleotide) RNA, 3×10^4 cpm/lane is sufficient for an overnight exposure. Therefore, a minimum of 4×10^4 cpm is necessary. This corresponds to 40 fmol of RNA at 10^6 cpm/pmol. This amount of radioactivity must be increased, if losses occur during the reaction of interest.

[7] T. E. England and O. C. Uhlenbeck, *Biochemistry* **17**, 2069 (1978).
[8] P. J. Romaniuk and O. C. Uhlenbeck, this series, Vol. 100, p. 52.

Purine Modification (A + G)

This reaction modifies A's more efficiently than G's, in about a 4:1 ratio, but can be used to analyze both purines in a single reaction.

Step 1. Purify the labeled RNA by gel electrophoresis. Use 12.5 μg of yeast RNA as carrier, when eluting the RNA from the gel slice.

Step 2. Take out the amount needed for the modification. Adjust the amount of yeast RNA in that aliquot to a total of 12.5 μg. Precipitate the RNA with ethanol, rinse the pellet thoroughly with 75% (v/v) ethanol, and dry.

Step 3. Redissolve the RNA pellet in 200 μl of buffer A. Add 2 μl of fresh DEPC. Vortex for 5 sec and incubate for 2.5 min at 90°. Open the lid on the tube before incubating, otherwise the CO_2 generated will pop the lid open and splash the contents of the tube.

Step 4. To stop the reaction, add 75 μl of 1 *M* sodium acetate, pH 4.5, and 750 μl of 100% (v/v) ethanol. Centrifuge for 20 min at 4° to pellet the RNA.

Step 5. Redissolve the pellet in 200 μl of 0.3 *M* sodium acetate, pH 3.8. Add 600 μl of ethanol and pellet again. The RNA is now ready to be used as a substrate in the reaction of interest.

Pyrimidine Modification

The pyrimidines can be modified either separately (U only or C only modifications) or together (U + C modification). With 3'-end-labeled RNA, all three types of modifications can be used successfully. With 5'-end-labeled RNA, only the U only modification reliably yields useful data; the U + C and the C only modifications yield unreadable ladders.

U Only Modification

Step 1. Purify the labeled RNA by gel electrophoresis. Use 12.5 μg of yeast RNA as carrier, when eluting the RNA from the gel slice.

Step 2. Take out the amount needed for the modification. Adjust the amount of yeast RNA in that aliquot to a total of 12.5 μg. Precipitate the RNA with ethanol, rinse the RNA pellet thoroughly with 75% (v/v) ethanol, and dry.

Step 3. Redissolve the RNA pellet in 10 μl of water and add 10 μl of anhydrous hydrazine. Hydrazine dissolves readily in water. Incubate for 10 min on ice.

Step 4. To stop the reaction, add 200 μl of 0.3 *M* sodium acetate, pH 3.8, and 750 μl of ethanol. Centrifuge for 20 min at 4° to pellet the RNA.

Step 5. Redissolve in 200 μl of 0.3 *M* sodium acetate, pH 3.8, add 600 μl of ethanol, and pellet again. The RNA is now ready to be used as a substrate in the reaction of interest.

U + C Modification

This modification should be performed exactly as that in the U only modification described in the previous section, except that, at step 3, the RNA should be redissolved in 20 μl of anhydrous hydrazine–0.5 *M* NaCl, and the incubation on ice should be for 30 min. The hydrazine–NaCl must be made up fresh, both for this and for the C only modifications described in the following section.

C Only Modification

This modification should be performed exactly as in the U + C reaction described in the previous section, except that, at step 3, the RNA should be redissolved in 20 μl of anhydrous hydrazine–3.0 *M* NaCl.

Cleavage of Modified RNA with Aniline and Gel Electrophoresis[4]

Step 1. Separate the products from the unreacted RNA after the reaction of interest. This often can be accomplished by gel electrophoresis. Ethanol precipitate the purified RNA. Wash with 75% (v/v) ethanol and dry.

Step 2. Redissolve the sample in 20 μl of 1 *M* aniline. Incubate for 20 min at 60° in the dark.

Step 3. To stop the reaction and remove the aniline, fill the Eppendorf tube with 1-butanol (about 1.4 ml) and vortex for at least 5 sec. Butanol precipitation is faster than the lyophilization step of Peattie.[4] Centrifuge for 10 min at room temperature to pellet the RNA. Carefully remove the butanol with a Pasteur pipet. The RNA pellet may not adhere to the side of the tube.

Step 4. Redissolve the RNA in 150 μl 1% (w/v) SDS.

Step 5. Fill the tube with butanol again, vortex, and centrifuge for 10 min at room temperature.

Step 6. Rinse the pellet with 1 ml of 100% (v/v) ethanol. Dry thoroughly. Incomplete drying of the pellet at this point will cause the RNA to smear on the gel.

Step 7. Redissolve in 5 μl of buffer B. Use of this loading buffer, rather than a formamide-based loading buffer, results in better resolution on the sequencing gel.

Step 8. Boil for 90 sec. Chill on ice immediately and load onto a sequencing gel. The sample should not be boiled longer than 90 sec or the urea will crystallize when put on ice. If this happens, centrifuge the tube briefly to bring any condensed water on the sides of the tube to the bottom or add 0.5 μl of water to the sample.

An aliquot of modified RNA should be reserved as the starting RNA. This sample will yield bands at every modified position and provides a standard to which RNA products can be compared. The comparison is simplest if the same amount of radioactivity is present in the ladders being compared. Thus, before loading the gel, each sample should be counted, and the approximate number of bands expected in each ladder of RNA should be calculated (often this number differs considerably in the RNA substrate and product). Load the same radioactivity per band onto each lane.

Purification of RNA

After both the transcription and the end-labeling reactions, the RNA must be separated from unincorporated nucleotide triphosphates. This can be done either by gel filtration chromatography or by electrophoresis through a denaturing polyacrylamide gel.

Rapid Column Chromatography[9]

Step 1. Remove the cap from a silanized Eppendorf tube and place inside a sterile 17 × 100-mm tube. Plug a 1-ml disposable syringe with silanized glass wool and place inside the tube.

Step 2. Fill the syringe with Sephadex G-25 that has been swollen in buffer C. Allow the beads to pack.

Step 3. Add more Sephadex G-25 until the packed volume is about 1 ml. Discard the excess buffer.

Step 4. Remove the void volume from the column by centrifugation at 1000 g for 5 min at room temperature.

Step 5. Apply the RNA (100 μl) to the column. Let the sample equilibrate at room temperature for 5 min.

Step 6. Recover the sample by centrifugation for exactly the same time at the same temperature as in step 4. The recovered

---

[9] T. Maniatis, E. F. Fritsch, and J. Sambrook, "Molecular Cloning: A Laboratory Manual," p. 466. Cold Spring Harbor Laboratory, Cold Spring Harbor, New York, 1982.

sample volume should be 95–105 μl. This procedure will remove about 98% of the unincorporated nucleotide triphosphates.

Polyacrylamide Gel Electrophoresis[4]

Step 1. Load the RNA onto a denaturing polyacrylamide gel. A 200-nt
RNA can be purified by electrophoresis through a 40-cm-long
6% acrylamide (w/v) sequencing gel at 30 V/cm for 1.5 hr.
Step 2. Locate the RNA in the gel either by autoradiography or by
fluorescence after the gel has been stained with ethidium bromide (EtBr). Cut out the gel slice containing the RNA.
Step 3. Soak the gel slice in 500 μl of buffer D for at least 2 hr at 37°. If
the RNA is unlabeled, 20 μg of glycogen is included as carrier;
if the RNA is ^{32}P labeled, 12.5 μg of yeast RNA is included as
carrier.
Step 4. Transfer the buffer containing the eluted RNA to a new tube
and precipitate with ethanol.
Step 5. Redissolve the RNA in 200 μl of 250 mM NaCl.
Step 6. Centrifuge for 10 min to remove debris.
Step 7. Transfer the RNA to a new tube and precipitate with ethanol a
second time.
Step 8. Rinse the pellet in 75% (v/v) ethanol, dry, and redissolve in
water, unless otherwise indicated.

Analysis of mRNA 3′ End Formation

The following example, an analysis of the requirements for 3′ end
cleavage of SV40 late precursor mRNA (pre-mRNA), is used to illustrate
the method.[1] The details of cleavage and polyadenylation *in vitro* are
discussed in detail elsewhere.[10,11]
   In this experiment, we analyzed a reaction in which a pre-mRNA is
cleaved at a specific nucleotide [the poly(A) site], producing two "half-
molecules."[1] End-labeled RNA modified with either DEPC or hydrazine
was incubated in HeLa cell nuclear extract[12] under conditions in which
cleavage occurs. RNA was recovered from the extract. The products of
the cleavage reaction were separated from the unreacted RNA by gel
electrophoresis. These RNAs and the starting RNA were treated with
aniline and were analyzed by PAGE.

[10] C. Moore, this series, Vol. 181, p. 6.
[11] C. L. Moore and P. A. Sharp, *Cell* **41**, 845 (1985).
[12] J. D. Dignam, R. M. Lebovitz, and R. G. Roeder, *Nucleic Acids Res.* **11**, 1475 (1983).

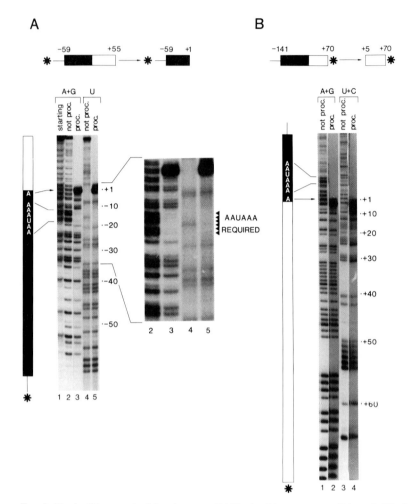

FIG. 3. Nucleotides required for cleavage. (A) Nucleotides upstream of the poly(A) site. 5′-End-labeled −59/+55 RNA was treated with DEPC (lanes 1–3) or with hydrazine (lanes 4 and 5) and was incubated in a nuclear extract, containing 0.5 m*M* EDTA to prevent polyadenylation. 5′ Half-molecules were isolated and treated with aniline. (Lane 1) DEPC-treated RNA that was not incubated in the extract. (Lane 2) DEPC-treated RNA that was not cleaved during incubation in the extract. (Lane 3) DEPC-treated RNA that was cleaved in the extract. (Lane 4) Hydrazine-treated RNA that was not cleaved in the extract. (Lane 5) Hydrazine-treated RNA that was cleaved in the extract. The sequence surrounding AAUAAA is enlarged to show detail. Modification of any nucleotide in AAUAAA prevents cleavage in the extract. The decreased intensity of the base at −36 (lane 3) is not reproducible. (B) Nucleotides downstream of the poly(A) site. 3′-End-labeled −141/+70 RNA was treated with DEPC (lanes 1 and 2) or with hydrazine (lanes 3 and 4) and was incubated in a nuclear extract, containing 0.5 m*M* EDTA. 3′ Half-molecules were isolated and treated with aniline. (Lane 1) DEPC-treated RNA that was not cleaved in the extract. (Lane 2) DEPC-treated RNA that was cleaved in the extract. (Lane 3) Hydrazine-treated RNA that was not cleaved in the extract. (Lane 4) Hydrazine-treated RNA that was cleaved in the extract. No single nucleotide modification downstream of the poly(A) site prevents cleavage.

The results are shown in Fig. 3. In the analysis of 5'-end-labeled RNA (Fig. 3A), the bases in the hexanucleotide AAUAAA are clearly absent from the processed RNA lanes (compare lanes 2 and 4 to lanes 3 and 5), consistent with results obtained by mutagenesis.[13] All other bases are present in the processed RNA, both in the 5' half-molecules (Fig. 3A) and in the 3' half-molecules (Fig. 3B). We conclude that the only single modifications that prevent cleavage lie in AAUAAA. We have performed comparable analyses of the formation of processing complexes and the addition of poly(A).[1] Rymond and Rosbash, who developed the modification interference method independently, have applied the method to the splicing of yeast pre-mRNAs.[2]

Other Modifications

Modification interference requires two steps: modification of specific bases with chemical reagents and the detection of those modified bases in substrate and product RNAs. Reagents other than DEPC and hydrazine satisfy these requirements. For example, dimethyl sulfate (DMS) has been used to examine G, A, and C residues.[14] Other potentially useful reagents include kethoxal (G specific), CMCT[15] (U and G specific),[14] and ethylnitrosourea (ENU) (phosphate specific).[16] Phosphorothioates can be incorporated using phosphorothioate containing nucleotide triphosphates. The modified positions can be identified by sensitivity to aniline cleavage, as is the case with the modifications described in this chapter, or by any other method that provides single-nucleotide resolution. For example, primer extension by reverse transcriptase is blocked by many base modifications and phosphodiesters containing a phosphorothioate are readily cleaved with iodoethane.[14,17]

[13] M. Wickens and P. Stephenson, *Science* **226**, 1045 (1984).
[14] D. Moazed and H. F. Noller, *Cell (Cambridge, Mass.)* **47**, 985 (1986).
[15] 1-Cyclohexyl-2-morpholinocarbodiimide metho-*p*-toluene sulfonate.
[16] P. Romby, D. Moras, M. Bergdoll, P. Dumas, V. V. Vlassov, E. Westhof, J. P. Ebel, and R. Giege, *J. Mol. Biol.* **184**, 455 (1985).
[17] G. Gish and F. Eckstein, *Science* **240**, 1520 (1988).

Section III

RNA Interactions

A. Cross-Linking
Articles 27 through 29

B. Other Methods
Articles 30 through 34

[27] Photoaffinity Cross-Linking Methods for Studying RNA–Protein Interactions

By MICHELLE M. HANNA

Photochemical cross-linking can be used to characterize RNA–protein interactions in ribonucleoprotein (RNP) complexes, which exist transiently during transcription, translation, splicing, or RNA processing or as stable structural and catalytic components of cells. Experiments can be designed to answer questions as simple as whether a RNA–protein interaction exists, or as complex as which nucleotide(s) or amino acid(s) are involved in a known RNP complex. Photochemical cross-linking "traps" weak or transient RNA–protein associations, which might not survive isolation procedures, such as immunoprecipitation, gel filtration, or filter binding, permitting detailed structural studies of these complexes.

There are a number of approaches that can be taken in photochemical crosslinking, and the choice of an approach depends on the specific question to be answered, the purity of the proteins being studied, the sequence and structure of the RNA, and the strength of the interaction between the RNA and protein. One method, which has been frequently used, involves the direct irradiation of a RNP complex with short wavelength ultraviolet (UV) light. This method relies on the direct excitation of nucleotides or amino acids, depending on which wavelength is used, to generate chemically reactive species. While this nonspecific labeling may show whether a protein–nucleic acid interaction exists, there can be problems with this method, the most serious being the degradation of some proteins by the irradiation itself.

An alternate approach involves the use of photosensitive cross-linking reagents. These reagents contain groups that are chemically inert in the absence of light, but can be easily converted to chemically reactive species by irradiation, sometimes at long wavelengths in the UV and visible spectrums, which will not nick protein or nucleic acid. Even when short-wavelength UV light must be used, the high reactivity of these groups can result in cross-linking, before considerable damage is done to the protein or nucleic acids.

METHODS IN ENZYMOLOGY, VOL. 180

Approaches to Protein–RNA Cross-Linking

Addition of Bifunctional Cross-Linking Reagent to Ribonucleoprotein Complex Containing Unmodified RNA and Protein

Bifunctional cross-linking reagents contain two cross-linking groups separated by linkers of about 5 –20 Å. Some linkers contain bonds that can be chemically cleaved by thiol or periodate. Both cross-linking groups may be photoreactive, or more commonly, these reagents contain one chemically reactive group and one photoreactive group. Most groups contain photoreactive aryl azides as the cross-linking functional group.[1] The half-life of the photolytically produced reactive species, the nitrene, is on the order of a millisecond, and its insertion reactions are highly nonspecific.[2] This nonspecificity is particularly useful, when probing the environment of a protein or nucleic acid, as there need not be a specific functional group within contact of the nitrene to achieve cross-linking. The wavelength required for photolysis of unsubstituted aryl azides is between 260 and 290 nm; however, substitutions can shift the excitation maxima to longer wavelengths. Addition of a nitro group to most aryl azides shifts the excitation maxima to between 300 and 460 nm. These substituted azides can undergo cross-linking at wavelengths that will not damage proteins or nucleic acids. Several of these bifunctional cross-linking agents are listed in Table I.[3-30]

[1] G. W. J. Fleet, R. R. Porter, and J. R. Knowles, *Nature (London)* **224**, 511 (1969).

[2] A. K. Schrock and G. B. Schuster, *J. Am. Chem. Soc.* **106**, 5228 (1984).

[3] P. A. S. Smith and B. B. Brown, *J. Am. Chem. Soc.* **73**, 2438 (1951).

[4] A. Reiser, H. M. Wagner, R. Morley, and A. Bowes, *Trans. Faraday Soc.* **63**, 2403 (1967).

[5] R. B. Mikkelsen and D. F. H. Wallach, *J. Biol. Chem.* **251**, 7413 (1976).

[6] W. L. Dentler, M. M. Pratt, and R. E. Stephens, *J. Cell Biol.* **84**, 381 (1980).

[7] R. V. Lewis, M. F. Roberts, E. A. Dennis, and W. S. Allison, *Biochemistry* **16**, 5650 (1977).

[8] Y. A. Vladimirov, D. J. Roshchupkin, and E. E. Fesenko, *Photochem. Photobiol.* **11**, 227 (1970).

[9] D. J. Kiehm and T. H. Ji, *J. Biol. Chem.* **252**, 8524 (1977).

[10] M. Haeuptle, M. L. Aubert, J. Djiane, and J.-P. Kraehenbuhl, *J. Biol. Chem.* **258**, 305 (1983).

[11] A. M. Tometsko and J. Turula, *Int. J. Pep. Protein Res.* **8**, 331 (1976).

[12] B. D. Burleigh, W.-K. Liu, and D. N. Ward, *J. Biol. Chem.* **253**, 7179 (1978).

[13] R. E. Galardy, L. C. Craig, J. D. Jamieson, and M. P. Printz, *J. Biol. Chem.* **249**, 3510 (1974).

[14] C. W. T. Yeung, M. L. Moule, and C. C. Yip, *Biochemistry* **19**, 2196 (1980).

[15] T. H. Ji, *J. Biol. Chem.* **252**, 1566 (1977).

[16] K. Sutoh and F. Matsuzaki, *Biochemistry* **19**, 3878 (1980).

[17] T. H. Ji and I. Ji, *Anal. Biochem.* **121**, 286 (1982).

[18] I. Ji, J. Shin, and T. H. Ji, *Anal. Biochem.* **151**, 348 (1985).

These reagents can be used to cross-link molecules that have not been modified with reactive groups and therefore exist in their native conformations, provided care has been taken in their isolation. Cross-linking reactions are prepared by combining unmodified protein and RNA (either as purified components or in cellular fractions), using conditions that are optimal for association of the RNA and protein. This allows the RNP complex to form in the absence of added chemical reagents not normally found in the system. The advantage of this method is that unmodified macromolecules are more likely to form complexes more similar to *in vivo* complexes than modified RNA or protein. After formation of the RNP complex, the bifunctional cross-linking reagent is added to the reaction (in dim light), and covalent linkages are formed between the RNA and protein, by adjusting reaction conditions for the chemical cross-linker and/or by excitation of the photoreactive group(s) with light. Cross-linker can be added directly as solid to the RNA–protein solution, if sufficiently soluble, or can be dissolved in a minimal amount of organic solvent [dimethyl sulfoxide (DMSO) or dimethylformamide (DMF)] and then can be added to the aqueous solution. Most of these bifunctional reagents hydrolyze rapidly in water and should not be stored as aqueous solutions. Most chemical modifications are complete in about 30 min at room temperature.

Disadvantages of this approach include the following.

1. Use of cross-linkers containing two photoreactive groups requires a trimolecular reaction to achieve cross-linking. The cross-linking reagent must become positioned as a bridge between the protein and RNA in such a way that one cross-linking group is within a few angstroms of each of these macromolecules. On irradiation, both groups must undergo suc-

[19] V. Chowdhry, R. Vaughan, and F. H. Westheimer, *Proc. Natl. Acad. Sci. U.S.A.* **73**, 1406 (1976).

[20] V. Chowdry and F. H. Westheimer, *Annu. Rev. Biochem.* **48**, 293 (1979).

[21] E. F. Vanin and T. H. Ji, *Biochemistry* **20**, 6754 (1981).

[22] J. U. Baenziger and D. Fiete, *J. Biol. Chem.* **257**, 4421 (1982).

[23] D. A. Zarling, J. A. Miskimen, D. P. Fan, E. K. Fujimoto, and P. K. Smith, *J. Immunol.* **128**, 251 (1982).

[24] S. M. Jung and M. Moroi, *Biochim. Biophys. Acta* **761**, 152 (1983).

[25] K. Ballmer-Hofer, V. Schlup, P. Burn, and M. M. Burger, *Anal. Biochem.* **126**, 246 (1982).

[26] H. Wollenweber and D. C. Morrison, *J. Biol. Chem.* **260**, 15068 (1986).

[27] S. H. Hixson and S. S. Hixson, *Biochemistry* **14**, 4251 (1975).

[28] M. Erecinska, *Biochem. Biophys. Res. Commun.* **76**, 495 (1977).

[29] R. B. Moreland, P. K. Smith, E. K. Fijimoto, and M. E. Dockter, *Anal. Biochem.* **121**, 321 (1982).

[30] C. K. Huang and F. M. Richards, *J. Biol. Chem.* **252**, 5514 (1977).

TABLE I
BIFUNCTIONAL PHOTOCROSS-LINKING REAGENTS

Reagent[a]	Specificity[b]	Cleavage	Advantages[c]	Reference(s)
DABP	Nonspecific			3–5
DAN	Nonspecific			3–5
DTBPA	Nonspecific	Thiols		3–6
ANB-NOS	Amines		Ex = 320–350 nm	7, 8
EADB	Amines	Thiols		9, 10
FNPA	Amines		Ex = 320–350 nm	11, 12
HSAB	Amines			13, 14
MABI	Amines			15, 16
NHS-ASA	Amines		Iodinatable	17, 18
PNP-DTP	Amines			19, 20
SADP	Amines	Thiols		21–24
SANPAH	Amines		Ex = 320–350 nm	22, 25
Sulfo-SADP	Amines	Thiols	Water soluble	21–23
SAND	Amines	Thiols	Water soluble	7
			Ex = 320–350 nm	
Sulfo-SANPAH	Amines		Water soluble	22, 25
			Ex = 320–350 nm	
SASD	Amines	Thiols	Water soluble	26
			Iodinatable	
APB	Thiols			27, 28
APTP	Thiols	Thiols		9, 29
DNCO	Thiols	Thiols	Ex = visible light	30

[a] ANB-NOS, N-5-Azido-2-nitrobenzoyloxysuccinimide; APB, p-azidophenacyl bromide; APTP, N-(4-azidophenylthio)phthalimide; DABP; 4,4'-diazidobiphenyl; DAN, diazidonaphthalene; DNCO, di-N-(2-nitro-4-azidophenyl)cystamine-S-S-dioxide; DTBPA, 4,4'-dithiobisphenylazide; EADB, ethyl 4-azidophenyl-1,4-dithiobutyrimidate-HCl; FNPA, 4-fluoro-3-nitrophenylazide; HSAB, N-hydroxysuccinimidyl-4-azidobenzoate; MABI, methyl-4-azidobenzoimidate; NHS-ASA, N-hydroxysuccinimidyl-4-azidosalicylic acid; PNP-DTP, p-nitrophenyl-2-diazo-3,3,3-trifluoropropionate; SADP, N-succinimidyl (4-azidophenyl)-1,3'-dithiopropionate; SAND, sulfosuccinimidyl 2-(m-azido-o-nitrobenzamido)ethyl-1,3'-dithiopropionate; SANPAH, N-succinimidyl 6-(4'-azido-2'-nitrophenylamino)hexanoate; SASD, sulfosuccinimidyl 2-(p-azidosalicylamido)ethyl-1,2'-dithiopropionate; sulfo-SANPAH, sulfosuccinimidyl 6-(4'-azido-2'-nitrophenylamino)hexanoate; sulfo-SADP, sulfosuccinimidyl(4-azidophenyldithio)propionate.

[b] Specificity refers to the group modified by the chemical cross-linking groups on heterobifunctional reagents. Nonspecific means that both cross-linking groups are photoreactive and react nonspecifically with many functional groups.

[c] Ex, Optimal wavelengths for excitation of the photocross-linking groups.

cessful cross-linking to the macromolecules, as opposed to solvent. As positioning of the cross-linking group is random, intramolecular cross-links within the protein and RNA molecules may be as likely as RNA–protein intermolecular cross-links. Such intramolecular cross-links can severely complicate characterization of cross-linked complexes. These reagents frequently give low cross-linking yields.

2. Use of bifunctional reagents, containing a chemical cross-linking group, can limit the types of molecular interactions that can be identified. These chemical cross-linkers are generally specific for some functional group, typically amines or thiols, and cannot be used to probe regions of molecules that do not contain accessible groups of the necessary type.

3. Chemical cross-linking groups generally require a narrow pH range for optimal reactivity. This is in the basic range (pH 7–11) for amine modifying groups and in the slightly basic range (pH 7–8) for thiol modifying groups. As most RNA molecules contain no thiol groups, and many proteins contain few, if any, free-thiol groups, amine modifying reagents may be the only choice for a given system. The pH required for efficient modification may not be that in which the RNP complex is stable or can exist in its native conformation.

4. Since bifunctional reagents cannot be targeted to the protein or RNA molecule of interest, this approach is generally useful only with purified components. Since these reagents can react with almost any molecule the reagents encounter, cross-linking in crude extracts can lead to formation of large aggregates, which complicates identification and isolation of the desired complex.

5. Most of the bifunctional cross-linkers are only slightly soluble in water and must be dissolved in organic solvents. Addition of such organic solvents to biological solutions may have some effect on normal RNP complex formation.

6. Buffers containing amines cannot be used with amine-modifying reagents. Tris- or glycine-containing buffers must be changed to acetate-, phosphate-, borate-, or citrate-containing buffers. Thiols, such as dithiothreitol (DTT) or 2-mercaptoethanol, cannot be used in buffers when thiol-modifying reagents or reagents containing disulfide bonds are used. These changes may affect normal RNA–protein interactions.

Placement of Photoreactive Cross-Linking Group on Purified Protein and Addition of This Modified Protein to Unmodified RNA

The reagents listed in Table I, which contain both chemical and photochemical cross-linking groups, can also be used to modify purified proteins or purified nucleic acids. The use of such reagents and their mechanisms of cross-linking have been thoroughly discussed by others.[31,32] The major limitation of this approach is that one must have both a purified protein and an assay for the activity of that protein. Photoreactive cross-linking groups can be placed on either thiol or amino groups. Cross-linker-to-protein ratios can be adjusted to add (on the average) only one cross-linking group or several per protein molecule.

Proteins must be dialyzed out of storage buffers, which contain amines and/or thiol, before reaction with the bifunctional cross-linker, as these reagents will react with buffer components as well. Although the cross-linking reagents may require solution conditions, which are not optimal for the protein, in many cases, normal conformations and activities will be regained, when the unreacted reagents are removed by dialysis of the modified protein back into its storage buffer.

The extent of protein modification can be determined by monitoring absorbance of the protein–cross-linker adduct, when a photoreactive group has been chosen that absorbs light above 300 nm. After covalent attachment of the cross-linker to the protein, unreacted reagent is removed by dialysis. An aliquot of the modified protein can then be scanned to determine absorbance at the absorption maxima for both the protein and the photoreactive cross-linking group. By knowing the extinction coefficient for both at these wavelengths, one can determine fairly accurately the number of cross-linking groups added per molecule. This requires a great deal of protein, however. When limited quantities are available, the extent of modification can be estimated by reacting proteins with fluorescent tags specific for the same group to which the photoreactive cross-linker was attached, optimizing conditions for complete protein modification, rather than protein stability. Comparison of the number of fluorescent groups added to cross-linker-modified and unmodified protein provides an estimation of the average number of cross-linking groups per molecule.

Disadvantages of this method include the following.

1. Except in the case in which there may be only one free thiol on a protein, it is impossible to target the cross-linker to one specific amino acid or to ensure that a protein will be modified by only one cross-linking group.

2. Unless large quantities of protein are available, determining the extent of modification can be difficult.

3. Conditions necessary for the modification of the protein may place the protein in a conformation that exposes groups normally buried in the

[31] H. Bayley and J. R. Knowles, this series, Vol. 46, p. 69.
[32] V. Chowdhry and F. H. Westheimer, *Annu. Rev. Biochem.* **48,** 293 (1979).

native structure. Modification of these groups may inhibit the correct refolding of the protein, even after dialysis back into its storage buffer.

4. Removal of thiol (and in some cases, certain amines) may result in irreversible loss of biological activity of the protein, even before exposure to chemical cross-linking reagents.

5. Modification of proteins may lead to loss of activity or loss of the ability to correctly associate with other proteins or assemble into a RNP complex. If chemical modification of the protein has resulted in loss of its normal activity, the contacts the protein makes when added back to RNA may be meaningless. One must assay to determine whether the chemically modified protein still has normal activity or binding properties before proceeding with photocross-linking.

Placement of Photoreactive Group within RNA Molecule and Addition of This Modified RNA to Unmodified Protein

By placing the photoreactive cross-linking group on the RNA molecule, it is possible to study RNA–protein interactions with proteins in their native conformations. Care must be taken, however, to maintain the native conformation of the RNA during introduction of the cross-linker. Photoreactive groups can be placed within RNA molecules by two means. The first is similar to method described previously, Placement of Photoreactive Cross-Linking Group on Purified Protein and Addition of This Modified Protein to Unmodified RNA, and involves the chemical attachment of bifunctional reagents to purified RNA molecules. Except in the case of tRNA, where 4-thiouridine can react with thiol-modifying reagents, the target for modification on RNA would be primary amine groups. Modification of 4-thiouridine with azidophenacyl bromide has been discussed elsewhere,[33,34] and the methods described in these papers can be adapted for other thiol-modifying photocross-linking reagents. Other RNA molecules contain no thiol groups, and the only primary amines available are on the bases themselves. These groups may be involved in Watson–Crick base pairs as part of RNA secondary structures, and these secondary structures may be important for both genetic regulation and RNP complex assembly. Amine modification on RNA will, therefore, not be discussed.

The second method for placing cross-linking groups on RNA involves the use of photoreactive nucleotide analogs. These can be incorporated into the RNA chain during transcription with purified RNA polymerase or in some crude transcription systems. Several types of nucleotide analogs

[33] I. Schwartz and J. Ofengand, *Proc. Natl. Acad. Sci. U.S.A.* **71,** 3951 (1982).
[34] I. Schwartz, E. Gordon, and J. Ofengand, *Biochemistry* **14,** 2907 (1984).

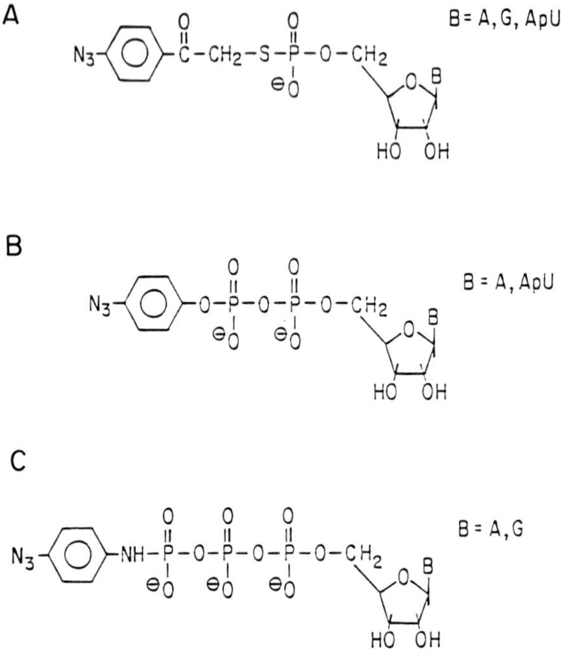

FIG. 1. Initiating photoreactive nucleotide analogs. Three photocross-linking nucleotide analogs that can be placed at the 5' end of a RNA chain are shown. These analogs can be used to initiate transcription with *Escherichia coli* RNA polymerase. These analogs can be used to probe interactions between proteins and the 5' end of a RNA chain. (A) α-[(4-Azidophenacyl)thio]nucleotide, (B) β-(4-azidophenyl)nucleotide, and (C) γ-(4-azidoanililyl)nucleotide.

have been characterized, both purines and pyrimidines, with different cross-linking groups, excitation maxima, distances between the nucleotide and the cross-linker, and cross-linker position on the nucleotide (base, sugar, or phosphate). The structures of a few of these are shown in Figs. 1–3. Although every photoreactive nucleotide is not shown, representatives of analogs that can be positioned at the 5' or 3' ends of a RNA chain, or at internal positions within a RNA chain, have been included.

Analogs cannot easily be placed at just one position within a RNA chain. There are exceptions to this, however. Interactions with the 5' end of a RNA molecule can be probed by using an analog that contains the photoreactive cross-linking group on the 5'-phosphate of a mono- or dinucleotide (Fig. 1). These analogs can be used to initiate transcription, and the cross-linking group will be found only at the 5' end of the RNA chain. Alternatively, one can place a cross-linking group at the 3' end of a

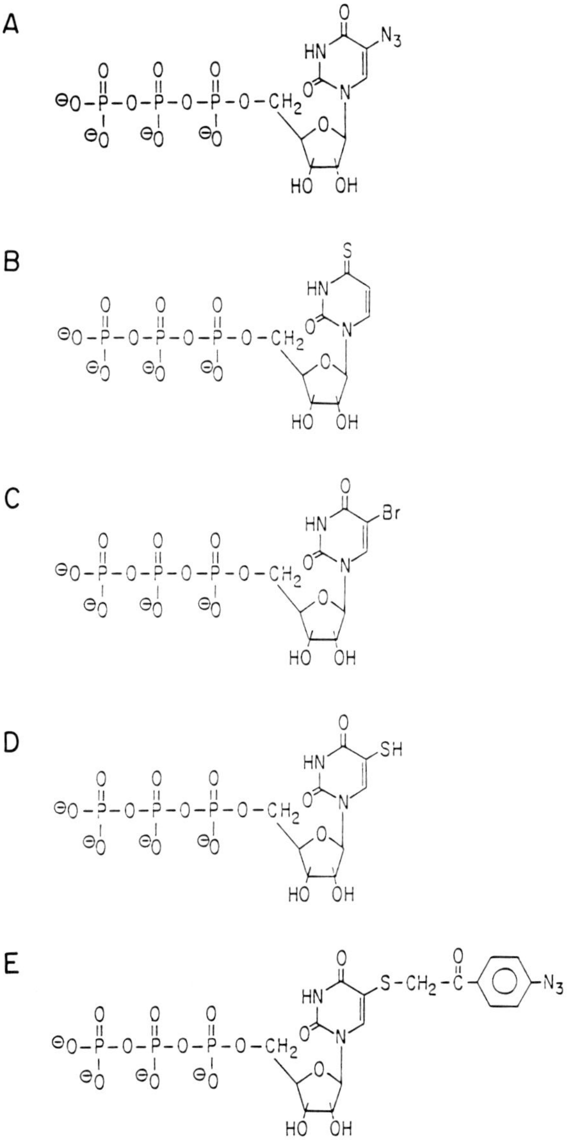

FIG. 2. Internal photoreactive nucleotide analogs. Five photocross-linking analogs that can be incorporated at internal positions within RNA chains are shown. All can be used with *Escherichia coli* RNA polymerase. Some can be used with other polymerases. These analogs can be used to probe interactions between proteins and internal sequences of a RNA chain. (A) 5-Azido UTP, (B) 4-thio-UTP, (C) 5-bromo-UTP, (D) 5-mercapto-UTP, and (E) 5'-[(4-azidophenacyl)thio]-UTP.

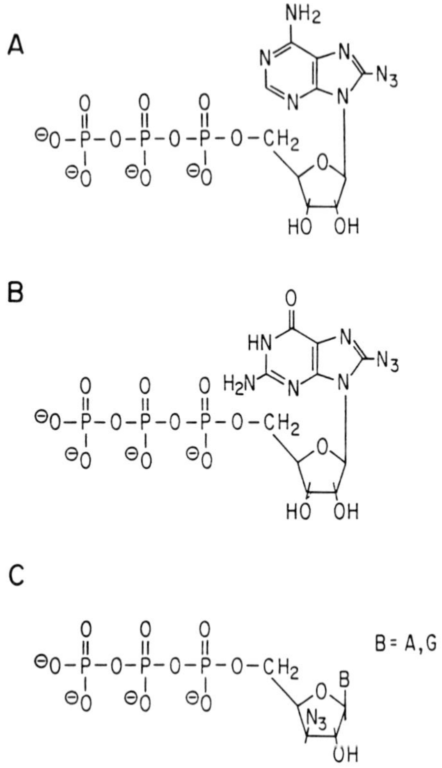

FIG. 3. Terminating photoreactive nucleotide analogs. Three photocross-linking analogs that can be incorporated at the 3'-terminal position of a RNA chain are shown. All act as RNA chain terminators with *Escherichia coli* RNA polymerase. These analogs can be used to probe interactions between proteins and the 3' end of a RNA chain. (A) 8-Azido-ATP, (B) 8-azido-GTP, and (C) 3'-azido-NTP.

RNA chain by using an analog that functions to terminate transcription (Fig. 3). Cross-linkers can easily be placed specifically at internal positions during transcription only for nucleotides that occur very close to the 5' end of the RNA molecule. While methods for specific placement of nucleotide analogs at internal positions of short RNA molecules have been described,[35,36] these are very complex and not easily applicable to larger RNA chains.

[35] M. Krug, P. L. de Haseth, and O. C. Uhlenbeck, *Biochemistry* **21,** 4713 (1982).
[36] W. L. Wittenberg and O. C. Uhlenbeck, *Biochemistry* **24,** 2705 (1985).

Even with these limitations, the introduction of photoreactive nucleotides into a RNA chain can be used for (1) identification of protein(s) (or other nucleic acids) that interact with or bind to a specific RNA, (2) identification of protein domain(s) or amino acid(s) involved in a specific RNA interaction, and (3) identification of RNA regions or nucleotides involved in a specific protein interaction.

Selection of Light Source

We have tested several light sources for efficiency of excitation of different photoreactive groups and for effects on protein and nucleic acid structure. These light sources have excitation maxima at 254, 302, or 337 nm. The three light sources used in our studies are (1) 254 nm: short-wavelength mercury vapor lamp (Gates, Long Island, NY), 63 μW/cm^2 at 1 m (from 225 to 313 nm); (2) 302 nm: medium-wavelength mercury vapor lamp (Spectroline, Westbury, NY, model XX-15B), 1800 μW/cm^2 at 15 cm (302 nm maximum); and (3) 337 nm: monochromatic nitrogen laser [Model LN1000, (Photochemical Research Company, Oakridge, TN)], 1300 μJ/pulse, 600-psec pulse, average power at 20 Hz–25 mW.

Light sources were tested for each analog at distances of 5 cm (254 nm), 1.5 cm (302 nm), or 1.0 cm (337 nm). We have found that the 254-nm light source rapidly degrades some proteins in less than a minute, unless samples are shielded.[37] We also see degradation of both DNA and RNA, but degradation occurs much less rapidly. RNA appears to be most stable to irradiation with this light source. Samples have been irradiated with and without cooling, and the degradation is not due to heating. Degradation can be eliminated by irradiating samples in borosilicate tubes, which absorb most wavelengths shorter than 290 nm. Even analogs that have absorption maxima below 290 nm can be activated while shielded, because of the high extinction coefficients of the cross-linking groups; however, longer irradiation times are required when samples are shielded.

The 302- and 337-nm light sources do not degrade protein, DNA, or RNA, even at irradiation times of 30 min. The 302-nm light source does, however, rapidly cleave some disulfide bonds.[37] This cleavage can act to degrade cross-linking reagents, which contain such bonds before covalent attachment of RNA to protein has occurred, or can cause photolytic separation of RNA and protein, which have been joined through the photocross-linking group of disulfide-containing cross-linkers. This disulfide bond breakage is not eliminated by shielding with borosilicate tubes. We

---

[37] M. M. Hanna, J. Ahdoot, H. Ngyuen, and A. Rafatjoo, unpublished (1987).

have found, therefore, that, whenever possible, the 254-nm mercury lamp (with shielding) or the 337-nm laser should be used with reagents having disulfide bonds. If one wishes to sequence proteins and nucleic acids involved in RNP complexes after cross-linking, it is important that a light source that does not degrade these macromolecules be chosen.

Selection and Synthesis of Nucleotide Analog

Important considerations for the use of the photoreactive nucleotide analogs shown in Figs. 1–3 will be discussed. Detailed synthetic procedures will be given only for those which have not been previously published. References to the syntheses of the remaining analogs will be provided. Cleavage methods will be described for those analogs that contain cleavable bonds. Figure 1 shows analogs that can be incorporated into the 5' end of a RNA chain, Fig. 2 shows analogs that can be incorporated at internal positions within a RNA chain, and Fig. 3 shows analogs that can be incorporated into the 3' end of a RNA chain.

Initiating Nucleotide Analogs

All of the photoreactive nucleotide analogs shown in Fig. 1 have been tested with *Escherichia coli* RNA polymerase. None has been tested with SP6, T7, or eukaryotic RNA polymerases, and none is commercially available. These analogs have proved very useful in defining the path of a nascent RNA chain through the *E. coli* RNA polymerase transcription complex.[38-43] The analogs should also be useful in examining the contacts made with the ribosomal components as the RNA leaves the surface of RNA polymerase.

α-[(4-Azidophenacyl)thio]nucleotides. These nucleotide analogs {5'-[(4-azidophenacyl)thio]phosphorylnucleotides} contain an aryl azide group at approximately the same position as the γ-phosphate on a normal nucleoside 5'-trisphosphate. The analogs do not serve as elongation substrates for RNA polymerase. Initiation with *E. coli* RNA polymerase is effective at or above 100 μM, if the normal initiating substrate is kept low (5 μM). If higher concentrations of the normal initiator are needed, the

[38] M. M. Hanna and C. F. Meares, *Proc. Natl. Acad. Sci. U.S.A.* **80**, 4238 (1983).
[39] M. M. Hanna and C. F. Meares, *Biochemistry* **22**, 3546 (1983).
[40] L. H. DeRiemer and C. F. Meares, *Biochemistry* **20**, 1606 (1981).
[41] M. A. Grachev and E. F. Zaychikov, *FEBS Lett.* **130**, 23 (1981).
[42] G. A. Nevinski, O. I. Lavrik, O. O. Favorova, and L. L. Kisselev, *Bioorg. Khim.* **5**, 352 (1980).
[43] M. A. Grachev, D. G. Knorre, and O. I. Lavrik, *Sov. Sci. Rev., Sect. D* **2N**, 107 (1981).

nucleotide analog concentration should be increased as well. The sulfur–phosphorus bond between the azide and the sugar is stable to acid, base, and thiol, but can be cleaved selectively with organomercurials. This type of cleavable bond is extremely useful, as the bond remains stable throughout transcription, electrophoresis, gel staining and destaining, and most steps involved in the isolation of a cross-linked RNP complex.[38] The azide absorption is at 300 nm ($E_{300\ nm} = 2 \times 10^4\ M^{-1}\ cm^{-1}$), but the 337-nm laser is quite effective in its photoexcitation. The 302-nm source can be used, as there are no photolabile disulfide bonds. Synthesis will be described for the mononucleotides. Detailed synthetic procedures for the dinucleotide are given in Ref. 39.

Synthesis of 5'-[(4-Azidophenacyl)thio]phosphoryladenosine or 5'-[(4-Azidophenacyl)thio]phosphorylguanosine

Synthesis should be carried out in dim light. Dissolve adenosine 5'-*O*-(thiomonophosphate) or guanosine 5'-*O*-(thiomonophosphate) (Boehringer Mannheim, Indianapolis, IN) in water to give a 13 m*M* solution. Combine 100 μl of this with 10 μl of 0.2 *M* sodium bicarbonate and 30 μl of methanol. Add to this 30 μl of 90 m*M* azidophenacyl bromide (Sigma, St. Louis, MO) in methanol and allow the reaction to proceed for 45 min at room temperature. Extract the reaction mixture three times each with isobutyl alcohol and then with ethyl ether and remove the ether under reduced pressure at room temperature. Product can be isolated by anion-exchange chromatography or high-performance liquid chromatography (HPLC),[39] if available. A volatile solvent should be chosen, such as triethylammonium bicarbonate, if using anion-exchange chromatography, so the solvent can be removed by lyophilization. The product should have absorption bands at both 260 nm (from the nucleotide) and 300 nm (from the azide). Product structure can be verified as described in Ref. 39 for the dinucleotide.

Once a RNA chain containing this analog has been cross-linked to a protein (or another nucleic acid) and the RNP complex has been isolated, the two covalently attached molecules can be separated by cleavage of the phosphorus–sulfur bond with phenylmercuric acetate. A saturated solution of phenylmercuric acetate in either water or 0.1% (w/v) sodium dodecyl sulfate (SDS) should be prepared immediately before use. Cleavage is complete after 24 hr at room temperature, though less time may be required.

β-(4-Azidophenyl)nucleotides. Analogs of this type have been prepared for both adenosine and the dinucleotide adenylyl(3'–5')uridine 5'-phosphate (pApU), and detailed procedures for the synthesis and characterization of these are given in Ref. 40. Synthesis occurs by way of a reactive *p*-azidophenyl phosphorimidazolidate, prepared by the coupling of *p*-nitrophenyl phosphate with *N,N'*-carbonyldiimidazole. Since the phosphorimidazolidate reacts with the terminal phosphates of nucleosides, synthesis of the guanosine, cytosine, and uridine adducts should also be possible. The absorption maximum for these compounds is at about 260 nm, with an extinction coefficient of $E_{260\,nm} = 2.56 \times 10^4\ M^{-1}$ cm^{-1}, resulting from the sum of the extinction coefficients for the nucleotide and the azide. This type of analog does not contain a chemically cleavable bond.

γ-(4-Azidoanililyl)nucleotides. These analogs (nucleoside 5'-triphosphate γ-azidoanilidates) serve both as initiation and elongation substrates for *E. coli* RNA polymerase[41]; however, the azide group is cleaved from the analog during incorporation into an internal position in a RNA chain. The azide group, therefore, will remain attached only to the initiating nucleotide. Synthesis of the adenosine, guanosine, and cytosine analogs have been described.[42,43] The absorption maximum for these compounds is at 260 nm. The nitrogen–phosphorus bond linking the azide and the sugar is acid-labile. This allows covalently attached RNA and protein molecules to be separated by treatment with 10% (v/v) acetic acid (2 hr at 37°). The major disadvantage of this type of analog is the instability of the RNA–protein cross-link to standard protein staining and destaining procedures, as acetic acid is frequently used to fix proteins and RNA in gels.

Internal Nucleotide Analogs

The photoreactive nucleotide analogs shown in Fig. 2 have all been incorporated into internal positions in RNA chains with *E. coli* RNA polymerase. Some have also been used with SP6, T7, or eukaryotic RNA polymerases. Only 5-bromo-UTP is commercially available (Sigma).

4-Thiouridine 5'-Triphosphate. The thioketo groups of thiopyrimidine derivatives can be modified with photoreactive cross-linking groups or can be directly activated with long-wavelength UV light to induce cross-linking. 4-Thiouridine and its corresponding nucleotides undergo photooxidation to electrophilic intermediates that react with nucleophiles under mild conditions.[44,45] As the cross-linking group is part of the pyrimidine, 4-thiouridine can be used to examine contacts with RNA within a few angstroms of uridine residues. The excitation maximum for this analog is between 320 and 340 nm. The nitrogen laser is an excellent photoexcitation source[37]; however, the 302-nm mercury lamp can be used as well because of the absence of disulfide bonds. 4-Thiouridine has been shown

[44] M. Pleiss, H. Ochiai, and P. A. Cerutti, *Biochem. Biophys. Res. Commun.* **34,** 70 (1969).
[45] M. G. Pleiss and P. A. Cerutti, *Biochemistry* **10,** 3093 (1971).

to be taken up and converted to a substrate for both prokaryotic[46] and eukaryotic RNA polymerases[47] *in vivo.* 4-Thio-UTP has been shown to be a substrate for HeLa RNA polymerase II,[48] T7,[49] *E. coli,*[50] and SP6[51] RNA polymerases *in vitro.*

While the original synthesis of 4-thio-UTP was chemical,[52] the availability of 4-thio-UDP (Sigma) allows for very simple and rapid enzymatic synthesis. A method involving transfer of a γ-phosphate from ATP using nucleoside diphosphate kinase has been published.[48] This method, however, results in the copurification of some ATP with the 4-thio-UTP. This can be a problem for some strategies used to place cross-linking groups at just one position in a RNA chain. An alternate synthesis that allows for the purification of 4-thio-UTP, which is free from all other nucleotides, involves the addition of a γ-phosphate using glycolytic enzymes.[49]

Synthesis of 4-Thiouridine 5'-Triphosphate

Synthesis is described for the preparation of 0.5 μmol of 4-thio-UTP; however, this synthesis has been scaled up 10-fold with similar yields. All steps should be carried out on ice or at 4°, unless otherwise specified. Sodium pyruvate and cysteine solutions must be prepared immediately before use. Aqueous solutions of L-α-glycerol phosphate and β-NAD can be stored at −20° for 6 months.

Prepare an ammonium sulfate suspension of enzymes (see Enzyme Stock) containing the following: 0.1 mg of glycerol-3-phosphate dehydrogenase, 0.001 mg of triose-phosphate isomerase, 0.1 mg of glyceraldehyde-3-phosphate dehydrogenase, 0.01 mg of 3-phosphoglycerate kinase, and 0.05 mg of lactate dehydrogenase. Enzymes can be purchased from Boehringer Mannheim as ammonium sulfate suspensions and should be mixed gently before use. Spin the enzyme solution for 5 min in a microcentrifuge and remove the supernatant solution by aspiration. The enzyme pellet should be stored on ice until just before use. Prepare a 60 mM cysteine stock solution, pH 8–9, by dissolving 10 mg of cysteine hydrochloride and 12.5 mg of Tris base in 1.0 ml of water. Prepare an enzyme–cysteine mixture (solution A) by resuspending the enzyme pellet in 0.333

[46] E. Hajnsdorf, A. Favre, and A. Expert-Bezançon, *Nucleic Acid Res.* **14,** 4009 (1986).
[47] W. T. Melvin, H. B. Milne, A. A. Slater, H. J. Allen, and H. M. Keir, *Eur. J. Biochem.* **92,** 373 (1978).
[48] B. Bartholomew, M. E. Dahmus, and C. F. Meares, *J. Biol. Chem.* **261,** 14226 (1986).
[49] N. K. Tanner, M. M. Hanna, and J. Abelson, *Biochemistry* **27,** 8852 (1988).
[50] F. Cramer, E. M. Gottschalk, H. Matzura, K.-H. Scheit, and H. Sternbach, *Eur. J. Biochem.* **19,** 379 (1971).
[51] M. M. Hanna, unpublished (1987).
[52] K. H. Scheit, *Chem. Ber.* **101,** 1147 (1968).

ml of 0.15 M Tris–HCl, pH 9.0, and then by adding 0.667 ml of 60 mM cysteine stock. This solution should be stored on ice until needed.

Prepare a 40 mM sodium pyruvate solution by placing approximately 5 mg of sodium pyruvate in a dry test tube and by dissolving in water to give a 4.4-mg/ml solution. All remaining steps should be carried out in dim light. Prepare 3.5 ml of solution B, as shown below. Add 0.875 ml of solution A and 4.4 ml of 0.6 mM phosphoric acid (HPLC grade) to solution B. Incubate the mixture for 1 hr at 25° and then extract twice with phenol to remove enzymes before purification.

Dilute the reaction mixture to 100 ml with water and load onto a 1.75 × 25-cm Whatman DE-52 column (at 4°), equilibrated in 0.1 M triethylammonium bicarbonate (TEAB), pH 8.0. Wash the column with 10 column volumes of 0.1 M TEAB and elute the product with a linear gradient from 0.5 to 1.0 M TEAB, pH 8.0 (500 ml of each). 4-Thio-UTP elutes at approximately 0.7 M. Locate the product fractions by determining absorbance at 330 nm, pool, and lyophilize to remove the solvent. The product should be dissolved in water and again lyophilized two or three times or until all visible salt is removed. Dissolve the product in water and determine the concentration by dilution of an aliquot into 0.01 M phosphate buffer, pH 7, using an extinction coefficient of $E_{330\,nm} = 21.2 \times 10^3$.[53] If the product has been dried for a long time and resists dissolution, 2-mercaptoethanol can be added on a molar basis to the 4-thio-UTP.

Enzyme Stock

Enzyme	Stock	Volume
Glycerol-3-phosphate dehydrogenase	10.0 mg/ml	10.0 μl
Triose-phosphate isomerase	2.0 mg/ml	0.5 μl
Glyceraldehyde-3-phosphate dehydrogenase	10.0 mg/ml	10.0 μl
3-Phosphoglycerate kinase	10.0 mg/ml	1.0 μl
Lactate dehydrogenase	5.0 mg/ml	10.0 μl

Solution B

Component	Final concentration	Stock concentration	Volume/3.5 ml
Tris–Cl, pH 9	0.15M	0.5 M	1.05 ml
DTT	0.015 M	1.0 M	0.052 ml
Magnesium chloride	0.035 M	1.0 M	0.122 ml
α-Glycerophosphate	0.34 mM	0.02 M	0.057 ml
4-Thio-UDP	0.14 mM	0.1 M	0.005 ml
β-NAD	1.43 mM	0.035 M	0.143 ml
Sodium pyruvate	2.8 mM	0.04 M	0.245 ml
Water			1.82 ml

5-Mercaptouridine 5'-Triphosphate. While 5-mercapto-UTP (Fig. 2) does not contain a highly photoreactive group, 5-mercapto-UTP does contain a thiol group, which can be modified with thiol-specific bifunctional photocross-linking reagents (Table I). 5-Mercapto-UTP is an excellent substrate for *E. coli* RNA polymerase.[51] Since the 5 position of the uracil is not involved in normal Watson–Crick base pairing, RNA containing this analog should form normal secondary and tertiary structures, even at high levels of substitution. RNA, containing complete substitution of 5-mercapto-UTP for UTP, forms normal ρ-independent transcription termination signals, as recognized by *E. coli* RNA polymerase.[51] Analog-containing RNA can be posttranscriptionally modified to place photocross-linking groups on accessible thiol groups. Residues that are accessible to these reagents are most likely to be accessible to regulatory or structural proteins as well. Even without posttranscriptional modification, there is some photocross-linking of 5-mercapto-UTP-containing RNA (not seen with UTP-containing RNA), when irradiated with long-wavelength UV light (wavelengths longer than 300 nm).

Synthesis of 5-Mercaptouridine 5'-Triphosphate

5-Mercapto-UTP is synthesized by a modification of the synthetic procedure for 5-mercaptouracil[54] and 5-mercapto-UMP.[55] Fresh reagents must be used. Dissolve 2.0 grams (4 mmol) of UTP in 125 ml of dimethylacetamide (4°) in a 500-ml-stoppered Erlenmeyer flask. Stir at 4° for at least 4 hr. A suspension of UTP will result. Prepare a fresh solution of methyl hypobromite[56] by placing 75 ml of anhydrous methanol (HPLC grade, −20°) into a 250-ml Erlenmeyer flask, containing a magnetic stir bar and capped with a stopper. Insert a drying tube and a 5-ml syringe (16- to 18-gauge needle) through holes in the stopper. Place the flask in a dry ice–ethanol bath to maintain the temperature at or below −15° and place the apparatus in dim light. Add 28.0 g (100 mmol) of silver carbonate and stir vigorously for 5 min. The solution should turn an olive-green color. While maintaining the temperature below −15° and stirring vigorously, add 5.5 ml (100 mmol) of bromine *dropwise* via the syringe at a *constant* rate of 1 ml/min. Stir this solution for 4 hr in the dark, maintaining the temperature below −15°. The color will change to a yellowish green.

[53] M. Pleiss, H. Ochiai, and P. A. Cerutti, *Biochem. Biophys. Res. Commun.* **34,** 70 (1969).
[54] T. J. Bardos and T. I. Kalman, *J. Pharm. Sci.* **55,** 606 (1966).
[55] Y. K. Ho, L. Novak, and T. J. Bardos, *in* "Nucleic Acid Chemistry" (L. B. Townsend and R. S. Tipson, eds.), Vol. 2, p. 813. Wiley, New York, 1976.
[56] R. Duschinsky, T. Gabriel, W. Tautz, A. Nussbaum, M. Hoffer, E. Grundberg, J. H. Burchenal, and J. J. Fox, *J. Med. Chem.* **10,** 47 (1967).

While the above solutions are stirring, prepare a fresh solution of sodium disulfide by dissolving 23 g (96 mmol) of sodium sulfide nonahydrate in 150 ml of boiling 100% (v/v) ethanol in a 250-ml beaker. Add 3.1 g (96 mmol) of sulfur and continue stirring until the volume is less than about 25 ml. This requires about 1 hr. Do not heat longer than needed. Cool the solution in an ice bath and a yellow solid will form. This solid must be broken up into small pieces as the solid forms to prevent formation of a large, insoluble solid. Store under vacuum and on dry ice until needed.

After the methyl hypobromite solution has stirred for 4 hr, filter the solution through Celite or through a medium glass-fritted funnel into a cold and dry 125-ml filtration flask, maintaining the temperature below −15° with dry ice. Rinse the flask, which contained the methyl hypobromite, four times each with 2 ml of dimethylacetamide and wash the filter with these rinses. *Immediately* add this solution to the suspension of UTP dimethylacetamide. A yellow slurry will form. Cover the reaction mixture with foil and continue stirring vigorously at 4° for 5–6 hr. The reaction mixture should change to a clear orange and should be used immediately. Add 23 g (80 mmol) of the freshly prepared sodium disulfide to the clear orange solution. Stir for at least 12 hr at 4°, in the dark, or until the solution turns and remains dark brown (no longer than 48 hr). The intermediate product bis(uridine 5′-triphosphate-5-yl) disulfide (bis-UTP) can be stored at 4° in the dark at this stage, if necessary, before removal of other reactants and side products.

Purify bis-UTP by adding 1 liter of cold water to the reaction mixture and by extracting three times each with 250 ml of water-saturated diethyl ether. A small amount of precipitate will form. Filter the aqueous phase first through Whatman 1 filter paper and then through a 0.2-μM filter. Wash the precipitate with water, combine the filtrate and water, and wash and dilute with water until the conductivity is equal to or below that of 0.02 M TEAB, pH 8.3, at room temperature (several liters). Load this onto a 1200-ml DE-52 column (at room temperature), equilibrated in 0.02 M TEAB, pH 8.3. Wash the column with 5 column volumes of 0.02 M TEAB and then elute the product with a 4-liter linear gradient from 0.02 to 1.5 M TEAB, pH 8.3. Locate bis-UTP by monitoring absorbance at 273 nm. The disulfide, containing bis-UTP, absorbs at 273 nm. When the disulfide bond is reduced, the 5-mercapto-UTP formed will also adsorb at 330 nm (at pH 8).

Pool the product fractions and remove the solvent under vacuum. Lyophilization for 3–5 days, with several reconstitutions with DEPC-treated water, is sufficient to remove salt. 5-Mercapto-UTP can be prepared for use in transcription by reduction of bis-UTP with DTT. Adjust the solution to pH 8 and determine the yield using $E_{330\ nm} = 8 \times 10^3\ M^{-1}$ cm^{-1} for 5-mercapto-UTP.[54,55]

5-Azidouridine 5'-Triphosphate. This UTP analog contains an azide group on the 5 position of the uridine ring (Fig. 2). Modification at this position should not interfere with formation of normal base pairs with other nucleotides. 5-Azido-UTP synthesis is described in Ref. 57. This analog is a substrate for *E. coli* RNA polymerase and has been used to photoaffinity label the active site of this enzyme in a ternary transcription complex.[58] The excitation maximum for this analog is 288 nm.

5-[(4-Azidophenacyl)thio]uridine 5'-Triphosphate. This UTP analog contains an aryl azide group about 10 Å from the 5 position of the uridine residue (Fig. 2), which should not interfere with normal Watson–Crick base pairing. The azide absorption is at 300 nm ($E_{300\ nm} = 2 \times 10^4\ M^{-1}$ cm^{-1}), and the 302-nm light may be used, because there are no disulfide bonds present. This analog is synthesized by the alkylation of 5-mercapto-UTP (5-SH-UTP) with azidophenacyl bromide.[59] The sulfur atom between the nucleotide and the azide can be removed by desulfurization with Raney nickel with accompanying bond scission, making this a reversible cross-linker. The sulfur–carbon bonds between the azide and the base are stable to acid, base, and thiol, however. This type of cleavable bond is extremely useful, as the bond remains stable throughout transcription, electrophoresis, gel staining and destaining, and most steps involved in the isolation of a cross-linked RNP complex. The analog can be incorporated into internal positions in RNA molecules by *E. coli* RNA polymerase and has been used to identify the contacts between this enzyme and the nascent RNA during transcription *in vitro.*[60]

Terminating Nucleotide Analogs

The photoreactive nucleotide analogs shown in Fig. 3 cannot be incorporated into internal positions in a RNA chain by *E. coli* RNA polymerase. Their use with other RNA polymerases has not been examined. These terminating analogs can be used to probe the active site of a RNA polymerase, as phosphodiester bond formation occurs at the 3' end of the RNA.

8-Azidonucleoside 5'-Triphosphates. The 8-azidopurines act as RNA

[57] R. K. Evans and B. E. Haley, *Biochemistry* **26**, 269 (1987).
[58] A-Y. M. Woody, R. K. Evans, and R. W. Woody, *Biochem. Biophys. Res. Commun.* **150**, 917 (1988).
[59] M. M. Hanna, S. Dissinger, B. D. Williams, and J. E. Colston, *Biochemistry,* in press (1989).
[60] S. Dissinger and M. M. Hanna, in preparation.

chain terminators, because the azide group on position 8 forces the nucleotide into a conformation that does not allow further chain elongation. Whether these analogs would act as terminators with other RNA polymerases is not known. The synthesis of 8-azido-ATP (Fig. 3) from 8-bromo-ATP is described in Ref. 61. This analog is also commercially available from Sigma. This analog has an absorption maximum at 281 nm at pH 7.4 with $E_{281 \text{ nm}} = 1.33 \times 10^4 \, M^{-1} \text{cm}^{-1}$.[61] The GTP analog can be synthesized in a similar manner. These compounds are stable, when stored in methanol at $-20°$ (or lower), showing 5–10% decomposition in 5–6 months. Samples can be photolyzed at 254 nm for times ranging from 15 sec to 5 min, depending on the intensity of the light source and the distance of the sample from the light source.[62] Irradiation should be done with shielding if proteins are to be analyzed. We have found, however, that both the 302- and 337-nm light sources are more effective for activation of 8-azido-ATP than the 254-nm light source, without the need for shielding to protect proteins and nucleic acids.[37] Detailed procedures for the use of 8-azidopurine analogs has been discussed in detail in an earlier volume of this series.[62]

3'-Azidonucleoside Triphosphates. The 3'-azidonucleotides (Fig. 3) act to inhibit transcription, because there is no hydroxyl group on the 3' position of the ribose to allow phosphodiester bond formation with the next nucleotide in the RNA chain. The syntheses of both the purine analogs are described in Ref. 63. These analogs should cause termination of transcription with all RNA polymerases for which these analogs may be substrates, similarly to other 3'-substituted ribonucleotides.[64] This type of analog allows one to examine the contacts made with the ribose in the active site of RNA polymerases.[65]

RNA–Protein Cross-Linking with Photoreactive Nucleotide Analogs

Synthesis of Analog-Containing RNA

All of the nucleotide analogs shown in Figs. 1–3 contain 5'-triphosphates and are best incorporated into RNA chains via transcription in the absence of light. Radioactivity can be incorporated into the RNA at internal positions with α-^{32}P-labeled nucleotides. If the analog-containing RNA will be sequenced after cross-linking to protein, the 5' end of the

[61] B. E. Haley and J. F. Hoffman, *Proc. Natl. Acad. Sci. U.S.A.* **71**, 3367 (1974).
[62] R. L. Potter and B. E. Haley, this series, Vol. 91, p. 613.
[63] D. Panka and D. Dennis, *J. Biol. Chem.* **259**, 8384 (1984).
[64] H. T. Shigeura and G. E. Boxer, *Biochem. Biophys. Res. Commun.* **17**, 758 (1964).
[65] V. W. Armstrong and F. Eckstein, *Biochemistry* **18**, 5117 (1979).

RNA can be labeled by incorporation of γ-^{32}P-labeled nucleotides (if a photoreactive-initiating nucleotide analog is not being used). Alternatively, one can initiate transcription with a dinucleotide, which does not contain a 5'-phosphate, and can add radioactivity with T4 polynucleotide kinase after isolation of the RNA. If a 5' analog is used, the RNA can be end labeled at the 3' end with RNA ligase. By making the RNA radioactive, one can identify proteins that become covalently attached to the RNA by cross-linking during irradiation, thereby becoming radioactively labeled.

Although all of the analogs shown are substrates for *E. coli* RNA polymerase, only 4-thio-UTP has been tested with SP6, T7, and eukaryotic RNA polymerases. It is probable that the uridine analogs substituted in the 5 position will be substrates for at least T7 RNA polymerase, as these analogs contain groups that are smaller than those found in biotinylated UTP, which does serve as a substrate for this enzyme.[66]

With the exception of 4-thio-UTP and 5-bromo-UTP, all of the analogs shown contain azide groups. Because azides are rapidly reduced by compounds containing two thiol groups, such as DTT,[67] one must eliminate such compounds from the transcription buffer and try to keep the pH of the solution below 8.0. Monothiols will also reduce azides, but less rapidly. It is therefore desirable to replace DTT with a monothiol, such as 2-mercaptoethanol, if thiol is absolutely required in the reaction. We have found, however, that thiol can be eliminated from the transcription buffer for *E. coli* RNA polymerase with no loss of activity for several hours.[68] It is also possible to completely remove thiol from the *E. coli* RNA polymerase storage buffer with no loss of transcriptional activity or specificity for up to 2 weeks, when the enzyme is stored at $-20°$.

In order to study interactions between RNA and RNA polymerase or transcription factors, it is not necessary to isolate the analog-containing RNA before cross-linking. Transcription complexes can be irradiated during active transcription, or transcription can be stopped without dissociation of the transcription complex by incorporation of terminating nucleotides.[38] To examine interactions between analog-containing RNA and proteins not involved in transcription, it is possible, in some cases, to add the purified protein or protein-containing fraction directly to the transcription reaction after synthesis of the RNA is complete. One must judge each system individually when deciding on an approach. It is frequently neces-

[66] P. R. Langer, A. A. Waldrop, and D. C. Ward, *Proc. Natl. Acad. Sci. U.S.A.* **78**, 6633 (1981).

[67] J. V. Staros, H. Bayley, D. N. Standring, and J. R. Knowles, *Biochem. Biophys. Res. Commun.* **80**, 568 (1978).

[68] M. M. Hanna, C. Bowser, and M. Hodge, unpublished (1987).

sary to isolate the RNA and then to add the RNA back to purified proteins or extracts containing these proteins. Most of the methods described below are applicable to all of these approaches.

Isolation of Analog-Containing RNA

All steps must be carried out in dim light. There are several ways in which RNA can be isolated. The method chosen may depend on the importance of the RNA secondary and tertiary structure to the RNA–protein interaction of interest. When working with a large molecule with extended secondary structure, it may be desirable to choose an isolation method that does not fully denature the RNA, as the correct structure may not reform after isolation. We have found the following methods to be most useful.

Precipitation of RNA. If working with a system that allows the preparation of analog-containing RNA with purified DNA and RNA polymerase where only one RNA transcript is formed, it is possible to simply remove the polymerase and DNA from such reactions. After transcription, the DNA template can be degraded by treatment of the reaction mixture with DNase, which is free of RNase (RQ1 DNase, Promega, Madison, WI). RNA polymerase and the DNase can be removed by extraction of the transcription reaction with phenol. The RNA can be precipitated two times with ammonium acetate–ethanol to remove unincorporated nucleotides[69] and can be resuspended in the appropriate buffer for formation of the RNP complex. (This buffer should not contain dithiols, if an azide cross-linker is being used.)

Isolation by Gel Electrophoresis. RNA can be isolated on nondenaturing or denaturing acrylamide or agarose gels, depending on the size of the RNA. Gels should be prepared and run according to standard procedures.[69] After electrophoresis, the radioactive RNA can be located by autoradiography (gels should not be fixed, stained, or dried). It is important to place radioactive marker dots or light spots all around the gel so that, after development, the film can be aligned accurately above the gel to allow excision of the RNA. If the RNA is very radioactive, one can expose the gel to film at room temperature. However, sometimes gels will need longer exposures, making freezing the gel and use of an intensifying screen desirable. When working with agarose or very low-percentage acrylamide gels (less than 10%), freezing and thawing of the gel can cause the gel to change size and become "mushy." We have had the best results by leaving the gels on one glass plate and by prefreezing the gel on a slab

[69] T. Maniatis, E. F. Fritsch, and J. Sambrook, "Molecular Cloning: A Laboratory Manual." Cold Spring Harbor Laboratory, Cold Spring Harbor, New York, 1982.

of dry ice before putting film on the gel. It is best to use two pieces of X-ray film, saving one as a record of the results of the experiment and using the other to directly overlay on the gel. It is easiest to cut directly through this second piece of film and the underlying gel with a razor blade. It is important that the gel remain frozen through all steps, again being placed on dry ice, while the gel piece is being cut from the gel.

The RNA can be recovered from the gel slice by one of several methods. For small RNA molecules, the gel piece can be crushed, and the RNA can be isolated by diffusion. An example for the isolation of analog-containing tRNA is given in Ref. 49. Alternatively, one can place the gel piece in a dialysis bag and isolate the RNA by electrophoresis. This is effective even for large RNA molecules and RNA–protein complexes. An example of this method is given in Ref. 38. Similarly, the RNA can be isolated by electrophoresis into a high-salt solution using an electroeluter apparatus (IBI, New Haven, CN). Last, dissolvable gels, which will be described in a later section, can be used.

Isolation by Affinity Chromatography. RNA can be separated from protein and DNA by passage over small columns containing covalently linked borate groups.[70–73] These groups have affinity for the *cis*-diol on the 3'-terminal ribose of RNA. Columns contain acetylated N-[N'-(m-dihydroxyborylphenyl)succinamyl]aminoethyl cellulose (DBAE-cellulose), available from Collaborative Research (Waltham, MA), or DBASE-polyacrylamide (Affi-Gel 601), available from Bio-Rad (Richmond, CA). RNA can be eluted from these columns with 0.05 M sodium acetate, pH 5.1, containing 0.2 M NaCl and 1 mM EDTA, and can be directly precipitated with ethanol. With proper storage, these columns can be regenerated and reused up to 60 times.[73]

Preparation of RNA–Protein Cross-Linked Complexes

Once a cross-linking reagent and a light source have been chosen, one must determine the optimal time for cross-linking empirically. If the RNA and protein associate to form a stable, long-lived complex, it is possible to irradiate for very long times and get cross-linking yields as high as 90%. If one is examining a transient interaction, however, it may be necessary to irradiate for shorter times, usually giving much lower cross-linking yields. It has recently been shown that salt can greatly affect the strength of protein–nucleic acid complexes. Substitution of glutamate for chloride ions has increased protein–DNA interactions more than 10-fold in some

[70] H. L. Weith, J. L. Wiebers, and P. T. Gilham, *Biochemistry* **9**, 4396 (1970).
[71] R. K. Goitein and S. M. Parsons, *Anal. Biochem.* **87**, 636 (1978).
[72] A. E. Annamalai, P. K. Pal, and R. F. Colman, *Anal. Biochem.* **99**, 85 (1979).
[73] S. Ackerman, B. Cool, and J. J. Furth, *Anal. Biochem.* **100**, 174 (1979).

cases.[74] It may be worthwhile to optimize the salt concentration for each RNP complex before photocross-linking.

There are several controls which must be included to ensure that photocross-linking data are meaningful. These include the following.

1. Always analyze a sample that has been treated identically to the photolyzed sample, except for irradiation. These dark controls are critical, as radioactively labeled nucleotides can become covalently attached to proteins nonenzymatically, even in the absence of light.[75] Although this labeling occurs at a low efficiency, this labeling can be of the same order of magnitude as some photocross-linking.

2. Set up parallel cross-linking reactions with a protein of similar size and charge, as the protein proposed to be involved in the RNP complex. This will help one to distinguish between specific and nonspecific binding to RNA.

3. Set up parallel cross-linking reactions with a RNA of similar size and secondary structure, if possible, as the RNA proposed to be involved in the RNP, for the same reason mentioned previously.

Isolation and Analysis of Cross-Linked RNA–Protein Complexes

In the simplest case, one may simply wish to ask whether a certain RNA and protein interact. This is done most easily by analyzing the cross-linking reactions (including controls) by SDS–polyacrylamide gel electrophoresis (PAGE) and autoradiography. As previously mentioned, if the RNA is radioactive and the protein is not, the formation of a radioactive protein molecule during irradiation of the RNP complex indicates an interaction between the RNA and the protein. Using electrophoresis, cross-linked complexes can be separated from free RNA and protein. A problem arises with large RNA molecules and small proteins, as both will tend to comigrate in these gels. The following method is useful for separation of free RNA from protein and RNA–protein complexes. It involves the simultaneous electrophoretic transfer of these molecules from the gel to overlayed nitrocellulose and nylon membranes, which can then be probed with antibodies or can be subjected to staining and autoradiography. Proteins and RNA–protein complexes will stick to the nitrocellulose and free RNA will bind to the nylon.

Transfer of Cross-Linked RNA–Protein Complexes to Nitrocellulose and Nylon Membranes

Unless otherwise stated, all procedures should be carried out in the dark with only a red light. All paper and membranes should be wetted in

---

[74] S. Leirmo, C. Harrison, D. S. Cayley, R. R. Burgess, and M T. Record, Jr., *Biochemistry* **26**, 2095 (1987).

[75] M. C. Schmidt and M. M. Hanna, *FEBS Lett.* **194**, 305 (1986).

transfer buffer (25 mM Tris, 192 mM glycine, and 20% (v/v) methanol). Tris–glycine solution should be prepared in DEPC-treated water and should be autoclaved before HPLC-grade methanol is added. Transfer buffers can be varied to favor transfer of a given RNA–protein complex. After separating components of the cross-linking reaction(s) by gel electrophoresis, transfer the gel from the glass plate to Whatman chromatography paper (Whatman 3MM). Place the gel on an electroblot apparatus (as used for western transfers). Layer onto the gel in the following order: a piece of Biotrace NT nitrocellulose [#66486S (Gelman Sciences, Ann Arbor, MI)], a piece of chromatography paper, a piece of Nytran nylon membrane [0.2-μm pore size (Schleicher and Schueller, Keen, NH)], and another piece of chromatography paper. Make sure there are no air bubbles between the layers. Enclose the layered sandwich of paper–gel–nitrocellulose–paper–nylon–paper in two larger, folded pieces of chromatography paper. Close the cassette assembly and place the cassette vertically in the blotting unit in the following orientation: negative electrode (black), black Scotch-Brite pad, paper, gel, nitrocellulose, paper, nylon, paper, white open-cell sponge pad, and positive electrode (red). Transfer should be carried out in the dark at 4° for 12–48 hr at 150 mA. Transfer is carried out in the dark so that photoreactive molecules, which may not have been activated during photolysis or are in the control reactions, will not become covalently linked to the gel or the membranes by room light during transfer.

After transfer, procedures can be carried out with the lights on. Remove the cassette assembly from the blotting unit and air dry the nylon membrane between Kimwipes and paper towels. Nitrocellulose can be probed with antibody against protein according to standard procedures. Both the nitrocellulose and the nylon membranes can then be subjected to autoradiography to locate RNA-containing bands. RNA–protein complexes should be bound to the nitrocellulose. Free RNA should be bound to the nylon. The gel should also be analyzed by staining and autoradiography to ensure that complete transfer has occurred. If a radioactive band is present on the nitrocellulose in the same place as a band visualized with antibody, then the band is inferred to correspond to a protein–RNA cross-linked complex.

Isolation of Cross-Linked RNA–Protein Complexes

To determine which domains or regions of a protein or RNA molecule, or more specifically, which amino acids or nucleotides, are involved in a RNA–protein complex, the covalently linked RNA–protein complex must be separated from free RNA and protein before beginning characterization of the cross-linked molecules. If antibody to the protein involved in the complex is available, the RNA–protein complex, along with free protein, can be separated from free RNA and other proteins in the reaction by immunoprecipitation. The free protein in the immunoprecipitated fraction can then be separated from the RNA–protein complex using a borate column, as described previously.[70-73]

Alternately, cross-linked RNA–protein complexes can be isolated by gel electrophoresis. When cross-linkers have been used which do not contain disulfide bonds, standard SDS–polyacrylamide gels containing methylenebisacrylamide can be used,[76] including thiol in the sample loading buffer. For very large RNA molecules cross-linked to small proteins, there is sometimes a problem in separating free RNA from RNA–protein complexes on such gels. Better resolution can sometimes be obtained by using polyacrylamide–urea gels.[77] If the two still cannot be separated, it may be necessary to partially digest the RNA before attempting separation. This will result in a small piece of RNA cross-linked to the protein, and the protein size will dictate mobility in these gels. When cross-linkers have been used which contain disulfide bonds, cross-linked proteins can be isolated on thiol-dissolvable acrylamide gels containing bisacrylcystamine.[78-80]

When reaction components have been separated by gel electrophoresis, it is necessary to recover the RNA–protein complex from the gel for further analysis. Gels should not be stained, destained, or dried after electrophoresis. It is advisable to run marker proteins in one lane of the gel, which can be sliced off for staining and destaining, to approximate the position of the proteins of interest in the unstained portion of the gel. RNA–protein complexes are located by autoradiography of the gel, as described in Isolation of Analog-Containing RNA. One must identify the band which is believed to correspond to the cross-linked complex by comparing the photolyzed test reaction to all of the control reactions. The RNA–protein complex can be recovered from the gel piece by any of the methods described in Isolation of Analog-Containing RNA.

Analysis of Cross-Linked RNA–Protein Complexes

Several questions can be answered once the RNA–protein complex has been isolated. If antibody to the protein is available, the identity of the protein(s) in such a complex can be determined or verified by the methods

[76] U. Laemmli, *Nature (London)* **277**, 680 (1970).
[77] C. W. Wu, F. V. H. Wu, and D. C. Speckhard, *Biochemistry* **20**, 5449 (1977).
[78] J. N. Hansen, *Anal. Biochem.* **76**, 37 (1976).
[79] J. N. Hansen, B. H. Pheiffer, and J. A. Boehnert, *Anal. Biochem.* **105**, 192 (1980).
[80] J. N. Hansen, *Anal. Biochem.* **116**, 146 (1981).

described previously. If antibodies are not available, the RNA can be cleaved from the protein, and both can be analyzed by gel electrophoresis or some other physical method for which characteristics of the putative cross-linked RNA or protein are known. If a cross-linker which contains a cleavable bond has been used, the RNA and the protein(s) can be separated simply by incubation of the complex with the appropriate cleavage reagent. If there is no cleavage procedure to separate the RNA and the protein, the RNA can be digested with ribonucleases or by treatment with alkali to leave a tag on the protein. Similarly, the protein can be digested with proteases, leaving a tag on the RNA.

To identify domains of the RNA and/or protein that are involved in the cross-link, RNA and/or protein sequencing must be used. If the RNA has been radioactively end labeled, the RNA cross-linked to protein can be sequenced by chemical[81,82] or enzymatic[83,84] means and can be compared to the sequencing pattern for RNA that has not been cross-linked. Sequencing patterns for both RNAs will be identical for all nucleotides between the radioactive label (whether 5' or 3') and the nucleotide that is cross-linked to protein. The nucleotide that is cross-linked to protein and all nucleotides beyond the bound protein and the labeled end will have an aberrant mobility on sequencing gels and will be missing from the sequencing ladders. If the RNA on the RNA–protein complex was partially digested to facilitate isolation of the cross-linked complex, the RNA can be relabeled by treatment with phosphatase to remove the 5'-phosphate group and by addition of a [^{32}P]phosphate group with polynucleotide kinase.[69]

To identify the domains of a protein that are involved in a cross-link, it is best to use uniformly labeled, whole-body [α-^{32}P]RNA for cross-linking. After isolation of the complex, the RNA can be digested, preferably chemically, so that only small pieces of radioactive RNA remain attached to the protein. This complex can then be treated with proteases, and the radioactively labeled fragment can be isolated and identified by standard protein-sequencing techniques.

Acknowledgments

Work from this laboratory was supported in part by an American Cancer Society grant (#NP544A) and by University of California Cancer Research Coordinating Committee funds.

[81] D. A. Peattie and W. Gilbert, *Proc. Natl. Acad. Sci. U.S.A.* **77,** 4679 (1980).
[82] D. A. Peattie, *Proc. Natl. Acad. Sci. U.S.A.* **76,** 1760 (1979).
[83] H. Donis-Keller, A. M. Maxam, and W. Gilbert, *Nucleic Acid Res.* **4,** 2527 (1977).
[84] H. Donis-Keller, *Nucleic Acid Res.* **8,** 3188 (1980).

[28] Ultraviolet-Induced Cross-Linking of RNA to Proteins in Vivo

By SERAFÍN PIÑOL-ROMA, STEPHEN A. ADAM, YANG DO CHOI, and GIDEON DREYFUSS

This chapter describes methods for RNA–protein cross-linking in living cells as a means of identification of RNA-binding proteins *in vivo*. The discussion focuses on heterogeneous nuclear RNA (hnRNA)- and messenger RNA (mRNA)-binding proteins, but the major considerations apply to any other polynucleotide. Heterogeneous nuclear ribonucleoprotein (hnRNP) particles and messenger ribonucleoprotein (mRNP) particles are the complexes of specific proteins with hnRNA or mRNA, respectively. Isolation of hnRNP and mRNP complexes and identification of their components has traditionally relied on cosedimentation of proteins with RNA in sucrose density gradients.[1-3] These fractionation methods, however, do not yield pure particles, since, as in any cellular fractionation procedure, some non-RNP proteins may become adventitiously associated with the RNA during the purification and some genuine RNP proteins may dissociate during such fractionation procedures. Other methods, such as purification of hnRNPs and mRNPs by oligo(dT) chromatography, while having some advantages over sedimentation methods, suffer from the fact that many non-RNP proteins bind oligo(dT). It thus becomes necessary to devise a strategy that identifies genuine RNP proteins *in vivo*. The most stringent definition of genuine RNP proteins is that these are proteins which are bound directly to the RNA of interest in the living cell. Experimentally, the identification of these proteins can be largely accomplished by ultraviolet (UV) light-induced cross-linking of the bound proteins to RNA *in vivo* and by subsequent isolation of these complexes under protein-denaturing conditions.[4-9]

[1] A. L. Beyer, M. E. Christensen, B. W. Walker, and W. M. LeStourgeon, *Cell (Cambridge, Mass.)* **11**, 127 (1977).

[2] J. Karn, G. Vidali, L. C. Boffa, and V. G. Allfrey, *J. Biol. Chem.* **252**, 7307 (1977).

[3] T. E. Martin, P. B. Billings, J. M. Pullman, B. J. Stevens, and A. J. Kinniburgh, *Cold Spring Harbor Symp. Quant. Biol.* **42**, 899 (1978).

[4] A. J. M. Wagenmakers, R. J. Reinders, and W. J. van Venrooij, *Eur. J. Biochem.* **112**, 323 (1980).

[5] C. A. van Eekelen, T. Riemen, and W. J. van Venrooij, *FEBS Lett.* **130**, 223 (1981).

[6] S. Mayrand and T. Pederson, *Proc. Natl. Acad. Sci. U.S.A.* **78**, 2208 (1981).

[7] S. Mayrand, B. Setyono, J. R. Greenberg, and T. Pederson, *J. Cell Biol.* **90**, 380 (1981).

[8] G. Dreyfuss, S. A. Adam, and Y. D. Choi, *Mol. Cell. Biol.* **4**, 415 (1984).

[9] G. Dreyfuss, Y. D. Choi, and S. A. Adam, *Mol. Cell. Biol.* **4**, 1104 (1984).

UV cross-linking of RNP complexes takes advantage of the fact that UV light of sufficient intensity generates highly reactive species of RNA, which react virtually indiscriminately with molecules, including proteins, with which the RNA is in stable, direct contact.[10,11] In principle, isolation of the RNA from cells that have been exposed to UV light will therefore result in the coisolation of at least some of the proteins that contacted the RNA in the cell. This covalent association of the RNA with proteins makes it possible to isolate the UV-cross-linked complexes under strongly denaturing and disulfide-bond-reducing conditions that ensure that only proteins cross-linked to the RNA are isolated with the RNA. The specific polynucleotides of interest can be isolated by hybridization to immobilized complementary polynucleotides. For most of the precursor mRNA (pre-mRNA) and the mRNA in the cell, one can take advantage of the fact that most of these sequences possess poly(A) tracts at their 3′ ends and can isolate polyadenylated RNA by chromatography on oligo(dT)-cellulose under protein-denaturing conditions. The cross-linked proteins can be released from the RNA by exhaustive RNase digestion and can be analyzed by gel-electrophoretic techniques. This method thus provides a powerful tool for the identification of bona fide RNP proteins. Fractionation of cells into nuclear and cytoplasmic components prior to the chromatographic step allows the distinction between hnRNA-containing RNP complexes (hnRNPs), from the nuclear fraction, and mRNA-containing RNP complexes (mRNPs), from the cytoplasmic fraction. The proteins can sometimes (depending on the amount of starting material) be visualized directly by staining with Coomassie blue or silver stain or by radiographic methods, if the proteins were radioactively labeled. To do so, cells can be prelabeled with radioactive amino acids, or the visualization of the proteins can be effected by *in vivo* labeling of the RNA with radioactively labeled ribonucleotide precursors. Residual radioactively labeled nucleotides, which survive the RNase digestion and which are covalently cross-linked to the proteins, provide a means of tagging the proteins by label transfer. In fact, this constitutes an affinity-labeling procedure for proteins that contact RNA *in vivo*. The proteins can then be separated by sodium dodecyl sulfate–polyacrylamide gel electrophoresis (SDS–PAGE) and can be visualized by fluorography.

The UV cross-linking method has several general limitations. Because of the denaturing conditions under which UV-cross-linked RNP complexes are isolated, these complexes are, for the most part, not suitable

---

[10] K. C. Smith, *in* "Photochemistry and Photobiology of Nucleic Acids" (S. Y. Wang, ed.), Vol. 2, p. 187. Academic Press, New York, 1976.
[11] J. R. Greenberg, *Nucleic Acids Res.* **8**, 5685 (1980).

for further biochemical studies. In addition, the efficiency of cross-linking for individual proteins to the RNA is highly variable; cross-linking is strongly dependent on both the specific protein and on the segment of RNA to which the protein is bound.[10] Thus, there may be some RNA-binding proteins which, due to chemical or steric considerations, do not cross-link well to the RNA. Furthermore, proteins that are genuine components of RNP complexes, but which associate with them primarily by protein–protein interactions, rather than direct protein–RNA base interactions, will not remain associated during the isolation procedure. The dependence of this method on oligo(dT) selection also does not allow examination of the proteins that are bound to poly(A)⁻ RNA; for these proteins, specific cDNAs need to be used with the result that the yields are very low. The specific yield of the UV-cross-linking reactions has not been determined, and the yield varies for different proteins. However, in general, the amount of cross-linked proteins obtained is rather small. Nevertheless, the specificity of the procedure and the ability to use the procedure on live cells renders this method a powerful tool for the identification of genuine RNP complex proteins. The UV-cross-linked complexes, as discussed in following sections, are also an excellent source of antigen for the generation of monoclonal antibodies to the cross-linked RNP proteins.

Procedure

Reagents. Oligo(dT)-cellulose (type 3) is from Collaborative Research (Lexington, MA). The following stock solutions are used at various steps in the procedure: 20% (v/v) Triton X-100; 5 M LiCl; 0.5 M ethylene-diaminetetraacetic acid (EDTA), pH 8.0; 10% (w/v) SDS; aprotinin, leupeptin, and pepstatin A from Sigma (St. Louis, MO); 10× MAGIK [10% (v/v) Tween 40 and 5% (w/v) deoxycholate]; and 200 mM vanadyl-adenosine (20×)[12] (see Preparation of Vanadyladenosine).

Buffer Solutions

Reticulocyte standard buffer (RSB): 10 mM Tris, pH 7.4, 10 mM NaCl, and 1.5 mM MgCl$_2$

Dulbecco's phosphate-buffered saline [PBS(+)]: PBS containing 1 mM CaCl$_2$ and 0.5 mM MgCl$_2$

Binding buffer for oligo(dT): 10 mM Tris–HCl, 1 mM EDTA, 0.5% (w/v) SDS, and 0.5 M LiCl

Elution buffer for oligo(dT): 10 mM Tris–HCl, 1 mM EDTA, and 0.05% (w/v) SDS

[12] S. L. Berger, this series, Vol. 152, p. 227.

Cell Culture and Labeling

The labeling protocol may vary depending on the cell type. HeLa cells are grown in monolayer culture in Dulbecco's modified Eagle's medium, containing 10% (v/v) fetal calf serum at 37° in an atmosphere of 5% CO_2. All cultures are supplemented with penicillin–streptomycin and are used at subconfluent densities. Cells are labeled overnight with [^{35}S]methionine at 10–20 μCi/ml in medium containing 10% (v/v) the normal amount of methionine and 5% (v/v) fetal calf serum. Alternatively, cells can be labeled with [^{35}S]methionine for 4 hr in methionine-free medium containing 2% (v/v) fetal calf serum. Labeling with ^{3}H-labeled nucleosides is carried out similarly in complete medium with 50 μCi of isotope/ml. ^{32}P labeling can also be carried out in phosphate-free medium with 50 μCi of ^{32}P$_i$/ml. Labeling with ^{32}P, however, also results in labeling of proteins directly due to protein kinase activity, and thus it becomes difficult to distinguish between labeling by virtue of label transfer from the RNA and labeling due to direct protein phosphorylation.

Ultraviolet Irradiation of Cell Monolayers

The cells, cultured on 10-cm tissue culture plates, are washed twice with cold (0–4°) PBS(+), the excess PBS(+) is removed, and 2 ml of PBS(+) is added to each 10-cm tissue culture plate. The plates are then placed without the lid under a 15-W germicidal lamp (Sylvania G15T8, GTE Sylvania Inc., Danvers, MA) at a distance of 4.5 cm. Several plates (up to three) can be placed next to each other under the lamp. The setup is covered with an aluminum foil "tent," and the cells are irradiated for 3 min. Under these conditions, the UV dose, as measured by a light intensity meter, is approximately 6.5×10^3 ergs/mm^2. Care should be taken at this point to protect the eyes and skin from exposure to the UV light. While longer irradiation times or higher doses of UV light increase the amount of proteins cross-linked to the RNA, these conditions result in nicking of the RNA and may increase nonspecific protein cross-linking. Furthermore, RNAs to which large amounts of proteins are cross-linked are less efficiently recovered in the oligo(dT)-cellulose chromatography step, presumably because the cross-linked proteins interfere with the hybridization (for example, see Adam et al.[13]). For different cell types, the optimal balance between these parameters needs to be determined experimentally. The amount of proteins and RNA recovered can be determined after labeling the cells with radioactive amino acids and nucleosides, respectively. We also find that different proteins cross-link with different

[13] S. A. Adam, Y. D. Choi, and G. Dreyfuss, J. Virol. **57**, 614 (1986).

efficiencies. For example, the hnRNP C group proteins cross-link very readily, whereas the hnRNP A and B group proteins do not cross-link to the RNA very efficiently.

Extraction of Cross-Linked Ribonucleoproteins

After UV irradiation, pour off and drain excess PBS(+). Replace with 1.75 ml of RSB, containing the RNase inhibitor vanadyladenosine at 10 mM, 0.5% (v/v) aprotinin, and 1 μg/ml each of leupeptin and pepstatin A. Float plates on an ice bath for 5 min to allow the cells to swell and then lyse the cells by adding 50 μl of 20% (v/v) Triton X-100 (0.5% final concentration). Add 200 μl of 10× MAGIK and scrape the cells with a rubber policeman. If the nuclei are fragile or if it is not essential to obtain nuclei completely clean of cytoplasmic contamination, the MAGIK can be omitted. Homogenize the cells by four passages through a 25-gauge needle, avoiding formation of bubbles. The nuclei are then pelleted by brief centrifugation at 3000 g, 4° (turn the centrifuge off as soon as the centrifuge reaches the appropriate speed). The supernatant is saved as the cytoplasmic fraction. The pellet contains the nuclear fraction. To the cytoplasmic fraction, add SDS to 1% (w/v), 2-mercaptoethanol to 1% (v/v), and EDTA to 10 mM and place on ice. Resuspend the nuclei in RSB, containing 10 mM vanadyladenosine, 0.5% (v/v) aprotinin, and 1 μg/ml each of leupeptin and pepstatin A, and vortex hard to break up the pellet. Add DNase I to 50 μg/ml, mix gently (do not vortex), and incubate at 37° for 15 min. After the digestion, add 3.5 ml of RSB with the protease inhibitors and bring to 0.5% (w/v) SDS, 1% (v/v) 2-mercaptoethanol, and 10 mM EDTA. This step is especially important for the nuclear fraction, since any residual DNase activity will digest the oligo(dT) column.

Poly(A)-Specific Fractionation by Chromatography on Oligo(dT)-Cellulose

Oligo(dT)-Cellulose Usage and Storage. To prepare oligo(dT)-cellulose for use, suspend the powder in eluting buffer. For four columns of 0.5 ml each, weigh out approximately 0.75 g of cellulose. Suspend and rock overnight. Pack 0.5-ml autoclavable, disposable polypropylene columns and wash extensively with elution buffer, followed by equilibration with binding buffer. It is advisable to get rid of fine particles, by allowing the cellulose suspension to stand for about 5 min and by decanting the supernatant, which contains the fine cellulose particles. Fine particles will interfere with the flow properties of the column. The columns can be rejuvenated as needed, by washing with 0.1 N KOH (wash with distilled water prior to and after the KOH wash in order to avoid precipitation of the SDS in the column).

Loading Procedure. For UV-cross-linked samples, the entire binding and eluting procedure must be carried out twice to ensure recovery and purity of the sample. For a 0.5-ml column, run 3- or 4-ml samples, which can be brought up to full volume with binding buffer. Vortex the samples and incubate in a 65- to 70°-water bath for 5 min. The sample, which has approximately 95% rRNA will be melted out, allowing only the mRNA to bind to the column. After heating, quench the samples by rapidly placing the samples in an ice bath for 5 min. After quenching, the sample salt concentration needs to be increased to binding conditions. Add 5 *M* LiCl to a final concentration of 500 m*M* and vortex the samples. Mix the sample with the column, cover the columns with Parafilm, and rock at room temperature for 15 min to ensure complete binding. Clamp the column to a column stand and let the sample run through the column twice. Wash the columns with 20–30 ml of binding buffer.

Elution. To elute for first run-through, run 3 ml of elution buffer through the column. Bring the eluate back to SDS-binding conditions, repeat the heating and chilling steps, bring the LiCl concentration back up to binding conditions, and repeat the binding to the column. For the second elution run, remove the column completely of binding buffer by tapping and collect 0.5-ml fractions. These fractions can then be assayed for OD_{260} or for radioactivity, and the peak fractions should be saved.

2-Butanol Sample Volume Reduction. Add 1.0 ml of 2-butanol to each of the saved fractions, vortex well for at least 30 sec, and spin in a microcentrifuge for 3 min. Draw off the alcohol phase (top) and pool the aqueous phases. This 2-butanol extraction may be repeated until a final combined volume of about 300 μl is obtained for the aqueous phase. With the 2-butanol volume reduction, the sample may precipitate in the aqueous phase. This is not a problem, although care must be taken that, when combining fractions, the entire sample (aqueous + precipitate) is recovered. After the volume reduction is completed, add LiCl to a final concentration of 0.2 *M* and precipitate the RNA–protein complexes with 2.5–3 volumes of ethanol by incubating at −20° overnight or by incubating in a dry ice–acetone bath for 1 hr. On solidification, spin the samples in the microcentrifuge for 15 min, draw off the entire supernatant, and dry in desiccator under vacuum until the sample is completely dry.

RNase Digestion of Ultraviolet-Cross-Linked RNA–Protein Complexes. The ethanol-precipitated, dry RNA–protein pellets are resuspended in 75 μl of RNase digestion cocktail (10 m*M* Tris, pH 7.4, 1 m*M* CaCl$_2$, 25 μg/ml of preboiled RNase A, and 400 U/ml of micrococcal nuclease) containing 0.5% (v/v) aprotinin, leupeptin, and pepstatin A. The samples are then digested for 1 hr at 37°. The ethanol-precipitation step of the sample is then repeated, and the dry pellets can then be resuspended in SDS sample buffer for gel electrophoresis.

After electrophoresis, the gels can be processed for autoradiography for the detection of the radioactive proteins. The cross-linked proteins normally exhibit a slightly lower mobility by SDS–PAGE, than their non-cross-linked counterparts, and also appear somewhat more fuzzy, due to the residual nucleotide(s), which survive the nuclease digestion step and remain covalently attached to the protein. Thus, this increase in apparent molecular weight is to be taken into account, when estimating the size of cross-linked proteins by SDS–PAGE.

Preparation of Vanadyladenosine

To prepare a 200 mM (20×) stock solution of vanadyladenosine follow these steps.

1. Make up 2 M VaOSO$_4$ (2.0 g in 4.6 ml) in H$_2$O.
2. Make up 250 mM adenosine in H$_2$O (i.e., 2.0 g in 30 ml).
3. Heat the adenosine solution with stirring on a hot, stirrer plate until the adenosine solution dissolves.
4. Add 3 ml of VaOSO$_4$ solution to 24 ml of hot adenosine solution with stirring (solution turns green-black).
5. Adjust the pH (initially ~2.5) to 6.0 with 10 N NaOH (a precipitate will form until the pH reaches ~6.0 and then will disappear and the solution will turn very dark). Then adjust the pH to 7.0 with 1.0 N NaOH.
6. Adjust the final volume to 30 ml. Keep hot enough and continue stirring while taking aliquots.
7. Store in 0.5-ml aliquots at −70°.

(Note) The vanadyladenosine solution has to be heated up to about 65° for the vanadyladenosine to go into solution on thawing. Discard the unused portions once thawed.

General Comments

1. It is essential, as a control for this procedure, to run parallel samples treated in exactly the same way, except without exposure to UV light.

2. The procedure can be readily adapted for cells grown in suspension culture. The cells are collected by centrifugation and washed with PBS(+) prior to the UV-irradiation step. The cells are then resuspended in a small

volume of PBS(+) (about 2 ml/10-cm plate) and are poured in culture dishes to form a very thin layer (1–3 mm) of cells for irradiation with UV light. The cells are collected by centrifugation, and the remaining steps are carried out as described in Ultraviolet Irradiation of Cell Monolayers, except that addition of the hypotonic RSB and of the detergents is carried out in the centrifuge tube.

3. Cross-linked RNP complexes have been used as antigens for the production of polyclonal and monoclonal antibodies against hnRNPs from HeLa cells,[9] hnRNPs and mRNPs from adenovirus-infected cells,[14] and the mRNA poly(A)-binding protein from yeast.[15]

4. This procedure has also been useful for the detection of changes in mRNA–protein interactions on treatment of cells with a variety of transcriptional modulators, as well as on viral infection.[8,13,14]

5. Under the conditions described in this chapter, UV-induced protein–protein cross-linking is not a problem. It needs to be kept in mind, however, that cross-linking may occur, and this should, if possible, be examined by the use of antibodies to specific proteins. The mobility of the cross-linked protein, by immunoblotting of SDS–PAGE gels, should be only slightly slower than that of the uncross-linked protein. The slower mobility, corresponding to an apparently higher molecular weight, is a characteristic of the cross-linked proteins and results from the residual cross-linked nucleotide(s).

6. UV light-induced cross-linking of proteins to RNA has also been successfully used for the identification of proteins contacting specific RNA molecules or RNA segments *in vitro*. The principle behind these procedures is essentially the same as that discussed previously for UV-induced cross-linking *in vivo*. The reaction mixtures usually contain extracts or protein mixtures, containing the RNA-binding proteins or the activity of interest, and continuously labeled RNA substrate (usually with ^{32}P), together with unlabeled, nonspecific competitor. The reaction is then placed in a minimal volume (an upside-down Eppendorf tube cap or a 24-well tissue culture plate are usually appropriate containers) under the UV light for different lengths of time, preferably on ice if the specific RNA–protein interaction allows this. We find that a 3- to 10-min exposure is sufficient in most cases. The reaction mixture is then digested with the RNase digestion cocktail, as described for RNP complexes cross-linked *in vivo,* and the cross-linked proteins, which become labeled by the transfer of residual radiolabeled nucleotides, can be analyzed by SDS–PAGE

[14] S. A. Adam and G. Dreyfuss, *J. Virol.* **61**, 3276 (1987).
[15] S. A. Adam, T. Nakagawa, M. S. Swanson, T. K. Woodruff, and G. Dreyfuss, *Mol. Cell. Biol.* **6**, 2932 (1986).

and can be visualized by autoradiography. Examples of this procedure can be found in the study of poly(A)-binding proteins in free mRNP and polyribosomes,[16] RNA–protein complexes in cell-free translation systems,[17] and in the identification of proteins interacting with specific RNA sequences, such as the polyadenylation signal AAUAAA.[18]

[16] B. Setyono and J. R. Greenberg, *Cell (Cambridge, Mass.)* **24**, 775 (1981).
[17] J. R. Greenberg and E. Carroll III, *Mol. Cell Biol.* **5**, 342 (1985).
[18] J. Wilusz and T. Shenk, *Cell (Cambridge, Mass.)* **52**, 221 (1988).

[29] Analysis of Ultraviolet-Induced RNA–RNA Cross-Links: A Means for Probing RNA Structure–Function Relationships

By ANDREA D. BRANCH, BONNIE J. BENENFELD, CYNTHIA P. PAUL, and HUGH D. ROBERTSON

An Overview of Ultraviolet-Induced Cross-Linking

Suitable RNAs for Cross-Linking Studies

In any group containing three or more RNAs, chances are good that one of the species could benefit from an analysis of the photoreactive sites lying within the species. To detect these sites, irradiation with ultraviolet (UV) light is used to produce a stable bond connecting two nucleotides. The susceptibility to UV cross-linking indicates that two nucleotides, which may be widely separated in the primary sequences, are very closely apposed in the native RNA. UV cross-linking takes advantage of photoreactive sites already built into the RNA and does not require any added chemical cross-linking agents, which might disrupt the native structure. Since non-Watson–Crick (tertiary) bonds are often essential components of UV-sensitive sites,[1-3] the mapping of photoreactive sites is one of the few techniques available for identifying elements of tertiary structure. In addition, the UV-induced covalent bond is a durable clamp, locking two parts of the molecule together, and thus indicating which of the many

[1] M. Yaniv, A. Favre, and B. G. Barrell, *Nature (London)* **223**, 1331 (1969).
[2] J. Atmadja, R. Brimacombe, H. Blöcker, and R. Frank, *Nucleic Acids Res.* **13**, 6919 (1985).
[3] A. D. Branch, B. J. Benenfeld, and H. D. Robertson, *Proc. Natl. Acad. Sci. U.S.A.* **82**, 6590 (1985).

possible secondary structures best depicts the native molecule.[4,5] Finally, UV cross-linking can be used to seek points of interaction between two RNA molecules, such as M1 RNA, the RNA subunit of RNase P, and tRNA precursors (pre-tRNA).[6]

Techniques are available for mapping UV-induced cross-links into a very wide variety of RNA molecules. The specific analytical procedures to be used will be dictated by both the size and the form of the RNA (whether the RNA is radiolabeled) and by the objective of the experiment.

To obtain maximum information about local bonding patterns, the UV-induced cross-link must be mapped to the nucleotide level. Direct RNA analysis will be needed. However, if cross-links are to be used only to determine which portions of the RNA are close together in the three-dimensional structure, indirect techniques, such as primer extension, may be employed. Our interest in the biologically active sites of RNA molecules and in the tertiary structures the sites often contain has led us to concentrate our UV studies on small RNAs because the requirements for direct RNA analysis are most easily met by RNAs smaller than 2000 bases. Furthermore, when working with relatively small RNAs, it is possible to use two-dimensional gel electrophoresis to purify the UV-cross-linked form of the RNA so that the impact of the UV-induced cross-link on RNA function can be studied. It is much more difficult to separate control and cross-linked forms when the RNAs are greater than about 2000 bases in length.

We have mapped cross-links in RNAs ranging in size from 55 to 1679 nucleotides. Our studies have included a number of infectious RNAs, such as viroids[3] and RNA of the agent responsible for δ hepatitis,[5] a RNA-processing site which is a substrate for cleavage by RNase III, many 5 S rRNAs, and a catalytic RNA, M1 RNA (which has been analyzed in collaboration with Drs. Sidney Altman and Cecelia Guerrier-Takada, Department of Biology, Yale University, New Haven, CT). UV irradiation has been conducted both on RNAs extracted from tissues or cells and on RNAs transcribed *in vitro*.

Although it is often most convenient to work with prelabeled RNAs, prepared by carrying out *in vitro* transcription in the presence of α-[32P]-labeled nucleoside triphosphates (NTPs) or by incubating cells in the presence of ortho[32P]phosphate, nonradioactive RNAs can be postlabeled at internal C residues by *in vitro* iodination or at termini by using

[4] W. Stiege, C. Glotz, and R. Brimacombe, *Nucleic Acids Res.* **11**, 1687 (1983).
[5] A. D. Branch, B. J. Benenfeld, B. M. Baroudy, F. V. Wells, J. L. Gerin, and H. D. Robertson, *Science* **243**, 649 (1989).
[6] S. Altman and C. Guerrier-Takada, personal communication.

polynucleotide kinase or RNA ligase. Indirect techniques can be applied to nonradioactive RNAs themselves. We provide examples of RNAs from a number of different categories.

While we typically irradiate RNA samples which have been freed of protein by prior extraction with phenol, photo-induced cross-linking can also be carried out on ribonucleoprotein (RNP) particles. The mapping of an extensive series of RNA–RNA cross-links introduced into rRNA, while the rRNA remained within intact ribosomal subunits, led to a detailed three-dimensional picture of 16 S rRNA.[2]

Technical Considerations

UV irradiation is used to introduce a new covalent bond into a RNA of interest. This bond marks a photoreactive site inherent to the RNA. The cross-link is stable to heating and all other treatments we commonly employ for RNA analysis.[7] Since UV irradiation can also cause nonspecific damage to RNA, it is best to minimize the exposure time. When care is taken to limit UV damage, the cross-linked derivative will typically constitute only 1–30% of the starting material, with the variation occurring because some photoreactive sites are much more susceptible to cross-linking than others.

The challenge of this procedure comes after the cross-link(s) have been introduced and their location must be determined. If the impact of the cross-link on RNA function is to be assessed, it will be necessary to purify the cross-linked form of the RNA as an intact molecule. For small RNAs, this purification can often be achieved by the use of two-dimensional gel electrophoresis. As indicated in Fig. 1, cross-linking transforms linear RNAs into α-shaped molecules. Cross-linking transforms circular RNAs, such as viroids, into figure 8-shaped molecules. Such changes in the overall geometry translate into altered gel mobility in gels containing urea, providing the basis for the use of two-dimensional gel electrophoresis[7,8] in the purification of cross-linked derivatives (see Two-Dimensional Gel Electrophoresis in the Preparation of Cross-Linked RNAs). Because two-dimensional gel electrophoresis is often one of the first techniques used to detect and study a cross-linked derivative, this procedure is described prior to the combination of methods needed to map the position of the UV-induced cross-link. We are presenting only the subset of mapping procedures we have used most commonly.

Many of an extremely wide variety of RNA analytical techniques may be needed to characterize fully a single cross-linked RNA. Both the struc-

---

[7] A. D. Branch, B. J. Benenfeld, and H. D. Robertson, *Nucleic Acids Res.* **13**, 4889 (1985).
[8] J. Schumacher, J. W. Randles, and D. Riesner, *Anal. Biochem.* **135**, 288 (1983).

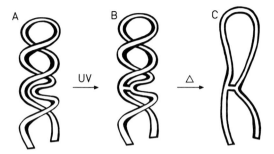

FIG. 1. UV-induced cross-linking within a tertiary structure. RNA containing a photo-reactive tertiary structure (A) is altered by the addition of a UV-induced covalent bond (B). The new bond creates a circular element. On denaturation, the molecule reveals its overall α-shaped geometry (C).

ture and the biological properties of an individual RNA must be considered in order to choose the most effective set of experiments for analyzing a cross-linked derivative. Since the functional tests to be carried out must be individually tailored to each RNA and thus comprise a tremendously diverse set of procedures, the tests could not be covered in this chapter. Instead, we have included references to a limited number of functional assays to illustrate an approach for using UV-cross-linking analyses to understand RNA structure–function relationships. The technique of RNA secondary analysis, an important component of our structural analyses (but one which must be adapted for each RNA), is treated in a similar manner.

Materials and Methods

The procedures described in this section are those which come into the preparation and analysis of UV-cross-linked RNAs at a number of different points. Methods which are more restricted in their application appear in later sections.

Preparation of Glassware

Many of the steps in a UV-cross-linking study require the analysis of small quantities of RNA. Glass tubes are used for most procedures because their clarity permits even minute amounts of RNA to be seen and followed. On request, 13 × 54-mm glass tubes can be purchased with a partial band of colored paint around their midsection (Dynalabs, Rochester, NY). This band provides a space for indelibly labeling the tube with a felt-tipped pen. To prepare the tubes for use, the tubes are first filled with 5% (v/v) dichlorodimethylsilane (Pierce Chemical Co., Rockford, Illinois) in chloroform, emptied, washed twice with distilled water, drained, and baked in a 150° oven for 2 hr. A similar procedure is used to prepare centrifuge tubes and dounce homogenizers.

Preparation of Carrier RNA

Carrier RNA [*Escherichia coli* tRNA (Sigma, St. Louis, MO)] is prepared by extracting 0.5-g aliquots with BDH phenol (Gallard-Schlesinger, Carle Place, NY) [saturated with 0.01 M Tris–HCl, pH 7.6, and 0.001 M ethylenediaminetetraacetic acid (EDTA)] three times, followed by two rounds of chromatography on Whatman cellulose CF11.[9] This carrier RNA is needed for many different steps. For example, the carrier RNA is used to facilitate ethanol precipitation of minute quantities of radiolabeled RNA and to prevent overdigestion during ribonuclease treatments. In addition, certain gel systems give more satisfactory results when samples contain at least 1 μg of RNA; this requirement is often filled by the addition of carrier RNA.

Autoradiography

Radioactive Ink. To make radioactive ink, all abandoned ^{32}P (usually about 1 mCi) which is considered to be unsuitable for further use in experiments, is pooled and added to 1 ml of Sheaffer black ink. Serial dilutions of the ink are made by mixing 0.1 ml of the ^{32}P-containing ink with 0.9 ml of ink from the stock bottle. A 10-μl Drummond microcap is filled with one of the dilutions and is placed into a small glass tube, which is then put into a scintillation vial. The Cerenkov counts are determined for a 0.1-min time interval. In this configuration, radioactive ink, which will be clearly visible on Fuji RX X-ray film after a 1-hr-long exposure, generates 10,000 counts/0.1 min. The intensity of all serial ink dilutions are calculated; inks are labeled and stored. Ink is applied to stickers with a superfine, flexible artists' pen [product number 9432, 102 Hunt Crowquill (Hunt Manufacturing Co., Statesville, NC)].

Exposure of Gels. For autoradiography of preparative-scale gels, one glass plate is removed at the end of electrophoresis. The gel and its supporting glass plate are wrapped in Saran Wrap and are mounted into a large manila folder (36 × 43 cm) using waterproof tape (Scotch 3M). Stickers, marked with ink bearing the appropriate level of radioactivity, are used to indicate the positions of the origin, dyes, and various addi-

---

[9] R. M. Franklin, *Proc. Natl. Acad. Sci. U.S.A.* **55**, 1504 (1966).

tional sites along the gel, which will be guides for precise gel band excision. Two pieces of film are placed over the gel and a weight is applied to ensure good contact. Eventually, bands of interest will be cut out of one film using a razor blade and then this film will be lined up over the preparative gel and will be used as a template for gel band excision. The other copy of the film will be retained as a record of the results of the experiment. We use either Kodak XAR or Fuji RX X-ray film. The Kodak film is about three times faster; the Fuji film survives storage better and gives cleaner backgrounds.

Gels carried out for strictly analytical purposes are often transferred from the bottom glass plate to a sheet of "junk" $35\frac{1}{2} \times 43$-cm X-ray film. To effectuate this transfer, the film is placed over the gel and then the assembly is inverted so that the gel falls onto the film. The glass plate is removed and then the gel is covered with Saran Wrap, using waterproof tape to seal the edges. The experiment is put under film in a cassette (Wolf) containing CRONEX HI-PLUS intensifying screens (Dupont, Wilmington, DE). The film is exposed at $-70°$ for up to 1 month.

Introduction of a Covalent Cross-Link by Irradiation with Ultraviolet Light

As a source of UV irradiation, we use the Model TS-15 of Ultra-Violet Products, Inc. (San Gabriel, CA), equipped with four shortwave tube bulbs (254-nm peak). This is a typical UV light box of the sort routinely used for looking at nucleic acids following staining with ethidium bromide (EtBr) and thus is a piece of equipment which is available in many laboratories. We first discovered a UV-sensitive site in viroid RNA because a cross-linked derivative was unavoidably produced during the purification procedure we used at the time, which involved the excision of EtBr-stained RNA gel bands. Considering that RNAs are often purified by cutting them out of gels under the guidance of UV light, it is interesting that the UV irradiation needed to identify the viroid RNA bands generated the cross-linked viroid at such a rapid rate that it was impossible to get the gel on and off the UV box fast enough to eliminate the photoreaction. Considering this history, it is important to point out that the UV-induced cross-linking we have studied is not dependent on EtBr.

Several different formats, each with its unique advantages, have been used for irradiation; nucleic acids can be placed on a sheet of Saran Wrap and can be irradiated in a drop, through the walls of a quartz cuvette, or while contained in a nondenaturing gel. To minimize heating during UV treatment, the surface of the UV box is prechilled by the application of plastic bags filled with ice and water. During all irradiations, UV exposure to the eyes and skin should be controlled by exercising care and by wearing UV-absorbing goggles and protective clothing.

Irradiation in a Drop

When this method is used, RNA samples are pipetted directly onto a sheet of Saran Wrap covering a prechilled UV box. For UV treatments longer than 1 min, 30-sec intervals of irradiation are alternated with 60-sec cooling periods, during which samples are removed to siliconized glass tubes and are maintained on ice.

One advantage of this format for UV treatment is that samples which are irradiated while resuspended in a drop of buffer or water can be directly tested for maintenance (or loss) of biological activity. This is important if functional studies of the cross-linked form of the RNA are planned for a later stage. Before going to the trouble of purifying a cross-linked derivative and before attempting to test its activity, it is beneficial to make sure in advance that the UV dose used to generate the cross-linked RNA did not inactivate the control molecules which are UV treated, but not cross-linked. As illustrated by the experiments shown in Figs. 2 and 3, it is possible to make this judgment by carrying out functional assays using a mixed population of control and cross-linked forms because the control molecules greatly outnumber the cross-linked ones.

Thus, M1 RNA, the catalytic subunit of *E. coli* RNase P,[10] was found to retain its capacity to cleave a pre-tRNA after 30 sec of irradiation in a drop of buffer containing magnesium.[11] Subsequently, two-dimensional gel electrophoresis revealed that the 30-sec treatment had produced a readily detectable level of a cross-linked derivative of M1 RNA (Fig. 3). Similar results were obtained when M1 RNA was irradiated in the presence of a Tris–acetate buffer,[12] although some loss of activity could be seen after 30 sec of irradiation, suggesting that it may be necessary to test a variety of buffers when seeking conditions for optimal cross-linking with minimal UV damage.

To carry out experiments in which loss of activity (due to nonspecific UV damage) and production of cross-linked derivatives must be monitored in parallel, it is useful to test the following irradiation times with the RNA at a concentration of 10 ng/μl: 0, 30, 90, 270, and 540 sec. Higher RNA concentrations will require longer irradiation times.

[10] C. Guerrier-Takada, K. Gardiner, T. Marsh, N. Pace, and S. Altman, *Cell* (*Cambridge, Mass.*) **35**, 845 (1983).
[11] A. D. Branch, B. J. Benenfeld, C. Guerrier-Takada, S. Altman, and H. D. Robertson, unpublished experiments.
[12] T. J. Morris and E. M. Smith, *Phytopathology* **67**, 145 (1977).

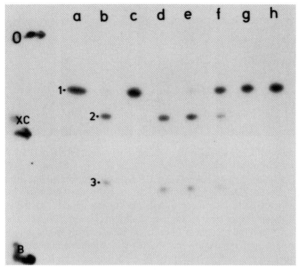

FIG. 2. Retention of the catalytic activity of *E. coli* M1 RNA after 30 sec of exposure to UV light. Samples of M1 RNA, the catalytic subunit of RNase P,[10] were resuspended in processing buffer (0.050 M Tris–HCl, pH 7.6, 0.060 M MgCl$_2$, and 0.10 M NH$_4$Cl), irradiated for various periods of time, and assayed for their ability to cleave a ^{32}P-labeled precursor of *E. coli* tRNA^{Tyr} (kindly provided by Drs. S. Altman and C. Guerrier-Takada), using conditions described previously.[10] At the end of a 30-min incubation at 37°, samples were fractionated in a 10% (w/v) polyacrylamide gel containing 7 M urea. The position of the pre-tRNA incubated in buffer alone is shown in lanes a and c; the position is marked band 1. The two products of bona fide M1 cleavage are identified in lane b as bands 2 and 3. The irradiation times of the M1 RNAs assayed in lanes d–h were as follows: lane d, 0 min; lane e, 30 sec; lane f, 90 sec; lane g, 270 sec; and lane h, 540 sec. O, The origin of electrophoresis; XC, the position of xylene cyanol dye marker; and B, the position of bromphenol blue dye marker.

As a final note, the cross-linked form of M1 RNA was recovered from two-dimensional gels (such as the one shown in Fig. 3B) and was found to be unable to cleave pre-tRNAs, while the UV-treated control M1 molecules carried out accurate cleavage.[11]

Irradiation in a Nondenaturing Gel

UV irradiation is often carried out, not as a drop of fluid as described previously, but after the first part of two-dimensional gel electrophoresis has been performed and the sample is immobilized in a gel strip. In this configuration, it is easier to prevent overheating during the UV exposure and, for samples requiring two-dimensional gel electrophoresis prior to any other analysis anyway (e.g., RNAs needing gel purification), no time

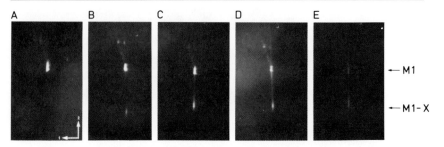

FIG. 3. Kinetics of UV-induced cross-linking of *E. coli* M1 RNA. M1 RNA (150 ng) was resuspended in 15 μl of M1-processing buffer (0.050 *M* Tris–HCl, pH 7.6, 0.060 *M* MgCl$_2$, and 0.10 *M* NH$_4$Cl) and then was exposed to UV light on a prechilled light box for (A) 0 sec, (B) 30 sec, (C) 90 sec, (D) 270 sec, or (E) 540 sec following a protocol in which 30-sec periods of irradiation were alternated with 1-min cooling periods. At the end of the UV treatment, 1-μl aliquots were assayed for their ability to cleave a ^{32}P-labeled pre-tRNA^{Try} (as described in Fig. 2) and the remainder of each sample was analyzed by two-dimensional gel electrophoresis. The horizontal and vertical arrows in A indicate the directions of first- and second-dimension separations, respectively. The positions of control and UV-cross-linked *E. coli* M1 RNA are identified by M1 and M1-X, respectively.

is lost by using this format for UV treatment. This approach to UV irradiation is described in detail as part of the method for two-dimensional gel electrophoresis presented in the following section.

Two-Dimensional Gel Electrophoresis in Preparation of Cross-Linked RNAs

Nondenaturing Gels for First-Dimensional Separation

Slab gels to be used for the first dimension of this procedure are made from a solution of acrylamide–bisacrylamide at a ratio of 30 : 1.5 (w/w), 0.040 *M* Tris–acetate, pH 7.2, 0.020 *M* sodium acetate, and 0.001 *M* EDTA.[12] For RNAs 50–100 bases long, gels containing 8% (w/v) polyacrylamide are used. For RNAs 100–400 bases long, gels containing 5% (w/v) polyacrylamide are used. For RNAs 400–1000 bases long, gels containing 4% (w/v) polyacrylamide are used. A standard slab gel is 16 × 14 × 0.15 cm. The sample wells are about 0.6 cm across and can accommodate up to 15 μg of nucleic acid. The following recipe can be used to make three to four gels, allowing about 30 ml for each gel, and will yield gels which are 5% (w/v) polyacrylamide/0.25% (w/v) bisacrylamide. To obtain gels of a different composition, the volumes of the 30% (w/v) acrylamide/1.5% (w/v) bisacrylamide stock solution and water are varied appropriately. The stock solution of acrylamide/bisacrylamide is filtered through

Whatman 3MM paper at the time the solution is made up and then is stored at room temperature. Once the gels have polymerized, the gels can be stored in a refrigerator for 1 week prior to use if kept in air-tight bags with enough water to provide moisture.

5% Nondenaturing Gels (120 ml Total Volume)

20 ml of 30% acrylamide/1.5% bisacrylamide [20 : 1 (w/w)]
12 ml of 10× Tris–acetate buffer: 10× Tris–acetate buffer contains 0.4 M Trizma base, 0.2 M sodium acetate, and 0.01 M EDTA, with glacial acetic acid added to give a pH of 7.2
87 ml of glass distilled water

After the solutions listed above are mixed, 1.0 ml of 10% (w/v) ammonium persulfate and 0.1 ml of TEMED are added in a fume hood

To prepare samples for gel electrophoresis, $\frac{1}{4}$ volume of a dye mix [0.002 M Tris–HCl, pH 7.6, 0.2% (w/v) bromphenol blue (BPB) dye, and 50% (w/v) sucrose] is added. For samples that contain only a small quantity of RNA, such as those with a very high specific activity, it is essential to add 1–5 μg of carrier RNA (E. coli tRNA, prepared as described in Materials and Methods). When loading samples onto the gel, it is important to leave at least 1 cm between the samples to allow space for later manipulations. Gels are run at 100 V, with a current of about 55–65 mA. In order to preserve native RNA structure, the gel temperature should not rise substantially above room temperature during the run.

Ultraviolet Treatment of RNA in a Gel Strip

Following electrophoresis, the top glass plate is removed from the slab gel, which is then carefully sliced into individual lanes using a fresh razor blade. The surface of the UV box is covered with Saran Wrap. The gel strip is placed on this protective layer and then covered with a second sheet of Saran Wrap. A plastic bag, filled with ice and water, is placed on top for cooling. Samples are irradiated for 1-min intervals, alternating with 30-sec cooling periods. Gel slices containing control samples are wrapped in Saran Wrap and are kept on ice, while others are irradiated.

Urea Gels for the Second-Dimension Separation

To prepare for the second-dimension separation, individual gel lanes are placed at the bottom of a glass plate, plastic side rails are placed in their customary positions, a top piece of glass is carefully applied, and then the assemblage is sealed into place with waterproof tape (Scotch 3M). Our standard gel is 16 × 14 × 0.15 cm. Also, similar to the gels used for the separation in the first dimension, the percentage of acrylamide in the gel used for the second dimension varies depending on the length and geometry of the RNA to be analyzed: 10% (w/v) polyacrylamide (acrylamide plus bisacrylamide) for RNAs 10–100 bases long, 10–6% (w/v) polyacrylamide for RNAs 100–200 bases long, 5% (w/v) polyacrylamide for RNAs 200–500 bases long, 4% (w/v) polyacrylamide for RNAs 500–1000 bases long, and 3.5% (w/v) polyacrylamide for RNAs 1000–2000 bases long. The optimal gel must often be determined by trial and error.

5% (w/v) Polyacrylamide Gels Containing 7 M Urea (100 ml of Total Volume)

42 g of urea
5 ml of 10× Tris–borate buffer: 10× Tris–borate buffer contains 0.9 M Trizma base, 0.9 M boric acid, and 0.03 M EDTA and has a pH of 8.2
12.5 ml of 38% acrylamide/2% bisacrylamide [19:1 (w/w)]
51.5 ml of distilled water

Once the ingredients listed above have been mixed and the urea has been dissolved, 0.1 ml of TEMED and 0.4 ml of 10% (w/v) ammonium persulfate are added in a fume hood

Immediately after pouring the second-dimension gel, gentle rapping is used to release any air bubbles, which might be attached to the first-dimension gel strip. Once the gel has polymerized, it is possible to supply additional dye markers by sticking a small amount of dye [supplied by a bit of cotton saturated with a solution of 99% (v/v) formamide and 0.01% (w/v) each of BPB and xylene cyanol (XC)] into the bottom of the urea gel. Fractionation in these gels is from bottom to top, requiring a reversal of the electrode leads. To provide an open electrical circuit, the integrity of the tape at the bottom of the gel must be breached; at least once a year, we forget to slit the tape. The ionic strength of the running buffer matches that of the urea gel and is ½× relative to the so-called 10× Tris–borate buffer, which is a widely used stock solution.[13] The lower ionic strength is used because the lower ionic strength often improves resolution of cross-linked RNAs. In order to achieve as much denaturation of the RNA as possible and thereby enhance the mobility differential between control and cross-linked forms, the gels are run with the accelerating voltage necessary to give the highest temperature compatible with intact gel plates. For a 5% (w/v) polyacrylamide–7 M urea gel, we use 300 V with 20–30 mA of current. When electrophoresis has been completed,

[13] A. C. Peacock and C. W. Dingman, *Biochemistry* **7**, 668 (1968).

RNAs are either stained with EtBr or are located by autoradiography (see Materials and Methods). Noting that cross-linked derivatives may migrate either more rapidly or more slowly than their non-cross-linked counterparts, RNAs of interest are excised and either are stored as gel bands at −20° in sealed containers or are directly extracted from the gels as described in Extraction of RNAs from Polyacrylamide Gels. To prepare a sample of a cross-linked RNA which is more than 90% pure, it is often necessary to fractionate the cross-linked RNA eluted from the two-dimensional gel in a third gel of a different composition, as illustrated in Ref. 7.

Extraction of RNAs from Polyacrylamide Gels

Extraction of RNAs Longer than 70 Bases

The RNA-containing gel band is placed into a siliconized dounce homogenizer (Bellco, Vineland, NJ) with 1 ml of 1× TSE (0.050 M Tris–HCl, pH 7.6, 0.1 M NaCl, and 0.001 M EDTA) and 0.5 ml of buffer-saturated BDH phenol (Gallard-Schlesinger, Carle Place, NY) (saturated with 0.010 M Tris–HCl, pH 7.6, and 0.001 M EDTA). In addition, radiolabeled RNA samples receive 10 μg of E. coli tRNA. Unlabeled samples do not. The gel piece is thoroughly crushed with a Teflon pestle and is pulverized by hand using 30 strokes with the pestle. If any gel pieces remain after 30 strokes, the process is continued. Using a siliconized Pasteur pipet, the sample is transferred to a 12-ml siliconized Sorvall centrifuge tube. The homogenizer and pestle are rinsed with 0.5 ml of 1× TSE by seven strokes with the pestle and then this solution is added to the sample in the 12-ml tube. The sample is then centrifuged for 5 min in a Sorvall GLC-3 tabletop centrifuge at 3000 rpm. The aqueous phase is transferred to a 30-ml siliconized Corex tube. The phenol and polyacrylamide layers are reextracted with 0.5 ml of 1× TSE and 0.2 ml of chloroform. The components are mixed by vortexing for 2 min and then are centrifuged as above. The aqueous phase is collected and pooled with that from the previous step. RNA is purified by chromatography on Whatman CF11 cellulose[9] as described in RNA Purification by Whatman CF11 Cellulose Chromatography.

Elution of Small RNAs

RNA fragments and very small RNAs (those less than 70 bases in length) can be recovered from polyacrylamide by soaking the gel band in 0.9 ml of 1× TSE or water overnight at 37°. Ethanol precipitation is an effective way to remove urea from RNAs eluted in this manner. Radiolabeled RNAs (in 0.9 ml of 1× TSE or water) are transferred by a siliconized Pasteur pipet to a siliconized 12-ml Sorvall centrifuge tube and 0.1 ml of 2 M sodium acetate, pH 5.2, 10 μg of tRNA, and 2.5 ml of chilled absolute ethanol are added. After 40 min at $-70°$ (and no longer, to avoid precipitation of the urea), RNAs are pelleted by centrifugation at 13,000 rpm for 30 min in a Sorvall SS-34 rotor. The RNA pellets are dried under vacuum and are resuspended in a convenient volume of water (usually 20–50 μl). This extraction procedure has proved extremely useful for the purification of ribonuclease-resistant oligonucleotides containing UV-induced cross-links.

RNA Purification by Whatman CF11 Cellulose Chromatography

To prepare a stock supply, 260 g of Whatman CF11 Fibrous Cellulose Powder (catalog number 1113) is pretreated with 2 liters of 0.1 M NaOH at room temperature for 30 min and then is washed 10 times in distilled water (taking care to eliminate the fine particles with each wash). The Whatman CF11 cellulose is stored beneath a large volume of 0.002 M EDTA at 4° or room temperature until used.

Small columns, with the capacity to bind a maximum of 50 μg of single-stranded RNA, are cast in GLASPAK (Becton-Dickinson, Rutherford, NJ) 1-ml sterile glass disposable tuberculin syringes. The plugs for these columns are made of little bundles of glass fiber (from a supply of glass fiber, which had previously been autoclaved with dry heat for 30 min). To cast the column, syringe barrels are first filled with glass-distilled water and then a thin slurry of cellulose CF11 is added and allowed to settle gently, yielding a final packed bed volume of 0.20 ml. To remove fine particles from the column, the column is washed successively with 1 ml of a degassed solution which is 50% ethanol/50% 1× TSE (1× TSE is 0.050 M Tris–HCl, pH 7.6, 0.1 M NaCl, and 0.001 M EDTA), 1 ml of 1× TSE, 1 ml of 50% ethanol/50% 1× TSE, and 8 ml of water. It is not always necessary to remove the fines thoroughly from the column; however, failure to do this may lead to RNA samples containing a considerable residue of CF11 cellulose.

To ready the column for RNA binding, the column is equilibrated with 50% ethanol/50% 1× TSE. RNA samples, in 1× TSE (salt is required for binding), are prepared for chromatography by the addition of an equal volume of absolute ethanol, to make the samples 50% ethanol/50% 1× TSE. Samples are loaded onto the column, and the flow-through fraction is collected into a scintillation vial. The column is then washed three times with 1 ml of 50% ethanol/50% 1× TSE and is allowed to run dry between each successive wash. The RNA is eluted into a 12-ml Sorvall centrifuge tube with 0.9 ml of distilled water and is concentrated by ethanol precipitation: to a 0.9-ml sample of a radiolabeled RNA, 10 μg of tRNA, 0.10 ml of 2 M sodium acetate, pH 5.2, and 2.5 ml of chilled absolute ethanol are added.

In some cases, the need for ethanol precipitation is circumvented by passing 1 ml of absolute ethanol (room temperature) over the column prior to eluting the sample with water, which is later removed by evaporation. The ethanol wash substantially reduces the amount of salt, which would otherwise be present in the eluate.

A similar protocol is used to purify RNA samples containing more than 50 μg of nucleic acid, except that the column size is adjusted upward in accordance with the rule that 1 cm³ of CF11 cellulose can bind 250 μg of single-stranded RNA. A bed containing 2 cm³ of CF11 cellulose can be cast conveniently in the barrel of a 5-ml syringe.

Mapping Ultraviolet-Induced Cross-Links

While additional procedures are usually required to locate a UV-induced bond in a large RNA molecule, a cross-link in a RNA smaller than 1000 nt can often be detected directly by a change the cross-link makes in the characteristic pattern of ribonuclease-resistant oligonucleotides. For example, as shown in Fig. 4, the cross-linked form of viroid RNA releases a novel RNase T1-resistant oligonucleotide not present in the one-dimensional pattern of oligonucleotides from control viroid. One- and two-dimensional oligonucleotide analyses have played an important role in the mapping of the UV-induced cross-links we have studied. One-dimensional oligonucleotide analysis is described in One-Dimensional Oligonucleotide Analysis Using Polyacrylamide Gels. Two-dimensional analysis (RNA fingerprinting), because of its greater complexity, is described in a separate chapter in this volume.[14] Both of these procedures are often coupled with RNA secondary analysis carried out as described by Barrell[15] and as discussed in RNA Secondary Analysis.

Generation of Ribonuclease-Resistant Oligonucleotides from Cross-Linked RNAs

General Considerations. Individual, disposable Drummond microcaps are used for measuring the ribonucleases. Enzyme stocks are made up in 0.01 M Tris–HCl, pH 7.6, and 0.001 M EDTA and are stored in a freezer designated for this purpose at −20°. Procedures involving ribonucleases

[14] A. D. Branch, B. J. Benenfeld, and H. D. Robertson, this volume [12].
[15] B. G. Barrell, *in* "Procedures in Nucleic Acid Research" (G. L. Cantoni and D. R. Davies, eds.), Vol. 2, p. 751. Harper, New York, 1971.

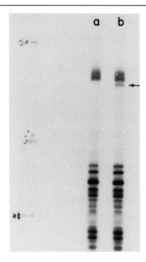

FIG. 4. Cross-linked viroid RNA contains a novel oligonucleotide. Circular (lane a) and UV-cross-linked (lane b) forms of potato spindle tuber viroid RNA were digested with RNase T1, were labeled with [γ-³²P]ATP and polynucleotide kinase, and then were fractionated in a 0.3-mm-thick 20% (w/v) polyacrylamide–7 M urea gel as described in the text. The arrow identifies an oligonucleotide found only in the digest of cross-linked viroid RNA. Radioactive ink marks, including one giving the position of the XC marker, were used to guide excision of the cross-link-specific oligonucleotide.

are conducted in a specified part of the laboratory, where only disposable glass- and plasticware are used. Digestions are carried out in the tips of drawn-out capillary tubes (Kimax, 0.8–1.10 × 100 mm) in 37° incubators: one reserved for RNase T1 and one reserved for pancreatic RNase A.

The conditions (salt, temperature, enzyme concentration, and time) used for ribonuclease digestion can affect the pattern of oligonucleotides released from a cross-linked RNA. We have found that ribonuclease digestion is often inhibited at a phosphodiester bond neighboring the UV-induced bond. Thus, it is frequently possible to purify a partial digestion product containing the UV-induced cross-link, greatly facilitating mapping. This phenomenon is exemplified by the cross-link-containing oligonucleotide in the RNase T1 digest of UV-cross-linked viroid RNA, which has the structure[3]

5'-GAAACCUG-3'
|
3'-GCCCAUCAUC-5'

As discussed in Special Conditions to Generate RNase T1 Partial Digestion Products, if necessary for mapping purposes, it is possible to manipulate the digestion conditions to obtain even larger fragments surrounding

the UV-induced bond. The exact protocol chosen for ribonuclease digestion is also strongly dependent on whether the RNA is already radiolabeled.

*Digestion of Radiolabeled RNAs. Digestion under standard con-
ditions.* For digestion under standard conditions, radiolabeled RNAs are
dried down in small silicon-coated glass tubes, almost always in the presence of 10 μg of tRNA to provide enough RNA to prevent overdigestion.
The RNA is resuspended in 2 μl of either 1 mg/ml of RNase T1 (Sankyo
Co., Calbiochem, San Diego, CA) (in a solution also containing 0.010 *M*
Tris–HCl, pH 7.6, and 0.001 *M* EDTA) or 0.25 mg/ml of pancreatic
RNase A (Worthington, Freehold, NJ) (also in 0.010 *M* Tris–HCl, pH 7.6,
and 0.001 *M* EDTA). RNase T1 digests are incubated for 45 min, while
pancreatic RNase digests are incubated for 10 min. To stop the digestion,
oligonucleotides to be separated in a polyacrylamide–urea gel are expelled into 4 μl of a formamide-dye mixture [99% (v/v) formamide and
0.01% (w/v) each of BPB and XC].

*Outcome of RNase T1 digestion carried out under standard
conditions.* When the standard conditions are used, RNase T1 cleaves
primarily after G residues and resolves almost all cyclic phosphate intermediates to 3′-phosphate groups. However, there are several exceptions
to this rule. First, certain A residues are nearly as susceptible to cleavage
by RNase T1 as standard G residues, leading to fragments ending with A
rather than G and providing some very confusing data at times.[16] Second,
as in the viroid cross-link, when the UV-induced bond actually involves a
G residue (the base specifying cleavage by RNase T1[17]), ribonuclease T1
may be reluctant to cleave the adjacent bond, giving rise to partial digestion products of the sort noted previously. Third, stable structural features may inhibit cleavage after a fairly large number (10 or more) of the G
residues contained within the features. In some cases, this can be very
helpful, yielding large partial digestion products containing the cross-link.
In the case of transcripts of the δ agent, this partial digestion product
generates such a large radioactive signal that one-dimensional oligonucleotide analysis can be carried out directly on the UV-treated sample,
obviating the need to first purify the cross-link-containing RNA molecule
(Fig. 5).

RNase T1 digestion at elevated temperature. The oligonucleotide constituents of the partial digestion product of δ RNA can be released by
altering the digestion conditions and by treating the RNA fragment with 1
mg/ml of RNase T1 at 65° for 1 hr in a sealed drawn-out capillary tube

[16] A. D. Branch, B. J. Benenfeld, and H. D. Robertson, unpublished experiments.
[17] K. Sato and F. Egami, *J. Biochem. (Tokyo)* **44,** 753 (1957).

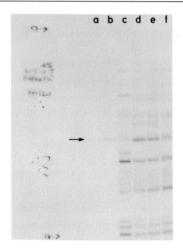

FIG. 5. A cross-link-specific oligonucleotide detected in UV-treated RNA transcripts of the δ agent. A partial cDNA clone of the δ hepatitis agent[5] was transcribed *in vitro* and mixed with tRNA, and then aliquots either were directly digested with RNase T1 at 37° (lane c) or were irradiated while in a small volume of water for various times (lane d, 1 min; lane e, 2 min; and lane f, 4 min prior to digestion with RNase T1). Faint markers for the cross-link-specific oligonucleotide appear in lanes a and b, as indicated by the arrow. Radioactive hieroglyphs were used to guide gel band excision.

(Fig. 6). If a tight structural element elsewhere in the molecule gives rise to a partial digestion product which interferes with the detection of a cross-link-specific oligonucleotide (e.g., because the two comigrate), the problem can be overcome by carrying out the digestion at 65°, providing that the higher temperature does not eliminate the possibility of identifying the cross-link-specific digestion product.

Outcome of pancreatic RNase A digestion under standard conditions. Just as the standard conditions for RNase T1 digestion are a compromise, so are the standard conditions for pancreatic RNase A digestion. The standard conditions minimize cleavages after A residues, while permitting efficient cleavage after C and U residues. However, some long A-rich oligonucleotides may suffer nicking and thus appear in less than their molar amounts. In addition, many of the oligonucleotides may occur as doublets with a 3'-phosphate on one component and a 2',3'-cyclic phosphate on the other, due to incomplete resolution of the cyclic phosphate intermediate. If either of these problems is acute, it may be necessary to raise or lower the ribonuclease concentration or to vary the digestion time. It is important to keep the presence of a cyclic phosphate

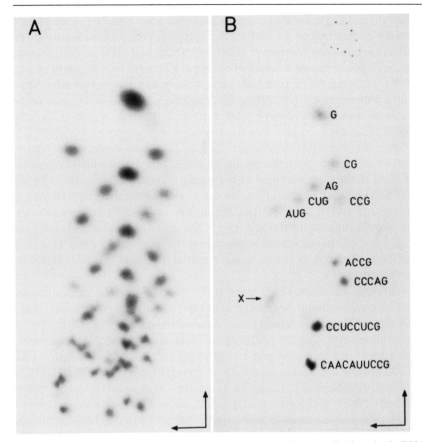

FIG. 6. Elevated temperature is needed to cleave near the cross-linking site in RNA transcripts of the δ agent. To provide material for RNA fingerprinting analysis, a cross-linked form of a δ RNA transcript was purified by two-dimensional gel electrophoresis,[5] digested with RNase T1 at 37° (as described in Fig. 5), and from the resulting fragments, a cross-link-specific partial digestion product was purified. This species was treated with RNase T1 at 65°, and then the constituents of the partial digestion product were fractionated into a two-dimensional pattern (B) and were compared to a fingerprint of the full-length cross-linked RNA transcript (A). Secondary analysis[15] was used to identify the oligonucleotides released from the cross-link-specific partial digestion product. X, Identifies the spot containing the UV-induced cross-link. The UV-sensitive site occurs in a conserved region of the genome of the δ agent and brings together two widely spaced segments (one including bases 703–735 and the other including bases 856–876).[5] The pairs of horizontal and vertical arrows in A and B indicate the directions of first- and second-dimension separations, respectively.[14]

group in mind, because the group may become a factor if secondary analysis is carried out as part of the mapping procedures. Because the 2',3'-cyclic phosphate group is resistant to both calf alkaline phosphatase and the 3'-phosphatase activity of nuclease P1, these two standard assays for the presence of a terminal phosphate group may give misleading results. Furthermore, paralleling the situation with RNase T1, pancreatic RNase A may fail to cleave after a C or U residue which is directly involved in a cross-link.

Special conditions to generate RNase T1 partial digestion products. In order to map cross-link-containing oligonucleotides which cannot be identified in digests produced either under standard conditions or by digestion at elevated temperature, we are exploring the use of conditions at the opposite end of the spectrum: conditions specifically designed to generate partial digestion products. Since the cross-link generally stabilizes the portion of the molecule surrounding the cross-link, there is a high probability that one of the prominent partial digestion products will contain the cross-link.

Although conditions will need to be worked out for each different RNA, we recommend beginning with the following protocol. Dry down the samples (the cross-linked form of the RNA and the irradiated, but not cross-linked, RNA) in the presence of 10 μg of carrier tRNA. Resuspend each sample in 3 μl containing 2 μl of water and 1 μl of 5× PRB (5× PRB is 1 M NaCl, 0.1 M MgCl$_2$, and 0.25 M Tris–HCl, pH 7.5[18]). Then add 2 μl of 33-mg/ml RNase T1. Digest at 0° for 1hr. At the end of the digestion, add 5 μl of 99% (v/v) formamide dye and, using a low voltage (to maintain a moderate temperature), fractionate the samples in a 1× Tris–borate and 15% (w/v) ultrathin polyacrylamide gel, containing 7 M urea, until the BPB has migrated 25–30 cm. Elute RNA from the gel bands which are more prominent in the digest of the cross-linked RNA and digest them with 1 mg/ml of RNase T1 at 65°. Separate the products in a 20% (w/v) polyacrylamide–7 M urea gel (as described later) or by RNA fingerprinting.[14] Apply RNA secondary analysis[15] as necessary to map the partial digestion products into the genome and to locate the UV-induced bond. A similar approach is used to generate partial digestion products of pancreatic RNase A.

Digestion and Kinase Labeling of Nonradioactive RNAs. Special conditions are needed for nonradioactive RNAs, whose oligonucleotides are to be labeled after digestion by incubation in the presence of polynucleotide kinase and [γ-^{32}P]ATP. Such RNAs are often available in only limited

[18] J. M. Adams, P. G. N. Jeppesen, F. Sanger, and B. G. Barrell, *Nature (London)* **223,** 1009 (1969).

amounts (far less than 1 μg) and thus require a reduced concentration of ribonuclease to avoid overdigestion. For example, 2 μl of 30-μg/ml RNase T1, 0.010 M Tris–HCl, pH 7.6, and 0.001 M EDTA are used to digest 5–500 ng of RNA. Incubation is carried out in the tip of a drawn-out capillary tube at 37° for 30 min. After digestion, samples in the 200- to 500-ng range are treated with 0.15 U of calf alkaline phosphatase [(Boehringer Mannheim, Indianapolis, IN) purified by chromatography on Sephadex G-75 prior to use] for 15 min at room temperature and then for 15 min at 37° to remove the 3′-phosphate groups produced by ribonuclease digestion. Samples are then sealed into drawn-out capillary tubes and are heated in a bath of boiling water for 3 min to inactivate the phosphatase.

Reactions to label the 5′ termini are carried out in 25-μl volumes containing the RNA digest, 15–20 U of polynucleotide kinase (Pharmacia, Piscataway, NJ), 0.010 M Tris–HCl, pH 8.0, 0.010 M MgCl$_2$, 0.015 M 2-mercaptoethanol, and 1–10 μCi of [γ-^{32}P]ATP (Amersham, Arlington Heights, IL). Reaction mixtures are incubated in a 37° waterbath for 45 min. A one-dimensional analysis of kinase-labeled oligonucleotides from control and cross-linked viroid RNA is shown in Fig. 4.

Samples in the 5- to 200-ng range do not require treatment with calf alkaline phosphatase, because even the relatively weak phosphatase activity provided by the kinase preparation is sufficient to completely remove the 3′-phosphate groups. The calf alkaline phosphatase treatment is needed for samples containing larger amounts of RNA to avoid a situation in which the weak phosphatase associated with the kinase preparation would partially remove the 3′-phosphate groups, causing the digestion products to be a mixture of the two forms.

Best results require compatible amounts of RNase T1, RNA, and [^{32}P]ATP. When it is not possible to obtain an accurate estimate of the RNA concentration, it may be necessary to test a variety of conditions.

One-Dimensional Oligonucleotide Analysis Using Polyacrylamide Gels

Ultrathin "sequencing" gels[19] produce excellent resolution of RNA oligonucleotides as illustrated in Figs. 4 and 5. These gels are cast between especially thick glass plates [5 mm (Swift Glass Co., Elmira, NY)] to prevent cracking during electrophoresis, which is carried out at an elevated temperature. The inside surface of one of the two glass plates must be siliconized before assembly to permit the manipulations which come later. To silicon-coat the glass plate, 1 ml of a solution containing 5% (v/v) dichlorodimethylsilane in chloroform is squirted onto a pre-cleaned plate in a fume hood and is spread around with a Kimwipe. The

[19] F. Sanger and A. R. Coulson, *FEBS Lett.* **87**, 107 (1978).

glass plate is then rinsed with distilled water and air-dried. Gels are poured between glass plates measuring 20 × 40 × 0.5 cm, which are separated by 0.3-mm-thick side rails. The comb and spacers are cut from plastic sheets [Slater's (Plastikard) Ltd., Matlock Bath, Derbyshire]. We make 16-slot combs. The gels themselves are 20 × 40 cm and 0.3 mm thick; the recipe given below can be used to make up two gels.

20% (w/v) Polyacrylamide Gels Containing 7 M Urea
(100 ml of Total Volume)

42 g of urea
5 ml of 10× Tris–borate buffer: 10× Tris–borate buffer contains 0.9 *M* Trizma base, 0.9 *M* boric acid, and 0.030 *M* EDTA and has a pH of 8.2
50 ml of 38% (w/v) acrylamide/2% (w/v) bisacrylamide [19 : 1 (w/w)]
14 ml of distilled water

The ingredients listed above are stirred until the urea is dissolved, and then 0.1 ml of TEMED and 0.4 ml of 10% (w/v) ammonium persulfate are added to the solution while the solution is in a fume hood. All components are mixed briefly.

When the samples are ready to be loaded onto the gel, the comb is removed, and the wells of the gel are immediately cleared by squirting them vigorously with running buffer (0.5× Tris–borate buffer) using a Pasteur pipet. Sample volumes are usually between 4 and 8 μl [1 part RNA digest and 3 parts a formamide-dye mixture which is 99% (v/v) ultrapure formamide (BRL, Gaithersburg, MD) and 0.01% (w/v) each of BPB and XC]. Samples are loaded onto the gel through drawn-out capillary tubes with particularly long tips. Gels are run in 0.5× Tris–borate buffer at 1400 V, usually until the BPB dye has migrated 30 cm. Autoradiography is carried out as described in Materials and Methods. Fragments are eluted from excised gel bands as described in Elution of Small RNAs.

Primer Extension: An Indirect Technique for Mapping
Ultraviolet-Induced Cross-Links

Primer extension is useful for mapping cross-links, especially when employed in combination with direct RNA analysis. However, we try to avoid relying on this, or other, indirect techniques as the sole means for mapping a cross-link because of the uncertainties tied to these methods. For example, reverse transcriptase may not copy precisely to the last nucleotide before a cross-link and/or the cross-linked form of a RNA may have a strong stop for reverse transcriptase which is located far from the actual cross-link but which results from the increased structural stability conferred to the RNA by the cross-link. The following protocol for primer extension is adapted from one developed by Sänger and colleagues for the synthesis of full-length copies of viroid RNA, a molecule notorious for its resistance to reverse transcriptase copying.[20]

The primer extension reactions are carried out in 1.5-ml microfuge tubes. For each primer extension reaction, an individual microfuge tube is set up that contains the following.

1 μl (0.020 μg) of a [^{32}P]kinase-labeled 20-base-long DNA primer
6.25 μl of 2× dNTPs: a 45-μl cocktail of 2× NTPs contains 7.2 μl of 20 mM dATP, 7.2 μl of 20 mM dCTP, 7.2 μl of 20 mM dGTP, 7.2 μl of 20 mM dTTP, and 16.2 μl of water
2.5 μl of 1 M Tris–HCl, pH 8.3
0.75 μl of water
2.0 μl (0.2 μg) of RNA

To bind the DNA to the RNA, the 12.5-μl sample is heated in an 80° water bath for 4 min and then is quenched immediately in ice water. To each microfuge tube add 2.5 μl of 0.1 M dithiothreitol (DTT), 1.5 μl of 0.1 M MgCl$_2$, 1.9 μl of 0.2 M KCl, 0.5 μl of Na$_4$P$_2$O$_7$, 0.25 μl of actinomycin D (10 ng/μl), 2.0 μl of reverse transcriptase [9 U/μl (Promega, Madison, WI)], and 3.85 μl of water. These additions bring the total reaction volume up to 25 μl. Samples are incubated at 37° for 2 hr and then the products are analyzed by gel electrophoresis.

RNA Secondary Analysis

RNA secondary analysis can be used to define the precise location of a UV-induced cross-link once the approximate location of the cross-link has been determined by other approaches, such as one- and two-dimensional oligonucleotide analyses, primer extension studies, and additional tests.[3] RNA secondary analysis is one of the techniques for RNA sequencing worked out by Sanger and colleagues.[21] In this method, treatment with a battery of enzymes is used to analyze purified oligonucleotides.[15] RNA secondary analysis relies heavily on knowledge of enzyme specifities. Some of the less well-known properties of the enzymes commonly used for RNA studies are presented in Digestion of Radiolabeled RNAs.

We illustrate in this section how RNA secondary analysis was used to map the position of a novel structural element in 5 S rRNA from HeLa

[20] M. Tabler, M. Schnölzer, and H. L. Sänger, *Biosci. Rep.* **5**, 143 (1985).
[21] F. Sanger, G. G. Brownlee, and B. G. Barrell, *J. Mol. Biol.* **13**, 373 (1965).

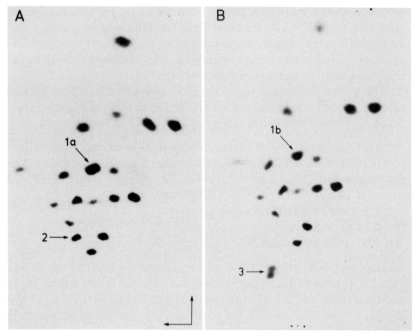

FIG. 7. Fingerprints of HeLa 5 S rRNA and its UV-cross-linked derivative. Pancreatic RNase A-resistant oligonucleotides of 5 S rRNA (A) and UV-cross-linked 5 S rRNA (B) were fractionated to give a two-dimensional array, as described.[3,14] RNA was eluted from the spots designated by the arrows [spots 1a and 2 (A) and spots 1b and 3 (B)] and was characterized by secondary analysis as described in the text. The horizontal and vertical arrows in A indicate the directions of high-voltage electrophoresis and ascending RNA homochromatography, respectively.

cells. Spots were eluted from the pancreatic RNase A fingerprints of 5 S rRNA shown in Fig. 7 and analyzed. The fingerprint of cross-linked 5 S rRNA was missing spot 2 (Fig. 7A), deduced to have the sequence GGGAAU by its position in the fingerprint and because RNase T1 digestion released AAU and G. Spot 1a (Fig. 7A), from the fingerprint of control 5 S rRNA, was found to contain both GAU and AGU. The analogous spot from cross-linked 5 S rRNA, spot 1b (Fig. 7B), contained only GAU and lacked AGU, an oligonucleotide that occurs only once in the sequence of HeLa 5 S rRNA.[22] RNase T1 products of spot 3 (Fig. 7B), which was specific to the fingerprint of cross-linked 5 S rRNA, included AAU, AG, G, and an additional product with an anomalous mobility. The release of AG indicated that the U residue of the oligonucleotide AGU

[22] B. G. Forget and S. M. Weissman, *J. Biol. Chem.* **244**, 3148 (1969).

was part of the cross-link, while the release of AAU implicated one of the three G residues from the oligonucleotide GGGAAU. Secondary analysis of spots from the RNase T1 fingerprint of control and cross-linked 5 S rRNA identified the third G residue as the one making up the second half of the UV-induced bond.[3]

Conclusions

We wish to stress three points about UV-sensitive sites in RNA and their application to structure–function studies. First, as already discussed in this chapter, the analysis of a UV-induced cross-link requires choices to be made among the many techniques available for RNA analysis. For example, enzymatically active RNAs, such as M1 RNA, might need one approach, while those with a structural role, such as 5 S rRNA, might require another approach. In addition, the protocols developed for mapping cross-links will differ depending on the sequence of the RNA and on the identity of the residues forming the UV-induced bond. In general, we have found that direct, RNA-level analysis gives the best chance for mapping the structural feature manifested by UV cross-linking to the nucleotide level.

The second point is as follows: the analyses of UV-sensitive sites— similar to all detailed studies of higher order structure in RNA molecules—are in their early stages. The similarities between the UV-sensitive sites in viroid and 5 S rRNAs[3] suggest that the tertiary elements in these two molecules have revealed a structural motif which may have a widespread distribution. To test this idea and explore the possibility that additional classes of UV-sensitive sites can be identified, UV-induced cross-links need to be evaluated in many more RNAs. A number of groups (including those of R. Brimacombe, C. Zwieb, T. R. Cech, and S. Altman and C. Guerrier-Takada) are working in this area, so progress can be expected soon. As of now, only one type of RNA molecule, tRNA, has been analyzed by X-ray crystallography. In that case, the previously known UV-sensitive site, which had originally indicated how the tRNA molecule is folded, was precisely mapped and its highly favored chemistry was explained. During the next 5 years, it is likely that a number of additional RNA molecules will have their structures solved by X-ray techniques, making it possible to place a number of the UV cross-links in their exact orientation in the native molecule.

Finally, improved techniques[23] are allowing the synthesis of milligram quantities of the RNA domains within which UV cross-linking takes

[23] J. F. Milligan, D. R. Groebe, G. W. Witherell, and O. C. Uhlenbeck, *Nucleic Acids Res.* **21,** 8783 (1987).

place.[24] This opens the way for precise chemical identification of the bonds formed at UV cross-linking sites and for physical studies of their structural context before and after the UV-cross-linking process. Using this approach, the analysis of UV-sensitive sites will provide detailed information about the intricate bonding patterns present in biologically active RNA molecules.

[24] A. D. Branch, B. J. Benenfeld, C. P. Paul, and H. D. Robertson, unpublished experiments.

[30] Analysis of Splicing Complexes and Small Nuclear Ribonucleoprotein Particles by Native Gel Electrophoresis

By MARIA M. KONARSKA

In recent years, native gel electrophoresis has become the standard technique for studying nucleic acid–protein interactions. Perhaps the most popular application of this technique has been the DNA–protein "gel shift" assay, in which detection of a DNA-binding protein associated with its cognate DNA sequence is based on the retarded mobility of the complex as compared to the free DNA fragment. The first quantitative assays for DNA-binding proteins were developed by Gardner and Revzin[1] and Fried and Crothers[2–4] utilizing the bacterial *lac* operon regulatory system as a model. A similar technique had been used previously for the electrophoretic separation of nucleosomes in agarose gels.[5,6] Subsequently, a modification of this method using composite agarose–acrylamide gels had been applied in studies of ribosomal particles.[7,8] Recently, a similar technique of electrophoresis in low-percentage native polyacrylamide gels has been successfully applied in analysis of the interaction between a RNA fragment and a protein factor.[9]

[1] M. M. Gardner and A. Revzin, *Nucleic Acids Res.* **9**, 3047 (1981).
[2] M. Fried and D. M. Crothers, *Nucleic Acids Res.* **9**, 6505 (1981).
[3] M. Fried and D. M. Crothers, *J. Mol. Biol.* **172**, 241 (1984a).
[4] M. Fried and D. M. Crothers, *J. Mol. Biol.* **172**, 263 (1984b).
[5] A. Varshavsky, V. V. Bakayev, and G. P. Georgiev, *Nucleic Acids Res.* **3**, 477 (1976).
[6] V. V. Bakayev, T. G. Bakayeva, and A. Varshavsky, *Cell (Cambridge, Mass.)* **11**, 619 (1977).
[7] A. C. Peacock and C. W. Dingman, *Biochemistry* **7**, 668 (1968).
[8] G. H. Goodwin and A. F. Dahlberg, *in* "Gel Electrophoresis of Nucleic Acids" (D. Rickwood and B. D. Hames, eds.), p. 199. IRL Press, Washington, D.C., 1982.
[9] E. A. Liebold and H. N. Munro, *Proc. Natl. Acad. Sci. U.S.A.* **85**, 2171 (1988).

In this chapter, we describe the application of native gel electrophoresis in studies of RNA–protein interactions involved in generation of large ribonucleoprotein (RNP) complexes during precursor mRNA (pre-mRNA) splicing. Involvement of small nuclear ribonucleoprotein (snRNP) particles in splicing of nuclear precursor RNA (preRNA) is now well established.[10–13] Splicing of preRNA occurs in a large multicomponent complex, termed the spliceosome.[14–16] Analysis of splicing complexes resolved by electrophoresis in native polyacrylamide gels has shown that the assembly of the spliceosome is based on sequential, multiple interactions of snRNP particles and preRNA[17–20] (Fig. 1). In particular, at least U2, U4, U5, and U6 snRNP particles interact stably with preRNA and/or each other during spliceosome formation. This process is assisted by various protein factors interacting with preRNA, including heterogeneous nuclear ribonucleoproteins (hnRNPs)[21] and probably multiple other proteins, which may provide the specificity necessary for splicing. The first stable, splicing-specific complex to form, A, is generated by binding of U2 snRNP to the region upstream of the 3′ splice site in pre-RNA. In the next step, a larger complex, B, is formed, apparently by association of complex A with a multi-snRNP particle containing U4, U5, and U6 small nuclear RNAs (snRNAs).[20] Improved resolution of large splicing complexes has revealed the presence of another complex, C, which contains the splicing RNA intermediates, the 5′ exon, and the lariat form of intervening sequence–3′ exon[22,23] (M. M. Konarska, unpublished observations). Conversion of complex B to C is apparently correlated with the release of the U4 snRNP particle from the spliceosome.[18,23,24] The final step of splicing results in release of the ligated exons and generation

[10] R. A. Padgett, P. J. Grabowski, M. M. Konarska, S. Seiler, and P. A. Sharp, *Annu. Rev. Biochem.* **55,** 1119 (1986).
[11] M. R. Green, *Annu. Rev. Genet.* **20,** 671 (1986).
[12] T. Maniatis and R. Reed, *Nature (London)* **325,** 673 (1987).
[13] C. Brunel, J. Sri-Widada, and P. Jeanteur, *Prog. Mol. Subcell. Biol.* **9,** 1 (1985).
[14] E. Brody and J. Abelson, *Science* **228,** 963 (1985).
[15] P. J. Grabowski, S. Seiler, and P. A. Sharp, *Cell (Cambridge, Mass.)* **42,** 345 (1985).
[16] D. Frendewey and W. Keller, *Cell (Cambridge, Mass.)* **42,** 355 (1985).
[17] C. W. Pikielny and M. Rosbash, *Cell (Cambridge, Mass.)* **45,** 869 (1986).
[18] C. W. Pikielny, B. C. Raymond, and M. Rosbash, *Nature (London)* **324,** 341 (1986).
[19] M. M. Konarska and P. A. Sharp, *Cell (Cambridge, Mass.)* **46,** 845 (1986).
[20] M. M. Konarska and P. A. Sharp, *Cell (Cambridge, Mass.)* **49,** 763 (1987).
[21] G. Dreyfuss, *Annu. Rev. Cell Biol.* **4,** 415 (1986).
[22] A. I. Lamond, M. M. Konarska, and P. A. Sharp, *Genes Dev.* **1,** 532 (1987).
[23] A. I. Lamond, M. M. Konarska, P. J. Grabowski, and P. A. Sharp, *Proc. Natl. Acad. Sci. U.S.A.* **85,** 411 (1987).
[24] S.-C. Cheng and J. Abelson, *Genes Dev.* **1,** 1014 (1987).

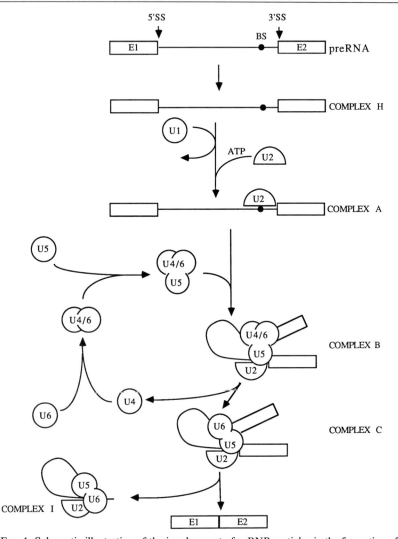

FIG. 1. Schematic illustration of the involvement of snRNP particles in the formation of splicing complexes. The preRNA substrate contains two exons (E1 and E2) separated by an intervening sequence. 5'SS, 3'SS, and BS indicate 5' and 3' splice sites and branch sites, respectively. Relative positions and molar ratios of snRNPs in various complexes are arbitrary.

of complex I, which contains the lariat form of the intervening sequence bound to U2, U5, and U6 snRNPs.[20]

As a substrate in our studies, we have used RNA containing the first two leader exons of the adenovirus type 2 major late transcription unit, separated by a shortened form of the first intervening sequence.[20] Another substrate used routinely in our laboratory contains the second and third exons of rabbit β-globin RNA separated by a shortened form of the second intron.[22,23] These RNAs, prepared by *in vitro* transcription of plasmid DNA templates with SP6, T3, or T7 RNA polymerases, are incubated under appropriate conditions with a nuclear extract of HeLa cells.[25] The splicing complexes formed on such preRNA substrates *in vitro* are separated in native polyacrylamide gels. Two methods of monitoring the complexes are used. First, a ^{32}P-labeled preRNA is used in the splicing reaction, and detection of the assembled complexes is simply achieved by autoradiography of the gel. Alternatively, an unlabeled preRNA serves as a substrate in the reaction. Splicing complexes are then resolved in a native gel, are transferred onto a nylon membrane, and are detected by Northern hybridization with a [^{32}P]RNA probe complementary to preRNA. This second method, although more laborious and time-consuming, offers greater experimental flexibility, allowing for the detection of all RNA components of analyzed complexes for which RNA probes are available. Specifically, the snRNP components of splicing complexes can be identified in this way. In addition, the distribution of endogenous snRNP complexes, not only those associated with preRNA, can be analyzed by this technique.[20,24]

Materials

Plasmids

pBSAd1: Containing the first two leader exons, the shortened form of the first intervening sequence, and part of the second intervening sequence[20]

pBSAL4: Containing the second and third exons of rabbit β-globin, separated by a shortened form of the second intervening sequence[22]

pBSU1, pBSU2b, pBSU4, and pBSU6a: Containing sequences complementary to snRNAs U1, U2, U4, and U6, respectively[20]

pSPU5b: Containing nucleotides 1–79 of human U5 snRNA, with an inserted G at position 14, cloned into pSP65 vector (kindly provided by Drs. Doug Black and Joan A. Steitz)

Restriction enzymes and T3, T7, and SP6 RNA polymerases: Commercially available enzymes are adequate for the described procedures

Nuclear extract of HeLa cells: Prepared as described by Dignam *et*

[25] J. D. Dignam, R. M. Lebovitz, and R. D. Roeder, *Nucleic Acids Res.* **11**, 1475 (1983).

al.,[25] except that MgCl₂ is omitted from the dialysis buffer. Both nuclear and whole-cell extracts from several other sources (BSC-1, F-9, WEHI, 70Z, and BJAB cells) have also been successfully analyzed by the native gel technique

Preparation of Sample

A typical reaction mixture (10 μl) contains 1 mM ATP, 5 mM creatine phosphate, 2 mM MgCl₂, 30 mM KCl, 6 mM HEPES, pH 7.6, 0.1 mM ethylenediaminetetraacetic acid (EDTA), 0.5 mM dithiothreitol (DTT), 6% (v/v) glycerol, and 20–30% (v/v) of HeLa nuclear extract prepared as described by Dignam *et al.*,[25] except that MgCl₂ is omitted from the dialysis buffer. Splicing reactions contain in addition a ³²P-labeled preRNA [typically 2 × 10⁴ cpm (counts per minute), ~5 × 10⁻⁴ pmol] for direct detection of complexes in a gel or unlabeled RNA (~0.18 pmol) for Northern analysis. The optimal length of the RNA substrate is estimated to fall between 100 and 500 nt (nucleotides). Splicing complexes formed on longer fragments are difficult to resolve in a gel. In some cases, when splicing complexes need to be separated from the endogenous pool of snRNPs, it is advisable to use RNA substrates longer than 300 nt. Under these conditions, splicing complex A migrates in a gel more slowly than the endogenous, unbound U2 snRNP or the U4/5/6 snRNP complex. RNA substrates, typically prepared by *in vitro* transcription with T3, T7, or SP6 RNA polymerases, are gel purified and are stored at −20°. The detailed protocols for RNA transcription reactions are presented elsewhere in this volume.[26]

Following an appropriate incubation (usually 10–60 min at 30°), the reaction mixture is loaded directly onto a native gel. No addition of loading buffer to the sample is necessary, since glycerol present in the extract provides sufficient density. However, the addition of xylene cyanol FF and bromphenol blue dyes [0.1% each (w/v)] loaded in the same or in parallel lanes of a gel is useful in monitoring the electrophoresis.

It should be noted that loading excessive amounts of nuclear extract proteins in a single well results in uneven distribution of RNP complexes due to precipitation of material at the top of the gel. Typically 2–3 μl (50–80 μg of protein) of a HeLa nuclear extract can be conveniently separated on a 1.5 × 5-mm gel slot; however, the exact amount of an extract that can be loaded on a gel must be determined experimentally.

[26] J. K. Yisraeli and D. A. Melton, this volume [4].

Treatment with Heparin

Nuclear extracts contain large amounts of proteins that bind to negatively charged polyanions, such as nucleic acids. In order to suppress this nonspecific electrostatic interaction, a polyanion, heparin, is added to splicing reactions. Heparin treatment is thought to destabilize various nonspecific interactions between proteins and nucleic acids, thus greatly improving separations of stable RNP complexes.[19,24] This treatment is similar to the addition of nonspecific DNA competitor frequently used in studies of DNA–protein interactions.[27–29] It is important to note, however, that, unlike the competitor DNA in such studies, heparin is not present during the reaction, but is added immediately prior to electrophoresis. For example, heparin treatment is very useful in resolving eukaryotic ribosomal particles in native polyacrylamide gels. Splicing complexes, once formed, are resistant to heparin treatment (5 mg/ml of final concentration, 5–10 min at 30° or on ice). The size of such heparin-treated complexes, as judged by glycerol-gradient centrifugation, is reduced, possibly due to removal of some protein components[19,30] (M. M. Konarska, unpublished data). However, the same set of snRNAs is found associated with native spliceosomes and their heparin-treated counterparts.[19,20,24,30] Addition of nonspecific RNA competitor to splicing reactions has also been used to improve the resolution of splicing complexes by a native gel electrophoresis.[18]

The effects of heparin treatment on the endogenous pool of snRNP particles in nuclear extracts are quite different. Most normally occurring snRNP–snRNP interactions are destroyed in the presence of heparin at 30°, such that U4/6 and U4/5/6 snRNP complexes are not detectable in heparin-treated reactions. Moreover, monomeric snRNP particles also seem to be greatly distorted. For example, the distribution pattern of U2 snRNP changes drastically on heparin treatment.[20] Thus, it is not surprising that addition of heparin to the splicing reaction prevents the assembly of spliceosomes. Furthermore, the heparin resistance of spliceosomes indicates a highly stable association of snRNPs in these complexes and probably reflects the induction of a conformational change in snRNPs on assembly into spliceosomes. Lowering the temperature of the heparin treatment may improve stability of endogenous snRNP complexes.[24]

Freeze–thawing of splicing reactions seems to generate effects similar

[27] F. Strauss and A. Varshavsky, *Cell (Cambridge, Mass.)* **37**, 889 (1984).
[28] H. Singh, R. Sen, D. Baltimore, and P. A. Sharp, *Nature (London)* **319**, 154(1986).
[29] R. W. Carthew, L. A. Chodosh, and P. A. Sharp, *Cell (Cambridge, Mass.)* **43**, 439 (1985).
[30] P. J. Grabowski and P. A. Sharp, *Science* **233**, 1294 (1986).

to those of heparin treatment on the pattern of splicing complexes separated in a gel. However, the behavior of the endogenous snRNPs under these conditions has not been studied in detail. Both heparin treatment and freeze–thawing of splicing reactions can be used to stop the incubation.[19,31] Incubations can also be conveniently stopped by placing on ice. This mild treatment is specially recommended, since it does not detectably change the observed pattern of snRNP complexes separated in a gel.

Standard Protocol for Preparation of Native Polyacrylamide Gels

The gel mixture consists of the following:

4% (w/v) Polyacrylamide (acrylamide-to-bisacrylamide weight ratio of 80 : 1)
50 mM Tris–glycine (50 mM Tris base and 50 mM glycine, pH ~8.8)
0.1% (w/v) Ammonium persulfate
0.1% (v/v) TEMED

The gel mixture is prepared from 10× concentrated stock solutions of 40% (w/v) acrylamide–bisacrylamide and 0.5 M Tris–glycine buffer. Such concentrated stock solutions can be stored at 4° for several months. It is not necessary to degas the gel mix before pouring the gel. Typically, 15 × 14 × 0.15-cm gels are prepared, using approximately 35 ml of the gel solution. Smaller minigels (8 × 10 × 0.15 cm) are also very convenient to use, specially for analytical purposes. In the analysis of a standard 10-μl reaction, optimal results were obtained using combs with teeth at least 0.5 cm wide. When polymerized, gels are preelectrophoresed for 10–30 min in 50 mM Tris–glycine running buffer at approximately 13 V/cm. This preelectrophoresis step does not critically influence the quality of separation and can be omitted, when necessary. After loading the samples and tracking dyes, the electrophoresis is carried out at the same voltage gradient, at a maximal power of 4 W. The buffering capacity of the Tris–glycine system eliminates the need for recirculation of the running buffer during the electrophoresis. Electrophoresis at room temperature was found to be optimal for separation of snRNP particles from crude nuclear extracts. The pattern of snRNP complexes separated at 4° is essentially identical; however, complexes have a tendency to stick to one another, decreasing resolution. Increasing the gel temperature above 25–30° during electrophoresis results in progressive dissociation of certain multi-snRNP complexes.

[31] G. Christofori, D. Frendewey, and W. Keller, *EMBO J.* **6,** 1747 (1987).

Typically, electrophoresis is carried out for 4–6 hr, until xylene cyanol FF dye migrates close to the bottom of the gel (~12 cm). These conditions give satisfactory separation of major endogenous snRNP particles, including fast-migrating monomeric forms of U1, U4, U5, and U6, slow-migrating forms of U2 and U5 snRNPs, and multi-snRNP U4/6 and U4/5/6 complexes. When better resolution of large splicing complexes B and C is desired, the electrophoresis should be continued for an additional 2–4 hr. Following electrophoresis, one of the glass plates is removed, and the gel is transferred onto a double layer of Whatman 3MM filter paper cut to size. Low-percentage polyacrylamide gels are very loose and sticky. Occasionally, during opening of the glass plates, the gel will stick to both plates, resulting in distortion of the gel or in the trapping of air bubbles under the gel. In such cases, the gel should be returned to its original shape by briefly soaking it in electrophoresis buffer or water prior to transferring it onto the filter paper. For reactions containing ^{32}P-labeled preRNA, such gels are simply dried under vacuum and are exposed. If separated RNP complexes are to be detected by Northern hybridization, gels require additional processing as described in Electrophoretic RNA Transfer and Northern Hybridization sections.

Modifications of the Gel System

Conditions of electrophoresis, as well as the type of running buffer, can be changed to accommodate various experimental needs. The ionic strength of the buffer in the gel can be varied from 25 to 100 mM Tris–glycine, pH 8.8, and the gel can be electrophoresed in 50 mM Tris–glycine buffer. Lowering the pH of the gel buffer does not affect the pattern of snRNPs; 25 mM Tris–192 mM glycine or 12.5 mM Tris–96 mM glycine, pH 8.3, gel buffers produce very similar results.

Similarly, a Tris–borate buffering system (45 mM Tris–borate and 1.2 mM EDTA, pH 8.3) has been used to separate splicing complexes. However, due to the apparently lowered gel capacity in this system, samples of nuclear extract required heparin treatment in order to cleanly resolve RNP complexes.[19]

In some cases, it may be useful to modify the porosity of the gel. This can be achieved by changing the concentration of acrylamide or by using composite acrylamide–agarose gels. For example, lowering acrylamide concentration from 4 to 3.75% results in a markedly improved resolution of large RNP complexes. Under these conditions, splicing complexes B and C can be separated from one another (see Fig. 2). However, the inconvenience of handling such low-percentage gels restricts the use of this method to specific cases. Another option is to use composite acrylamide–agarose gels.[17,18] This type of gel was originally used in electrophoretic separation of nucleosomes and was adapted for resolution of ribosomes and ribosomal subunits.[7,8] For the separation of snRNP complexes, we have tested a combination 3% (w/v) polyacrylamide–2% (w/v) agarose gel run either in a Tris–glycine or in a Tris–borate buffer system. In both cases, the resulting pattern of snRNP distribution was nearly analogous to that on a 4% (w/v) polyacrylamide gel.

Other modified electrophoretic systems have been employed in several laboratories in the analysis of splicing[17,18,24,31–33] and polyadenylation complexes.[34]

Electrophoretic RNA Transfer

Conditions for transfer of RNA to nylon membrane and for Northern hybridization using native RNP gels are essentially the same as those recommended in standard protocols.[35] Handling low-percentage polyacrylamide gels, however, requires additional care to avoid distortion of the gel shape and the trapping of air bubbles between the gel and membrane. Briefly, the gel is placed on two layers of Whatman 3MM paper and is soaked for at least 10–15 min in the following transfer buffer:

10 mM Tris base
5 mM Sodium acetate
0.5 mM EDTA (final pH 7.8)

A piece of nylon membrane, two additional pieces of filter paper, and two fiber pads are also equilibrated in the transfer buffer. Other low-ionic-strength buffers, including the 50 mM Tris–glycine electrophoresis buffer, give comparable results. However, even when using the same buffer system for both the electrophoresis and transfer, the equilibration step should not be omitted or shortened to less than 10 min. Several types of nylon membranes are suitable for Northern hybridization analysis of native gels, including Zeta Probe (Bio-Rad, Richmond, CA) and GeneScreen (DuPont NEN, Wilmington, DE). Transfers onto nitrocellulose are generally more difficult, since low-percentage polyacrylamide gels tend to stick tightly to this membrane.

The transfer "sandwich" is prepared by placing the presoaked membrane on top of the gel plus filter paper, covering it with the two layers of presoaked Whatman 3MM filter paper and inserting the whole assembly

[32] M. Zillmann, S. D. Rose, and S. M. Berget, *Mol. Cell. Biol.* **7**, 2877 (1987).
[33] M. Zillmann, M. L. Zapp, and S. M. Berget, *Mol. Cell. Biol.* **8**, 814 (1988).
[34] H. Skolnik-David, C. L. Moore, and P. A. Sharp, *Genes Dev.* **1**, 672 (1987).
[35] T. Maniatis, E. F. Fritsch, and J. Sambrook, "Molecular Cloning: A Laboratory Manual." Cold Spring Harbor Laboratory, Cold Spring Harbor, New York, 1982.

between the two presoaked fiber pads in a gel holder. The critical step in this operation is the placing of the membrane on the gel. Care must be taken to eliminate both air bubbles and excess buffer from between the two surfaces. The gel holder is then placed in a transfer cell, filled with prechilled transfer buffer, and the transfer is carried out overnight, at ~200 mA at 4°. The actual time required for complete transfer of most RNAs under these conditions is only 2–4 hr. However, longer transfers are usually very convenient and are equally effective.

After transfer, the membrane is separated from the gel, covered with plastic wrap and ultraviolet light (UV) irradiated,[36] by placing the membrane directly on the surface of a 300-nm UV lamp for 15–30 min. This treatment fixes the transferred RNA to the membrane and is particularly recommended, when the single filter is to be processed through several rounds of hybridization. The membrane is then air-dried (or vacuum dried for 1 hr at 80° for longer storage) and used for hybridization.

Northern Hybridization

The membrane is prehybridized in the hybridization buffer consisting of the following:

  50% (v/v) Formamide
  0.1% (w/v) Poly(vinylpyrrolidone) (MW 40,000)
  0.1% (w/v) Bovine serum albumin (BSA)
  0.1% (w/v) Ficoll (MW 400,000)
  5× SSC
  50 mM Sodium phosphate, pH 6.5
  1% (w/v) Sodium dodecyl sulfate (SDS)
  2.5% (w/v) Dextran sulfate (MW 500,000)
  0.1 mg/ml Salmon sperm DNA

  1× SSC is 0.15 M NaCl and 15 mM trisodium citrate, pH 7.0

The prehybridization is carried out in a sealed bag for 1–5 hr at 42°. The radioactive RNA probe is added to the prehybridization mixture in 2–3 ml of the same buffer, and hybridization is carried out for 16 hr at 42°, with constant agitation.

RNA probes complementary to snRNAs are produced by transcription of appropriate DNA templates with T3, T7, or SP6 RNA polymerases.[20] After transcription, the reaction is treated with DNase, and RNA is purified over a small Sephadex G-50 spun column to remove unincorporated nucleotides. Typically 1–2 × 10^5 cpm/ml of the appropriate RNA probe is used for hybridization.

---

[36] G. M. Church and W. Gilbert, *Proc. Natl. Acad. Sci. U.S.A.* **81,** 1991 (1984).

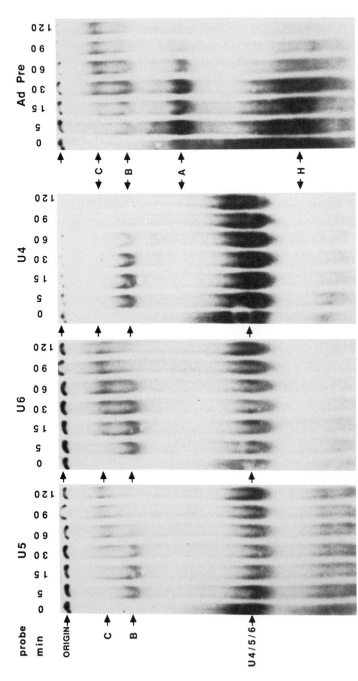

FIG. 2. Electrophoretic separation of spliceosomes and endogenous snRNP complexes. Splicing reactions (100 μl), containing a HeLa nuclear extract and ~1.8 pmol of unlabeled preRNA (Ad Pre, prepared by transcription of a *Sau*3AI-cleaved pBSAd1 plasmid with T3 RNA polymerase), were carried out in the presence of ATP. At times indicated at the top, 10-μl aliquots were withdrawn and were placed on ice. After 2 hr of incubation, samples were loaded onto a 3.75% (w/v) polyacrylamide and 50 m*M* Tris–glycine gel. The electrophoresis was carried out for 8 hr at the maximal power of 4 W. The separated products of the reaction were transferred onto a nylon membrane and were analyzed by hybridization with [32P]RNA probes complementary to U5, U6, and U4 snRNAs and to preRNA. Positions of splicing

Following hybridization, the membrane is washed three times for 30 min in $0.5\times$ SSC and 0.1% (w/v) SDS at room temperature. The membrane, still moist, is then covered tightly with plastic wrap and is autoradiographed at $-70°$ with an intensifying screen. Typically, short exposures (1–6 hr) are sufficient for the detection of abundant complexes. If rehybridization is planned, special care should be taken to avoid drying the membrane. The probe is stripped off by boiling the membrane in several changes of a large volume of 10 mM Tris–HCl, pH 7.5, 1 mM EDTA, and 0.1% (w/v) SDS. To ensure that the labeled probe has been completely removed by this treatment, the membrane is autoradiographed prior to rehybridization.

Results

An example of the analysis of splicing complexes by native gel electrophoresis combined with detection of RNAs by Northern hybridization is shown in Fig. 2. In this case, preRNA transcribed from pBSAd1 plasmid cut with Sau3AI restriction enzyme was used as a substrate in the splicing reaction. A time course of the assembly of splicing complexes is shown; the reactions were stopped at the indicated times by placing on ice. In order to obtain better separation between splicing complexes B and C, a 3.75% (w/v) [rather than 4% (w/v)] polyacrylamide gel was used for the separation of products. A standard, 50 mM Tris–glycine buffer was used, and the gel was electrophoresed for ~8 hr at maximal power of 4 W. RNP complexes were then transferred onto a nylon GeneScreen membrane and were probed sequentially with snRNA probes (hybridization with U5, U6, and U4 probes is shown). In addition, the same membrane blot was hybridized with a RNA probe complementary to the preRNA, revealing a complete pattern of the complexes assembled on the precursor. The resulting picture allows for the identification of snRNPs participating in formation of various splicing complexes, schematically illustrated in Fig. 1.

Acknowledgments

I am grateful to Phil Sharp, in whose laboratory this work was completed, for his help and continuous interest. I am also grateful to Sharon Seiler, Mariano Garcia-Blanco, and Anna Gil for helpful comments on the manuscript. M. M. Konarska is a Lucille P. Markey Scholar. The results presented in this work were supported by a grant from the Lucille P. Markey Charitable Trust (No. 87-19) to M. M. Konarska and by a grant from the National Institutes of Health (No. GM34277) and partially by a National Cancer Institute core grant (No. CA14051) to Phillip A. Sharp.

[31] Determination of RNA–Protein and RNA–Ribonucleoprotein Interactions by Nuclease Probing

By KATHERINE A. PARKER and JOAN A. STEITZ

Introduction

Processing signals in precursor RNAs (preRNAs) are protected from ribonuclease digestion by their interaction with bound factors. Two sorts of experiments have exploited this phenomenon to delineate regions of a RNA that are specifically recognized by a particular protein or ribonucleoprotein (RNP).

The first type of experiment involves only protection from RNase. A radiolabeled model preRNA is incubated with an active extract or fraction and is digested with RNase, and the total surviving RNA is isolated and analyzed on a gel. Partial digestion fragments that appear in the presence of extract, but not in the naked RNA, locate the binding site(s) of the factor(s). However, unless highly purified fractions are used, the identity of the responsible factor cannot be determined.

The second technique adds an immunoprecipitation step after the RNase digestion, which selects for RNA fragments that are tightly associated with a factor recognized by an antibody. Thus, interactions of a particular factor can be characterized, if one is fortunate enough to have available an antibody directed against that protein or RNP.

The latter technique, in which a preRNA is incubated with extract and is immunoprecipitated either during or after digestion with RNase, will be described in detail in this chapter. Methods for RNase digestion alone[1-3] are very similar, except that the immunoprecipitation step is omitted.

Extracts or Fractions

A radiolabeled precursor is incubated with an extract or fraction that contains the RNP or protein being tested for binding. If possible, an active extract should be used under conditions that allow the processing reaction of interest to occur. Although binding experiments have been successfully performed using purified small nuclear ribonucleoprotein (snRNP) fractions (e.g., recognition of the 5' splice site by U1 snRNPs[4]) or with puri-

[1] B. Ruskin and M. R. Green, *Cell (Cambridge, Mass.)* **43**, 131 (1985).

[2] A. Bindereif and M. R. Green, *Mol. Cell. Biol.* **6**, 2582 (1986).

[3] A. Krämer, *J. Mol. Biol.* **196**, 559 (1986).

[4] S. M. Mount, I. Pettersson, M. Hinterberger, A. Karmas, and J. A. Steitz, *Cell (Cambridge, Mass.)* **33**, 509 (1983).

fied protein fractions (e.g., recognition of the 3' splice site by a 70-kDa protein[5]), many interactions (such as the recognition of the intron branch point by U2 snRNPs[6,7]) require the participation of multiple components present in an active extract. A high-salt extract of nuclei, prepared by the method of Dignam *et al.*,[8,9] is active for splicing, polyadenylation, and histone precursor mRNA (pre-mRNA) processing and has been used for protection–immunoprecipitation experiments. Such extracts can be stored in aliquots at $-70°$ in buffer D [20 mM HEPES, pH 7.9, 20% (v/v) glycerol, 100 mM KCl, 0.2 mM ethylenediaminetetraacetic acid (EDTA), and 0.5 mM dithiothreitol (DTT)] for several months with no loss of activity. When highly purified fractions are substituted for crude extract, it may be necessary either to add exogenous bulk RNA[5] or to decrease the amount of ribonuclease[1,2] to avoid overdigestion.

Precursor RNA (PreRNA)

Rabiolabeled, model preRNAs are transcribed from gene sequences cloned downstream of a promoter recognized by a commercially available RNA polymerase. SP6, T7, and T3 polymerases all incorporate a GpppG cap at the 5' end, which increases both transcript stability and the efficiency of some processing reactions. If the processing signals have been well defined, decreasing the size of the preRNA to less than a few hundred relevant nucleotides will facilitate analysis of the immunoprecipitated bands (and may give enhanced processing or protection).

High-specific-activity transcripts are required, because usually less than 1% of the total counts per minute (cpm) added are immunoprecipitated; this is probably because complexes are lost during the nuclease step, even though the original binding may be quite efficient. The transcription protocol[10] for the plasmid template (previously cut with an appropriate restriction enzyme) is standard, except that 75–150 μCi of [α-^{32}P]GTP at 400 Ci/mmol is dried down, is resuspended in water, and is added to a 20-μl reaction mixture with 0.5 mM GpppG cap in the absence of cold GTP. If RNase T1 is to be used to analyze protected fragments, it is important that the labeled nucleotide(s) initially include GTP to ensure that all resulting oligonucleotides (except G monomers which are not adjacent to a G) will be labeled. The specific activity of the transcript can

[5] V. Gerke and J. A. Steitz, *Cell (Cambridge, Mass.)* **47**, 973 (1986).
[6] D. L. Black, B. Chabot, and J. A. Steitz, *Cell (Cambridge, Mass.)* **42**, 737 (1985).
[7] B. Ruskin, P. D. Zamore, and M. R. Green, *Cell (Cambridge, Mass.)* **52**, 207 (1988).
[8] J. D. Dignam, R. M. Lebovitz, and R. G. Roeder, *Nucleic Acids Res.* **11**, 1475 (1983).
[9] A. Krämer and W. Keller, this series, Vol. 181, p. 3.
[10] J. K. Yisraeli and D. Melton, this volume [4].

be further increased by using 1 mCi of [α-^{32}P]GTP at 4000 Ci/mmol (again in the absence of cold GTP) or by labeling with more than 1 nt (nucleotide) (CTP and UTP are successfully incorporated and greatly facilitate secondary analysis of protected fragments). However, high-specific-activity transcripts undergo rapid autoradiolysis and must be used within a few days. Transcripts can be purified on a 4% (w/v) polyacrylamide gel and can be eluted with 0.4 ml of 0.3 M sodium acetate, 1 mM EDTA, 0.1% (w/v) sodium dodecyl sulfate (SDS), and 10 mM Tris–HCl, pH 7.5, then precipitated with 1 ml of ethanol after phenol extraction.

To synthesize ^{3}H-labeled standards for double-label fingerprinting,[11] the transcription reaction is scaled-up 5-fold to a total volume of 100 μl, and 250 μCi of [8-^{3}H]GTP (11 Ci/mmol) dried down and resuspended in water, is added in the absence of cold GTP to yield transcripts of specific activity 18.1 × 10^6 dpm (disintegrations per minute)/μg. Addition of 1000 cpm of [α-^{32}P]GTP to the transcription reaction allows the RNA to be gel purified, but does not interfere with subsequent fingerprinting. The RNA is eluted, is precipitated with ethanol, is resuspended in 100 μl, and is stored at $-20°$; 5–10 μl are cofingerprinted with protected fragments as described later.

Incubation of Radiolabeled PreRNA with Extract

Although some protection–immunoprecipitation reactions (e.g., recognition of the 5' splice site by the U1 snRNP[4,6] or the hairpin loop of histone pre-mRNA by an unidentified factor[12]) occur on ice, most require incubation at higher temperatures (usually 30°). For example, the intron branch point cannot be immunoprecipitated by anti-(U2)RNP antibodies,[6] or the downstream purine-rich box of histone pre-mRNAs cannot be immunoprecipitated by anti-Sm antibodies[12] (see Fig. 2A), unless the substrate is incubated under processing conditions for 60 or 30 min, respectively. Therefore, various times of incubation should be tested to ensure optimal binding.

Digestion with RNase and Immunoprecipitation

Although our laboratory has performed RNase digestion at 4° in an effort to stabilize RNA–factor complexes (even transient dissociation, when RNase is present, permits digestion of the region of interest), other laboratories have successfully carried out the digestion at 30°.[7,13] The

[11] K. A. Parker, J. P. Bruzik, and J. A. Steitz, Nucleic Acids Res. 16, 10493 (1988).
[12] K. L. Mowry and J. A. Steitz, Mol. Cell. Biol. 7, 1663 (1987).
[13] A. Bindereif and M. R. Green, EMBO J. 6, 2415 (1987).

RNase chosen should specifically cleave after a single (or at most two) base: more promiscuous RNases give very heterogeneous arrays of protected bands. RNase T1, which cleaves 3' to G residues, has been most widely used, because RNase T1 gives the simplest digestion patterns; but pancreatic RNase, which cleaves 3' to pyrimidines, is sometimes preferable.[7] For maximum reproducibility, the enzyme should be made up or diluted from a concentrated stock just prior to use. Use of both T1 and pancreatic RNases (in separate experiments) gives the best idea of the true extent of protection; T1 alone may greatly overestimate the site size, especially if the region has a paucity of Gs that are single stranded.

Although the amount of RNase required must be determined empirically, usually at least 75% of the transcript should be digested to completion. This is assessed by analyzing an aliquot of the reaction mixture at the end of the digestion period, as detailed later.

Antibodies

The specificity and titer of the antibody used for immunoprecipitation are very important. For most protection–immunoprecipitation experiments, either mouse monoclonal antibodies or human sera derived from patients with autoimmune diseases, have been used. Control antibodies are other mouse monoclonals of the same subclass or serum derived either from normal lab personnel (young males are best) or from an autoimmune patient whose serum recognizes a different protein or RNP.[14]

To maximize the signal, the antibody must be titered[14] and added in excess over the protecting protein or RNP. In many protection–immunoprecipitation protocols, purified immunoglobulin[15] or crude antiserum or ascites fluid[4,6,13,16,17] was added directly to the reaction mixture. To enable the antigen–antibody complexes to be pelleted, protein A [either in the form of formalin-fixed *Staphylococcus aureus* cells[4] (pansorbin) or conjugated to Sepharose beads[6,13,15,16,17] was added 0–30 min later. In some more recent experiments, the antibody has been prebound to protein A–Sepharose (PAS) beads, washed, then added to the reaction.[11,12,18] The latter procedure is preferable for two reasons: (1) crude antiserum or ascites fluid contains varying levels of nucleases, which are washed away during prebinding of the antibody, and (2) free antibody, which might

---

[14] J. A. Steitz, this volume [32].
[15] J. E. Stefano and D. E. Adams, *Mol. Cell. Biol.* **8**, 2052 (1988).
[16] C. Hashimoto and J. A. Steitz, *Cell (Cambridge, Mass.)* **45**, 581 (1986).
[17] B. Chabot, D. L. Black, D. M. LeMaster, and J. A. Steitz, *Science* **230**, 1344 (1985).
[18] K. L. Mowry and J. A. Steitz, *Science* **238**, 1682 (1987).

compete with the PAS-bound antibody for binding to the RNA–factor complex, is removed.

Although human immunoglobulin G (IgG) binds protein A with high affinity, IgGs from some species do not. In such cases, a second antibody, which itself can bind PAS and which is directed against the Fc portion of the first antibody, must be used to obtain precipitation.

Order of Addition of RNase and Antibody

In some protection–immunoprecipitation protocols, antibody is not added until 30 min after the RNase.[7,12,18] Although it is theoretically possible that antibody binding to the RNP or protein factor could either destabilize the complex or alter the digestion pattern by sterically hindering the nuclease, this does not appear to be a practical problem, and other protocols add antibody and RNase simultaneously[5,6,11,15,16,17] or add RNase 30 min after antibody[4,13]; the advantage here is that longer times of incubation will increase the efficiency of immunoprecipitation. Although major differences are not generally observed, different orders of addition should be tested initially to discover which is optimal. Deviation from an established protocol may require retitration of the amount of nuclease needed to achieve adequate digestion of the precursor.

Typical Protocol for Protection–Immunoprecipitation Experiment

That given in detail has been used to study the interaction between the U3 RNP and precursor rRNA (pre-rRNA).[11] Similar protocols have been used to study processing signals for pre-mRNA splicing,[4–7,13] polyadenylation,[15,16] and histone pre-mRNA 3'-end formation.[12,18]

Prebinding of antibodies to PAS is performed as follows. For an experiment with 10 separate reactions, add 40 mg of PAS (Pharmacia, Piscataway, NJ) to 5 ml of NET-2 [150 mM NaCl, 50 mM Tris–HCl, pH 7.5, and 0.05% (v/v) Nonidet P-40 (NP-40)]. After swelling for >5 min, aliquot (while mixing frequently as PAS settles very quickly) 0.5 ml into 1.5-ml Eppendorf tubes. Add 20 μl of antiserum or ascites fluid and mix gently at room temperature for 1–6 hr or overnight at 4°. To wash, spin in a microfuge 10 sec, then carefully remove the supernatant (the pellet can be unintentionally aspirated!) with a Pasteur pipet. Add 1 ml of NET-2, invert the tube three times to mix, and repeat three times. After the last wash, add 0.1 ml of NET-2 and store at 4°. The antibody–PAS mixture is stable for at least 2 days.

If the same processing reaction is to be immunoprecipitated with several antibodies, it is best to perform a single large reaction, then to aliquot a portion onto each type of antibody–PAS mixture. A protocol for a reaction performed in a total volume of 50 μl, which can be scaled-up appropriately, is as follows.

1. Add 30 μl of Dignam extract in buffer D.[8,9]
2. Add 2 μl of a 1 M solution of KCl, which will contribute 40 mM to the final concentration of KCl; recall that Dignam buffer D (undiluted) contains 100 mM KCl and thus contributes 60 mM, making the total KCl concentration 100 mM.
3. Add 2.5 μl of a 100 mM solution of MgCl$_2$ to a final concentration of 5 mM.
4. Add 2.5 μl of a 100 mM solution of DTT to a final concentration of 5 mM.
5. Add 2 μl of a 10 mM solution of ATP to a final concentration of 0.4 mM.
6. Add 1 μl of a 0.5 M solution of creatine phosphate to a final concentration of 10 mM.
7. After combining the first six reagents on ice, add 10 μl of the radiolabeled preRNA present in aqueous solution as described above. The mixture should contain either $1-10 \times 10^6$ cpm (Cerenkov) of preRNA, if the transcript was labeled with [α-^{32}P]GTP at 400 Ci/mmol to a specific activity of the RNA of 3.3×10^8 cpm/μg, or $1-10 \times 10^7$ cpm, if the transcript was labeled with [α-^{32}P]GTP at 4000 Ci/mmol to a specific activity of the RNA of 3.3×10^9 cpm/μg. (This represents, in both cases, about 3–30 ng of precursor or approximately $0.5-5 \times 10^{-13}$ mol for a 200-nt RNA.) For each reaction, however, the amount of precursor should be titered empirically to maximize the signal-to-noise ratio.
8. After incubation at 30° for 15 min, place the mixture on ice and add 20 μl of a 30,000-U/ml solution of T1 RNase [specific activity 3000 U/mg (Calbiochem, San Diego, CA)] or 20 μl of a 2-mg/ml solution of pancreatic RNase [specific activity 5500 U/mg (Worthington, Freehold, NJ)].
9. After 5 min, aliquot 70 μl onto the prebound antibody–PAS mixture (in 100 μl of NET-2) and nutate the samples at 4° for 60 min. Alternatively, the antibody–PAS mixture may be added after 30 min, then may be incubated at 0° with occasion mixing for an additional 30 min. Spin 10 sec in a microfuge at room temperature, then transfer 2 μl of the supernatant into a separate tube containing 300 μl of NET-2, and immediately phenol extract; this represents the "total" and provides a measure of the amount of digestion that has occurred. Most of the preRNA should be digested to limit oligonucleotides (lane 1 of Fig. 1a). The remainder of the supernatant is carefully aspirated with a Pasteur pipet and is discarded. After

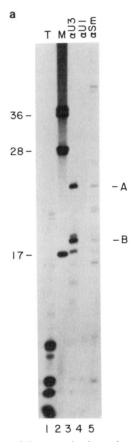

FIG. 1. RNase T1 protection of the α-sarcin site region of 28 S rRNA and immunoprecipitation by anti-(U3)RNP antibodies. (a) Protection–immunoprecipitation. A 268-nt model rRNA transcript containing nucleotides 4542–4810 of 28 S (Ref. 19) was radiolabeled with [α-^{32}P]GTP and was incubated with Dignam extract as described in the text. The sample was digested with T1 RNase and was immunoprecipitated with a mouse monoclonal anti-(U3)RNP antibody [αU3 (lane 3)], a patient anti-(U1)RNP antibody [αU1 (lane 4)], or a mouse anti-Sm monoclonal antibody [αSm (lane 5)]. The resulting RNA was electrophoresed on a 20% (w/v) polyacrylamide–8 M urea gel. The RNA fragments present in 1/80 of the total anti-(U3)RNP antibody sample at the end of the digestion period, just prior to washing the immunoprecipitate, are shown [T (lane 1)]. Markers, obtained by filling in a HpaII digest of pBR322, are also shown [M (lane 2)]. (b) Fingerprint of fragment A and autoradiograph without a fluor. Fragment A was eluted from the gel, was ethanol precipitated, and was mixed with 2 × 10^5 cpm of [^{3}H]GTP-labeled total transcript. The mixture was fingerprinted by electrophoresis on cellulose acetate in the first dimension (CA), followed by homochromatography in the second dimension (HC). This autoradiograph shows the fingerprint after autoradiography in the absence of a fluor. (c) Fingerprint of fragment A and autoradiograph with a fluor. The fingerprint in B was sprayed with Enhance and was reautoradiographed. The oligonucleotides from fragment A (shown in Fig. 1b) were identified by virtue of which spot the oligonucleotides comigrated with; this constellation can only

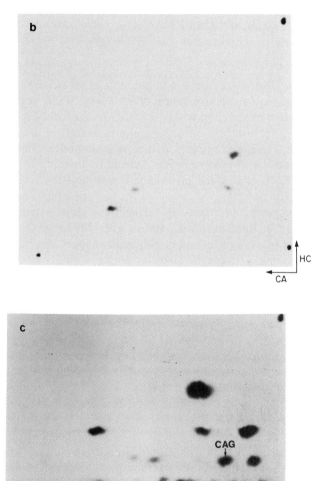

FIG. 1. (*continued*)

originate from nucleotides 4570–4590 of human 28 S rRNA,[19] which is immediately adjacent to the α-sarcin site. Fragment B contains nucleotides 4575–4590. The spots in the extreme upper right and both lower corners of b and c represent ^{32}P-labeled markers that were placed on the completed fingerprint prior to the bland first autoradiography. (From Parker *et al.*[11])

washing the pellets four to six times with cold NET-2, the RNA is phenol extracted as described[14] and is ethanol precipitated in the presence of 20 μg of carrier RNA.

Analysis of Protected Fragments

Fragments are fractionated on a high-percentage gel (15–20%), which gives good resolution in the size range of 10–50 nt. If the protection–immunoprecipitation experiment was successful, bands will be seen in the appropriate lanes, but not the controls. Examples include both fragments derived from a 28 S rRNA substrate[19] that were immunoprecipitated with anti-(U3)RNP antibodies[11] (shown in Fig. 1a) and fragments derived from a histone pre-mRNA that were precipitated with anti-Sm antibodies[12] (shown in Fig. 2A).

To identify the protected fragments, further analysis is essential. Even if bands produced at different times and/or precipitated with different antibodies appear to comigrate on a one-dimensional gel, the bands may not represent the same fragment. Usually, length variants of the same region, which differ by the presence of one or more oligonucleotides, will be observed. For example, fragment A in Fig. 1a contains AACCG in addition to all the oligonucleotides present in fragment B. The ratio of such variants can vary slightly between experiments.

Since, in general, bands derived from a protection–immunoprecipitation experiment are not digested to completion, a second digestion with the same nuclease will generate a series of limit products that can be further analyzed to allow definitive determination of the origin of the fragment. Three different methods have been used for analyzing immunoprecipitated bands: (1) RNA fingerprint analysis[20]; (2) fractionation of RNase T1 digestion products on a 15% (w/v) polyacrylamide gel, followed by secondary analysis of each eluted band using pancreatic RNase and paper electrophoresis[12,17]; or (3) fractionation of RNase T1 digestion products on a 10–12% (w/v) polyacrylamide gel, followed by secondary analysis of each eluted band using pancreatic RNase and two-dimensional thin-layer chromatography on polyethyleneimine (PEI) plates.[7,13,21]

For the initial analysis, we strongly recommend RNA fingerprinting.[20] Although secondary analyses of oligonucleotides, separated on a one-dimensional gel after primary digestion, has been used successfully to initially identify immunoprecipitated fragments,[7,13] this method might fail

[19] I. L. Gonzalez, J. L. Gorski, T. J. Campen, D. J. Dorney, J. M. Erickson, J. E. Sylvester, and R. D. Schmickel, *Proc. Natl. Acad. Sci. U.S.A.* **82**, 7666 (1985).
[20] A. D. Branch, B. J. Benenfeld, and H. D. Robertson, this volume [12].
[21] G. Volckaert and W. Fiers, *Anal. Biochem.* **83**, 228 (1977).

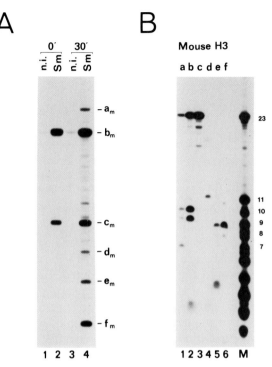

Fig. 2. T1 protection of the mouse histone pre-mRNA and immunoprecipitation with anti-Sm antibodies. (A) Protection–immunoprecipitation. A 249-nt model mouse H3 histone pre-mRNA, which contained sequences flanking the site of 3′-end formation, was radiolabeled with [α-^{32}P]GTP, [α-^{32}P]CTP, and [α-^{32}P]UTP, was incubated with nuclear extract under histone 3′-end-processing conditions, was digested with RNase T1, and was immunoprecipitated with a human nonimmune [n.i. (lanes 1 and 3)] control serum or a mouse anti-Sm [Sm (lanes 2 and 4)] monoclonal antibody, and the RNA fragments were electrophoresed on a 15% (w/v) polyacrylamide–8 M urea gel. Fragments, which were immunoprecipitated from a sample that was not incubated at 30° [0′ (lanes 1 and 2)], are shown, and the samples, which were incubated for 30 min at 30° [30′ (lanes 3 and 4)], are also shown. Fragment b_m derives from the hairpin loop just upstream of the cleavage site, fragment c_m derives from the region just upstream of the hairpin loop, and fragment a_m derives from an extension of the hairpin loop to include part of the purine-rich box where the U7 snRNA binds. Fragments e_m and f_m originate from the extreme 5′ end of the precursor, and fragment d_m originates from a region about 60 nt from the 5′ end of the transcript. (B) One-dimensional gel analysis of the bands shown in A. The RNA fragments shown in A were eluted from the gel, were digested to completion with RNase T1, and were electrophoresed on a 20% (w/v) polyacrylamide–8 M urea gel (lanes 1–6). The T1 oligonucleotides contained in fragments a_m–f_m of A are compared with those from a T1 digest of the entire mouse histone H3 transcript (lane M) whose sizes are indicated on the right of B. (From Mowry and Steitz.)[12]

if several small oligonucleotides did not separate on a one-dimensional polyacrylamide gel [although oligonucleotides of 10 nt or more usually separate well on a one-dimensional gel, shorter oligonucleotides of the same or even some of similar lengths (such as a 7-mer and 8-mer) may not be distinguished]. Once a fragment has been definitively identified by RNA fingerprinting, however, one-dimensional fractionation of the primary digestion products in parallel with a total digest of the transcript (see Fig. 2B), followed by secondary analysis of the eluted bands, can be used in subsequent experiments.

Protected Fragment Analysis by Double-Label Fingerprinting

We have recently developed a modified fingerprint technique, which greatly facilitates the identification of fragments generated in protection–immunoprecipitation experiments.[11] α-^{32}P-labeled fragments are produced as described previously, are separated on a 20% (w/v) polyacrylamide–8 M urea gel (as shown in Fig. 1a), and are eluted. Each fragment is then mixed with 2×10^5 cpm of H-labeled total transcript, and 20 μg of carrier RNA is added. The mixture is dried down, is resuspended in 2 μl of RNase T1 at 3.5 U/μl, is digested at 37° for 30 min, and is fingerprinted.[20]

This technique was used to identify fragment A (shown in Fig. 1a). The first autoradiograph shows only the ^{32}P-labeled oligonucleotides derived from fragment A (see Fig. 1b), as ^{3}H requires a fluor to be visualized. The fingerprint is then sprayed with the fluor Enhance (New England Nuclear, Wilmington, DE) and is reexposed (see Fig. 1c). This gives a total fingerprint, which is identical to the total ^{32}P-labeled fingerprint; spots can be easily identified by their relative mobilities. The first and second exposures are then aligned, and the spots from the ^{32}P-labeled fragment are definitively identified without secondary analysis. Initial identification by secondary analysis of all spots from a ^{32}P-labeled total fingerprint is of course required.

Determination of the Nature of the Protecting Factor

Once it has been established that immunoprecipitation of a protected fragment is specific, several types of experiments can be performed to probe the factor–RNA interaction and, in particular, to determine if a RNA component is intimately involved. These experiments include (1) pretreatment of an extract or fraction with micrococcal nuclease, (2) selective destruction of RNAs with RNase H, or (3) immunoprecipitation with antitrimethylguanosine (TMG) antibodies.

Microccal Nuclease Pretreatment

The activity of micrococcal nuclease is dependent on calcium, which in turn can be specifically chelated with ethylene glycolbis(β-aminoethyl ether)tetraacetic acid (EGTA). To test the requirement for a RNA component in the protecting factor, the first six reagents of a protection–immunoprecipitation experiment are mixed (see Typical Protocol for Protection–Immunoprecipitation Experiment) and 1 μl of a 30 mM solution of CaCl$_2$ (final concentration 0.75 mM) and 1 μl of a 50,000-U/ml solution of micrococcal nuclease [final concentration 1,250 U/ml (Worthington)] are added and incubated at 30° for 5 min. One microliter of a 100 mM solution of EGTA is added (final concentration of 2.5 mM), the reaction is mixed well, and 10 μl of prerna [reagent 7 see Typical Protocol for Protection–Immunoprecipitation Experiment]) is added, and the reaction is completed precisely as outlined previously.

Figure 3 illustrates that protection–immunoprecipitation of a region of 28 S rRNA near the α-sarcin site by U3 RNPs is completely abolished on pretreatment of the extract with micrococcal nuclease.[11] It is critical to show that the loss of protection–immunoprecipitation activity is not due to a failure to inactivate the micrococcal nuclease. This can be tested by adding calcium and EGTA first, and then adding the micrococcal nuclease; as shown in lane 4 (+pre-EGTA), this order of addition has no effect on the reaction. Individual components, such as calcium (lane 2) or micrococcal nuclease (lane 3) should also be tested. These testing procedures are essential, since calcium-activated proteases could destroy the protecting factor.

RNase H Pretreatment

An excellent way to test whether a particular RNP is involved in the protection–immunoprecipitation reaction is to target its RNA for cleavage, using a complementary oligodeoxynucleotide and RNase H, which cleaves the RNA component of a RNA–DNA hybrid.[22] If activity is abolished, as occurs for the immunoprecipitation of the intron branch point after selective destruction of U2,[6] that RNP is most likely involved, although it is always possible that the oligonucleotide targeted destruction of another less-abundant RNA responsible for the activity.

Protection Experiments Using Antitrimethylguanosine (TMG) Antibodies

One distinguishing characteristic of RNAs contained in Sm snRNPs is that the RNAs have a TMG cap at their 5' end. Anti-TMG antibodies can

---

[22] A. Kramer, this series, Vol. 181.

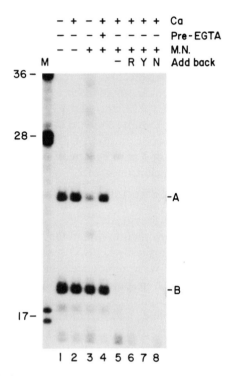

FIG. 3. Effect of micrococcal nuclease on the RNase T1 protection of the α-sarcin site region of 28 S and immunoprecipitation by anti-(U3)RNP antibodies. The first six reagents of a protection–immunoprecipitation experiment were prepared as described in the text. The samples shown then received the following: (lane 1) 2 μl of water; (lane 2) 1 μl of $CaCl_2$ (+Ca) at 30 mM and 1 μl of water; (lane 3) 1 μl of micrococcal nuclease (+M.N.) at 50,000 U/ml and 1 μl of water; (lane 4) 1 μl of $CaCl_2$, 1 μl of EGTA (+pre-EGTA) at 100 mM, and then, after mixing, 1 μl of micrococcal nuclease; (lanes 5–8) 1 μl of $CaCl_2$ and 1 μl of micrococcal nuclease. All samples were incubated for 5 min at 30°, then 1 μl of EGTA was added to each, except sample 4, which had already received the EGTA. After chelation of $CaCl_2$, the sample in lane 6 received 50 μg of 16 S rRNA (R), the sample in lane 7 received 50 μg of yeast small RNAs (Y), and the sample in lane 8 received 6 μg of human nuclear RNA (N). After mixing well, the radiolabeled rRNA transcript was added, and the binding reaction was performed as described previously, using a mouse monoclonal anti-(U3)RNP antibody for the immunoprecipitation. (Lane M) Markers obtained by filling in a HpaII digest of pBR322. (From Parker et al.[11])

be used in protection–immunoprecipitation experiments to determine if such a RNA is involved. Although immunoprecipitation of a fragment (e.g., the 5′ splice site,[17] the 3′ splice site,[5,17] or the intron branch point[17]) with anti-TMG antibodies implies that a TMG-containing RNA either directly or indirectly interacts with the processing signal, a negative result does not necessarily mean that a TMG-containing RNA is not involved. If a highly accessible stretch of RNA lies between the 5' end and the RNA sequence that recognizes a processing signal, the accessible region is likely to be cleaved during RNase digestion, leading to an inability of the anti-TMG antibody to immunoprecipitate the protected fragment.

Interpretating Significance of Protected–Immunoprecipitated Fragments

The most difficult aspect of protection experiments is deciding whether an immunoprecipitated band is significant. The first question which must be answered is whether the band is specific for the antibody. Overexposure of gel-fractionated fragments in any immunoprecipitate will eventually reveal bands. The most obvious way to determine if a band is specific is to directly compare the pattern of protected fragments given by the antibody with one or more controls. If controls never immunoprecipitate a fragment [as is the case for 5' splice site immunoprecipitation by anti-(U1)RNP antibodies[6]], the interaction is definitely specific.

Difficulties in interpretation arise if a fragment can be immunoprecipitated to a lesser extent with control antibodies. There are two potential explanations for the same band appearing in varying amounts with different antibodies. One possibility is that there is a "specific" and "nonspecific" component to the immunoprecipitation and that any differences are significant. The second possibility is that different antibodies have different capacities for nonspecific binding. It is possible that either the protein component alone or the intact RNP binds RNAs nonspecifically in order to scan for a specific signal; thus the relative abundance of the RNP could determine the relative precipitation of nonspecific fragments. In such cases, an interaction is considered likely to be specific if a large number of controls immunoprecipitate the same bands <5% as efficiently as the cognate antibody. Moreover, multiple sources of the same antibody specificity should be used and shown to give the same protection pattern. However, further experiments are still required to prove specificity.

The most powerful tools for proving specificity in protection–immunoprecipitation experiments are mutant transcripts, which no longer process but maintain, identical RNase digestion patterns. For example, the 3' splice site fragment is strongly precipitated by anti-Sm antibodies and very weakly (<2%) by controls.[5,17] A point mutation (AG to GG) in this region completely abolishes splicing, but shortens the protected fragment by only 1 nt. A dramatic decrease (>90%) in the precipitation of this mutant region by anti-Sm antibodies[5,17] argues strongly that the immunoprecipitation of the wild-type sequence is functionally significant. Analysis of mutant transcripts in a protection–immunoprecipitation assay may not always be definitive, however, as some mutations might allow binding of the factor to occur, but may prevent the reaction from proceeding further.

Once it is established that the immunoprecipitation of a particular fragment is specific, it is still necessary to determine if the protection–immunoprecipitation is relevant to the processing reaction being investigated. For example, purified U1 snRNPs can protect 5' splice site consensus sequences contained in 16 S rRNA.[23] In other cases, protection–immunoprecipitation results are clearly related to function. Ultimately, if the factor is a RNP, genetic suppression experiments[24,25] can be employed to establish which sequences in the RNA component of the factor recognize a specific processing signal in the substrate RNA.

Acknowledgments

We would like to thank Kim Mowry for careful reading of the manuscript and several helpful suggestions. K. A. Parker is a recipient of a National Institutes of Health Physician Scientist Award and J. A. Steitz is supported by the Howard Hughes Medical Institute and GM grant 26154.

[23] S. M. Mount and J. A. Steitz, unpublished observations.
[24] Y. Zhuang and A. M. Weiner, Cell (Cambridge, Mass.) 46, 827 (1986).
[25] R. Parker, P. Siliciano, and C. Guthrie, Cell (Cambridge, Mass.) 49, 229 (1987).

[32] Immunoprecipitation of Ribonucleoproteins Using Autoantibodies

By JOAN A. STEITZ

Introduction

The development of a simple immunoprecipitation technique led to the original realization that small ribonucleoproteins (RNPs) with distinct RNA and protein components exist in mammalian cells. These complexes display remarkable variety in both their RNA and protein moieties. Well-characterized small RNPs have sedimentation values up to about 15 S and contain from one to nine polypeptides and usually one RNA molecule—between 60 and 300 nt (nucleotides) long. Small RNPs range in abundance from 10^8 to as few as 10^3 particles/cell. Their RNA components can be encoded by either the host or by a viral genome; to date, all known small RNP proteins are host encoded. With a few exceptions, small RNPs occupy discrete cellular compartments: small nuclear RNPs (snRNPs) are found in the nucleoplasm, small cytoplasmic RNPs (scRNPs) are found in the cytoplasm, and small nucleolar RNPs (snoRNPs) are found in the nucleolus. Even mitochondria and chloroplasts contain small RNPs.

An unexpected source of probes for characterizing the structures and cellular functions of small RNPs is the sera of patients suffering from autoimmune disease. A large fraction of such sera contains relatively high titers of autoantibodies of the immunoglobulin G (IgG) class directed against determinants carried on small RNPs. Autoantibodies always occur in multiple individuals; yet, some specificities are much more common than others (10- to 100-fold more frequent), and frequencies do vary depending on the racial background of the population. Furthermore, there are acknowledged correlations between the specificity of a patient's autoantibodies and the clinical manifestations of disease, despite the fact that the etiology of the autoimmune response is not understood. Since most patient sera contain a mixture of autoantibody specificities, multiple sera must be screened and compared to ensure that any selected for biochemical use are relatively monospecific.

Virtually every known mammalian small RNA–protein complex, whatever its initial method of detection, is known to be targeted by some patient autoantibody (see Table I[1-22]). With only rare exceptions, the

[1] M. R. Lerner and J. A. Steitz, *Proc. Natl. Acad. Sci. U.S.A.* **76,** 5405 (1979).

[2] T. Mimori, M. Hinterberger, I. Pettersson, and J. A. Steitz, *J. Biol. Chem.* **259,** 560 (1984).

[3] J. Wilusz and J. D. Keene, *J. Biol. Chem.* **261,** 5467 (1986).

[4] K. L. Mowry and J. A. Steitz, *Science* **238,** 1682 (1987).

[5] K. A. Montzka and J. A. Steitz, *Proc. Natl. Acad. Sci. U.S.A.* (in press).

[6] S. I. Lee, S. C. S. Murthy, J. J. Trimble, R. C. Desrosiers, and J. A. Steitz, *Cell (Cambridge, Mass.)* **54,** 599 (1988).

[7] M. R. Lerner, J. A. Boyle, J. A. Hardin, and J. A. Steitz, *Science* **211,** 400 (1981).

[8] J. P. Hendrick, S. L. Wolin, J. Rinke, M. R. Lerner, and J. A. Steitz, *Mol. Cell. Biol.* **1,** 1138 (1981).

[9] J. Rinke and J. A. Steitz, *Cell (Cambridge, Mass.)* **29,** 149 (1982).

[10] J. C. Chambers, M. G. Kurilla, and J. D. Keene, *J. Biol. Chem.* **258,** 11438 (1983).

[11] C. Hashimoto and J. A. Steitz, *J. Biol. Chem.* **258,** 1379 (1983).

[12] M. G. Kurilla and J. D. Keene, *Cell (Cambridge, Mass.)* **34,** 837 (1983).

[13] M. R. Lerner, N. C. Andrews, G. Miller, and J. A. Steitz, *Proc. Natl. Acad. Sci. U.S.A.* **78,** 805 (1981).

[14] J. G. Howe and M.-D. Shu, *J. Virol.* **62,** 2790 (1988).

[15] M. D. Rosa, J. P. Hendrick, M. R. Lerner, J. A. Steitz, and M. Reichlin, *Nucleic Acids Res.* **11,** 853 (1983).

[16] M. B. Mathews, M. Reichlin, G. R. V. Hughes, and R. M. Beinstein, *J. Exp. Med.* **160,** 420 (1984).

[17] J. A. Hardin, D. R. Rahn, C. Shan, M. R. Lerner, S. L. Wolin, M. D. Rosa, and J. A. Steitz, *J. Clin. Invest.* **70,** 141 (1982).

(Footnotes continued on page 471)

TABLE I

MAMMALIAN SMALL RNPs RECOGNIZED BY AUTOANTIBODIES[a]

RNP class	Antibody	RNA component(s)	Initial reference(s)
Sm snRNPs	anti-Sm	U4	1
		U6	1
		U5	1
	anti-(U2)RNP	U2	1, 2
	anti-(U1)RNP and		
	anti-(U1)RNA	U1	1, 3
		U7	4
		U11–...	5
		HSUR 1–5	6, D. Wassarman and
		(*Herpesvirus saimiri*)	S. Lee (personal
			communication)
Ro scRNPs	anti-Ro	Mouse Y1–Y2	7
		Human Y1–Y5	8
La RNPs	anti-La	Ro RNAs (see above)	
(mostly snRNPs)		Rat 4.5S$_1$	8
		Mouse or hamster	7
		4.5S	
		tRNA precursors	9
		pre-5S rRNA	9
		Other class III RNAs	10, 11
		(e.g., pre-7SL,	
		pre-7-2)	
		VAI,II (adenovirus)	7
		VSV leader RNA	12
		(vesicular stomatitis	
		virus)	
		EBER 1,2	13
		(Epstein-Barr virus)	
		HVP 1,2	14
		(*Herpesvirus papio*)	
Jo scRNP	anti-Jo (synthetase)	tRNA^{His}	15
scRNP	anti-PL7 (synthetase)	tRNA^{Thr}	16
scRNP	anti-tRNA1	tRNA^{Ala}	16, 17
	(synthetase)		
scRNPs	anti-LL, anti-SU	Other tRNAs	18
snoRNP	anti-5S RNP	5S rRNA	17
snoRNPs	anti-Th (To)	7-2	11, 19
		8-2 (RNase P)	20
snoRNPs	anti-(U3)RNP	U3	21
		U8	K. Tyc (personal
			communication)
		U13	K. Tyc (personal
			communication)
scRNP	anti-SRP	7SL	22
snRNP	anti-7SK	7SK	K. Mowry (unpub-
			lished observation)

[a] All antibodies listed are directed against protein epitopes, except for anti-(U1)RNA and perhaps some of the "other tRNA" specificities indicated.

recognized determinants are in the protein(s), rather than the RNA components of the particles. The RNA–protein associations in most small RNPs are relatively strong; many particles survive exposure to mild denaturants and detergents. Therefore, immunoprecipitation techniques can be used to identify the RNA and protein components of small RNPs present in various types of cell extracts. Moreover, because an antigenic protein often interacts (either directly or indirectly) with more than one small RNA, autoantibodies provide a way of classifying small RNPs according to common protein components (or epitopes).

In this chapter, immunoprecipitation procedures developed to identify and characterize the components of small RNPs using patient polyclonal or mouse monoclonal autoantibodies are outlined. These procedures are based on the original protein A procedure of Kessler,[23] as adapted for use with autoantibodies by Lerner and Steitz[1] and as later modified by Matter et al.[24] and Mimori et al.[2] For representative immunoprecipitation patterns of RNAs and proteins, see Refs. 1, 2, and 24.

Antibodies

Crude patient serum is prepared from whole blood (drawn without heparin) in the following way. Transfer the blood to a glass tube and incubate at 4° overnight to allow a clot to form. Spin for 10 min at 2500 rpm in a clinical centrifuge and carefully aspirate the clear serum. The concentration of IgG in the prepared serum is usually about 10–20 mg/ml; often autoimmune sera have higher protein concentrations than normal sera. Sera are usually stored frozen at −70° in aliquots; patient autoantibodies can remain active for years and also can survive several cycles of freezing and thawing with only a moderate loss of activity. Sera can also be stored at 4° or can be shipped unrefrigerated after the addition of 0.1%

[18] Ciechanover, S. L. Wolin, J. A. Steitz, and H. F. Lodish, *Proc. Natl. Acad. Sci. U.S.A.* **82,** 1341 (1985).

[19] R. Reddy, E. M. Tan, D. Henning, K. Nohga, and H. Busch, *J. Biol. Chem.* **258,** 1383 (1983).

[20] H. A. Gold, J. Craft, J. A. Hardin, M. Bartkiewicz, and S. Altman, *Proc. Natl. Acad. Sci. U.S.A.* **85,** 5483 (1988).

[21] M. A. Lischwe, R. L. Ochs, R. Reddy, R. G. Cook, L. C. Yeoman, E. M. Tan, M. Reichlin, and H. Busch, *J. Biol. Chem.* **260,** 14304 (1985).

[22] W. H. Reeves, S. K. Nigam, and G. Blobel, *Proc. Natl. Acad. Sci. U.S.A.* **83,** 9507 (1986).

[23] S. W. Kessler, *J. Immunol.* **115,** 1619 (1975).

[24] L. Matter, K. Schopfer, J. A. Wilhelm, T. Nyffenegger, R. F. Parisot, and E. M. DeRobertis, *Arthritis Rheum.* **25,** 1278 (1982).

(w/v) sodium azide. For certain biochemical uses, however, the presence of azide in the sera may not be acceptable, since azide may inhibit *in vitro* reactions. Alternatively, antibodies in patient sera can be precipitated three times at room temperature with 40% (w/v) ammonium sulfate, as described by Garvey *et al.*,[25] followed by dialysis against cold 17.5 mM sodium phosphate buffer, pH 6.3, and then by centrifugation to remove cryoglobulins. Such a step can serve to concentrate the IgGs.

Monoclonal antibodies, either in the form of mouse ascites fluid or cell culture supernatants, can also be used for immunoprecipitation of small RNPs. After drawing from the animal, ascites fluid is incubated at 37° for 2–3 hr or at room temperature overnight and is centrifuged at 2500 rpm for 10 min in a clinical centrifuge. The supernatant is then recentrifuged at 8000 rpm for 5 min to remove any remaining cells. Ascites fluid can be further prepared by ammonium sulfate precipitation as described above. Cell culture supernatants have significantly lower concentrations of autoantibodies, compared to patient serum or ascites fluid (often >100-fold lower). Monoclonal antibodies may not be as stable to freezing and thawing as patient sera; each antibody should therefore be tested and stored at 4° (with 0.1% (w/v) sodium azide) if necessary.

Cell Extracts

Tissue culture cells are the preferred source of extracts from which to immunoprecipitate small RNPs. Tissues can be used, but as tissues are bathed in serum before removal from the animal, ribonuclease levels are usually high and can result in RNA degradation during the immunoprecipitation procedure. Moreover, cultivated cells can be easily labeled with $^{32}PO_4$ for analysis of RNAs, can be labeled with [^{35}S]methionine or other labeled amino acids (or $^{32}PO_4$) for analysis of proteins (or phosphoproteins), or can be used unlabeled for nonradioactive analysis of the RNA components of small RNPs.

For $^{32}PO_4$ labeling of tissue culture cells, logarithmically growing cells are suspended in phosphate-free minimal essential medium (MEM), supplemented with 10% (v/v) dialyzed calf serum and 60 μg/ml of penicillin and 100 μg/ml of streptomycin, at 2 × 10^5 cells/ml. After the addition of $^{32}PO_4$ at 0.1 mCi/10 ml of cell suspension, the cells are incubated at 37° for about ½ of 1 doubling time. (HeLa cells, for example, have a doubling time of about 20 hr; therefore, an 8- to 16-hr labeling time is normally used.) If

[25] J. S. Garvey, N. E. Cremer, and D. H. Sussdorf, *in* "Methods in Immunology" (J. S. Garvey, N. E. Cremer, and D. H. Sussdorf, eds.), 3rd ed., p. 218. Benjamin, Reading, Massachusetts, 1977.

longer labeling times are to be employed, it may be necessary to supplement the medium with 1–2% (w/v) phosphate, especially in the case of fast-growing cells.

For [³⁵S]methionine labeling, 4 μCi/ml is added to 2 × 10⁵ cells/ml in methionine-free MEM. Incubation is for $\frac{3}{4}$–1 cell doubling time (usually 16–18 hr in the case of HeLa cells).

Regardless of how the cells are labeled, a whole-cell extract is prepared in the following way. Cells are collected by centrifugation in a clinical centrifuge (5 min at 1500 rpm) and are washed with $\frac{1}{5}$ volume of cold TBS (150 mM NaCl and 40 mM Tris–Cl, pH 7.4). All following steps are conducted at 0–4°. The cells are suspended in an "appropriate" volume of either TBS or NET-2 [150 mM NaCl, 50 mM Tris–Cl, pH 7.4, and 0.05% (v/v) Nonidet P-40 (NP-40, RNase free)]. An appropriate volume is usually about 1 ml/20–80 ml of labeled cell culture (although higher numbers of cells can be used); the total volume should be ≥2 ml. Cells are broken by sonication for 3 × 30 sec at setting three of a Branson sonifier. The extract is incubated on ice for ≥20 sec between sonications. The cell extract is then cleared by centrifugation at 14,000 g for 10 min (10,000 rpm for 10 min in a SS34 rotor). The supernatant is the prepared whole-cell extract.

Because some RNPs are unstable to sonication and heat, it may be necessary to sonicate for shorter periods, using lower sonication power, and to exercise extreme caution that the sample temperature does not rise above 0°. Extracts are usually used immediately; however, some RNPs are stable in extracts for hours on ice or even resist freezing and thawing (although the backgrounds in immunoprecipitates will usually rise).

Alternatively, cell or nuclear extracts, prepared according to other protocols (e.g., a Dignam nuclear extract[26]), can be used as a source of antigen for immunoprecipitation. What is important is to be able to calculate the approximate number of cells' worth of material represented in the preparation, so that the appropriate amount of antibody can be used. If the cell–antibody ratio is too low, high backgrounds will result.

Immunoprecipitation

The preferred procedure is to preadsorb IgGs present in the serum onto protein A–Sepharose CL-4B [PAS (Pharmacia, Piscataway, NJ)] and then to add the cell extract containing the target RNPs. Ribonucleases that are present in crude sera are removed (to some extent) by the washing of the PAS-bound antibody, as are other serum proteins that may

[26] J. D. Dignam, R. M. Lebovitz, and R. G. Roeder, *Nucleic Acids Res.* **11**, 1475 (1983).

produce high backgrounds in the immunoprecipitate. Alternatively, immune complexes can be preformed by incubation of the serum with the cell extract and then can be collected by exposure to pansorbin (or PAS) in a second step. Pansorbin (Calbiochem-Behring, San Diego, CA) is a formalin-fixed preparation of *Staphylococcus aureus* cells; pansorbin is prepared according to the manufacturer's instructions.

For the PAS-prebinding procedure, swell 2.5 mg of PAS dry resin in 0.5 ml NET-2 for each serum to be tested. PAS (2.5 mg) has a bed volume of about 10 μl and contains about 2 mg of protein A/ml bed volume. Since 1 mg of protein A binds 10 mg of IgG and human serum contains 10–20 mg/ml of IgG, 10 μl of swollen PAS should completely bind the IgGs present in 10 μl of serum. After swelling for about 5 min, each 0.5-ml aliquot of PAS suspension is transferred to a 1.5-ml Eppendorf-type microtube; during transfer, the PAS–NET-2 suspension should be mixed frequently to avoid settling of the resin. Add 10 μl of crude serum or ascites fluid (or about 0.3 mg of purified IgG) to each 0.5-ml aliquot of swollen PAS. Nutate either overnight at 4° or for 1–6 hr at room temperature. (Preparation of the cell extract can be done during this period.) Wash the PAS-bound antibody by spinning out the resin in a microfuge (5 sec), remove the supernatant carefully with a Pasteur pipet, and resuspend the pellet in 1 ml of NET-2. Repeat the wash three times. Do not add NET-2 to the resin pellet after the final wash, if the pellet is to be used immediately. Alternatively, the washed PAS–antibody mixture can be stored in NET-2 for at least 2 days at 4°.

Use the PAS-bound antibodies to precipitate RNPs as follows. Adjust the total volume of the cell extract, so that each 0.5-ml sample represents between 2 × 10^6 and 1 × 10^7 cells. (To ensure quantitative immunoprecipitation of abundant particles such as the Sm snRNPs, the lower concentration should be used.) Save 50–100 μl of the extract for a total control. [This extract may be frozen or preferably immediately phenol extracted (for RNA) or boiled in sample buffer (for proteins).] Add 0.5 ml of cell extract to each washed PAS–antibody resin pellet and nutate for 1–2 hr at 4°. Spin out the resin, discard the supernatant (carefully, if labeled), and wash the resin a minimum of four times with 1 ml of cold NET-2 as described previously.

If pansorbin is to be used instead of PAS, the cell extract should first be cleared of nonspecific binding material by incubating the cell extract for 20 min on ice with an equal volume of pansorbin in NET-2 and by centrifuging to remove the pansorbin. The precleared extract is then incubated with serum (or other antibody preparation) for ≥15 min on ice. Pansorbin is added at an extract–pansorbin ratio of 2 : 1, and incubation is continued for ≥15 min. The pansorbin–immune complex precipitates are collected by centrifugation and are washed a minimum of four times in NET-2. If RNases in the serum appear to be degrading the RNAs present in the extract during the immunoprecipitation, the addition of carrier RNA (50 μg of yeast tRNA/10 μl of serum) to the cell extract may be helpful.

The NET-2 washing procedure can be advantageously substituted by higher salt washes, if the RNP is stable (e.g., most Sm snRNPs appear to be stable to high salt,[2] but the U3 RNP is not[27]). To achieve a significantly lower background in the analysis of both RNAs and protein, IPP [500 mM NaCl, 10 mM Tris–Cl, pH 8.0, and 0.1% (v/v) NP-40] can be used instead of NET-2 in the preparation of both the PAS-bound antibody and the PAS-bound antigen–antibody complexes. Obviously, both the RNA–protein interactions in the RNP and the antigen–antibody interactions must be stable to the higher salt concentration; thus, this procedure should always be checked in parallel with NET-2.

The above immunoprecipitation protocol can be adjusted for use with non-IgG antibodies or with IgGs from species that are not recognized by protein A. In this case, a second antibody that is bound by protein A (and also recognizes the specific antibody) is first bound to PAS. The procedure is exactly as described previously. Then the washed PAS–second antibody is used to adsorb the antibody of interest, simply by repeating the binding protocol. It is essential to prebind sufficient second antibody, so that the specific antibody will be quantitatively adsorbed. The antibody ratio must therefore be calculated from the known capacity of the second antibody and the concentration of the specific antibody or (preferably) must be titered. Alternatively, it may be possible to use protein G, which recognizes all mouse IgG subclasses.

Analysis of Immunoprecipitated RNAs

If the immunoprecipitates are to be analyzed for the RNA components of small RNPs, the washed protein A–immune complex pellets should be immediately phenol extracted as follows. To each sample, add the following in order: 300 μl of NET-2, carrier (1–2 μl of 10-mg/ml yeast RNA, if the sample is ^{32}P labeled, or 2 μl of 5-mg/ml glycogen, if the sample is nonradioactive and is to be visualized by staining or is to be 3' end labeled with [^{32}P]pCp), 30 μl of 10% (w/v) sodium dodecyl sulfate (SDS), and 300 μl of PCA [phenol–chloroform–isoamyl alcohol (50:50:1) plus 0.1% (w/v) 8-hydroxyquinoline, saturated (preferably) with SSC (0.15 M NaCl and 0.015 M sodium citrate)]. To enhance the recovery of RNA (espe-

---

[27] K. A. Parker and J. A. Steitz, *Mol. Cell. Biol.* **7**, 2899 (1987).

cially from large numbers of cells), the mixture can be warmed to 37° and can be incubated in a water bath for 15 min, vortexing every 5 min. Separate the layers by spinning for 5 min in a microfuge and remove the aqueous fraction to a fresh tube. Add to the aqueous layer 40 μl of 3 M sodium acetate plus 1 ml of 95% (v/v) ethanol and allow the precipitate to form on dry ice for ≥20 min or to form overnight at −20°. Spin out the pellet for 10 min in a microfuge. Wash with cold (−20°) 75% (v/v) ethanol and dry the pellet under vacuum. Resuspend each pellet in water or in 10 μl of TBE–urea sample buffer {10 M urea, 0.025% (w/v) bromphenol blue, and 0.025% (w/v) xylene cyanol FF in TBE [90 mM Tris–Cl, 90 mM borate, and 1 mM ethylenediaminetetraacetic acid (EDTA), pH 8.6]}. Heat to 70° for 3 min and fractionate 5 μl/lane on a polyacrylamide–7 M urea gel; a 10% (w/v) polyacrylamide gel is optimal for fractionating RNAs in the size range of 60–300 nt.

Enough RNA can be precipitated, using this protocol, to yield an autoradiograph of abundant [32P]RNAs (≥10^5/cell) in 16 hr, using an intensifying screen on a dried gel. If the immunoprecipitated RNAs have been previously well characterized, the RNAs need not be labeled and can be visualized instead by staining either with ethidium bromide (EtBr) or with silver. Forman et al.[28] have reported a procedure that can detect even the Ro RNAs, each present in humans at about 10^4 copies/cell. For EtBr staining, the gel is incubated in a 0.5-μg/ml solution in TBE for 5 min and is destained in TBE for 5 min. RNA bands are observed by exposure to high-intensity ultraviolet (UV) light (300 nm). For silver staining,[29] the gels (optimally <1 mm thick) are washed in 50% (v/v) methanol and 12% (v/v) acetic acid for 1–2 hr, are incubated for 5 min in an oxidizing reagent (0.0034 M potassium dichromate and 0.0032 M nitric acid in water), are washed four times for 30 sec in glass-distilled water, and are immersed in an aqueous solution of silver nitrate (0.012 M) for 30 min. During the first 5 min of exposure to the latter solution, the glass vessel, containing the gel, is placed on an X-ray viewer to provide uniform exposure to fluorescent light. The gels are rinsed twice with water, are rinsed once in developer (0.28 M sodium carbonate and 0.5 ml/liter of formalin), and then are shaken gently in developer until the desired stain intensity is achieved. Stained gels can be fixed in 1% (v/v) acetic acid. (The gels should be soaked in 5% (v/v) glycerol prior to drying down to avoid cracking.)

[32P]RNAs can be excised from gels and can be analyzed by standard techniques. For the analysis of low-abundance RNAs, sensitivity can be

[28] M. S. Forman, M. Nakamura, T. Mimori, C. Gelpi, and J. A. Hardin, *Arthritis Rheum.* **28,** 1356 (1985).

[29] C. R. Merrill, D. Goldman, S. A. Sedman, and M. H. Ebert, *Science* **211,** 1437 (1981).

increased by postlabeling unlabeled RNAs at their 3' ends with T4 RNA ligase and [^{32}P]pCp before gel fractionation. However, since the efficiency of labeling can vary as much as 100-fold from one RNA to another, relative yields cannot be directly assessed. Alternatively, unlabeled RNAs can be fractionated and then can be detected by Northern blotting. The peril in this method is that the background level of immunoprecipitation of nonspecific RNAs will not be visualized. If using Northern analysis, it is therefore essential to also probe for an abundant control RNA (e.g., 5S or U1) and to also analyze the supernatant from the protein A–antibody–antigen pellet to assess the specificity of the immunoprecipitation.

Analysis of Immunoprecipitated Proteins

^{35}S-Labeled proteins, contained in washed protein A–antibody–antigen pellets, are extracted by heating at 90° in SDS gel sample buffer [62.5 mM Tris–Cl, pH 7, 2% (w/v) SDS, 5% (v/v) mercaptoethanol, 10% (v/v) glycerol, and 0.005% (w/v) bromphenol blue] for 10 min. Fractionation is on Laemmli gels,[30] with polyacrylamide concentrations designed to resolve proteins in the size range of interest. Gels can be enhanced by soaking for 30 min in 500 mM sodium salicylate before drying and autoradiography.

If phosphoproteins are to be analyzed, RNAs contained in the immunoprecipitate must first be destroyed or the RNAs will produce an intolerably high background in the Laemmli gel. Cells labeled with ^{32}PO$_4$, as described previously for RNA, are immunoprecipitated, as described previously for ^{35}S-labeled proteins. After washing, the PAS-bound precipitates are incubated for 15 min at 25° with RNases A and T1, at final concentrations of 2 mg/ml and 3000 U/ml, respectively, based on the assumption that a PAS pellet contains about 30 μl. The PAS-bound precipitates are then loaded directly on the gel after boiling in sample buffer. To ensure that a resulting ^{32}P-labeled band is indeed a protein, a control sample should be digested with proteinase K (2 mg/ml) for 10 min before being loaded on the gel. (See Ref. 27, for examples.)

Immunoprecipitation will detect all proteins in a RNP complex, whether the proteins are antigenic or not. The antigenic protein(s) themselves can be detected by transfer from the gel to nitrocellulose and by Western blotting.[31] However, it is important to remember that autoanti-

---

[30] U. K. Laemmli, *Nature* (*London*) **227**, 680 (1970).

[31] I. Pettersson, M. Hinterberger, T. Mimori, E. Gottlieb, and J. A. Steitz, *J. Biol. Chem.* **259**, 5907 (1984).

bodies seem to be directed against "native" protein epitopes. Thus, the relative titer of a particular autoantibody can appear to be as much as 1000-fold lower, when reacting with a denatured protein on a blot relative to a native RNP in a cell extract. For the same reasons, a blot can over-represent the amount of reactivity with a protein, whose antigenicity may not contribute at all to the RNA profile obtained with a particular serum.

Often proteins which are not components of the RNP whose RNA has been identified by immunoprecipitation will appear either in labeled protein immunoprecipitates or in Western blots using a particular serum. Thus, multiple sera should be compared, and proteins which cofractionate (by sedimentation or chromatographic procedures) with the RNA should be analyzed in order to ascertain which polypeptides are actual RNP components.

Determination of Relative Concentrations of Antibodies Directed against RNPs in Mixed Sera

Patient sera are virtually never truly monospecific. Autoantibodies against one kind of RNP are almost invariably accompanied by antibodies against other RNPs. Certain pairs frequently recur, especially when protein targets are present in the same multimolecular complexes, e.g., anti-Sm and anti-(U1)RNP, anti-Ro and anti-La. Often, clues that a particular patient serum contains a mixture of autoantibodies can be obtained from the immunofluorescence pattern. Since most small RNPs are specifically localized in one or another cell compartment (i.e., nucleoplasm, nucleolus, or cytoplasm), staining of several compartments suggests mixed specificity. (Note, however, from Table I, that different classes of small RNPs can be identically localized and that some classes occupy multiple cellular compartments.) Yet, if after analysis, the titer of one autoantibody is judged to be ≥10-fold higher than any other specificity, the serum can usually be utilized successfully for biochemical studies.

To determine the relative titers of serum autoantibodies directed against different small RNPs, the immunoprecipitation assay must be adjusted to ensure that the antibodies, rather than the antigenic RNPs, are limiting. (Note that this is opposite to the strategy described previously in which the antibody is in excess, so as to precipitate the RNP quantitatively and thereby to generate optimal amounts for analysis.) Because the RNP targets of any mixed specificity serum may be overlapping, it is necessary to analyze several points in an antibody dilution assay. Previously, Pettersson et al.[31] have used 20- and 400-fold dilutions of patient sera to assess the relative titers of anti-(U1)RNP and anti-Sm present. The ratio of U1 to U2 RNA in the immunoprecipitates at the highest dilution provides an estimate of the relative titers in that particular serum.

Autoantibody Reactions with Small Rnps from Nonmammalian Species

Patient autoantibodies are remarkably cross-reactive with their target antigens from nonhuman sources. This cross-reactivity is in contrast to antibodies deliberately raised against (denatured) proteins, which can often be highly species specific. Within mammals, virtually all autoantibodies against RNPs are completely cross-reactive, suggesting that the autoantibodies target epitopes that are extremely highly conserved. Similarly, most patient autoantibodies will react with the corresponding RNPs from other vertebrates. Cross-reactions become weaker with RNPs from nonvertebrate metazoans, but successful use of anti-Sm and anti-(U1)RNP with insect,[32] nematode,[33] sea urchin,[32] and even malarial[34] and dinoflagellate[35] antigens has been reported. An occasional patient serum cross-reacts with the corresponding yeast[36] and plant[36] small RNPs.

When searching for autoantibodies reactive with small RNPs from distant species, two points should be kept in mind. First, patient sera, which are polyclonal in nature, are more likely to cross-react than monoclonals, which target a single epitope that may not be present in the distantly related antigen. Second, proteins detected in Western blots using patient autoantibodies are even more likely to be irrelevant to the RNP in question than when a mammalian extract is immunoblotted. Thus, immunoprecipitation techniques, which allow immune complex formation with the native antigen, must be used to analyze both the RNA and the protein components of the small RNP.

Interpretation of Immunoprecipitation Data

As mentioned several times in this chapter, the biggest problem with the use of autoantibodies to identify components of small RNPs is distinguishing true immunoprecipitation from background. Some cellular pro-

[32] M. R. Lerner, J. A. Boyle, S. M. Mount, S. L. Wolin, and J. A. Steitz, *Nature (London)* **283**, 220 (1980).

[33] J. D. Thomas, R. C. Conrad, and T. Blumenthal, *Cell (Cambridge, Mass.)* **54**, 533 (1988).

[34] A. M. Francoeur, C. A. Gritzmacher, C. L. Peebles, R. T. Reese, and E. M. Tan, *Proc. Natl. Acad. Sci. U.S.A.* **82**, 3635 (1985).

[35] R. Reddy, D. Spector, D. Henning, M.-H. Liu, and H. Busch, *J. Biol. Chem.* **258**, 13965 (1983).

[36] D. Tollervey and I. W. Mattaj, *EMBO J.* **6**, 469 (1987).

teins and small RNAs seem to have higher than average affinities for PAS (or pansorbin) or for IgGs and therefore produce high backgrounds, which can be mistaken for specific immunoprecipitation. Inclusion of the following controls in immunoprecipitation experiments should mitigate against false conclusions.

1. Always include a "total" RNA (or protein) lane and be certain that the pattern of the immunoprecipitate is not simply a lighter version of the total. RNA analysis by Northern blotting and protein analysis by Western blotting are disadvantageous because the total profile is not assessed.

2. Totals from newly opened cells should be compared to totals from the final extract to ensure that degradation products are not being generated during the procedure.

3. Always include a no-antibody and a normal human serum (or nonimmune IgG) lane to check for background sticking. Profiles obtained with autoantibodies of different specificity also provide useful controls.

4. Base conclusions on results with multiple sera, if possible. If a second serum produces the same RNA profile, but additional bands in the protein profile, the second serum is likely to be of mixed specificity.

5. Since spurious background bands can be generated if the ratio of PAS–antibody to cell extract is too high, several relative concentrations should be tested. High backgrounds can often be eliminated by using higher salt concentrations to wash the PAS–antibody–antigen precipitate, but be certain to check that the RNP is stable to higher salt concentrations.

6. If a desired autoantibody is present in a mixed serum, affinity purification can be used to select those IgGs directed against a particular protein antigen. Elution of antibodies from a specific band on an immunoblot is the ideal procedure,[37] but yields are often low, and some antibodies do not survive the pH 3 glycine elution step.

7. Immunoprecipitation analysis across a gradient or column is useful for determining whether RNAs and proteins coexist in the same physical particle. Backgrounds are also usually lower with fractionated, compared to crude, cell extracts.

Depletion of Small RNPs from Extracts Using Autoantibodies

One use of autoantibodies is to demonstrate that a small RNP has a particular biological activity. *In vitro* activity is always easier to inhibit by depletion of the RNP, rather than by simple addition of autoantibodies to

---

[37] D. E. Smith and P. A. Fisher, *J. Cell Biol.* **99,** 20 (1984).

an active system. On the other hand, the ability to achieve adequate depletion is highly dependent on the kind of extract subjected to the depletion protocol and can even vary from one sample to another of the same type of extract. Hence, it is always essential to check the extent of depletion achieved and to be aware that depletion almost invariably involves dilution of the activity originally present in the extract.

To obtain an extract depleted of a particular small RNP,[2] the extract is incubated for three successive 1-hr periods, each with a fresh preparation of PAS-bound antibody, usually in the buffer utilized in the *in vitro* system. Free antibody that may become released into the extract can be removed by exposure for 20 min to pansorbin, as described previously. After removal of pansorbin by centrifugation, the depleted extract can be tested for activity and its level of depletion can be checked by Western blotting for proteins or by staining or Northern blotting for RNAs. A parallel sample, exposed to PAS-bound normal serum or nonimmune IgG, provides a control for dilution and activity losses inherent to the manipulations. Note that some *in vitro* activities withstand exposure to PAS, but not pansorbin, or vice versa.

Acknowledgments

    The author is most grateful to many former and current colleagues for the development of the methods described in this chapter and to John Hardin and his colleagues for providing many novel autoantibodies. Susanna Lee, Karen Montzka, Jim Bruzik, Mei-Di Shu, and Kathy Parker made helpful comments on the manuscript; Lynda Stevens provided expert typing. Support was provided by grants GM21654 and CA16038 from the National Institutes of Health.

[33] Electron Microscopy of Ribonucleoprotein Complexes on Nascent RNA Using Miller Chromatin Spreading Method

By Yvonne N. Osheim and Ann L. Beyer

Introduction

The chromatin-spreading method, developed by Miller and colleagues about 20 years ago,[1,2] has proved extremely useful for the study of many aspects of genetic activity, including replication, transcription, and trans-

[1] O. L. Miller, Jr. and B. R. Beatty, *Science* **164**, 955 (1969).
[2] O. L. Miller, Jr. and A. H. Bakken, *Acta Endocrinol.* (*Copenhagen*) *Suppl.* **168**, 155 (1972).

lation. The chromatin-spreading method is particularly valuable, because the method allows analysis of *in vivo* events on an individual gene basis, after the gene has been unwound gently and rapidly from its compact *in situ* structure. We have used the technique to study the formation of ribonucleoprotein (RNP) complexes on the nascent transcripts of a gene, thus allowing comparison of RNP structure on multiple RNA molecules of the same sequence.[3] Several previous articles have reviewed information that has been gained from use of the chromatin-spreading method,[4–6] and other articles have included details of the method.[2,7–10] Two recent reviews are particularly comprehensive with regard to the technique.[9,10] However, as discussed by Trendelenburg and Puvion-Dutilleul,[10] the technique varies somewhat with different specimens and in different hands, such that it is not possible to describe a detailed "standard" procedure. In this chapter, we outline the methods in use in our laboratory for the visualization of transcriptionally active chromatin from *Drosophila* oocytes and embryos. We then describe how we analyze RNP structure and RNA-processing events on nascent RNA transcripts and conclude with prospects for improving the information yield of the approach.

Miller Chromatin-Spreading Method

Overview

The "Miller" chromatin-spreading technique for electron microscopy (EM) allows visualization of dispersed nuclear chromatin and its accompanying transcription. Nuclear chromatin is hypotonically disrupted from its compact three-dimensional *in situ* structure to a dispersed array that is

[3] A. L. Beyer, O. L. Miller, Jr., and S. L. McKnight, *Cell* (*Cambridge, Mass.*) **20,** 75 (1980).
[4] B. A. Hamkalo and O. L. Miller, Jr., *Annu. Rev. Biochem.* **42,** 379 (1973).
[5] O. L. Miller, Jr., *J. Cell Biol.* **91,** 15s (1981).
[6] M. F. Trendelenburg, *Hum. Genet.* **63,** 197 (1983).
[7] B. A. Hamkalo, *Methods Cell Biol.* **19,** 287 (1978).
[8] A. H. Bakken and B. A. Hamkalo, *in* "Principles and Techniques of Electron Microscopy" (M. A. Hayat, ed.), Vol. 9, p. 84. Van Nostrand-Reinhold, Princeton, New Jersey, 1978.
[9] H. Zentgraf, C.-T. Bock, and M. Schrenk, *in* "Electron Microscopy in Molecular Biology: A Practical Approach" (J. Sommerville and U. Scheer, eds.), p. 81. IRL Press, Washington, D.C., 1987.
[10] M. F. Trendelenburg and F. Puvion-Dutilleul, *in* "Electron Microscopy in Molecular Biology: A Practical Approach" (J. Sommerville and U. Scheer, eds.), p. 101. IRL Press, Washington, D.C., 1987.

centrifuged onto an EM grid and displayed two dimensionally. Although higher order nuclear interactions are disrupted, DNA remains in a nucleosomal configuration, and nascent RNA occurs in a RNP configuration. The procedure was developed empirically by Miller and Beatty and was first applied to the visualization of the amplified ribosomal genes of amphibian oocytes.[1] The two critical factors for optimal dispersal of eukaryotic chromatin are an alkaline pH (pH 8.5–9) and a very low ionic strength. The common dispersal solution, "pH 9 water," is glass-distilled water that has been adjusted to pH 8.5–9. Low concentrations of detergent are used, when manual disruption is not possible and when hypotonic shock alone is insufficient to destabilize cellular membranes. Biochemical fractionation of tissues or cells is generally avoided before chromatin spreading, since exposure to the salt and pH conditions in typical solutions renders subsequent spreading more difficult (but see Ref. 10). Large oocytes and embryos of amphibia and insects are particularly amenable to chromatin spreading, because these cells can be manipulated and broken manually. Different cell types from many organisms have been spread, however, including mammalian tissues and tissue-culture cells, bacteria, algae, slime molds, yeast, and higher plants.[4,5,9,10] The procedure is rapid; one can go from living cells to material fixed on a grid in 30–40 min. Very little cellular material and no specialized equipment is needed, other than an electron microscope and a vacuum evaporator.

Preparation of Carbon Support Film on Grids

Before chromatin is deposited on a grid, the grid must be covered with a support film. Carbon films are preferable to plastic support films, because of their purity and because carbon films can be made thinner, allowing higher resolution analysis of the specimen. We routinely use thin carbon films (30–50 nm) made by depositing carbon on a glycerin-coated glass slide, as previously described[7,8] and reproduced in the following list with some modifications. An alternative to carbon deposition on a glycerin-coated slide is deposition on freshly cleaved mica.[11] Both glycerin and mica provide a very smooth surface for forming the carbon film, which, in either case, is subsequently floated on water and is used to coat EM grids.

1. Fill a Coplin staining dish with a freshly prepared solution of 40% (v/v) glycerol (Mallinckrodt, St. Louis, MO), made in glass-distilled water.

2. Wipe a precleaned glass microscope slide with a Kimwipe and

[11] L. W. Coggins, *in* "Electron Microscopy in Molecular Biology: A Practical Approach" (J. Sommerville and U. Scheer, eds.), p. 1. IRL Press, Washington, D.C., 1987.

pass the slide back and forth through a Bunsen burner flame five times. Turn the slide over and repeat. Repeat the warming cycle again.

3. Holding the warm slide with forceps, dip the slide into the glycerol solution, so that most of the slide, excluding the top 1–2 cm, is submerged. Gently move the slide back and forth about 10 times in the glycerol solution. Remove the slide from the glycerol and hold the slide in a vertical position above (not in) a Bunsen burner flame. Gently move the slide back and forth for about 6–8 sec. Stand the slide on its short bottom edge against a rack to drain onto bibulous paper. Prepare four to six slides. The earlier slides will drain longer than those prepared later. Allow the last slide to drain about 1 min. Heating the slide causes the glycerol solution to drain rapidly, leaving a thin uniform coating. Drainage should occur from the sides of the slide to the middle and from the top to bottom, resulting in a smooth surface bisected vertically by a thin line of a thicker glycerin deposit down the center of the slide (Fig. 1a). Discard slides that

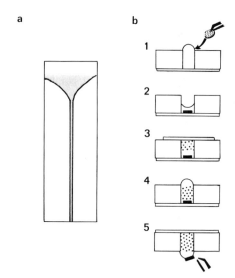

FIG. 1. (a) Desired drainage pattern of a glycerin-coated slide. The clear areas on either side of the shaded Y are suitable surfaces for deposition of a thin carbon film. (b) Steps in the spreading procedure shown on a transverse section of a microcentrifugation chamber. (1) The chamber, which is sealed on the bottom with a bonded cover slip, is overfilled with sucrose–formalin solution, and the rinsed grid is inserted in the convex meniscus. (2) Most of the sucrose–formalin solution is removed from the chamber after the grid has floated to the bottom. (3) The dispersing chromatin is layered over the sucrose–formalin solution, and the chamber is sealed on the top with a cover slip. (4) After centrifugation, the top cover slip is removed and the sucrose–formalin solution is again added to form a convex meniscus. (5) The grid chamber is inverted, and the grid floats to the surface of the hanging drop from which the grid is picked up with forceps.

display alternative drainage patterns. This procedure works best on dry, sunny days; the glycerol drains less efficiently on humid days.

4. Remove the glycerol from the back of the slide, which is the side that was leaning against the support, by wiping with Kimwipes. Place four to six coated slides on the baseplate of a vacuum evaporator on a sheet of 15-cm white filter paper, centered ~6 inches below the carbon source. The carbon source consists of two 3-mm-diameter spectroscopically pure graphite rods, one of which has been sharpened to an ~2-mm-long point and has been placed end-to-end, using a spring loaded clip, with the ~1-mm-indented end of the other rod. Before carbon deposition, the glycerol surfaces are glow discharged.[11] Plasma cleaning of the surface is essential for a uniform and strong carbon film. We pump down the chamber of a vacuum evaporator (Edwards 306) to 0.1–0.2 Torr with the mechanical pump and glow at maximum current for 1.5–2 min.

5. Immediately after glow discharge, evacuate the apparatus to at least 2×10^{-4} Torr and increase the current until the carbon rods just start to evaporate. Continue evaporation at this current until the desired thickness is obtained, which is estimated by the color change of the white paper on which the slides are placed. A very light brown or beige color corresponds to a thin film.

6. Set up a glass dish with a black background. We use a 170 × 190-mm round Pyrex dish covered on the outside with black tape. Place a 100-ml beaker in the dish and fill the dish to overflowing with glass-distilled water. Place a small piece of wire mesh or screen, measuring about 4 × 9 cm, across the top of the beaker. The size of the dish and beaker are not important, only that the support for the screen (e.g., the beaker) must be below the water surface by several centimeters. The carbon film will be floated on the water and lowered onto the grids sitting on the screen. Skim the water surface with a glass rod to remove dust particles.

7. Take a carbon-coated slide and lower the slide slowly into the water, carbon side up, at an angle of 35–40°, with the back of the slide supported on the side of the glass dish. The carbon film will float on the water surface, and the glycerin will dissolve. The dark background of the dish improves visibility, when used in combination with a gooseneck desk lamp.

8. Holding the grids with anticapillary tweezers (see Materials), dip the grids individually in 95% (v/v) ethanol for a few seconds and then place the grids close together on the wire mesh in a pattern roughly approximating the shape of the intact film, being careful not to disturb the floating carbon film. We use 300-mesh copper grids with one shiny side and one dull side (for orientation purposes) and put them on the screen with the shiny side up.

9. In the final step, position the floating film over the grids and lower the water level by aspiration, bringing the film down on top of the grids. We guide the film to the correct position using two pairs of tweezers (one in each hand). Aspirate water from the bowl, until the film is deposited on the grids. The grids are left on the wire mesh and are allowed to dry overnight in the refrigerator in a covered petri dish.

10. The strength of the film and evenness of grains can be checked in the EM the following day. Assess the extent of film coverage by viewing the dull side of the grid with reflected light at a 30–40° angle in a dissecting microscope. Sort the grids with respect to film coverage (good, 75–100%; fair, 60–75%) and store the grids in petri dishes on filter paper in the refrigerator. The grids can be stored for up to 3 months without apparent deterioration.

Solutions for Spreading Drosophila Cells

All solutions are made in deionized, glass-distilled water, which is stored in a glass container. The pH of the solutions is adjusted by the dropwise addition of pH 10 buffer standard for pH meters. We are currently using Fisher brand (Pittsburgh, PA) (Cat. No. SB116-500), which is a 0.05 M solution of potassium carbonate, potassium borate, and potassium hydroxide. All of the solutions are filtered through 0.2-μm pore size disposable Nalgene filter units and are stored and used at room temperature, unless otherwise noted.

1. pH 9 water adjusted to pH 8.5–9. The pH adjustment should require between 1 and 3 drops of pH 10 buffer/100 ml of water. Make 100 ml of pH 9 water just prior to use.
2. Sucrose–formalin containing 10% (v/v) formalin (i.e., 100 ml of 37% (w/w) formaldehyde solution/100 ml) and 0.1 M sucrose (RNase-free), adjusted to pH 8.5–9. Make the solution fresh weekly, but adjust the pH daily, if necessary.
3. Kodak Photo-Flo 200 [0.4% (v/v)] solution, pH 8.5–9. Make 100 ml of Photo-Flo fresh daily and store in a 100-ml beaker for convenient rinsing of grids.
4. Various detergent solutions, as discussed in Adapting to Other Systems, adjusted to pH 8.5–9 and made fresh daily.
5. Phosphotungstic acid [4% (w/v) PTA] stain solution. Microfilter and store in a brown glass bottle for up to 1 year.
6. Uranyl acetate [4% (w/v) UA] stock stain solution. Microfilter and store in a brown glass bottle for up to 1 year. (Caution: this solution contains a radioactive and toxic heavy metal salt).

Materials

1. Microcentrifugation chambers. These chambers are made by slicing plexiglass rods into disks of 5–8 mm in height. The diameter of the rod is not critical, provided that the disks will fit into the appropriate swinging bucket rotor for centrifugation. Standard 50- or 100-ml tube buckets of an International or DuPont Sorvall centrifuge are frequently used. A convenient diameter for the plexiglass rod is 25 mm. A hole of 3.5–4 mm in diameter, which is large enough to hold an EM grid, is drilled into the center of each disk, and a round cover glass is bonded to the bottom surface with Epon. In bonding the cover glass, care must be taken to form a complete seal around the hole with Epon, while avoiding Epon extrusion into the central hole. An alternative and simpler arrangement uses 10-mm-high disks with central holes drilled 7–8 mm into the disk.[10] The advantage of the first design is that the contents of the chamber can be viewed in an inverted microscope, which is occasionally useful for assessing the degree of dispersal or the presence of bubbles or large debris.
2. Dissecting microscope and light.
3. Microscope slides.
4. Double-stick tape.
5. Tweezers, Dumont #5, sharpened stainless steel (e.g., Ernest F. Fullam, Cat. No. 11040).
6. Tweezers, anticapillary, self-closing, Dumont pattern N4 (e.g., Ernest F. Fullam, Latham, NY, Cat. No. 14140).
7. Long forceps, 10–12 inches.
8. Slide with one to three concavities ("depression slide") for chromatin dispersal (e.g., Extra-thick hanging drop slide, Fisher Cat. No. 12-565B).
9. Pasteur pipets.
10. Ross optical lens tissue.
11. Bibulous paper.
12. Microtiter plate (96-well).
13. Small glass petri dish filled with 95% (v/v) ethanol.
14. Wash bottles, one with distilled H_2O and the other with 95% (v/v) ethanol.

Drosophila Chromatin Spreading

Drosophila as Experimental System. Drosophila is representative of eukaryotes in general, regarding transcription and RNA processing. The splicing mechanism in *Drosophila* is closely related, if not identical, to that of higher eukaryotes.[12] Thus, *Drosophila* is an appropriate system in which to study RNA processing, while being much more amenable to spreading than mammalian cells. There are two reasons why *Drosophila* is preferable to a mammalian system. The first is the ease of obtaining large, manipulable, actively transcribing oocytes and embryos, whose chromatin can be dispersed under very mild conditions with retention of the RNP structure. The second is the size of the *Drosophila* genome [1.6×10^5 kb (kilobases)], which is one-twentieth the size of a typical mammalian genome, resulting in a much higher probability of recognizing an active gene. As discussed in Electron Microscopy, the rate-limiting step in the method is scanning the grids searching for active genes among the inactive nucleosomal chromatin.

Embryo Collection. To obtain sufficient embryos or egg chambers at the appropriate stages of development, it is important that the flies be healthy and well fed. We typically use wild-type Oregon—R *Drosophila melanogaster* that are raised in uncrowded conditions in half-pint milk bottles, containing a cornmeal medium.[13] We rear flies on two-staggered 2-week cycles, which ensures a steady supply of fecund females. (Flies emerge 11–12 days after eggs are laid. New adults are transferred to a fresh bottle and are used for embryo collection for 3–9 days post-eclosion.)

Embryos are used at 3–4 hr of development. One embryo is needed per grid, and typically no more than 20-grid preparations are made per day. Embryo collection can be done conveniently on a small scale by transferring flies to an empty half-pint milk bottle and by capping the bottle with a moist laying surface, i.e., the bottom of a 60 × 15-mm petri dish that has been half-filled with 2.5% (w/v) agar containing 1% (w/v) sucrose. A small amount of yeast paste is spread thinly on the agar surface prior to use. (Both the agar plates and the yeast paste can be made ahead and can be stored in the refrigerator, but both should be at room temperature when used. The yeast paste is a smooth, thick paste made with dry baker's yeast and water.) The bottle containing the flies is inverted, so that the bottle is sitting on the laying plate, and is left undisturbed for 1 hr, at which time a fresh laying plate is substituted. We usually do three to four consecutive 1-hr lays early in the day. The first lay is discarded, because the lay may contain embryos of more advanced

[12] D. C. Rio, *Proc. Natl. Acad. Sci. U.S.A.* **85,** 2905 (1988).
[13] D. B. Roberts, *in* "Drosophila: A Practical Approach" (D. B. Roberts, ed.), p. 1. IRL Press, Washington, D.C., 1986.

stages, since females will sometimes hold eggs internally until fresh food is found.

The freshly laid eggs are allowed to develop at room temperature (25°) for 2 hr, at which time, the embryos range in age from 2 to 3 hr. Embryos (3-hr) are undergoing cellularization, at which stage the nuclei have migrated to the periphery of the embryo, and cell membranes have begun inward growth, but have not yet completely surrounded the nuclei.[14] It is an excellent time to spread embryos for visualization of active genes, because transcriptional activity increases markedly at this stage and because the nuclear contents are very accessible for dispersal using mild conditions, since the nuclear contents are not yet within cell membranes or tissues.[15]

To determine the developmental stage of an embryo, the embryo must be dechorionated. The chorion (eggshell) is manually removed, while viewing in a dissecting microscope. Using tweezers, the embryos are lifted individually or in small groups from the laying plate and are deposited carefully on a piece of double-stick tape on a standard microscope slide. The chorion is removed by touching the upper surface of the embryo with the tweezers and then by slightly rolling the embryo in either direction. The chorion will break and remain on the tape, and the embryo will readily stick to the tweezers. The embryos are still encased in a vitelline membrane, but the embryos are quite fragile and subject to desiccation and must be handled gently.

Once dechorionated, the embryos are staged by viewing in an inverted compound microscope. A well slide (made by boring a $\frac{1}{4}$-inch-diameter hole in a standard microscope slide and by covering the hole with a cover slip attached with melted paraffin) is filled with Photo-Flo solution, and the embryos are deposited on the bottom of the well. Using the 20 or 40× objective, one can see peripheral nuclei, which form a discrete monolayer, and can observe the extent of inward cell membrane growth. Embryos exhibiting about 10–12 μm of inward membrane growth are at an excellent stage for spreading.[15] Older embryos will have begun to gastrulate and are more difficult to spread; younger embryos are not transcriptionally active.

If an inverted microscope is not available, shorter egg lays (of ~20 min) will also allow fairly precise timing of development. At 25°, spreads should be done between 170 and 190 min after laying, but this time will vary for only slightly different temperatures.[13]

---

[14] V. E. Foe and B. M. Alberts, *J. Cell Sci.* **61,** 31 (1983).

[15] S. L. McKnight and O. L. Miller, Jr., *Cell (Cambridge, Mass.)* **8,** 305 (1976).

Procedure for Spreading Embryo Chromatin

1. Using tweezers, transfer an embryo of the desired stage to the well of a depression slide, which has, immediately prior to this, been filled with 60–80 μl of pH 9 water. Embryos will stick to the tines of the tweezers or will be in the liquid adhering to the tines.

2. Observing through the dissecting microscope and using two tweezers, pull the embryo apart, macerating and stirring until the embryonic contents have dispersed. Remove the insoluble translucent vitelline membrane with tweezers. Stir the solution continuously for at least 1 min and then pipet up and down with either a mouth-controlled "braking pipet," a Pasteur pipet with a pulled-out bore, or a micropipettor, being careful not to create bubbles.

3. After dispersing for at least 5 min with occasional stirring (or up to 30 min), add an equal volume of the sucrose–formalin solution and mix, either by stirring or pipetting. While the chromatin is dispersing for an additional 10 min or so, prepare the microcentrifugation chamber.

4. Using a Pasteur pipet, overfill a clean, dry microcentrifugation chamber with the sucrose–formalin solution to obtain a convex meniscus, making certain that no air bubbles are trapped in the chamber (Fig. 1b, step 1).

5. Using anticapillary tweezers, pick up a 300-mesh carbon-covered EM grid, holding onto the grid by the outer edge, so as not to break the film. Swish the grid for 1 min in 95% (v/v) ethanol in a small, open dish (which renders the grid hydrophilic), and rinse the grid well with the sucrose–formalin solution (about two refills of a Pasteur pipet). Without allowing the grid to dry, slide the grid under the rounded up meniscus on the microcentrifugation chamber with the shiny side up (Fig. 1b, step 1). Release the grid, and the grid should gently float to the bottom of the chamber.

6. Using a Pasteur pipet, remove most of the sucrose–formalin solution from the chamber, leaving the chamber $\frac{1}{4}$–$\frac{1}{3}$ full (Fig. 1b, step 2).

7. Gently mix the dispersing chromatin by pipetting and then carefully layer the chromatin over the sucrose–formalin solution, completely filling the chamber. Place a round, 18-mm cover slip on top of the microcentrifuge chamber and press down firmly to seal the cover slip (Fig. 1b, step 3). Blot the excess liquid with bibulous paper. Typically, four chambers containing chromatin from four different embryos are prepared during the same period, with the first embryos dispersing for longer periods of time than the latter embryos.

8. Place the microcentrifugation chambers into the buckets of an appropriate swinging bucket rotor. We typically use a Sorvall GLC-1

tabletop centrifuge equipped with 100-ml swinging buckets, which have been fitted with rubber pads to form a flat inside surface. Long-handled (10–12 inches) forceps are used to place the chambers on the rubber pads. The chambers are then centrifuged at 2000–2500 g for 6 min at room temperature without braking. Alternatively, a Sorvall centrifuge equipped with a HB-4 swinging bucket rotor can be used. Standard black plastic bottle caps with an outside diameter of about 2.8 cm are used as adaptors. A microcentrifuge chamber is placed inside a bottle cap, and the tines of the long-handled forceps are pressed against the inside walls of the cap, as the cap is slowly lowered to the bottom of the bucket.

9. After centrifugation, remove the cover slip from the grid chamber and add sucrose–formalin to round up the meniscus over the well (Fig. 1b, step 4). Invert the chamber, hold the chamber at eye level, and tap the chamber gently with the forceps. The grid should float to the surface of the drop from which the grid is picked up with the anticapillary tweezers (Fig. 1b, step 5).

10. Holding the grid securely by the edge, gently swish the grid in a beaker of Photo-Flo solution for about 30 sec. Lift the grid from the Photo-Flo and gently blot the excess liquid with lens tissue, touching only the edge of the grid and the area between the tines of the tweezers. The grid should air-dry uniformly. The Photo-Flo reduces surface tension during the drying process. The grid can now be stained, or grids can be accumulated for later staining.

Procedure for Spreading Oocyte Follicle Cell Chromatin. The developing *Drosophila* oocyte is surrounded by a monolayer of follicle cells, which are responsible for laying down the chorion. We have used these cells extensively for the visualization and analysis of the preferentially amplified chorion genes.[16,17] Follicle cell chromatin is spread in essentially the same manner as embryo chromatin. To be assured of having egg chambers at all stages of oogenesis, we transfer young flies to fresh food bottles, supplemented with a small amount of baker's yeast, for 3 consecutive days prior to use.

1. Put a well-fed female fly in a small dish, over ice, containing ice-cold pH 9 water. Dissect out the ovary and pull apart the ovarioles under water. Egg chambers of various late stages should be visible.
2. Rapidly select the desired stage (typically stage 10–14) and dissect the desired stage out. Still working under the pH 9 water, grasp the egg chamber with the tweezers in the middle and pinch off the end

[16] Y. N. Osheim and O. L. Miller, Jr., *Cell (Cambridge, Mass.)* **33,** 543 (1983).
[17] Y. N. Osheim, O. L. Miller, Jr., and A. L. Beyer, *Cell (Cambridge, Mass.)* **43,** 143 (1985).

containing the nurse cells. (See Mahowald and Kambysellis[18] for photographs and descriptions of the various stages.)

3. Using tweezers, transfer the remaining portion of the egg chamber to the well of a depression slide, filled with 100 μl of either pH 9 water or the detergent solution of choice (see later), and macerate the egg chamber using two tweezers, being especially certain that the chorion–vitelline membrane surrounding the egg chamber is well pulled apart.

4. Stir the dispersal solution for at least 2 min and gently pipet the solution up and down a few times, being careful not to create bubbles.

5. Let the solution disperse for 20–40 min with occasional mixing, then add 25 μl of the sucrose–formalin solution, and allow 5–10 min for additional dispersal.

6. Continue with steps 4–10 of Procedure for Spreading Embryo Chromatin.

We previously spread follicle cells in a solution of 0.05% (v/v) Joy detergent and 0.05% (w/v) sodium deoxycholate (pH 8.2)[16] and embryos in a solution of 0.05% (v/v) Joy detergent (pH 8.5).[3] However, in the past year, we have had more success in spreading both cell types in pH 9 water. We were told by Proctor and Gamble that the chemical composition of the dishwashing detergent Joy was modified in early 1987, and we feel that this modification has adversely affected the spreading capacity of the detergent.

Contrast Enhancement

Staining Material on Grids. We routinely double-stain grids with PTA to stain proteins and with UA to stain nucleic acids. Staining is done in 96-well microtiter plates in 1% (v/v) ethanolic solutions made from the 4% (w/v) stock solutions.

1. Fill a small beaker with 95% (v/v) ethanol for rinsing the grids between stains and have the 0.4% (v/v) Photo-Flo solution in a beaker at hand.

2. Immediately before staining, prepare one well of each stain solution by first adding 9 drops of 95% (v/v) ethanol (from a 5¾-inch Pasteur pipet) and then 3 drops of the stock stain solution. Mix by pipetting. Since the ethanol evaporates rapidly, concentrating the

[18] A. P. Mahowald and M. P. Kambysellis, *in* "The Genetics and Biology of *Drosophila*" (M. Ashburner and T. R. F. Wright, eds.), p. 141. Academic Press, New York, 1980.

stain, make up fresh diluted solutions after staining every two grids.

3. Pick up a grid to be stained by the edge, using anticapillary, self-closing tweezers.

4. Lower the grid into the well containing the PTA solution, being certain that the grid is completely submerged, but not touching the bottom or the sides of the well. Gently agitate the grid for 30 sec.

5. Immediately immerse the grid in the beaker of 95% (v/v) ethanol and swish the grid gently for 15–20 sec.

6. Remove the grid from the ethanol and put the grid into a well containing UA. Agitate the grid gently for 1 min.

7. Rinse the grid in the beaker of ethanol for about 30 sec.

8. Dip the grid into the Photo-Flo solution for a few seconds.

9. Blot the grid on its side with lens tissue and air-dry as before. The grids can now be viewed in the electron microscope, or to improve contrast, lightly shadowed with platinum.

Rotary Metal Shadowing. Metal shadowing is not necessary and does not improve the quality of a spread. However, given that grids must be carefully scanned at relatively high magnifications to find active transcription units, the increased contrast enhancement, afforded by shadowing, can be very helpful. We have used vacuum evaporaters, equipped with an electron gun evaporation system (Balzers BAE 080 T) or a heat resistance evaporation system (JEOL JEE-4X), for metal shadowing. Either type can be used to deposit a thin coating of platinum onto the grids. The important factors are that the grids are all in one plane at a 6–9° angle to the platinum source and that the grids rotate at about 30–60 rpm during platinum deposition. Additional details can be found in Coggins.[11] We prefer a light shadow of about 15–20 Å. It is important that grids remain as flat as possible during the spreading and staining procedures, so that the grids can be uniformly shadowed.

Electron Microscopy

We view grids in a JEOL 100 CX transmission electron microscope operated at 80 kV with either a 40- or 60-μm objective aperture. Initially, we assess the grid by completely scanning several grid squares in different regions of the grid, using the binoculars and at a magnification of 5,000–10,000, to determine if chromatin is present and to assess how well the chromatin has spread.

Chromatin distribution on the grid is frequently not uniform. Thus, if no chromatin is seen by this approach, a quicker scan of a larger area is possible by dispensing with binoculars and searching for light gray regions on the grid. On inspection with binoculars, these regions sometimes reveal chromatin masses that are only well dispersed at their periphery. To see well-spread regions, it is normally necessary to use the binoculars and magnifications of 5,000–10,000. If the initial assessment of the grid indicates a good chromatin spread, we carefully examine each grid square in an orderly sequence looking for active genes. This procedure may take 2 days or so and is the rate-limiting step in the method. Even on grids with a considerable amount of well-spread chromatin from an organism with a small genome at a transcriptionally active stage (e.g., cellular blastoderm stage *Drosophila* embryos), very few active genes in a well-dispersed configuration suitable for RNP mapping may be seen (0–20/grid). On such grids, examples of the tandemly repeated ribosomal RNA genes of nucleoli (Fig. 2) are typically spotted several times, as are many additional nonribosomal genes in regions where the chromatin is too dense to allow mapping of the transcripts. In our experience, it is very rare to find a nonribosomal gene in a mammalian tissue culture cell spread, presumably due to the larger size of the genome and the low percentage of genes that are active at any time.

We scan grids at rather low illumination to minimize film breakage and contamination. To minimize beam damage to the specimen, we photograph potentially interesting (usually all) genes as soon as the genes are seen. We normally take at least two photographs, one photograph at a lower magnification to record the nature and transcriptional activity of the general region (5,000–10,000) and then one photograph at a higher magnification of the specific gene (10,000–25,000). We routinely check the magnification indicator on the microscope with a diffraction grating, so measurements can be accurate.

The quality of a spread can vary considerably across a grid, so it is essential that different regions of a grid be examined before a grid is discarded. If one makes eight grids, a high success rate would be to have chromatin on five to six of the grids with one grid being exceptionally good, two grids being acceptable, and the remaining two to three grids being marginal. A good spread is characterized by nucleosomal chromatin, which is laying in a relaxed manner and in a fairly low density across the grid. Genes should have nascent fibrils that are unwound, with distinct RNP structures, and lying on both sides of the deoxynucleoprotein (DNP) template. Marginal grids may contain regions where the chromatin is stretched or where nascent transcripts are all oriented in one direction, as if the chromatin was under tension. Other marginal grids will have chromatin that is not well dispersed, so that, although one can detect active genes by the presence of "polymerase backbones," ultrastructural details associated with the nascent transcripts are obscured by overlapping chromatin. Particles normally found on transcripts can also be amorphous instead of distinct, and, on certain grids, the transcripts themselves are not well dispersed. Grids in which no nucleosomal chromatin is apparent will frequently have regions, roughly spherical, which appear to contain large globular cables of nucleoprotein. Occasionally, intact nuclei are also observed. More often, however, grids will be devoid of material.

Adapting to Other Systems

When attempting to spread chromatin from other cell types, it may help to vary the detergent, the detergent concentration, cell concentration, length of dispersal (5–60 min), and temperature (4–25°) during dispersal. It generally requires numerous attempts before anything approaching a reasonable spread is obtained, and conditions must be worked out empirically for each cell type. Protocols that include mono- and divalent cations should be avoided, if possible, and during the spreading procedure, one should never take a preparation from a low-salt to a higher salt solution. The more quickly a cell can be lysed, the easier it will be to spread the cell successfully. The problems most frequently encountered when attempting to spread other cell types are (1) difficulty in breaking the cell wall and/or nuclear envelope and (2) difficulty in dispersing the chromatin. As a first step, hypotonic shock in pH 9 water should be attempted, in the maximum amount of water that will still allow a reasonable concentration of chromatin on the grid. Vortexing or homogenizing is sometimes useful at this stage. If nuclei are still intact, the next step typically involves varying the type and amount of detergent(s) used, such that nuclei will break, and chromatin will be dispersed, yet still in a nucleosomal configuration (i.e., not deproteinized). The dishwashing detergent Joy was originally chosen as the detergent of choice after a considerable amount of trial and error and has been used at concentrations ranging from 0.05%[3] to 0.5% (v/v).[2] However, as mentioned in Procedure for Spreading Oocyte Follicle Cell Chromatin, the composition of Joy was changed by Proctor and Gamble in 1987, and the detergent now seems less useful for spreading. Various investigators have used other detergents successfully, including Triton X-100 at 0.1–0.5% (v/v),[7,19] Nonidet P-40 (NP-40) at 0.2–0.5% (v/v)[7,19] sarkosyl at 0.01–0.1% (w/v),[10] and digitonin at 0.1% (w/v).[20] Frequently, two or more detergents are used in combination. For example, *Drosophila* follicle cells were spread with 0.025% (w/v) sodium deoxycholate (NaDOC) and 0.05% (v/v) Joy,[16] and *Chironomus* salivary gland cells were spread with 0.5% (v/v) Joy, 0.05%

---

[19] R. Tsanev and I. Tsaneva, *Methods Achiev. Exp. Pathol.* **12,** 63 (1986).
[20] M. Jamrich and O. L. Miller, Jr., *Chromosoma* **87,** 561 (1982).

(w/v) NaDOC and 0.1% (w/v) sarkosyl NL-97.[21] *Drosophila* salivary gland cells were spread with 0.6% (w/v) CHAPS {3-[(3-cholamido-propyl)dimethylammonio]-1-propane sulfonate}.[22] Mammalian tissue culture cells were lysed in 0.5% (v/v) Joy and then were quickly diluted and allowed to spread in ~0.2% (v/v) Joy.[10]

For some cells, prior treatment is necessary before membranes are accessible to hypotonic shock or detergent treatment. For example, yeast cell walls have been digested with zymolyase,[23] bacterial cells with lyso-zyme,[24] and mammalian tissues with collagenase.[10] These treatments should be as rapid and gentle as possible and all salt-containing buffers should be removed or diluted out while lysing nuclei. An alternative method for yeast cell disruption for chromatin spreading involves sectioning a frozen cell pellet with a cryostat microtome.[25]

Once the nuclear envelope has been disrupted, there is no guarantee that chromatin will readily disperse. At this stage, detergents may facilitate spreading by partially extracting proteins from the chromatin. Sarkosyl NL-30 at concentrations ranging from 0.01 to 0.10% (w/v) selectively removes histone H1 and may improve spreading, while maintaining chromatin in a nucleosomal form; however, at concentrations above 0.2% (w/v), DNA is deproteinized.[26] Yeast RNA at 100 μg/ml added to a preparation may improve spreading by removing histone H1 and by inhibiting endogenous nuclease activity.[27] A reducing agent, 2-mercaptoethanol, may improve spreading by preventing disulfide bonds from cross-linking chromatin.[8] Addition of ethylenediaminetetraacetic acid (EDTA) (up to 2 mM) often facilitates chromatin dispersal.[9] Another approach to improving chromatin spreads is to cut the chromatin using DNase I,[28] micrococcal nuclease, or restriction enzymes.[16,29] However, the slightly higher salt levels needed to get a minimum number of cuts from most restriction enzymes results in RNP fibrils that are more compacted and cannot be mapped.

[21] C. Francke, J.-E. Edstrom, A. W. McDowall, and O. L. Miller, Jr., *EMBO J.* **1**, 59 (1982).

[22] E. J. Hager, Ph.D. Dissertation, University of Virginia, Charlottesville (1987).

[23] J. B. Rattner, C. Saunders, J. R. Davie, and B. A. Hamkalo, *J. Cell Biol.* **92**, 217 (1982).

[24] S. French, K. Martin, T. Patterson, R. Bauerle, and O. L. Miller, Jr., *Proc. Natl. Acad. Sci. U.S.A.* **82**, 4638 (1985).

[25] L. Saffer and O. L. Miller, Jr., *Mol. Cell. Biol.* **6**, 1148 (1986).

[26] U. Scheer, *Cell (Cambridge, Mass.)* **13**, 535 (1978).

[27] V. E. Foe, L. E. Wilkinson, and C. D. Laird, *Cell (Cambridge, Mass.)* **9**, 131 (1976).

[28] J. B. Rattner, A. Branch, and B. A. Hamkalo, *Chromosoma* **52**, 329 (1975).

[29] R. H. Reeder, T. Higashinakagawa, and O. L. Miller, Jr., *Cell (Cambridge, Mass.)* **8**, 449 (1976).

One can also vary the amount of centrifugation. If plasmids are being spread, one frequently needs centrifugal speeds of up to 25,000 g.[30] Occasionally, it is necessary to dilute a preparation considerably, with a concomitant reduction in the amount of material visible on the grid. Centrifuging preparations at higher speeds and for longer times will increase the amount of material on the grid. (See Refs. 9 and 10 for a description of an EM grid adaptor that fits a Beckman Airfuge and will achieve speeds of 150,000 g.)

We routinely swish grids in 95% (v/v) ethanol for 1 min to make them hydrophilic. If chromatin adsorption to the grid is a problem, other methods can be tried, such as glow discharge, as discussed by Coggins.[11]

Analysis of the Electron Microscopic Data

Identification of Polymerase I, II, and III Genes

Nascent RNA transcripts are identified in chromatin spreads as lateral fibrils extending from a chromatin strand. The transcripts typically occur in groups representing an actively transcribing gene. These genes display a polarity, in that one end, near the promoter, has short, newly initiated transcripts, which generally increase in length in going from the initiation to the termination end of the gene. Our interest is in the RNP structure of precursor mRNA (pre-mRNA), which is the product of genes transcribed by RNA polymerase II (Pol II). Thus, it is important to ultrastructurally distinguish Pol I, II, and III transcription units. Pol I [precursor rRNA (pre-rRNA)] genes, as seen in Fig. 2, are readily identified in chromatin spreads, because Pol I genes are multiple-copy and tandemly repeated in a head-to-tail orientation in *Drosophila* and most other eukaryotic organisms.[5,31] Pol I genes are characteristic in length for a given organism (~2.6 μm in *Drosophila*) and are typically very heavily transcribed, with RNA polymerase density near a maximum (~35–40/μm). The products of Pol III genes, such as 5 S RNA and tRNA, are very short [<300 nt (nucleotides)]. These genes typically go unnoticed in chromatin spreads (but see Ref. 32), because their nascent transcripts are too short to extend from the polymerase complex on the template. Thus, we classify all non-Pol I genes seen as Pol II genes. As expected for pre-mRNA transcripts, Pol II

---

[30] A. H. Bakken, G. Morgan, B. Sollner-Webb, J. Roan, S. Busby, and R. H. Reeder, *Proc. Natl. Acad. Sci. U.S.A.* **79**, 56 (1982).
[31] U. Scheer and H. Zentgraf, *in* "The Cell Nucleus" (H. Busch and L. Rothblum, eds.), Vol. 11, p. 143. Academic Press, New York, 1982.
[32] U. Scheer, *Biol. Cell.* **44**, 213 (1982).

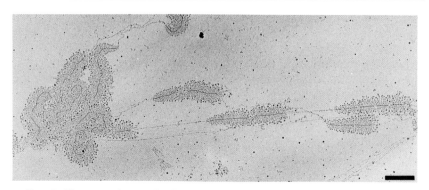

FIG. 2. Electron micrograph of a portion of dispersed nucleolar chromatin showing several pre-rRNA (Pol I) genes from *Drosophila*. These genes typically exhibit a high RNA polymerase density and occur in a tandem head-to-tail orientation. Bar, 1 μm.

genes are typically not tandemly repeated and vary considerably in length and polymerase density. Some of them, like the chorion genes, have a polymerase density as high as pre-rRNA genes,[17] while others consist of a single transcript, and the majority are transcribed at an intermediate level (Fig. 3).

RNP Fibril Mapping

If a gene exhibits several transcripts, one can compare RNP structure on multiple nascent transcripts from a single initiation site, and thus on RNA molecules of the same sequence. We use "RNP fibril mapping" to analyze and display this information and have applied this method to both pre-mRNA[3,17,33–35] and pre-rRNA genes.[35] RNP fibril maps display schematized RNA transcripts from a single transcription unit, such that similar RNA sequences are aligned and RNP structures occurring on the transcripts are shown at the appropriate positions. These two features are shown in Figs. 4 and 5.

The first step in RNP fibril mapping is measuring the contour length of the RNA transcripts and their position on the template with respect to the first (typically the shortest) transcript on the gene. We do length measurements on a Numonics graphics calculator from a print of the electron micrograph at a magnification of at least 50,000. In tracing the transcripts,

---

[33] A. L. Beyer, A. H. Bouton, and O. L. Miller, Jr., *Cell* (*Cambridge, Mass.*) **26,** 155 (1981).
[34] A. L. Beyer and Y. N. Osheim, *Genes Dev.* **2,** 754 (1988).
[35] Y. N. Osheim and A. L. Beyer, *UCLA Symp. Mol. Cell. Biol.* **26,** 277 (1985).

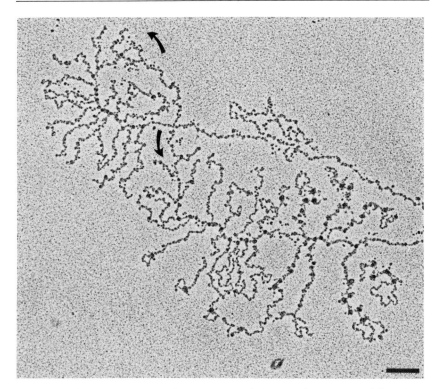

Fig. 3. Electron micrograph of pre-mRNA (Pol II) genes from a *Drosophila* embryo. Two divergently transcribed unidentified genes are shown on a single chromatin strand, with arrows indicating the direction of transcription. The shorter gene in the upper left portion of the figure has a higher RNA polymerase density $(17/\mu m)$ than the longer gene in the lower right portion $(13/\mu m)$. The transcripts of the two genes also differ in their RNP structure, with those of the shorter gene being mainly fibrillar, while those of the longer gene have RNP particles and loops superimposed on the basic fibril. Bar, 0.2 μm.

the position and type (see later) of RNP structures are recorded. Figure 4 illustrates the two methods that are then used to align RNA sequences graphically on the RNP fibril maps,[3,33] depending on whether the transcripts have been "processed" while nascent. This is determined by linear regression analysis of RNP fibril length for each gene. That is, the contour length of each transcript is plotted as a function of its position on the DNA template with respect to the first fibril of the array, and the linear equation best describing this distribution is derived by the least-squares method (Fig. 4b and b'). If a single line represents a good fit, as determined by the value of the correlation coefficient (>0.9), then it is assumed

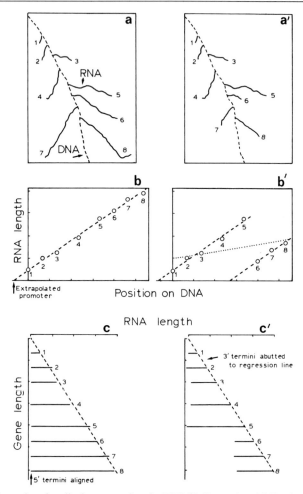

FIG. 4. Procedure for aligning transcripts for RNP fibril mapping. (a) Tracing of a theoretical gene whose transcripts increase in length along the gene. (a') Tracing of a theoretical
gene whose more mature transcripts are shorter than expected for their position on the
template. To simplify the figure, the RNP structures which typically accompany this abrupt
transcript shortening (RNP particles and loops) are not shown. (b and b') RNA length as a
function of DNA template position for the genes in a and a', respectively. The dashed lines
represent the best-fit linear regression lines. The dotted line in b' represents the regression
line, describing all of the transcripts in that gene as one population rather than two. (c and c')
Approximate graphical alignment of RNA sequences in preparation for RNP fibril mapping.
RNA sequences are aligned in c by 5'-terminal alignment and in c' by 3'-terminal abutment
to the regression line for the first (unprocessed) fibril length gradient. The spacing of the
transcripts on the vertical axis corresponds to their position on the chromatin template.

that the transcripts are initiated at a single site and have not been processed or broken while nascent.[3,36] In these cases, which represent the majority, transcripts are aligned by aligning 5' termini to the left in the fibril map (Fig. 4c). For some genes, however, the RNP fibril length gradients display abrupt transitions, and the transcript data are fit significantly better by two or three linear equations (with similar slopes and different intercepts) than the data are fit by a single line (Fig. 4b'). The 5'-terminal alignment procedure cannot be used, because many of the transcripts have lost sequences from near their 5' ends. (As we now know, these abrupt transitions in transcript length correspond to intron removal during the process of splicing, as discussed later.) In these cases, the 3' ends of the transcripts are abutted to a sloped line which represents the best-fit line to the first gradient of fibril length (Fig. 4c').

RNA sequences are not perfectly aligned by either of these methods because individual transcripts vary in their RNA compaction (i.e., nucleotides of RNA per microgram) when dispersed by the spreading method. The native configuration of these transcripts is most likely a compact granule, which is disrupted and decondensed to varying degrees by the technique. The degree of variation in RNA compaction is greater from one grid or one grid region to another[34,37] (ranging from 1.5 to 18 kb/μm),[3] than the degree of variation is among the transcripts of a given gene ($\pm 15\%$ of the "best-fit" transcript at a given template position).

Types of Ribonucleoprotein Structure Seen

There are only a few types of RNP structure seen when transcripts are prepared for microscopy by the Miller spreading technique. These are shown in Fig. 5 and presumably represent a limited heterogeneity in native RNP structure.

The basic structure of pre-mRNA transcripts is an irregular RNP "fibril" that averages about 5 nm wide and is sensitive to both ribonuclease and protease digestion. The fibril frequently appears to have an underlying repeating or beaded structure (Fig. 5b). The protein(s) responsible for this structure, which are probably heterogeneous nuclear RNPs (hnRNPs),[38] are deposited on the nascent transcript immediately at the DNA–RNA transcription fork and are presumably responsible for compacting the RNA, such that the RNA is typically one to four times shorter than the length of DNA that served as its template.

[36] A. L. Beyer, *Mol. Biol. Rep.* **9**, 49 (1983).

[37] U. Skoglund, K. Anderson, B. Bjorkroth, M. Lamb, and B. Daneholt, *Cell (Cambridge, Mass.)* **34**, 847 (1983).

[38] S. Y. Chung and J. Wooley, *Proteins: Struct., Funct., Genet.* **1**, 195 (1986).

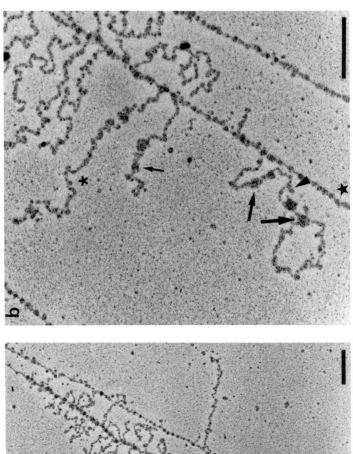

FIGS. 5a and 5b.

Superimposed on this fibril are RNP particles, which are categorized as small (<20 nm), medium (average 25 nm), and large (average 40 nm) (Fig. 5b). (These size measurements are done on metal-shadowed preparation and thus reflect relative rather than absolute sizes.) The medium-sized particles form a discrete size class and occur at nonrandom positions on nascent transcripts (e.g., Figs. 5 and 6). These positions are characteristic for the transcripts on a given gene, but are different from one gene to the next, as expected for RNA-processing signals. It can be readily determined from transcript length measurements and linear regression analyses that formation of these particles on the RNP fibril does not result in any further compaction of the fibril, at least in its dispersed configuration.[33] Thus, the amount of RNA in a 25-nm RNP particle is on average the same amount of RNA contained in 25 nm length of the RNP fibril. Due to differences in RNP compaction in different preparations, this ranges from about 70 to 300 nt.[3,17] (See below for converting length

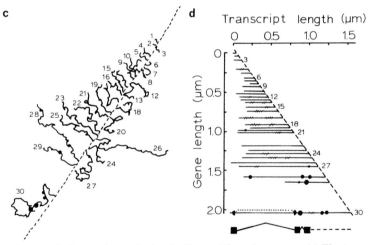

FIG. 5. RNP fibril mapping analysis of a *Drosophila* embryo gene. (a) Electron micrograph of an unidentified gene. Bar, 0.2 μm. (b) Higher magnification view of transcripts near the termination end of the gene showing the various RNP structures seen. (Small arrow) <20-nm RNP particle, (medium arrow) ~25-nm RNP particle, (large arrow) ~40-nm RNP particle at base of open RNP loop, (arrowhead) hairpin loop, (asterisk) ~5-nm RNP fibril, and (star) nucleosomal DNA. Bar, 0.2 μm. (c) Interpretive tracing of EM. (----) Template chromatin, (——) RNP fibrils, and (●) RNP particles. (d) RNP fibril map. Transcripts are linearized and aligned by abutting 3′ termini to regression line. RNP particles are shown as closed circles, hairpin loops are shown as slashed regions, and the open loop on transcript #30 is shown as a dotted line connecting the two "half-particles" at the points of intramolecular contact. At the bottom is shown the proposed exon–intron structure of the transcripts. (■) Exons, (——) intron, and (---) difficult to assign as intron or exon. Transcript #29 has lost the open (intron) loop by splicing of the nascent transcript. It is a general observation that the 25-nm particle at the 5′ end of the intron loop is not deposited until the 3′ end of the loop (3′ splice site) is synthesized.[34] (Adapted with permission from Ref. 35.)

measurements to molecular estimates.) The useful resolution for mapping particle position on a transcript thus varies from about 100 to 400 nt, with the additional uncertainty contributed by the variation in RNP compaction between transcripts on a given gene. Although this resolution seems poor in molecular terms, the resolution is quite reasonable for ultrastructural analysis of long transcripts.[3,33,34]

The small (<20-nm) RNP particles occur more randomly than the 25-nm particles and may represent fibril irregularities due to differing degrees of transcript dispersal. In a few cases, the irregularities occur at the same position as the 25-nm particles and appear to be a smaller form of these particles.

The large, ~40-nm particles form by the stable association, and apparent coalescense, of two 25-nm particles, looping out the intervening RNP fibrillar region (Fig. 5). The two 25-nm particles involved in forming a 40-nm particle are always nearest-neighbor particles, but can be spaced by less than 100 nt of RNA or up to 25 kb, resulting in widely variable loop sizes on different genes or gene regions. These "open loops" are one of two secondary RNP structures seen on nascent transcripts. We refer to the other type as "hairpin loops," because of their appearance as short, lateral projections (of approximately double-fibril thickness) from the basic RNP fibril [arrowhead (Fig. 5b)]. The lengths of these projections indicate that the projections encompass 100–400 nt of RNA, with those less than ~100 nt presumably not distinguishable from other fibril irregularities. The projections are most likely not hairpin loops in the typical base-paired sense, because (1) although the projections tend to occur in general regions of the transcript, their position varies more than particle position varies; (2) the projections are found on chorion gene transcripts, although these transcripts are known not to have any large inverted repeat sequences; and (3) double-stranded RNA formation is rare *in vivo* on pre-mRNA,[39] presumably due to the helix-destabilizing properties of hnRNP.[40] These considerations have led us to propose that the projections may be the result of topological constraints during fibril unwinding.[17] The hairpin loops are shown (linearized) on RNP fibril maps as slashed regions, while open loops are also linearized and indicated by a dotted line above the fibril connecting the two loop base sites (Fig. 5d).

Structures Related to Splicing on Nascent Transcripts

Both direct and indirect evidence have led us to conclude that the 25-nm particles represent spliceosome assembly intermediates, that the 40-

[39] D. Solnick and S. I. Lee, *Mol. Cell. Biol.* **7,** 3194 (1987).
[40] F. Cobianchi, R. L. Karpel, K. R. Williams, V. Notario, and S. H. Wilson, *J. Biol. Chem.* **263,** 1063 (1988).

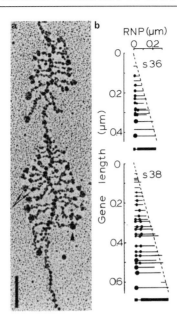

FIG. 6. RNP fibril mapping of two *Drosophila* chorion genes from oocyte follicle cells. (a) Electron micrograph of the tandemly linked *s36* (top) and *s38* chorion genes of *D. melanogaster*. The transcripts generally display either two 25-nm RNP particles (thin arrows) or a single larger particle (arrowhead) near their 5' ends. Bar, 0.2 μm. (b) RNP fibril map of the two chorion genes. When compared to the known gene structure (as shown below each map), it can be seen that the 25-nm particles occur at or near the 5' and 3' splice sites and that the more mature transcripts display a single 40-nm particle, resulting from the stable association of the two 25-nm particles. Intron loops are not seen on these transcripts due to their small size (91 and 228 nt). (Adapted with permission from Ref. 17.)

nm particles represent spliceosomes, and that the open loops formed between spliceosome intermediates represent intron loops. The direct evidence supporting this conclusion is the observation of these particles specifically at 5' and 3' splice junction sites on several *Drosophila* chorion genes (Fig. 6),[17] which are some of the few genes that have been identified in chromatin spreads.[16] The temporal sequence of events, which is seen on many unidentified genes, is that 25-nm particles form first at 3' and then 5' splice sites, followed by their stable coalescence into a 40-nm particle with concomitant loop formation, followed by removal of the particle and loop.[17,34] This sequence can be inferred, because the multiple transcripts on a gene show a series of increasingly mature RNA-processing intermediates in going from the initiation to the termination end of the gene. This ultrastructure, sequence of events, and range in loop length are exactly as expected for splicing, based on many biochemical studies *in vivo* and *in vitro*. A thorough discussion of the phenomenon and of the indirect evidence supporting the identity of the observed process as splicing can be found in Beyer and Osheim[34] and Beyer *et al.*[41] It is the phenomenon of loop removal, or splicing, that is responsible for the abrupt transitions in fibril length that are observed with some frequency on nascent transcripts and necessitate the second type of RNP fibril map (Fig. 4c').

Converting Length Measurements to Molecular Estimates

It is generally of interest to consider gene or intron size in kilobases rather than in micrometers of contour length measurement. Since DNA compaction (i.e., DNA kilobases per micrometer) varies over a smaller range and is more reproducible than RNA compaction in these spreads, we use the DNA template measurements to estimate the molecular size of the gene and to deduce the transcribed RNA lengths. The conversion of micrometer measurements of DNA to kilobases is based on the observations that nucleosome density[42] and DNA contraction[43] are inversely proportional to RNA polymerase density.[43] The assumption is made that DNA contraction varies on a linear continuum between known values at either extreme of RNA polymerase density. At one end is the contraction ratio of the fully transcribed rDNA and chorion genes in these preparations (3.3 kb/μm),[44] and at the other end is the contraction ratio of inactive nucleosomal DNA in Miller spreads (4.8 kb/μm).[42] The equation $y = -0.04x + 4.87$, where x is RNA transcripts per micrometer of DNA and y is DNA kilobases per micrometer, describes this empirical relationship.[33]

Specific Applications to Improve Information Yield of Technique: A Progress Report

Overview

All of the approaches listed in the following sections could significantly improve the information yield of the technique by allowing identification of specific genes, specific proteins, and specific nucleic acid sequences in chromatin spreads. Varying, but limited, degrees of success have been achieved in all of these approaches, as briefly reviewed in the next three sections. It is important to keep in mind, however, that a

---

[41] A. Beyer, M. Jamrich, and Y. Osheim, *UCLA Symp. Mol. Cell. Biol.* **94**, 133 (1989).
[42] S. L. McKnight and O. L. Miller, Jr., *Cell (Cambridge, Mass.)* **17**, 551 (1979).
[43] M. F. Trendelenburg and J. B. Gurdon, *Nature (London)* **276**, 292 (1978).
[44] Y. N. Osheim, O. L. Miller, Jr., and A. L. Beyer, *EMBO J.* **5**, 3591 (1986).

successful Miller spread is obtained by rapidly disrupting unfractionated cells, exposing the contents briefly to no salt, high pH conditions, and immediately depositing and fixing the chromatin on a grid. These optimal conditions are incompatible with incubations in physiological or harsher buffers required for immunological reactions or nucleic acid hybridization. It is hoped that ongoing research in several laboratories will make one or all of the following approaches more generally applicable than the approaches are at the present time.

Immunoelectron Microscopy

Reaction of specific antibodies with the nucleoprotein material in chromatin spreads could be very informative regarding the composition and function of various structures. There have been two general approaches, involving either antibody reaction in the chromatin lysate in solution or reaction on the spread chromatin on a grid. In general, when the reaction is done in solution, there is a trade-off between the efficiency of the reaction and the degree of dispersal, because of the different ionic conditions optimal for the two processes. However, when the antigen in question is present at many sites, even a low-efficiency reaction can be useful. For example, using antibodies to ribosomal proteins, Chooi and Leiby[45] maintained excellent ultrastructure and achieved high specificity, but achieved low-efficiency labeling of nascent rRNA transcripts. Due to the large number of transcripts per gene (~100) and multiple-gene copies, Chooi and Leiby were able to use a statistical approach to determine the initial point of protein assembly on the transcripts. Fakan *et al.*[46] got high-efficiency and high-specificity labeling of nascent transcripts, but the bound antibodies may have prevented good dispersal of those same transcripts. Fakan *et al.* were, nevertheless, able to make the important observation that both hnRNPs and snRNPs associate with nascent transcripts.

We are pursuing immunological reactions done on the grid after chromatin dispersal, because our interest is in identifying specific snRNP and hnRNP in specific structures on well-dispersed transcripts. With this approach, we have found that there is a trade-off between the specificity of the reaction and the preservation of ultrastructural detail. The carbon film and nucleoprotein material on the grid are both "sticky," and rather stringent blocking and washing conditions are required to decrease nonspecific adsorption of antibodies and electron-dense markers. These

[45] W. Y. Chooi and K. R. Leiby, *Proc. Natl. Acad. Sci. U.S.A.* **78**, 4823 (1981).
[46] S. Fakan, G. Leser, and T. E. Martin, *J. Cell Biol.* **103**, 1153 (1986).

blocking and washing conditions lead to both obscuring of the specimen and specimen loss from the grid, resulting in poorly preserved structure. We are linking the primary antibodies directly to colloidal gold particles in order to minimize the number of incubations and are also experimenting with various blocking and washing protocols.

In Situ Hybridization

In situ hybridization of specific nucleic acid probes could provide an opportunity to identify and analyze specific active genes, as well as the configuration of specific segments of the transcripts. Hamkalo and colleagues have pioneered the application of in situ hybridization using biotinylated probes on Miller chromatin spreads.[47,48] Hamkalo and coworkers have been very successful at the level of whole-mount metaphase chromosomes (prepared for EM by a modified Miller technique), on which these investigators have been able to localize moderately repetitive sequences with high resolution. Attempts to extend the resolution to the level of individual active transcription units have again met with the problem of incompatibility between optimal conditions for hybridization versus spreading. Although specific hybridization has been obtained using rDNA genes as a model system, specimen protein loss is a problem, and details of the gene structure are not retained.[48] Ongoing efforts are directed at improving morphological resolution.

Identification of Specific Genes in Chromatin Spreads

The routine identification of specific active genes with known sequence and intron–exon structure would be extremely valuable for many studies, as exemplified by the information we have obtained on Drosophila chorion gene amplification,[16,49] transcription,[16,49] RNA processing,[17] and termination.[44] If other genes can be made ultrastructurally identifiable, for example, by introducing the genes into cells on multicopy plasmids, harsher techniques for gene identification, such as in situ hybridization, can be avoided, and excellent transcript morphology can be preserved. This is particularly important to our goals of studying the in vivo splicing reaction in a minimally disrupted RNP configuration.

A few specific Pol II genes have been identified in Miller spreads by

[47] N. J. Hutchison, P. R. Langer-Safer, D. C. Ward, and B. A. Hamkalo, J. Cell Biol. 95, 609 (1982).

[48] S. Narayanswami and B. A. Hamkalo, in "Electron Microscopy in Molecular Biology: A Practical Approach" (J. Sommerville and U. Scheer, eds.), p. 215. IRL Press, Washington, D.C., 1987.

[49] Y. N. Osheim, O. L. Miller, Jr., and A. L. Beyer, Mol. Cell. Biol. 8, 2811 (1988).

virtue of ultrastructural peculiarities. These genes have been highly active and/or multicopy insect genes, which were identified in the particular tissues at the particular developmental stages when the genes were most active.[16,50,51] A mammalian virus gene has also been identified in dispersed chromatin by virtue of its location on multicopy extrachromosomal viral genomes.[52] However, this approach is generally unreasonable for a single-copy mammalian gene, which is contained in a ~20-fold larger genome, even if the gene product is extremely abundant.

There have been a few reports on the visualization of specific cloned genes on plasmids after injection into *Xenopus* oocytes,[30,35,43,53] although the limited number is an indication that the procedure is not yet routinely useful. The great majority (typically estimated at >90%) of injected plasmids remain transcriptionally inactive, and the search for an active gene in a favorably dispersed configuration becomes the limiting factor. Trendelenburg *et al.*[53] suggested that the low efficiency of initiation might be due to their use of a heterologous promoter, but we have found the same results using the *Xenopus TFIIIA* gene promoter (unpublished). We and others[53] have also noted the inefficiency of Pol II transcription termination in *Xenopus* oocytes, resulting in plasmids that are completely filled with transcripts of varying lengths, such that it is not possible to determine the initiation site of the gene.

Rather than attempting to find more efficient *Xenopus* promoter and terminator signals, we are focusing our current efforts on visualizing cloned genes on plasmid vectors after injection into *Drosophila* embryos, using promoters that are either active or inducible during early embryogenesis. Since these embryos can be dispersed for EM using very mild conditions, as described in this chapter, retention of the RNP structure on nascent transcripts should be excellent.

Acknowledgments

We thank Oscar Miller, Jr. for training us in the spreading technique and for continued support and encouragement. We thank Martha Farrell for sharing her film-making expertise with us. The authors' research is supported by grants from the National Institutes of Health and the ACS. A. Beyer is the recipient of an American Cancer Society Faculty Research Award.

[50] M. M. Lamb and B. Daneholt, *Cell (Cambridge, Mass.)* **17**, 835 (1979).
[51] S. L. McKnight, N. L. Sullivan, and O. L. Miller, Jr., *Prog. Nucleic Acids Res. Mol. Biol.* **19**, 313 (1976).
[52] A. L. Beyer, A. H. Bouton, L. D. Hodge, and O. L. Miller, Jr., *J. Mol. Biol.* **147**, 269 (1981).
[53] M. F. Trendelenburg, D. Mathis, and P. Oudet, *Proc. Natl. Acad. Sci. U.S.A.* **77**, 5984 (1980).

[34] Genetic Methods for Identification and Characterization of RNA–RNA and RNA–Protein Interactions

By ROY PARKER

The Problem

The biogenesis and function of RNA molecules require a large number of specific RNA–RNA and RNA–protein interactions. A frequent experimental goal is to identify and determine the nature of such interactions. For example, in the splicing of nuclear precursor mRNA (pre-mRNA), one aspect of the problem is as follows: what are the RNA–RNA and RNA–protein interactions that enable the branch point sequence to be recognized by the splicing machinery? Furthermore, because the basic rules of RNA base pairing are well established, the specific base pairs forming in RNA–RNA interactions can often be defined. There are now several experimental strategies useful in attacking this problem; other chapters in this volume outline biochemical techniques to identify both RNA–RNA and RNA–protein interactions, as well as the powerful approach of identifying RNA–RNA interactions through phylogenetic comparisons. The focus of this chapter is to discuss genetic approaches that can be used, both in the testing of specific RNA base-pairing models and in general open-ended scenarios to identify either RNA–RNA or RNA–protein interactions.

The Strategy

There are two basic levels in the application of genetic analyses to a RNA interaction. The first step is to define regions within the RNA that are involved in an interaction. In general, this goal is achieved by defining sequences, by deletion or point mutagenesis, which are required for the particular process under study. A subsequent genetic strategy to identify specific interactions occurring at such sequences is the analysis of pseudorevertants, revertants (or suppressors) which occur at a site different from the original lesion. The cornerstone in this rationale is that mutations, which disrupt important contacts within or between macromolecules, can sometimes be corrected by a compensating change in the other component of the interaction.[1,2] Two versions of this strategy have been useful in defining RNA interactions. Suppressors obtained *in vivo*

---

[1] J. Jarvik and D. Botstein, *Proc. Natl. Acad. Sci. U.S.A.* **70**, 2046 (1973).
[2] J. Jarvik and D. Botstein, *Proc. Natl. Acad. Sci. U.S.A.* **72**, 2738 (1975).

have identified both specific RNA base pairs and potential RNA–protein interactions.[3-5] More recently, the use of oligomutagenesis to synthesize compensatory changes in predicted RNA helices has been repeatedly successful in demonstrating both intra- and intermolecular base pairing.[6-8]

The remainder of this discussion will describe the application of these general strategies in the specific example of the analysis of nuclear pre-mRNA splicing in yeast. Clearly, the genetic techniques available in yeast allow sophisticated approaches not possible in all experimental organisms. However, in any experimental system in which genes altered *in vitro* can be introduced *in vivo,* it is possible to test the results of mutations, both individually and in combination with compensatory changes. For example, by the synthesis of compensatory changes, the base pairing of the U7 small nuclear RNA (snRNA) to the 3' end of histone pre-mRNA has been demonstrated in *Xenopus* oocytes, an organism with little potential for classical genetics.[8] Indeed, in the analysis of catalytic RNAs, the consequences of compensatory changes can often be directly determined *in vitro,* without the use of *any* experimental organism.[9-11]

A Genetic System

In order to fully exploit the genetic strategies available, it is necessary to have a biological phenotype dependent on the process under investigation. For the analysis of mRNA splicing in yeast, we constructed a gene fusion between the yeast actin gene and the *HIS4* gene, such that expression of the HIS4 gene product requires proper splicing of the actin intron (see Fig. 1).[12] This fusion provides a biological phenotype by allowing strains deleted for *HIS4* to grow on media containing the histidine precursor, histidinol. In addition, by direct analysis of the fusion transcripts produced, it is possible to biochemically assay the proper splicing of the actin intron.

[3] B. Weiss-Brummer, J. Holl, R. J. Scheweyen, G. Rodel, and F. Kaudewitz, *Cell (Cambridge, Mass.)* **33,** 195 (1983).
[4] K. W. Anderson and J. D. Smith, *J. Mol. Biol.* **69,** 349 (1972).
[5] J. R. Couto, J. Tamm, R. Parker, and C. Guthrie, *Genes Dev.* **1,** 445 (1987).
[6] Y. Zhaung and A. Weiner, *Cell (Cambridge, Mass.)* **46,** 827 (1986).
[7] R. Parker, P. G. Siliciano, and C. Guthrie, *Cell (Cambridge, Mass.)* **49,** 229 (1987).
[8] F. Schaufele, G. M. Gilmartin, W. Bannwarth, and M. L. Birnstiel, *Nature (London)* **323,** 777 (1986).
[9] R. B. Waring, P. Towner, S. J. Minter, and R. W. Davies, *Nature (London)* **321,** 133 (1986).
[10] J. M. Burke, K. D. Irvine, K. J. Kaneko, B. J. Kerker, A. B. Oettgen, W. M. Tierney, C. L. Williamsin, A. J. Zaug, and T. R. Cech, *Cell (Cambridge, Mass.)* **45,** 167 (1986).
[11] A. Jacquier and F. Michel, *Cell (Cambridge, Mass.)* **50,** 17 (1987).
[12] R. Parker and C. Guthrie, *Cell (Cambridge, Mass.)* **41,** 107 (1985).

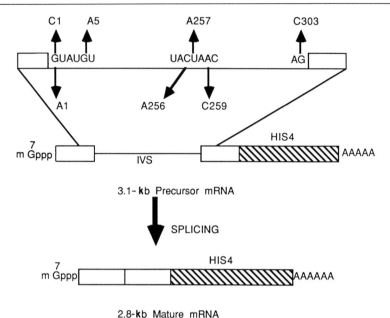

FIG. 1. Intron mutations in an actin–*HIS4* fusion. The structure of the fusion transcripts and several point mutations inhibit splicing to varying degrees. (□) Actin sequences. (▨) *HIS4* sequences. Mutations are named as the mutant nucleotide, followed by the position within the intron. In the C303 mutation, a AG dinucleotide immediately 3' of the splice junction is changed to AC as well. IVS, intervening sequence.

Starting Mutations

The initial starting point in any genetic analysis is the isolation of mutations that affect the process under investigation. These mutations, in addition to defining sites of interactions, are to provide the starting point for the generation and analysis of suppressors. To this end, we sought point mutations within the intron that affected splicing by two approaches. First, utilizing the biological phenotype, we looked directly *in vivo* for lesions that inhibited splicing (i.e., failed to grow on histidinol media[12]). In addition, based on phylogenetic conservation and deletion analyses of sequences located at the 5' splice site, the branch site, and the 3' splice site, we synthesized a number of point mutations and determined their effect on splicing of the intron by assaying both biological and biochemical phenotypes.[13]

[13] U. Vijayraghavan, R. Parker, J. Tamm, Y. Iimura, J. Rossi, J. Abelson, and C. Guthrie, *EMBO J.* **5**(7), 1683 (1986).

In conjunction with work from a number of laboratories, this collection of mutations (shown in Fig. 1) identified important interactions occurring at the 5' splice site, the branch point, and the 3' splice site. Two general strategies are available to determine the nature of these interactions. First, in cases with specific base-pairing models (either in cis or in trans), it is possible to synthesize compensatory changes to directly test the model. The major advantage of testing a specific model is that, if successful, a specific base-pairing interaction has been proposed and has been *rapidly* confirmed. Alternatively, by utilizing the biological phenotype, suppressors can be isolated *in vivo*. The clear advantage of identifying suppressors *in vivo* is that *no* limitations are placed on the types of possible suppressors; no starting model is required. A potential disadvantage is that a substantial time investment may be required to understand the details of the molecular interaction (see later).

Synthesis of Compensatory Changes *in Vitro*

An example of the use of the synthesis of compensatory changes to test base-pairing models is the analysis of interactions between the branch sequence, the so-called TACTAAC or UACUAAC box in yeast, and the U2 snRNA.[7] Based on the sequences of yeast U2 (also called snR20 and LSR1) and yeast branch points, it was possible to hypothesize a specific 7-bp (base pair) interaction between the branch sequence and the U2 snRNA (see Fig. 2). Moreover, though fewer base pairs were formed, a similar interaction could be drawn between the corresponding metazoan branch point and U2. In general, since the success of testing specific interactions is dictated in a large part by the starting model, phylogenetic comparisons are an invaluable tool, both for suggesting and discarding proposed base-pairing models.

Two of the intron mutations, A256 and A257 (see Fig. 1), were predicted to disrupt this base pairing. Therefore, to directly test this model, compensatory changes, predicted to restore base pairing to these mutations in the UACUAAC box, were introduced into the yeast U2 analog by oligomutagenesis (shown for the A257 mutation in Fig. 2). These potential suppressor U2 molecules were then subcloned onto a yeast centromere plasmid and were introduced into cells containing the corresponding mutant form of the actin–*HIS4* fusion. Analyses of these strains demonstrated that both of the consequences of the starting lesion, inability to grow on histidinol media and an inhibition of splicing, were alleviated.

Once a compensatory change has been demonstrated to suppress either the biochemical and/or biological defects of the original starting lesions, it is also important to establish that this suppression is due to the

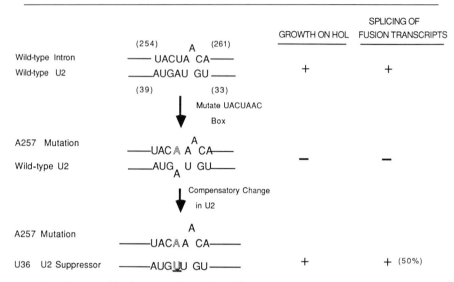

FIG. 2. Base pairing between the UACUAAC box and the U2 snRNA in mutant and wild-type combinations. In each base pairing, the top line represents the intron, and the bottom sequence represents U2 sequences. The A257 mutation is shown as an outlined letter. The U36 suppressor mutation is outlined, shadowed, and underlined. Growth on Hol illustrates the ability of the corresponding combinations to grow on media supplemented with histidinol. Note that, even in the presence of the U36 suppressor, splicing of the A257 fusion transcripts is only restored to 50% of wild type, as judged by steady-state levels of precursor and mature mRNA.[7]

restoration of base pairing per se. For example, it could be argued that the suppressor mutation results indirectly in a general stimulation of the process under consideration. Alternatively, and perhaps more likely, the suppressor mutations may (indirectly) reduce the specificity of the interaction and thus, such as certain "omnipotent" nonsense suppressors,[14] may gain the ability to recognize altered sequences. To address this issue, it is useful to examine the allele specificity of any suppressor. If the mechanism of action is by Watson–Crick base pairing, then the suppressors should be specific for those target mutations for which base pairing is restored. To this end, we examined the effect of each suppressor U2 molecule on three mutations within the UACUAAC element. For example, the suppressor designed to interact with the A257 mutation, U36 (see Fig. 2), was introduced into strains carrying fusions with either the A256

[14] F. Sherman, in "The Molecular Biology of the Yeast Saccharomyces" (J. N. Strathern, E. W. Jones, and J. R. Broach, eds.), Vol. 2. Cold Spring Harbor Laboratory, Cold Spring Harbor, New York.

or C259 mutations. In each case, the suppressor molecules suppressed only the defects in the mutation in which the suppressors were predicted to restore base pairing. These results allowed the conclusion that the nature of this interaction was direct Watson–Crick base pairing.

Note that, in several instances, it has been observed that certain positions within a proposed base pairing are more suppressible than others.[6,7] This difference is likely to reflect a requirement for additional features besides strict base pairing in some positions. In this light, it is advantageous to start with several target mutations and the corresponding bank of suppressor mutations. Testing each suppressor against each starting mutation, not only guards against some positions having non-base-pair sequence requirements, but provides important information as to the allele specificity of each suppressor.

Selection of Suppressors *in Vivo*

The open-ended strategy of identifying suppressors directly *in vivo* has also been applied to the collection of starting mutations described previously. Though suppressors were obtained, by selecting for growth on histidinol media, for all seven mutations shown in Fig. 1, only in the case of the C259 mutation was any suppressor demonstrated to specifically restore splicing of the intron containing the starting mutation. In general, since there are many possible mechanisms of suppression, it is critical to identify rapidly the specific class of the suppressors of interest. The salient features of the approach utilized in this instance are discussed later.

There are several kinds of mutations that can give rise to a "suppressor" phenotype. For example, overproduction of the target gene product may lead to biological suppression, particularly if the original lesion is "leaky." Indeed, our initial suppressor isolations yielded a set of recessive suppressors, which appeared to increase the expression of the fusion transcript independent of splicing.[15] (In this case, the selections were designed to identify suppressors that also conferred a temperature-sensitive phenotype. An additional phenotype greatly facilitates subsequent analysis of the suppressor gene product.) Suppressor mutations, which compensate for the original lesion by a change in an interacting gene product (i.e., interactive suppressors), have gained a new function, the ability to recognize an altered substrate. As such, this class of suppressors should be dominant or semidominant in nature. For this reason, selections were preformed in diploid cells in which dominant suppressors are more likely to arise. In addition, the use of diploid starting strains allows the recovery of recessive lethal suppressors.

[15] R. Parker, Ph.D. Thesis, University of California, San Francisco, 1985.

In analyzing the interactions with the splicing signals, extragenic revertants were of particular interest. To rapidly identify trans-acting suppressors, independent revertants of the original lesion were first selected by their ability to grow on histidinol media. In the second step, extragenic revertants were identified by their ability to suppress the phenotype of the same intron mutation in an actin–*lacZ* gene fusion. Because the starting mutations inhibit splicing, strains carrying the *lacZ* fusion are white or faint blue (if the original mutation is leaky), suppressors are identified as making bluer colonies on X-Gal indicator plates.[16] In the case of the C259 lesion, the majority of the revertants obtained (505 out of 521) failed to suppress the actin–*lacZ* fusion and were discarded.

Finally, to identify suppressors involved directly in splicing, a final screen was performed. The actin–*lacZ* fusion, present on a plasmid, was replaced with a nearly identical construct in which the actin intron had been precisely deleted. In this construct, β-galactosidase activity does not require splicing. Therefore, any suppressor, whose function is specific for intron removal, should not affect the expression of the intronless fusion. Suppressors of the class described previously, which increase gene expression by some other mechanism, enhance the production of *lacZ* from this plasmid. Of the sixteen trans-acting suppressors of the C259 mutation only one, R8, failed to increase the expression of this fusion. This suppressor was therefore identified as a trans-acting splicing-specific suppressor. To confirm this, classical genetic segregation analysis was used to show that the R8 suppressor was in a single nuclear gene, since defined as RNA16,[5] unlinked to the starting fusion. Critically, direct analysis of the fusion transcripts, produced in this revertant, demonstrated that this suppressor does increase the splicing of the C259 mutant substrate.

If the R8 suppressor interacts with a specific sequence within the intron, the suppressor should not restore the splicing of other mutations within the intron. Indeed, by examination of the splicing of the various mutant introns (see Fig. 1), the R8 suppressor was shown to affect only the C259 substrate. This allele specificity of the R8 suppressor argues for a specific interaction between the RNA16 gene product and the branch nucleotide, the site of the C259 mutation. Note that the absence of allele specificity does not necessarily indicate that two gene products do not interact. Suppressor analysis of λ repressor binding argues that suppressors can arise by increasing the number of nonspecific contacts with the DNA backbone.[17] In principle, such types of mutations could arise in RNA binding proteins as well.

[16] M. Rose, M. J. Casadaban, and D. Botstein, *Proc. Natl. Acad. Sci. U.S.A.* **78**, 2460 (1981).
[17] H. C. M. Nelson and R. T. Sauer, *Cell (Cambridge, Mass.)* **42**, 549 (1985).

Summary

Two general strategies using pseudorevertants have been useful in identifying RNA–RNA and RNA–protein interactions, direct synthesis of compensatory changes to test models or *in vivo* selection of suppressors. The most successful application of these approaches has been in the elucidation of RNA–RNA interactions, in a large part because the rules for RNA base pairing are, at least partially, understood. Beyond the initial identification of interactions, suppressors also provide important starting points for further analysis. In the case of an unknown gene product (i.e., the RNA16 gene product), the suppressor provides a route to cloning and analysis of the gene. In addition, the construction of a specific suppressor in a RNA molecule can provide a starting point for the dissection of the biogenesis and function of that RNA. This aspect is most important when the RNA in question normally provides an essential cellular role or is encoded by a multigene family. For example, the construction of a message-specific 16 S RNA in *Escherichia coli* should, in principle, allow for a genetic dissection of rRNA biogenesis and function.[18]

[18] A. Hui and H. A. DeBoer, *Proc. Natl. Acad. Sci. U.S.A.* **84,** 4762 (1987).

Section IV

Appendix

[35] Compilation of Small Nuclear RNA Sequences

By RAM REDDY

Figures 1 through 9 are a compilation of small nuclear RNA (snRNA) sequences for U1–U8 RNA and yeast. The abbreviations used for base and sugar modifications are as follows: m_3G, $N^{2,2,7}$-trimethylguanosine; A3, $2'$-O-methyladenosine; C3, $2'$-O-methylcytosine; G3, $2'$-O-methylguanosine; U3, $2'$-O-methyluridine; F, pseudouridine; A6, N^6-methyladenosine; G2, 2-methylguanosine; and N, unidentified nucleotide. The nucleotides that are identical to those in the sequences shown in the top line are indicated by dashes, and only the differences from the sequences on the top line are shown. Dots in the middle of a sequence indicate the absence of any nucleotide. The references are cited in the figure legends. In the case of capped small RNAs, the cap nucleotide is not included in the numbering system. Partial sequences, where available, are included in the figure legends and in the references, but are not included in the figures. Minor variants of some small RNAs are not included. Some of the RNA sequences are derived only from DNA sequencing of the genomic clones and the actual RNAs have not been sequenced. The 5′ and 3′ ends of snRNAs in these instances are derived from comparison with corresponding RNA sequences from related species.

```
                    1                           20                     40                      60                     80              90
Human   U1   m3GpppA3U3ACFFACCUGGCAGGGAG AUACCAUGAUCACGAAGGUG GUUUCCCAGGGCGAGGCUU AUCCAUUGCA3CUCCGAUGU GCUGACCCCU
Rat     U1   --------------------------- ------------------- ------------------ ------C------------- ----------
Chick.U1     --------------------------- ------G-C----------- ------------------ ---CC--------------- -G--------

                    1                           20                     40                      60                     80              90
Mouse U1-A1  m3GpppA3U3ACFFACCUGGCAGGGAG AUACCAUGAUCACGAAGGUG GUUUCCCAGGGCGAGGCUU AUCCAUUGCA3CUCCGGA.UGU GCUGACCCCU
Mouse U1-A2  --------------------------- ------------------- -----C------------ ------------------- ----------
Mouse U1-B1  --------------------------- ------------------- -----C------------ -C-------UU--GG---- ----------
Mouse U1-B2  --------------------------- ------U------------ -----C------------ -C-------UU--GG---- ----------
Mouse U1-B4  --------------------------- ------------------- -----C------------ -C-------UU--GG---- ----------
Mouse U1-B5  --------------------------- ------G------------ -----C------------ -C-----..UU--GG---- ----------
Mouse U1-B6  --------------------------- ------------------- -----C------------ -C-------UU--GG---- ----------

                    1                           20                     40                      60                     80              90
Frog  U1-A   m3GpppA U ACUUACCUGGCAGGGAG AUACCAUGAUCACGAAGGUG GUUCUCCCAGGGCGAGGCUC AGCCAUUGCA CUCCGGCUGU GCUGACCCCU
Frog  U1-B   m3GpppA U ACUUACCUGGACGGGGUC ------------------- ------------------- ----------------- -C--------

                    1                           20                     40                      60                     80              90
F.Fly U1     m3GpppA3U3ACFFACCUGGCGUAGAGG UUAACCGUGAUCACGAAGGC GGUCCUCCGAGUGAGGCU UGGCCAUUGCA CCUCGGCUG AGUGACCUC
Bean  U1     m3GpppA U ACUUACCUGGACGGGGUC AAUGGAUGAUCUAUAAGGUC CAUGGCCUAGGAAGUGACC UUCAUUGCACU CAGAAGGGG UGCCUACUCUA


                    100                        120                      140                    160
Human   U1   GCGAUUUCCC CAAAUGUGGGAAACUCGACU GCAU4AUUGUGGUAGUGGG GGACUGCGUUCGCGCUUUCC CCUG-OH (164)
Rat     U1   ---------- -------------C------ ------------------ -------C------------ ----- (164)
Chick.U1     ---------- -------------C------ ------------------ -------C------------ ----- (164)

                    100                        120                      140                    160
Mouse U1-A1  GCGAUUUCCC CAAAUGCGGGAAACUCGACU GCAUAAUUGUGGUAGUGGG GGA.CUGCGUUCGCGCUCUCC CCUG-OH (164)
Mouse U1-A2  ---------- ------------------- ------------------ -------------------- ----- (164)
Mouse U1-B1  ---------- ------------------- ------------------ -------------------- ----- (165)
Mouse U1-B2  ---------- ------------------- ------------------ -------------------- ----- (165)
Mouse U1-B4  ---------- ------------------- ------------------ -------------------- ----- (165)
Mouse U1-B5  ---------- ------------------- ------------------ -------------------- ----- (165)
Mouse U1-B6  ---------- ------------------- ------------------ ---G---------------- ----- (166)

                    100                        120                      140                    160
Frog  U1-A   GCGAUUUCCC CAAAUGCGGGAAACUCGACU GCAUAUUUCUGGUAGUGGG GGACUGCGUUCGCGCUUUCC CCUG-OH (164)
Frog  U1-B   ---------- --------U-----G----- ------------------ --G-------G-C------- -----

                    100                        120                      140                    160
F.Fly U1     UGCGAUUAUU CCUAAUGUGAAUAACUCGUG GCAUAUUUGGUAGUGGG GGAAUGGCGUUCGCGCCCGUC CCGA-OH (164)
Bean  U1     AGGUCUGUCC AAGUGAUGGAGCCUACGUCA UAAUUUGGUAGUGGGGGC CUGCGUUCGCGCGCCCCUU AC (162)
```

```
Yeast U1    m3GpppAUACUUACCUUAAGAUAUCA GAGGAGAUCAAGAAGUCCUA CUGAUCAAACAUGCGCUUCC AAUAGUAGAGGACGUUAAG CAUUUAUCAUUGAACUAUAA   100
UUGUUCAUUGAAGUCAUUGA UGCAAACUCCUUGGUCACAC ACACAUACGCGCGGAAGGC CUUCUGCCUGGAGAAGUUUG GUGUUGCUGACGUUUCCAU UCCCUGUUUCAAUCAUUGG UUAAUCCUUGAUUCCUUUG   220
GGGAUUUUUGGGUUAAACUG AUUUUUGGGGCCCUUUGUUU CUUCUGCCUGGAGAAGUUUG ACACCCAAAUUCAAAUUGGU GUAGGGGAGCUGGGGCCUU UCAAAAGAGAGCUUUGGUAGA   340
GGCAUUCUUUUGACUACUU UUCUCUAGCGUGCCAUUUUA GUUUUUGACGGAGAUUCGAA UGAACUUAAGUUUAUGAUGA AGGUAUGGCUGUUGAGAUUA UUUGGUCGGGGAUUGUAGUUU   460
GAAGAUGUGCUCUUUUGAGC AGUCUCAACUUGCUCGCUC CCGUUAUGGGAAAAAUUUUG GAAGGUCUUGGUAGGAACGG GUGGAUCUUUAUAAUUUUGA UUUAUUUU-OH   568
```

FIG. 1. U1 RNA. Human, HeLa cells[1]; human, placenta.[2] Rat, Brain[1]; rat, Novikoff hepatoma.[3] Mouse, Sperm[4]; Friend erythroleukemia and K1 cells[5]; F9 embryonal carcinoma.[6] Chicken, Liver[1]; chicken, oviduct.[7,8] Frog, Late gastrula-stage embryos[9]; frog, *Xenopus laevis*.[10,11] F. Fly, *Drosophila melanogaster*.[12,13] Silk worm, *Bombyx mori* (partial).[14] Yeast, *Saccharomyces cerevisiae*.[15,16] Bean, *Phaseolus vulgaris*.[17] Pea, *Pisum sativum* (partial).[18]

```
          1                 20                    40                        60                         80             •      100
Human  U2  m3GpppA3U3CGCUUCUCG3G3CCUUUUG3G  CUAAG3AUCAA3  GUGUAGUAU⊃3  FGUUCUU AUCAGUUAAUAU  C3UGAUACGUCCUCUAUCCGA  GGACAAUAUUAUUAAAUGGAU
Rat    U2  ------------------------FF-----  --!•--------  --6--F-F-F--  --FF---3----F--F--  ----------------------  -------------------F--
Mouse  U2  -------- ! ----------- ! -------  ------------  U-------  -------------------  ----------------------  -------------------
Chick.U2  -------- ! ----------- ! -------  ------------  U-------  -----------------G--GA--  ----UU-----------C---
Frog   U2  ------------------------------  ------------  U-------  ------------------U-G---  ------C-----------C---

          1                 20                    40                        60                         80             •      100            110
F.FLY U2A  m3GpppAUCGCUUCUCGGCCUUAUGG  CUAAGAGAAAGUGUAGUAU  CUGUUCLUAUCAGCUUAACA  UCUGAUAGUUCCUCCAUUGG  AGGACAACAAAUGUAAACUG  AUUUUUGGAA
F.Fly U2B  ----------------------------  ------------------  -----------AG----  ----------------------  -----------C---
F.Fly U2C  ----------------------------  ------------------  ----------------  ----------------------  --------------C---

          1               •     20                 40                      60                      80            •       100
Amoeba U2  NpppAGAUAAAGAAGAGAAACAAAGA  UUUAUAUGUCGCACAAAGUU  JAGUCGAUGUCCUACGGGGU  AAGUUGACAAUGAAAACUUU  UGUUGAAAAAUAUUUGGAGAC
Trypanosome U2 m3GpppAUAUCUUCGGCUAUUUAG  CUAAGCAAGUUAUUAAAC  AGAGUAACUCAGUAGUAACUC  CUGAUACGGCCUUUGGCCC  AAGGAUCAAAACUGUUGCCU

          1                20                    40                         60                     80                     100
Bean    U2  m3GpppAUACCFF3CFCGGCCFUFU63G  CUAAG3AUC3AA3GUGFAGFA3FC3  FGFFCUUAUUAGUUUAAFAF  CUGAUAUGUGGGCCAAUGGC  CCACACGAUAUUAAAUUUAU
Crucifer   m3GpppAUACCUUUCUCGGCCUUUUG  G  CAUUG AUC AA GUGUAGUA UC  UGUUCCUUAUCAGUUUAAAUAU  CUGAUAUGUGGGGCCAUCGGC  CCACACGAUAUUAACUCUA  100
U2.2,3,4,5,7 and 9

          •          120                   •     140                    •     160                      180
Human  U2  UUUUUGGAGCAGGGAGGAUGGA  AUAGGAGCUUGCUCCGUCCA  CUCCACGCAUCGACCUGGUA  UUGCAGUACCUCCAGGAACG  GUGCACC
Rat    U2  ------A-UA----U-----  -------------------  -------------------  -------------------  ------(A)-OH  (187,188)
Mouse  U2  ------AGUA----U----  -------------------  -------------------  -------------------  ------------
Chick. U2  ------GCGC-----U----  CCC---------C---G  ------G------U--C---  -----G--G---C---  ------------
Frog   U2  ------A----------  -G-A--------U---  ------G------U---  ------U-------C---  ------U------
```

524

```
                  •               130                    •               150                    •               170                    •               190
F.Fly U2A    UCAGACGGAGUGCUAGGAGC   UUGCUCCACCUCGUCGCGG   GUUGGGCCGGUAUUGCAGUA   CCGCCGGGAUUUUGGCCCAA   C   (191)
F.Fly U2B    --------------------   -------G-----------   --------------------   ----------C---------   -   (191)
F.Fly U2C    ----------------G---   ----------A--------   --------------------   ----------C---------   -   (191)

                  •               120                    •               140                    •               160                    •               180
Amoeba U2         UUUUGAGUAUUUAAAGGGUG   AGUGGCUAUAGAUCCUCUGU   UCUCAUGAUCGAUUAAUAUG   AUCCACCCAAUCAAUACUCG   AACACCC-OH   (187)
Trypanosome U2    GUCCCGCGUUCGUUUCGGGU   UCCACUUGUCCGACGGAGC    GCGACGGU.                                                           (148)
Bean U2           UUCUUGAGGGGAAGAGGCCA   CCACAGUAGCUUGCUAUUGG   GUCCUUACGUGUCGCUCUU    GCGUUGCACUAUAGCAAUUG   CUGGCGCACCCCA(C)-OH   (193,194)

Crucifer U2.3     UUUUUUAAGGGAGAAAGCCC   ACUAAGAUAGCUUGCUAUCU   GGGCUUUCACGAGUCGCCCA   UGCCGUUGCACUACUGCACGG   GCCUGGCUCAACCCGC   (196)
Crucifer U2.2     -C------------------   G---U--------------    ----C--------------    ----------------------   ----------------
Crucifer U2.4     --------------------   -------------------    -------A-----------    ---------U------------   ----------------
Crucifer U2.5     --------------G-----   GU-U---------------    ----------U--------    -------U--------------   ----------------
Crucifer U2.7     ------------A-------   ------G------------    ----------U--------    -------U---GA---------   --U-------------
Crucifer U2.9     --------------------   GU-----------------    -------------G-----    ---------G------------   ----------------

Yeast U2     m3GpppACGAAUCUCUUUGCCUUUUG   GCUUAGAUCAAGUCUAGUAU   CUGUUCUUUCAGUGUAACA   ACUGAAAUGACCUCAUGAG   GCUCAUUACCUUUUAUUUG      100
UUACAAUACACAUUUUUUGG   CACCCAAAAUAAUAAAAUUGG   ACGGGAAGAGACUUUUUAAG   UGCUUGUUGAGAUGACGGGU   GUCAUGAUUUUUCUUCCUCA   UUUUUAUUUUAUUUUUAGUCU   220
UCCUGUUUCUCCUUAGGUUG   GCUUUUUGCUCUUACUCUCU   CCCUGUUUCCGACCGCGGU   UAUGUCCAACGCGGGAUUUG   GUUUUCUCUGGGGAUUUUU   UCUGGUGGCGUCGCAGAGG   340
GAUUUUGGUUGCAAGGAA   AGGUGUCUUACGCGAUUCUUU   CCAUUGUUAAUUAAAUCG   CCUUUACUAUUAGUGGCGG   UUUUCAUUGAGGAGAGGUUCC   UUUUUCAUUGAGGAGAGGUAU   580
GAGGGGUUAGGGGUGGGGU   UGGUCUACAAUAAGAGUGU   GAGGAGAUGUGUGGAGGU   AUCUGGAGCGUUUUUGG   CAGAGGCGUUUUUUGGA   UUUUCAUUGAGGAGAGGAAGG   700
AGCUCUCGCAGGAUAUA   AUGGAGGUGUUUCAAAGGG   GAGGAGAUGUGUUGAGGU   AUCUGGAGCGUUUUUGG   CAGAGGCGUUUUUUGGA   AUGGAAGGCGUUUGCUGGGA   820
AAAGAGAGGCCAUGACUG   CAUCUGUGUUUUCAAGGCCA   GUUUUAUUAACCGCCUAUGU   CAUAGAGGCGUUUUUUUUGG   AGGAUUUGGAAGAACGCCCG   GCGGCAUCAAGAAACGGACU   1060
UGAUGGUUGACGCCUGUUUU   UAAAGUAUAGAGACGUCGCGA   CCCUCGCACUUUGUGGAGUCG   UUCCUGAUGUUUUUCUGCGU   UCCCGGUUCGUCUCUU   1175
```

FIG. 2. U2 RNA. Human, Placenta[19-21]; human, HeLa cells (partial);[22] Rat, Novikoff hepatoma[23]; rat, brain (partial).[22] Mouse, Liver.[24] Bird, Chicken liver[25]; bird, pheasant liver (partial).[22] F. Fly, *D. melanogaster*.[26,27] Frog, *X. laevis*.[28] Amoeba, *Dictyostelium discodeum*[29]; *Trypanosome brucei gambiense*.[30] Yeast, *S. cerevisiae*.[31] Plants, Wheat (partial)[32]; broad bean, *Vicia faba*[33]; crucifer, *Arabidopsis thaliania*.[34] Human U2 RNA has other posttranscriptional modifications not yet localized. The homology between U2 (Dd8) RNA of amoeba and U2 RNA of higher eukaryotes is not established.

```
        1                    20                     40                        60                      80                    100             110
Human U3   m3GpppA A GACUAUACUUUCAGGGAU CAUUUCUAUAGUGUGUUACU AGAGAAGUUUCUGAACGU GUAGAGCACCGAAAACCACG AGGAAGAGGUAGCGUUU CUCCUGAGCG

Rat U3B    m3GpppÂA3GACUAFACUFUCAGGGAU CAUUUCUAUAGUUCGUUACU AGAGAAGUUUCUGACUGU GUAGAGCACCGAAACCACG AGGACGAGACAUAGCGUCCC CUCCUGAGCG
Rat U3C    ------------------------- ------------------- ----------------- ------------------- ------------------- ----------
Rat U3A    ------------------------- ------------------- ----------------- --------G-----U-U --------G-----U-U ----------
Rat U3D    ------------------------- ------------------- ----------------- --------G-----U-U --------G-----U-U ----------

        1                    20                     40                        60                      80                    100             110
Dict. U3   NppppA U GACCAAACUCUUAGGAUC AUUUCUAGAGUAUCGUCUAU UAAAAUAUUCAUCAAUAAU UUUUCCUCUUUUCAUAGCUAG GAUGAUGAUAUACAUCACU AUACGAAAGC

        120                  140                    16C                      180                    200             217
Human U3   UGAAGCCGGC UUUCUGGCGUUGCUUGGCUG CAACUGCCGUCAGCCAUUGA UGAUCGUUCUUCUCUCCGUA UUGGGGAGUGAGAGGGAGAG AACGCGGUCUGAGUGGU-OH   (217)

Rat U3B    UGAAGCCGGC UCUAGGUGCUGCUUCGUGUGC AGCUGCCUCUUGCCAUUGAU GAUCGUUCUUCUCUCCUUCG GGGGGUAAGAGGGAGGGAA CGCAGUCUGAGUGGA-OH   (215)
Rat U3C    ---------- ----------------- ---------A--- ----------------- -------A---- ----------U-OH   (215)
Rat U3A    ---------- ---UA---U----- -A----UAU-G--- CGGUC-UC--U -A--UUCG------ ----A---- ----------U-OH   (214)
Rat U3D    ---------- ---UA---U----- -A----UAU-G--- CGAGC-UC--U -A--UUCG------ ----A---- ----------U-OH   (214)

        120                  140
Dict. U3   GUGAAACCGU UAUUAUCGAAUGAUUCAUUU AUUUGUUAUUAAUUAACAUUGAAUG ACCGUCUAAAUUCAGGGAUGA AUUGGUUGGUGGUGGGAUU CGUACUGGCU-OH   (210)

Yeast U3A (SNR17A)  m3GpppAACACAUUCUCACAGUAGGAU CAUUUCUAUAGGAAUGCGUCA CUCUUUGACUCUUCAAAAGA GCCACUGGAUUCCAACUUGGU UGAUGAGUCCCAUAACCUUU  100
Yeast U3B (SNR17B)  ---------A--------------- -------------- ---------------- -----C------ ---------

        GUACCCCAGAGAGAAACC GAAAUUGAAUCUAAAAUUAGC UUGGUUGAAUCUAAAUUUC GUUCGGCCAUCUAAAAUUUU GAAUAAAAAUUUUCCUUUGCC GUUGCAUUUGUAGUUUUUC  220
        ----------------G--AU ----C-----U------ ------U---G------ -------U---G---- -------A----GA-AAC-----

        CUUUGGAAGUAGUAAUUUACAAUA UUUUAUGGCGCGAUGAUCU GACCCAUCCUAUGUACUUCU UUUUGAAGGAAGGAUGAGGGCUC UAUGGGGUGGUUACAAAUGGC AGUCUGAC  328
        -----U-UG--G-G---- ---A--------- --------C A------G-A------A-U C-------
```

FIG. 3. U3 RNA. Human, Placenta[35]; HeLa cells (partial).[35,36] Rat, Novikoff hepatoma[37]; liver.[38] Amoeba, *D. discodeum.*[39] Silk worm, *B. mori* (partial).[40] Yeast, *S. cerevisiae.*[14] Plants, Broad bean (partial).[41]

```
                              1                    20                       40                         60                       80                          100
Human  U4A   m3GpppA3G3CFUUGC3GCAGUGGCAGUA UCGUAGCCAAUGAGGUUUAU CCGAGGCGCGAUAUAUUGCUA AUUGA3AAACUUFUCCCAAFA CCCCGCCGUGACGACUUGCA
Human  U4B   --------------------------------- -------------------- --------------------- --------------------- --------A----------A-
Rat    U4A   --------------------------------- -------------------- --------------------- --------------------- --------A----------A-
Rat    U4B   --------------------------------- -------------------- --------------------- --------------------- --------A----------A-
Mouse  U4A   --------------------------------- -------------------- --------------------- ---------A----------- --------A----------A-
Mouse  U4B   --------------------------------- -------------------- --------------------- ---------A----------- --------A----------A-
Chick. U4A   --------------------------------- -------------------- --------------------- ---------A----------- --------A----------A-
Chick. U4B   --------------------------------- -------------------- --------------------- ---------U----------- --------A----------A-
Chick. U4X   -----------U------ --------------- -------------------- --------------------- G--------U----------U- --------A----------A-

                              1                    20                       40                         60                       80                          100
F.Fly  U4A   m3GpppA3GCUUAGCGCAGUGGCAAUA CCGUAACCAAUGAAGCCUCC CUGAGGUGCGGUUAUUGCUA GUUGA3AAACUUUAACCAAC.C CACGCCAUG.GGACGUGAAAU
F.Fly  U4B   - U---U-----A---G--- U--------------------- --------A------------ ------------------UA- -C----------G-------
F.Fly  U4C   - U--U-----U------- --------U-CU--------- --------A------------ ------------------UA- -C----------G-------

                              120                              140
Human  U4A   AUAUAGUCGGCAUUGGCAAU UUUUGACAGUCUCUACGGAG ACUG(G)-OH (144,145)    Chick. U4A   AUAUAGUCGGCAUUGGCAAU UUUUGACAGUCUCUACGGAG ACUG(G)-OH (144,145)
Human  U4B   -------------------- -------------------- -            (141)         Chick. U4B   -------------------- -------------------- -            (141)
Rat    U4A   -------------------- -------------------- -            (144,145)    Chick. U4X   -------------------- ------G--C---CC----- G            (141)
Rat    U4B   -------------------- -------------------- -            (141)         F.Fly  U4A   ACCGUCCACUACGGCAAUUU UUGGAAGCCCFFACGAGGGC UAA          (143)
Mouse  U4A   -------------------- -------------------- -            (144,145)    F.Fly  U4B   -------------------- -----G-,----------C- -            (142)
Mouse  U4B   -------------------- -------------------- -            (144)         F.Fly  U4C   -------------------- -----G-,------U C--- -            (143)

Yeast  U4    m3GpppAUCCUUAUGCACGGGAAAUA CGCAUAUCAGUGAGGAUUCG UCCGAGAUUGUGUUUUUGCU GGUUGAAAUUUAAUUAUAAA CCAGACCGUCCUCCAUGGU 100
             CAAUUCGUGUUCGCUUUGAA UACUUCAAGACCUAUGUAGG GAAUUUUGGAAUACCUUU-OH 159
```

FIG. 4. U4 RNA. Human, HeLa cells.[42] Rat, Brain.[42] Novikoff hepatoma.[43] Mouse, Kidney lymphoma.[44] Chicken, Liver[42]; oviduct.[45] Insect, D. melanogaster.[46,47] Slime mold, Physarum polycephalum (partial).[48] Yeast, S. cerevisiae.[49]

```
                  1                    20                     40                        60                      80                       100
                  •                    •                      •                         •                       •                        •
Human  U5A  m3GpppA3U3ACUCUGGUUUCUCUUCAG  AUCGCAUAAAUCUUUCG3CCU   U3UFAC3FAAAGAUFUCCGUGG   AGAGGACAACUCUGAGUCU   UAACCCAAUUUUUUAG.CCU
Human  U5B  -----------------------------  ---------------------   -------------U-------    -------U-----------   ---G-U-------------
Rat    U5A  -----------------------------  ------U--------------   ---------------------    -------------------   -------A-----------
Mouse  U5A  -----------------------------  ------U--------------   ---------------------    -------U-----------   -------A--------G---
Mouse  U5B  -----------------------------  ------U--------------   ---------------------    -------------------   -----A----------GUUC
Chick. U5A
Frog   U5   m3GpppA U ACUCUGGUUUCUCUUCAA    AUUCGAAUAAAUCUUUCG CCU  U UUAC UAAAGAUUUCCGUGG     AGAGGACGACCAUGAGUU    UCGUUCAAUUUUUGAAGCCU


                  1                    20                        40                        60                       80                       100
                  •                    •                         •                         •                        •                        •
Dinof. U5   m3GpppA6U3C3ACAGUGFUCACUUCAAC  CGAAUCAAFCUFU3CG3CCUU3U   FAC3FAAAGGUUGCCGUGAAU    GGACCACAUCAAUGUGAAUC   UCUCAAUUUUUGAGGGCUCU
Tetra. U5   NpppA3U3CA CAGAACUCAGCUCAAU      ACGCUUAAAUUUFUCG3C: UU3   UFAC3FAAAGAUFACCGUGGG    CUGGGUUCUACAAUGUGAAU   UAUUAAAAUUUUGAGGAUU
Pea    U5A  m3GpppG A GC CGUGUGAUGAUGACAU     AGCGCAACUAU3CUFUCG3CUFU3  UFAC3FAAAGAAAUACFGUGUC   AGCGUCACAAUUAGCGGCAU   ACCGUAGUUUUUGGAGAGU
            110


Human  U5A  UGCCUUGGCAAGGCUA-OH     (116)        Chick. U5A  UGUUCCGGCAAGGCUA-OH        (116)
Human  U5B  ---UCC-A-----------     (116)        Frog   U5   GGUUCACCAGGUA              (116)
Rat    U5A  --UUCC-------------     (116)        Dinof. U5   GCCCCAC-OH                 (107)
Mouse  U5A  --U----------------     (117)        Tetra. U5   GUGUGAAUCCUA-OH            (112)
Mouse  U5B  ---UCCAA-----------     (117)        Pea    U5   UCUCAAGUUUUGAGGGCUCU G-OH   (121)


                                                                •
Yeast  U5 (SNR7-L)  m3GpppAAGCAGCUUUACAGAUCAAU  GGCGGAGGGAGGGUCACAUC  AAGAACUGGGCCUUUUAU  UGCCUAUGAAGAACUUAUAACG  AACAUGGUUCUUGCCUUUUA  100
                    CCAGAACCAUCCGGGUGUUG  UCUCCAUAGAAACAGGUAAA  GCGUGUCCGUUACUGUGGGCU  JUCUGCCCUUUUUCUCAAUG  AGUAAGGAGGGCGU-OH  214
```

FIG. 5. U5 RNA. Human, HeLa cells.[50] Rat, Brain[50]; Novikoff hepatoma.[51] Mouse, Kidney.[52] Bird, Chicken liver[50]; pheasant and duck liver (partial).[53] Frog, *X. laevis*.[54] Dinoflagellates, *C. cohnii*.[55,56] Amoeba, *Tetrahymena*.[53] Yeast, *S. cerevisiae*.[57] Pea, *P. sativum*.[18] Sequences of minor variants are not shown.

```
            1                      .          20                              .           40                                .              60                        .                                     80                             .           100
Rat   U6  XpppGUGCUCGCUUCGGCAGCACA UAUACUAAAAFUGGAACGAF ACA6GAGA3AGAUUA3G3CAU6GC3 CC3C3UGCGC AA3GG2AUGAC3ACG CAAAUFCGUGAAGCGUUCCA
Mouse U6  ------------------------ -------------------- ------------------------- ---------- --------------- --------------------
Human U6  ------------U----------- ------------------U- ------------------------- -------U-- --------------- -----------U--------
Frog  U6  -------U---U------------ ------------------U- ------------------------- -------U-- --------------- -------U-U----------
F.Fly U6  -------F---U------A------ -------------------F ------------------3------ -------3-- --------------- -------AF-----------

Physarum  XpppUUUUGUAUCACAUAUACUAA AAUGGCGCUAGCGA UUAAG C CGGC CGGUUGCUC AGUACC G  U GAGACGCUCG AGCGAAGCG  UUUGCGUUUUUUGUAAA (98)
Bean  U6  XpppCUUCGGGACAUCCGAUAAA AFUGGAACGACACA6GAGA3AG AUUA3GCAUGGCCC3CUGCGC3A A3GGAUGACACG3CACAAAUCG3 AGAAFGGUCCAAAUUUU-OH (98)

Rat   U6  UAUUUU(U)-OH (106,107)        Frog  U6  UAUUUU(U)-OH (106,107)
Mouse U6  ------------                  F.Fly U6  C------- (107)
Human U6  ------------

Rat   U6  UAUUUU(U)-OH (106,107)
Mouse U6  ------------
Human U6  ------------
```

FIG. 6. U6 RNA. Rat, Novikoff hepatoma.[58] Mouse, Liver.[59-60] Human, HeLa cells.[62,63] Frog, *X. laevis*.[64] F. Fly, *D. melanogaster Canton S*.[65,66] *Physarum microplasmodia*.[67] Broad bean, *V. faba L*.[68] Dinoflagellates, *C. cohnii* (partial).[56] X is an unidentified nonnucleotide.

```
                1                     20                        40                           •
Human      U7  m3GpppNNGUGUUACAGCUCUUUUAG AAUUUGUCUAGUAGGCUUUC UGGCUUUUCACCGGAAAGCC CCU-OH  (63)
Mouse      U7  m3GpppA(U)A----------------- --------------C---U---- --A---C.GGU-----A-G  ------  (62,63)
Sea urchin U7  m3GpppAUCUUUCAAGUUUCUCUAGA AGGUCUCCGUCCGAAGUC GGAGGCGAGUGCCCAAC        (57)
```

FIG. 7. U7 RNA. Human, HeLa cells.[69] Mouse.[70] Sea urchin, *P. miliaris*.[71,72]

```
            1                  20              U  •              40                   60               •               80                                 100
Rat   U8  m3GpppA3U3CGUCAGGAGGGUUAAUCCF UACCUGUCCCUCCUUUCGGA GGGCAGAUAGAAAAUGAUGA UUGGGAGCCUUGCAUGAUCUGC UGAUUAUAGCAUUUCCGUGU
Mouse U8  ------------------------------ ------------------- -------•----------- ------------------- -------••...----A--C

            120                    140
RAT   U8  AAUCAGGACCUGACAACAUC CUGAUUGCUUUCUAUCUGAUU-OH  (140)
Mouse U8  -------------------- --G------C-------------  (137)
```

FIG. 8. U8 RNA. Rat, Novikoff hepatoma.[73] Mouse, Friend leukemia.[74]

```
Yeast SNR3   m3GpppAACUUUGUCCUAAAGUACUA AUCCACCGCAUUAGACAGUA CGAAGAUCGAGCUAUUUAUU UGAACACUCGGUCUUAUUCG UGAUAAGCGUAAUGUGGAGA
GAUCAAUUUCCGGGUCAUUU AUAAGAACUCGAGUGGAUUG CUAGUGUUUUGAUUUA3CUG AAUGAGACUCGAGUGUCAGA AGAUGACUAUAUUU-OH  (194)
```

FIG. 9. SNR3 RNA Yeast. *Saccharomyces cerevisiae*.[75]

References for Figures 1 through 9

[1] C. Branlant, A. Krol, J. P. Ebel, E. Lazar, H. Gallinaro, M. Jacob, J. Sri-Widada, and P. Jeanteur, *Nucleic Acids Res.* **8**, 4143 (1980).

[2] T. Manser and R. F. Gesteland, *Cell (Cambridge, Mass.)* **29**, 259 (1982).

[3] R. Reddy, D. Henning, and H. Busch, *Biochem. Biophys. Res. Commun.* **98**, 1076 (1981).

[4] W. F. Marzluff, D. T. Brown, S. Lobo, and S. Wang, *Nucleic Acids Res.* **11**, 6255 (1983).

[5] N. Kato and F. Harada, *J. Biol. Chem.* **260**, 7775 (1985).

[6] E. Lund, B. Kahan, and J. E. Dahlberg, *Science* **229**, 1271 (1985).

[7] J. Earley, K. Roebuck, and W. Stumph, *Nucleic Acids Res.* **12**, 7411 (1984).

[8] D. R. Roop, P. Kristo, W. E. Stumph, M. J. Tsai, and B. W. O'Malley, *Cell (Cambridge, Mass.)* **23**, 671 (1981).

[9] D. J. Forbes, M. W. Kirschner, D. Caput, J. E. Dahlberg, and E. Lund, *Cell (Cambridge, Mass.)* **38**, 681 (1984).

[10] A. Krol, E. Lund, and J. E. Dahlberg, *EMBO J.* **4**, 1529 (1985).

[11] R. Zeller, M. Carri, I. W. Mattaj, and E. M. DeRobertis, *EMBO J.* **3**, 1076 (1984).

[12] S. Mount and J. A. Steitz, *Nucleic Acids Res.* **8**, 6351 (1981).

[13] J. C. Wooley, R. D. Cone, D. Tartof, and S. Chung, *Proc. Natl. Acad. Sci. U.S.A.* **79**, 6762 (1982).

[14] D. C. Adams, R. J. Herrera, R. Luhrmann, and P. M. Lizardi, *Biochemistry* **24**, 117 (1985).

[15] L. Kretzner, B. C. Rymond, and M. Rosbash, *Cell (Cambridge, Mass.)* **50**, 593 (1987).

[16] P. G. Siliciano, M. H. Jones, and C. Guthrie, *Science* **237**, 1484 (1987).

[17] V. L. Van Santen and R. A. Spritz, *Proc. Natl. Acad. Sci. U.S.A.* **84**, 9094 (1987).

[18] A. Krol, J. Ebel, J. Rinke, and R. Luhrmann, *Nucleic Acids Res.* **11**, 8583 (1983).

[19] S. W. Van Arsdell and A. M. Weiner, *Mol. Cell. Biol.* **4**, 492 (1984).

[20] G. Westin, J. Zabielski, K. Hammarström, H. Monstein, C. Bark, and U. Pettersson, *Proc. Natl. Acad. Sci. U.S.A.* **81**, 3811 (1984).

[21] G. Westin, E. Lund, J. T. Murphy, U. Pettersson, and J. E. Dahlberg, *EMBO J.* **3**, 3295 (1984).

[22] C. Branlant, A. Krol, J. Ebel, E. Lazar, B. Haendler, and M. Jacob, *EMBO J.* **1**, 1259 (1982).

[23] R. Reddy, D. Henning, P. Epstein, and H. Busch, *Nucleic Acids Res.* **9**, 5645 (1981).

[24] H. Nojima and R. D. Kornberg, *J. Biol. Chem.* **258**, 8151 (1983).

[25] G. M. Korf and W. E. Stumph, *Biochemistry* **25**, 2041 (1986).

[26] A. Alonso, J. L. Jorcano, E. Beck, and E. Spiess, *J. Mol. Biol.* **169**, 691 (1983).

[27] A. Alonso, E. Beck, J. L. Jorcano, and B. Hovemann, *Nucleic Acids Res.* **12**, 9543 (1984).

[28] I. W. Mattaj and R. Zeller, *EMBO J.* **2**, 1883 (1983).

[29] S. Kaneda, O. Gotoh, T. Seno, and K. Takeishi, *J. Biol. Chem.* **258**, 10606 (1983).

[30] C. Tschudi, F. Richards, and E. Ullu, *Nucleic Acids Res.* **14**, 8893 (1986).

[31] M. Ares, *Cell (Cambridge, Mass.)* **47**, 49 (1986).

[32] J. M. Skuzeski and J. J. Jendrisak, *Plant Mol. Biol.* **4**, 181 (1985).

[33] T. Kiss and F. Solymosy, *Nucleic Acids Res.* **15**, 1332 (1987).

[34] P. Vankan and W. Filipowicz, *EMBO J.* **7**, 791 (1988).

[35] D. Suh, H. Busch, and R. Reddy, *Biochem. Biophys. Res. Commun.* **137**, 1133 (1986).

[36] L. B. Bernstein, S. M. Mount, and A. M. Weiner, *Cell (Cambridge, Mass.)* **32**, 461 (1983).

[37] R. Reddy, D. Henning, C. Subrahmanyam, L. Rothblum, D. Wright, and H. Busch, *J. Biol. Chem.* **260**, 5715 (1985).

[38] I. L. Stroke and A. M. Weiner, *J. Mol. Biol.* **184**, 183 (1985).

[39] J. A. Wise and A. M. Weiner, *Cell (Cambridge, Mass.)* **22**, 109 (1980).

[40] J. M. X. Hughes, D. A. M. Konings, and G. Cesareni, *EMBO J.* **6**, 2145 (1987).

[41] T. Kiss, M. Toth, and F. Solymosy, *Eur. J. Biochem.* **152**, 259 (1985).

[42] A. Krol, C. Branlant, E. Lazar, H. Gallinaro, and M. Jacob, *Nucleic Acids Res.* **9**, 2699 (1981).

[43] R. Reddy, D. Henning, and H. Busch, *J. Biol. Chem.* **256**, 3532 (1981).

[44] N. Kato and F. Harada, *Biochem. Biophys. Res. Commun.* **99**, 1477 (1981).

[45] M. L. Hoffman, G. M. Korf, K. J. McNamara, and W. E. Stumph, *Mol. Cell Biol.* **6**, 3910 (1986).

[46] E. Myslinski, C. Branlant, E. D. Weiben, and T. Pederson, *J. Mol. Biol.* **180**, 927 (1984).

[47] J. Saba, H. Busch, D. Wright, and R. Reddy, *J. Biol. Chem.* **261**, 8750 (1986).

[48] D. S. Adams, D. Noonan, T. C. Burn, and H. B. Skinner, *Gene* **54**, 93 (1987).

[49] P. G. Siliciano, D. A. Brow, H. Roiha, and C. Guthrie, *Cell (Cambridge, Mass.)* **50**, 585 (1987).

[50] A. Krol, H. Gallinaro, E. Lazar, M. Jacob, and C. Branlant, *Nucleic Acids Res.* **9**, 769 (1981).

[51] R. Reddy and H. Busch, *in* "The Cell Nucleus" (H. Busch, ed.), Vol. 8, p. 261. Academic Press, New York, 1981.

[52] N. Kato and F. Harada, *Biochem. Biophys. Res. Commun.* **99**, 1468 (1981).

[53] C. Branlant, A. Krol, E. Lazar, B. Haendler, M. Jacob, L. Galego-Dias, and C. Pousada, *Nucleic Acids Res.* **11**, 8359 (1983).

[54] M. Kazmaier, G. Tebb, and I. W. Mattaj, *EMBO J.* **6**, 3071 (1987).

[55] M. Liu, R. Reddy, D. Henning, D. Spector, and H. Busch, *Nucleic Acids Res.* **12**, 1529 (1984).

[56] R. Reddy, D. Spector, D. Henning, M. Liu, and H. Busch, *J. Biol. Chem.* **258**, 13965 (1983).

[57] B. Patterson and C. Guthrie, *Cell (Cambridge, Mass.)* **49**, 613 (1987).

[58] P. Epstein, R. Reddy, D. Henning, and H. Busch, *J. Biol. Chem.* **255**, 8901 (1980).

[59] G. Das, D. Henning, D. Wright, and R. Reddy, *EMBO J.* **7**, 503 (1988).

[60] F. Harada, N. Kato, and S. Nishimura, *Biochem. Biophys. Res. Commun.* **95**, 1332 (1980).

[61] Y. Ohshima, N. Okada, T. Tani, Y. Itoh, and M. Itoh, *Nucleic Acids Res.* **9**, 5145 (1981).

[62] G. R. Kunkel, R. L. Maser, J. P. Calvet, and T. Pederson, *Proc. Natl. Acad. Sci. U.S.A.* **83**, 8575 (1986).

[63] J. Sri-Widada, J. P. Liautard, C. Assens, and C. Brunel, *Mol. Biol. Rep.* **8**, 29 (1981).

[64] A. Krol, P. Carbon, J. Ebel, and B. Appel, *Nucleic Acids Res.* **15**, 2463 (1987).

[65] G. Das, D. Henning, and R. Reddy, *J. Biol. Chem.* **262**, 1187 (1987).

[66] H. Saluz, Y. Choffat, and E. Kubli, *Nucleic Acids Res.* **16**, 1202 (1988).

[67] H. B. Skinner and D. S. Adams, *Nucleic Acids Res.* **15**, 371 (1987).

[68] T. Kiss, M. Antal, and F. Solymosy, *Nucleic Acids Res.* **15**, 543 (1987).

[69] K. L. Mowry and J. A. Steitz, *Science* **238**, 1682 (1987).

[70] M. Cotten, O. Gick, A. Vasserot, G. Schaffner, and M. L. Birnstiel, *EMBO J.* **7**, 801 (1988).

[71] M. DeLorenzi, U. Rohrer, and M. L. Birnstiel, *Proc. Natl. Acad. Sci. U.S.A.* **83**, 3243 (1986).

[72] K. Strub, G. Galli, M. Busslinger, and M. L. Birnstiel, *EMBO J.* **3**, 2801 (1984).

[73] R. Reddy, D. Henning, and H. Busch, *J. Biol. Chem.* **260**, 10930 (1985).

[74] N. Kato and F. Harada, *Biochim. Biophys. Acta* **782**, 127 (1984).

[75] D. Tollervey, J. Wise, and C. Guthrie, *Cell (Cambridge, Mass.)* **35**, 753 (1983).

[36] Sequences and Classification of Group I and Group II Introns

By John M. Burke

Group I and group II[1] introns represent distinct intron types. Some, but not all, group I and group II introns have been shown to be self-splicing *in vitro*.[2-6] Both groups of introns are widespread in nature. Group I introns have been identified in nuclear rRNA genes, in mitochondrial rRNA and protein genes, in chloroplast rRNA and tRNA genes, and in protein-coding genes of bacteriophage (Table I). Group II introns have been found in mitochondrial protein genes and in chloroplast genes for protein and tRNA (Table II).

Splicing of precursor RNAs (preRNAs) containing group I introns proceeds by a two-step transesterification mechanism, initiated by attack of the 3'-hydroxyl group of a free guanosine nucleotide on the 5' splice site.[2,7] Exon ligation liberates group I introns as linear species, which may subsequently cyclize. Splicing of preRNAs containing group II introns is initiated by the attack on the 5' splice site by the 2'-hydroxyl group of a particular adenosine residue near the 3' end of the intron.[4-6] Exon ligation liberates group II introns in the form of lariats. Nuclear precursor mRNA (pre-mRNA) introns are also excised as lariats with branch-point adenosine residues, and other similarities between group II introns and nuclear pre-mRNA introns have been noted.[8-10]

Both group I and group II introns have characteristic structural features, showing conserved primary, secondary, and tertiary structural ele-

---

[1] Group I introns have also been termed class I introns. Group II introns have also been termed class II introns.

[2] K. Kruger, P. J. Grabowski, A. J. Zaug, J. Sands, D. E. Gottschling, and T. R. Cech, *Cell* (*Cambridge, Mass.*) **31**, 147 (1982).

[3] T. R. Cech, *Gene* **73**, 259 (1989).

[4] C. L. Peebles, P. S. Perlman, K. L. Mecklenburg, M. L. Petrillo, J. H. Tabor, K. A. Jarrell, and H.-L. Cheng, *Cell* (*Cambridge, Mass.*) **44**, 213 (1986).

[5] R. van der Veen, A. C. Arnberg, G. van der Horst, L. Bonen, H. F. Tabak, and L. A. Grivell, *Cell* (*Cambridge, Mass.*) **44**, 225 (1986).

[6] C. Schmelzer and R. J. Schweyen, *Cell* (*Cambridge, Mass.*) **46**, 557 (1986).

[7] T. R. Cech, *Science* **236**, 1532 (1987).

[8] T. R. Cech, *Cell* (*Cambridge, Mass.*) **34**, 713 (1983).

[9] T. R. Cech, *Cell* (*Cambridge, Mass.*) **44**, 207 (1986).

[10] P. A. Sharp, *Science* **235**, 766 (1987).

A

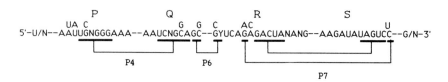

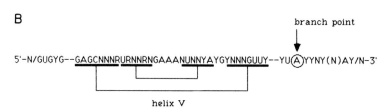

helix V

FIG. 1. Sequences characteristic of group I and group II introns. Splice sites are indicated by a slash. For group I introns (A), conserved sequence elements, P, Q, R, and S, are shown, and three of the nine base-pairing interactions characteristic of group I introns are indicated (P4, P6, and P7). For group II introns (B), the branch-point adenosine residue is circled. One of the six base-pairing interactions characteristic of group II introns is indicated (helix V). N, Unidentified nucleotide; Y, pyrimidine; R, purine.

ments.[11-14] Genetic studies have shown that a number of these structural elements are important for splicing *in vivo* and *in vitro*.[15,16] The folded structures of the introns must provide specific sites for substrate binding and catalysis.

Group I Introns

The structure of group I introns has been reviewed in some detail.[3,16,17] Group I introns have only a single, conserved nucleotide at each splice site, U preceding the 5′ splice site and G preceding the 3′ splice site (Table I, Fig. 1). Four conserved internal sequence elements, P, Q, R, and S, are

[11] F. Michel, A. Jacquier, and B. Dujon, *Biochimie* **64**, 867 (1982).
[12] R. W. Davies, R. B. Waring, J. A. Ray, T. A. Brown, and C. Scazzocchio, *Nature (London)* **300**, 719 (1982).
[13] J. M. Burke and U. L. RajBhandary, *Cell (Cambridge, Mass.)* **31**, 509 (1982).
[14] F. Michel and A. Jacquier, *Cold Spring Harbor Symp. Quant. Biol.* **52**, 201 (1987).
[15] A. Jacquier and F. Michel, *Cell (Cambridge, Mass.)* **50**, 17 (1987).
[16] J. Burke, *Gene* **73**, 273 (1989).
[17] R. B. Waring, R. W. Davies, T. A. Brown, and C. Scazzocchio, *Gene* **28**, 277 (1984).

always found in group I introns.[3] These sequence elements always occur in the same order and base pair with one another in the folded structure of the preRNA (Fig. 1).[3,16,18] Elements R [consensus sequence (C/G)YUCA(GA/AC)GACUANANG] and S (consensus sequence AAGAUAUAGUCY) are the most highly conserved sequences within group I introns and serve as convenient "landmarks" for the identification of group I introns (Fig. 1).

All group I introns are capable of folding into similar secondary structures composed of nine characteristic base-paired elements, termed P1–P9.[18] *In vivo* splicing of some, and perhaps all, group I introns requires the participation of specific proteins.[16] It is likely that proteins facilitate the splicing reactions by assisting the preRNA to fold into a conformation which is conducive to splicing.

The 5' splice site of group I introns is selected by binding to an "internal guide sequence" (IGS), located near the 5' end of the intron.[19,20] An overlapping region of the IGS has been implicated in the selection of 3' splice sites of some group I introns on the basis of comparative sequence data,[12,17] but this proposed pairing has not been demonstrated experimentally.

Two subclasses of group I introns have been identified. Group IA introns contain additional sequences between P7(5') and P3(3').[18] The sequences of elements R and S of group IA and IB introns vary in a characteristic fashion (Table I). In addition, several families of group I introns have been identified that have extensive structural similarities. One such family includes the *Tetrahymena thermophila* nuclear large rRNA introns, the *Neurospora crassa* mitochondrial NADH dehydrogenase subunit 4L intron, and the first intron of the *Podospora anserina* mitochondrial NADH dehydrogenase subunit I gene.[21,22]

Group II Introns

Keller and Michel[23] have formulated a set of rules that group II introns are likely to obey. These include consensus sequences GUGYG and YUAYYNY(N)AY at the 5' and 3' splice sites, respectively, and a char-

[18] J. M. Burke, M. Belfort, T. R. Cech, R. W. Davies, R. J. Schweyen, D. A. Shub, J. W. Szostak, and H. F. Tabak, *Nucleic Acids Res.* **15**, 7217 (1987).
[19] M. D. Been and T. R. Cech, *Cell* (*Cambridge, Mass.*) **47**, 207 (1986).
[20] R. B. Waring, P. Towner, S. J. Minter, and R. W. Davies, *Nature* (*London*) **321**, 133 (1986).
[21] F. Michel and D. J. Cummings, *Curr. Genet.* **10**, 69 (1985).
[22] M. A. Nelson and G. Macino, *Mol. Gen. Genet.* **206**, 318 (1987).
[23] M. Keller and F. Michel, *FEBS Lett.* **179**, 69 (1985).

TABLE I
Group I Introns†

Intron	SG	5' SS	R	S	3' SS	Structure	Sequence
An mit COB	IB	agguuuAUACA...	UCUCAGAGACUACAUG...	AUGAUAUAGUCC...	AAAUGcucu	a	b
An mit LSU	IA	agggauUUUUA...	GUUCAACGACUAAA...	AAGAAAUAGUUU...	AAUUGaaca	a	c
An mit OX1	IB	uauggcAAAAU...	CCUCAGAGACUACACG...	AAGACAUAGUCC...	AUUAGguuu	d	d
An mit OX2	IB	gccgguAAACA...	CUUCAGAGACUAUACG...	AAAAUAUAGUCC...	AAUAGggua	d	d
An mit OX3	IB	agagguCAAAA...	CCUCAGAGACUUUAUG...	AAGAUAAAGUCC...	UUUUGuuau	d	d
Ce chl LSU5	IB	aacgguUAAUU...	CCUCAGAGACUACACG...	AAGAUAUAGUCC...	UAUUGccua	e	e
Cm chl LSU1	IA	cgggauUUUAA...	GUUCAACGACUAGGGC...	AAGACAUAGUCU...	AAUUGaag	e	e
Cm chl LSU2	IA	uuugcuCUUUU...	GUUCAACGACUAGAAA...	AAGACAUAGUCU...	AAUUGuaga	e	e
Cm chl LSU3	IA	caggcuUAGUU...	GUUCAACGACUAGCGA...	AUGACAUAGUCU...	AACUGcaa	e	e
Cm chl LSU4	IA	agggauAGAAC...	GUUCAACGACUAGGAG...	AAGACAUAGUCU...	UACUGaaca	e	e
Cm chl LSU5	IA	cgucguAAACU...	CCGUAACGACUCUGUA...	AUGGUAUAGUCU...	AAUUGgaga	e	e
Cm chl psbA1	IB	gcucauAAAAA...	GUUCAGAGACUAAUGG...	AUGAUAUAGUCC...	AAAAGgg	e	e
Cm chl psbA2	IB	acgcucCAAAU...	CCUCAACGACUACACG...	AUGAUAUAGUCU...	AAAAGcauu	e	e
Cm chl SSU	IA	cgcgguCUGGC...	CCGUAACGACUUUUCU...	AAGGUAUAGUCU...	AAUCGaaua	e	e
Cr chl LSU	IA	aaacguAAAUA...	CCGUAGAGACUUAUA...	AAGACAUAGUCC...	UCAUGcgug	f	f
Kf mit ATP9	IA	uaggauUAAUU...	GUUCAACGACUAGAUA...	AUGACAUAGUCU...	UAAAGucgcu	g	g
Kt mit LSU	IA	aggauAAAUA...	GUUCAACGACUAGAUA...	AUGACAUAGUCU...	AAUUGaaca	h	h
Mp chl tRNAL	IB	ggacuuAAUUU...	GUGCAGAGACUAAAG...	AAGAURGAGUCC...	AAAUGaaaa	i	i
*Nc mit COB1	IB	cuggguUAUCA...	CCGUACAGACUGGGUC...	AAUGUACAGUCG...	GUACGuaug	j	k
Nc mit COB2	IA	uugaguCAAAA...	GUUCAACGACUAAUGA...	AAGAUAUAGUCU...	UAUUGucau	j	l
Nc mit LSU	IA	aggauGUUUG...	GUUCAACGACUAUAAG...	AAGAAAUAGUCU...	AGUUGaaca	a	m
Nc mit ND1	IB	gaggcuACURA...	CUUCAGAGACUAGACG...	AAGGUAUAGUCC...	UUUUGgaau	n	n

Nc mit ND4L	IB	uagauuAAUUA....GUUCAUCGACUAAAUG....AAGAUAUAGUCA....AACCGaaga	o	o
Nc mit ND5,1	IB	auccuCGGGG....GAGCAACGAGUAGACG....AAGAUGUACUCU....AAAAGcaua	o	o
Nc mit ND5,2	IB	gaagguUUUUU....CCUCAGAGAUUACAAG....AAGAUAUAGUCC....UUUUGccua	o	o
Nc mit OX2	IB	....CCUCAGAGACUACA....AAGAUAUAGUCC.... AUG	p	p
Nc mit OX3	IB	....CUUCAGAGGCUAUA....AAGAUAUGGUCC.... UUG	p	p
Nc mit OX4	IB	....CCUCAGAGACUCUA....AAUAUAAAGUCC.... UUG	p	p
Nc mit ATP6,2	IB	aaugguUAACA....CCUCAGAGACUACAAG....AAGAUAUAGUCC....UGCUGuccu	q	q
Nt chl tRNAL		ggacuuAAUUG....GUGCAGAGACUCAAUG....AAGAUAGAGUCC....AGAGGaaaa		r
*Pa mit ND1,1	IB	guagguCGCAC....GAUCAUCGACUAGACG....AAGAUAUAGUCA....AUUCGuauu	s	s
Pa mit ND1,2	IB	auaccuUAUGU....GAGCAACGAGCAGACG....AAGGUGUCUCU....AUAAGuaug	s	s
Pa mit ND1,3	IA	cuggauUUUAU....GUUCAACGACUAGUAU....AAGACAUAGUCU....AUUUGgaag	t	t
Pa mit ND1,4	IB	gaggcuGGGUU....CCUCAGAGACUAUACG....AAGAUAUAGUCC....AUAUGgaac	t	t
Pa mit OX7	IB	uuuucuUUAAA....CACUACAGACUAGUUC....AAUGUAUAGUCG....AAAUGgauu		u
Pa mit OX8	IB	uucgguCGAGA....CUUCAGAGACCAUACG....AAGAUGGGUCC....UGUUGcauc		u
Pa mit OX9	IB	uucgguCAAUA....UUUCAGAGGCCAUACG....AAGAGAUGGUCC....AAAAGuaua		u
Pp nuc LSU1	IB	agggauUGAUA....CUUCAACGACUGGAAA....AAGGUGCAGUCC....AUCGGaacu	v	w
Sc mit COB3	IB	ugggguUAAUA....AUUCAGAGACUUUAUA....AAGAUAUAGUCC....AAAAGcuca	x	x
Sc mit COB4	IB	uuagguCAAAA....CUUCAGAGACUACACG....AAGAUAUAGUCC....AAAAGcauc	a	y
*Sc mit COB5	IA	auugaaUUAA....GUUCAACGACUAGAAA....AUGACGUAGUCU....AAUUGugu	n	aa
*Sc mit LSU	IA	agggauAAUUU....GUUCAACGACUAGAUG....AUGAUAUAGUCU....AUUUGaaca	n	bb
*Sc mit OX3	IB	uuugguAACCA....CCUCAGAGACUACACG....AAGAUAUAGUCC....AAAUGaccu	n	cc
Sc mit OX4	IB	uuugguCAAAC....CCUCAGAGACUACAAG....AAGAUAUAGUCC....ACAAGcacc	n	cc
*Sc mit OX5a	IB	augauuAAUAA....GCGUAGAGAUUAGACG....AAGGUAUAAUCC....AAAAGagcu	a	dd
Sc mit OX5β	IB	cacgauAUUAA....UUGUAGAGACUAAACG....AAGAGAUUAAACG....UUAUGacuu	a	dd
Sp mit OX2	IB	agagguCAAUU....CCUCAGAGACUUUACG....AAAAUAAAGUCC....UAUUGuuau	ee	ee
Sp mit OX2a	IB	uuugguCAACG....CCUCAGAGACUAUAUG....AAGAGAUAGUCC....AUUUGcauc	ff	ff

(continued)

TABLE I (continued)

Intron	SG	5' SS	R	S	3' SS	Structure	Sequence
Sp mit OX3	IB	caugauAUUAA....	CCGUAGAGACUAAACG....	AAGUUAUAGUCC....	UUCAGaccu	ff	ff
*Sp phg g31	IA	gagccuAAAGA....	GUGCAACGACUAUC.....	AAGAUAUAGUCU....	UGUAGaacg	gg	gg
*T4 phg nrdB	IA	ugcguAAAAU....	GUUCAACGACUAGUCU....	GUGAUAUAGUCU....	AVACGguac	hh	ii
*T4 phg sunY	IA	augaguUAACG....	GCCCAAGACUAUACC....	AAGAUAUAGUCU....	CAUUGauga	hh	hh
*T4 phg td	IA	uugggguUAAUU....	GCCUAAGACUAUCCC....	GAGAUAUAGUCU....	UAAUGcuac	hh	jj
Tc nuc LSU	IB	cucucuAAAUU....	GUUCACAGACUAAAUG....	AAGAUAUAGUCG....	AAUCGuaag	kk	kk
Th nuc LSU	IB	cucucuAAAUU....	GUUCACAGACUAAAUG....	AAGAUAUAGUCG....	AAUCGuaag	kk	kk
Tm nuc LSU	IB	cucucuAAAUU....	GUUCACAGACUAAAUG....	AAGAUAUAGUCG....	AAUCGuaag	kk	kk
Tp nuc LSU	IB	cucucuAAAUU....	GUUCACAGACUAAAUG....	AAGAUAUAGUCG....	AAUCGuaag	kk	ll
Ts nuc LSU	IB	cucucuAAAUU....	GUUCACAGACUAAAUG....	AAGAUAUAGUCG....	AAUCGuaag	kk	kk
*Tt nuc LSU	IB	cucucuAAAUA....	GUUCACAGACUAAAUG....	AAGAUAUAGUCG....	ACUCGuaag	mm	nn
Vf chl tRNAL	IB	ggacuuAAUUG....	GUGCAGAGACUCAAUG....	AAGAUAGAGUCC....	AUCGGaaaa	oo	oo
Zm chl tRNAL	IB	ggacugGAAUG....	GUGCAGAGCACUCAAUG....	AAGAGAGAGUCC....	AAAGGaaaa	pp	pp
			C				
CONSENSUS	IA	.....u....	GYUCAACGACUANANN....	AAGAUAUAGUCU....	.......G....		
			G				
CONSENSUS	IB	.....u....	CYUCAGAGACUANANG....	AAGAUAUAGUCC....	.......G....		

† Sequences at the 5' splice site (5' SS), sequence element R, sequence element S, and the 3' splice site (3' SS) are shown. Lowercase letters indicate exon nucleotides; uppercase letters indicate intron nucleotides. Introns are identified by organism, genome, and gene. For example, Ce chl LSU5 is the fifth intron of the *Chlamydomonas eugametos* chloroplast large subunit rRNA gene. Introns are identified by subgroup (SG) as belonging to either subgroup IA or IB. Reference footnotes are given for structural diagrams and sequences for each intron. Introns that have been shown to be self-splicing are indicated with an asterisk (*). An, *Aspergillus nidulans*; ATP, subunit of ATP synthetase; Ce, *C. eugametos*; chl, chloroplast; Cm, *Chlamydomonas moewussi*; COB, apocytochrome *b*;

Cr, *Chlamydomonas reinhardtii*; g31, gene 31 product (DNA polymerase); Kf, *Kluyveromyces fragilis*; Kt, *Kluyveromyces thermotolerans*; LSU, large subunit rRNA; mit, mitochondrion; Mp, *Marchantia polymorpha* (liverwort); Nc, *Neurospora crassa*; ND, subunit of NADH dehydrogenase; nrd, subunit of ribonucleotide reductase; Nt, *Nicotiana tabacum* (tobacco); nuc, nucleus; OX, cytochrome oxidase subunit 1; Pa, *Podospora anserina*; phg, bacteriophage; Pp, *Physarum polycephalum*; psb, subunit of photosystem II; Sc, *Saccharomyces cerevisiae*; Sp, *Schizosaccharomyces pombe*; SP, SPO1 bacteriophage; SSU, small subunit rRNA; sunY, split gene of unknown function; T4, T4 bacteriophage; Tc, *Tetrahymena cosmopolitanis*; td, thymidylate synthase; Th, *Tetrahymena hyperangularis*; Tm, *Tetrahymena malacensis*; Tp, *Tetrahymena pigmentosa*; Ts, *Tetrahymena sonneborni*; Tt, *Tetrahymena thermophila*; Vf, *Vicia faba*; Zm, *Zea mays*.

[a] R. B. Waring and R. W. Davies, *Gene* **28**, 277 (1984).
[b] R. B. Waring, R. W. Davies, C. Scazzochio, and T. A. Brown, *Proc. Natl. Acad. Sci. U.S.A.* **79**, 6332 (1982).
[c] R. Netzker, H. G. Kochel, N. Basak, and H. Kuntzel, *Nucleic Acids Res.* **10**, 4783 (1982).
[d] R. B. Waring, T. A. Brown, J. A. Ray, C. Scazzocchio, and R. W. Davies, *EMBO J.* **3**, 2121 (1984).
[e] C. Lemieux, V. Durocher, A. Gauthier, G. Bellamare, M. Turmel, and J. Boulanger, personal communication.
[f] J. D. Rochaix, M. Rahire, and F. Michel, *Nucleic Acids Res.* **13**, 975 (1985).
[g] B. Dujon, L. Colleaux, A. Jacquier, F. Michel, and C. Monteilhet, *in* "Extrachromosomal Elements in Lower Eukaryotes" (R. B. Wickner, A. Hinnebusch, A. M. Lambowitz, I. C. Gunsalus, and A. Hollaender, eds.), p. 5. Plenum, New York, 1986.
[h] A. Jacquier and B. Dujon, *Mol. Gen. Genet.* **192**, 487 (1983).
[i] K. Umesono, H. Inokuchi, Y. Shiki, M. Takeuchi, Z. Chang, H. Fukuzawa, T. Kohchi, H. Sirai, K. Ohyama, and H. Ozeki, *J. Mol. Biol.* **203**, 299 (1988).
[j] M. H. Citterich, G. Morelli, and G. Macino, *EMBO J.* **2**, 1235 (1983).
[k] R. A. Collins, C. A. Reynolds, and J. Olive, *Nucleic Acids Res.* **16**, 1125 (1988).
[l] J. M. Burke, C. Breitenberger, J. E. Heckman, B. Dujon, and U. L. RajBhandary, *J. Biol. Chem.* **259**, 504 (1984).
[m] J. M. Burke and U. L. RajBhandary, *Cell (Cambridge, Mass.)* **31**, 509 (1982).
[n] G. Burger and S. Werner, *J. Mol. Biol.* **186**, 231 (1985).
[o] M. A. Nelson and G. Macino, *Mol. Gen. Genet.* **206**, 318 (1987).
[p] R. A. Collins, personal communication.
[q] G. Morelli and G. Macino, *J. Mol. Biol.* **178**, 491 (1984).
[r] K. Shinozaki, M. Ohme, M. Tanaka, T. Wakasugi, N. Hayashida, T. Matsubayashi, N. Zaita, J. Chunwongse, J. Obokata, K. Yamaguchi-Shinozaki, C. Ohto, K. Torozawa, B. Y. Meng, M. Sugita, H. Deno, T. Kamogashira, K. Yamada, J. Kusuda, F. Takaiwa, A. Kato, N. Tohdoh, H. Shimada, and M. Sugiura, *EMBO J.* **5**, 2043 (1986).
[s] F. Michel and D. J. Cummings, *Curr. Genet.* **10**, 69 (1985).

(continued)

Footnotes to TABLE I (continued)

[t] D. J. Cummings, J. M. Domenico, and F. Michel, Curr. Genet. 14, 253 (1988).

[u] T. Karsch, U. Kuch, and K. Esser, Nucleic Acids Res. 15, 6743 (1987).

[v] F. Michel and B. Dujon, EMBO J. 2, 33 (1983).

[w] H. Nomiyama, Y. Sakagi, and Y. Takagi, Proc. Natl. Acad. Sci. U.S.A. 78, 1376 (1981).

[x] J. Holl, C. Schmidt, and R. J. Schweyen, in "Achievements and Perspectives in Mitochondrial Research" (E. Quagliariello, ed.), p. 227. Elsevier, Amsterdam, 1985.

[y] P. Q. Anziano, D. K. Hanson, H. R. Mahler, and P. S. Perlman, Cell (Cambridge, Mass.) 30, 925 (1982).

[z] F. Michel, A. Jacquier, and B. Dujon, Biochimie 64, 867 (1982).

[aa] C. A. Bonjardin and F. G. Nobrega, Braz. J. Med. Biol. Res. 17, 17 (1984).

[bb] B. Dujon, Cell (Cambridge, Mass.) 20, 185 (1980).

[cc] S. G. Bonitz, G. Coruzzi, B. E. Thalenfeld, and A. Tzagoloff, J. Biol. Chem. 255, 1927 (1980).

[dd] L. A. M. Hensgens, L. Bonen, M. de Haan, G. van der Horst, and L. A. Grivell, Cell (Cambridge, Mass.) 32, 379 (1983).

[ee] B. F. Lang, EMBO J. 3, 2129 (1984).

[ff] H. Trinkl and K. Wolf, Gene 45, 289 (1986).

[gg] D. Shub, personal communication.

[hh] D. A. Shub, J. M. Gott, M.-Q. Xu, B. F. Lang, F. Michel, J. Pedersen-Lane, and M. Belfort, Proc. Natl. Acad. Sci. U.S.A. 85, 1151 (1988).

[ii] B.-M. Sjöberg, S. Hahne, C. Z. Mathews, C. K. Mathews, K. N. Rand, and M. J. Gait, EMBO J. 5, 2031 (1986).

[jj] F. K. Chu, G. F. Maley, D. K. West, M. Belfort, and F. Maley, Cell (Cambridge, Mass.) 45, 157 (1986).

[kk] H. Nielsen and J. Engberg, Nucleic Acids Res. 13, 7445 (1985).

[ll] M. A. Wild and R. Sommer, Nature (London) 283, 693 (1980).

[mm] J. M. Burke, M. Belfort, T. R. Cech, R. W. Davies, R. J. Schweyen, D. A. Shub, J. W. Szostak, and H. F. Tabak, Nucleic Acids Res. 15, 7217 (1987).

[nn] N. C. Kan and J. Gall, Nucleic Acids Res. 10, 2809 (1982).

[oo] G. Bonnard, F. Michel, J. H. Weil, and A. Steinmetz, Mol. Gen. Genet. 194, 330 (1984).

[pp] A. Steinmetz, E. J. Gubbins, and L. Bogorad, Nucleic Acids Res. 10, 3027 (1982).

TABLE II
Group II Introns†

Intron	5' SS	Branch	3' SS	Structure	Sequence
Ba chl tRNAV	uuacacGUGCG....AUCGUUUUUACcga				a
Eg chl psaA1	uuggagUUGCG....UCUAGUUUUAUguu				b
Eg chl psaA2	auguuuGUAGC....CUUAAUUUUAUuca				b
Eg chl psaA3	aauguuGUAUG....UUUAAUUUUAUuuu				b
Eg chl psaB1	uugcagGUGCG....UCGAGUUUUAUguu				b
Eg chl psaB2	uuuuuuGUGUG....UCGAGUUUUAUuca				b
Eg chl psaB3	augauaGUGUG....UCAAUUCUAAUaau				b
Eg chl psaB4	cacugaGUGCG....CCUAUUUUUACuac				b
Eg chl psaB5	uauggcGUGCG....AUCUAUUUUAUcaa				b
Eg chl psaB6	ucggaaGUGCG....UCUAUCUUUAUuga				b
Eg chl psbA1	auauaaGUGCG....AUCUAUUUUACuua				c
Eg chl psbA2	gaauggGUGUG....CUUUAGUUUAUuug				c
Eg chl psbA3	uuaaacGUGCG....GCUAGUGUUAUggu				c
Eg chl psbA4	guuaugGUGCG....UUUAGUUUAACcac	d			c
Eg chl psbC2	ugggugGUGUG....ACUAAAUUAAGcuc				c
Eg chl psbC4	GUGCG....UCUAGUUUAAU				c
Eg chl psbC5	GUGUG....AACUACUUUAU				c
Eg chl psbE1	gauauuGUGCG....UUUAAUUUCAUauu				e
Eg chl psbE2	auucauGUGUG....UUUAGUUUUAUagu				e
Eg chl psbF	aauaaaGUGCG....AUUUACCUAAUgau				e
Eg chl rbcL1	aaagcgGUGUG....UUUAGUUUUAUggu				f
Eg chl rbcL2	caagugGUGUG....UGCAAUUUUAUuca				f
Eg chl rbcL3	ccugccGUGUG....UUUAAUUUUAUgaa				f
Eg chl rbcL4	aggucgUUGUG....UCUAAUUUUAUuug				f
Eg chl rbcL5	gaugaaGUGCG....UUUAGUUUUAUaau				f
Eg chl rbcL6	gagaugGUGUG....CUUAAUUUUAUuac				f
Eg chl rbcL7	aaugcuGUGUG....UUCUAUUUUAAcua				f
Eg chl rbcL8	aaugcuGUGUG....CUUAAUUUUAUcca				f
Eg chl rbcL9	caagcuGUGCG....UUUAAUUUUAUucg				f
Eg chl rps3,1	uaacuaUUGAG....UCUAUUUUCUUagu				g
Eg chl X1	gguaaaGUGCG....UCUAGUUUAAUuau				d
Eg chl X2	cuaugcGUGUG....AACUACUUUAUguu				d
Mp chl atpF	gagugUGUGCG....UCUACUUUCAUuaa				h
Mp chl ndh1	cucuacGUGUG....UCGACUAUAAUuau				h
Mp chl ndh2	gaaggaGUGCG....UCGACUCUAACuca				h
Mp chl ORF135	cuagagGUGUG....CCUAUCCUAAUauu				h

(continued)

TABLE II (continued)

Intron	5' SS	Branch	3' SS	Structure	Sequence
Mp chl ORF167	gagaugGUGUG....ACAACCGUAACgaa				h
Mp chl ORF203,1	cguauaAUGCG....AUUAUCUUAAUcaa				h
Mp chl ORF203,2	cacgcuUUGUG....UAUUUAUUAAUagg				h
Mp chl petB	auggguGUGCG....CCUAUCUCAAUaaa				h
Mp chl petD	gggaguGUGUG....UCUACCUUAAUaac				h
Mp chl rpl16	cuuaguGUGUG....UCGACUAUAACccu				h
Mp chl rpl2	cuuugaGUGCG....AUCUACUUCAAcca				h
Mp chl rpoC1	ugcgauGUGUG....CCUAUCCCAAUcuc				h
Mp chl rps12,1	guguauGUGCG.//.UCAACUUUUCCacu	i			h
Mp chl rps12,2	guucuaGUGCG....AUCCACCCUACaau	i			h
Mp chl tRNAA	uugcaaUUGGG....GCUUACCCUGUggc				h
Mp chl tRNAG	guaaaaGUGCG....UCGACUAUAACccu				h
Mp chl tRNAI	ugauaaUUGCG....AUUUACUUCACggg				h
Mp chl tRNAK	uuuuaaGUGCG....AUCUACUUCAUccg	i			h
Mp chl tRNAV	uuacacGUGCG....AACUGUUUUACcga				h
Nt chl atpF	gaguguGUGCG....UCUACUUUCAUuaa				j
Nt chl ndh1	cucuacGUGUG....UCGACUAUGAUuau				j
Nt chl ndh2	gaaggaGUGCG....UCGACUCUGACucu				j
Nt chl petB	augaquGUGUG....CCUAUCUCAAUaaa				j
Nt chl petD	gggaguGUGUG....CCUAUCCCAAUaac				j
Nt chl rpl16	cuuaguGUGUG....UCAACUAUAACccc				j
Nt chl rpl2	cuuugaGUGCG....AUCUACUUCAAccg				j
Nt chl rpoC1	ugcgauGUGUG....CCUAUCCCAAUuuu				j
Nt chl rps12,1	guguauGUGCG.//.UCAACUUUUCCacu				j
Nt chl rps12,2	guucuaGUGCG....AUCCACCCUACaau				j
Nt chl rps16	agcaacGUGCG....UCAAUCCCAAUgag				k
Nt chl tRNAA	uugcaaUUGGG....GUUUACCCUGCggc				j
Nt chl tRNAG	guaaaaGUGUG....UCGACUAUAACccc				j
Nt chl tRNAI	ugauaaUUGCG....AUUUACUUCACggg				j
Nt chl tRNAK	uuuuaaGUGCG....AUCUACUCCAUccg				l
Nt chl tRNAV	uuacacGCGCG....ACCUGUUUUACcga				j
Pa mit OX1	uugcagGUGCG....GCUUAUCCUACaua	m			m
Pa mit OX4	ucugauGUGCG....AUCGACCCUACuau	n			n
Sa chl tRNAK	uuuuaaGUGCG....AUCUACUCCAUcgg	o			o
*Sc mit COB1	gacagaGUGAG....CCUAUCACAAUugu	p			q
Sc mit OX1	uuaaugGUGCG....UGCUAUUUCAUcuu	r			s

(continued)

TABLE II (continued)

Intron	5' SS	Branch	3' SS	Structure	Sequence
Sc mit OX2	uuuucuGUGCG....UGCUACUCUACucu			r	s
*Sc mit OX5g	auuuucGAGCG....CCUAUCGGGAUacu			p	s
So chl atpF	gaguguGUGCG....UCUACUUUCAUuaa				t
Sp mit COB1	auaaaaUUGCG....CUUAGUUCAAUgau			u	u
Ta chl atpF	gaguguGUGCG....UCUACUUUCAUuaa				v
Ta chl tRNAG	guaaaaGUGUG....UCGACUAUAACccc				w
Zm chl petB	augaguGUGUG....CCUAUCUCAAUaaa				x
Zm chl petD	gggaguGUGUG....CCUAUCCCAAUaac				x
Zm chl tRNAA	uugcaaUUGGG....GUUUACCCUGUggc			r	y
Zm chl tRNAI	ugauaaUUCGU....AUUUACUUCACggg			r	y
Zm chl tRNAV	uuacacGUGCG....AUCUGUUUUACcga				z
Zm mit OX2	aucggaGUGCG....GUCGACCCAACcuu			r	aa
CONSENSUS	......GUGYG......YUAYYNYAY...	(N)			

† Sequences at the 5' and 3' splice sites (5' SS and 3' SS, respectively) are shown. Lowercase letters indicate exon nucleotides; uppercase letters indicate intron nucleotides. Likely sites of branch-point formation are underlined. Introns are identified as described in Table I. Reference footnotes are given for structural diagrams and sequences for each intron. Introns that have been shown to be self-splicing are indicated with an asterisk (*). atp, Subunit of ATP synthetase; Ba, barley; chl, chloroplast; COB, apocytochrome b; Eg, *Euglena gracilis;* mit, mitochondrion; Mp, *Marchantia polymorpha* (liverwort); ndh, subunit of NADH dehydrogenase; Nt, *Nicotiana tabacum* (tobacco); ORF, open reading frame; OX, cytochrome oxidase subunit I; Pa, *Podospora anserina;* pet, subunit of photosynthetic electron transport complex; psa, subunit of photosystem I; psb, subunit of photosystem II; rbcL, large subunit of ribulose-1,5-bisphosphate carboxylase; rpl, large subunit ribosomal protein; rpo, RNA polymerase subunit; rps, small subunit ribosomal protein; Sa, *Sinapis alba* (mustard); Sc, *Saccharomyces cerevisiae;* Sp, *Schizosaccharomyces pombe;* Ta, *Triticum aestivum* (wheat); X, unidentified open reading frame; Zm, *Zea mays.*

[a] G. Zurawski and M. T. Clegg, *Nucleic Acids Res.* **12,** 2549 (1984).
[b] J. C. Cushman, R. B. Hallick, and C. A. Price, *Curr. Genet.* **13,** 159 (1988).
[c] G. D. Karabin, M. Farley, and R. B. Hallick, *Nucleic Acids Res.* **12,** 5801 (1984); M. Keller and E. Stutz, *FEBS Lett.* **175,** 173 (1984).
[d] M. Keller and F. Michel, *FEBS Lett.* **179,** 69 (1985).
[e] J. C. Cushman, D. A. Christopher, M. C. Little, R. B. Hallick, and C. A. Price, *Curr. Genet.* **13,** 173 (1988).
[f] J. C. Gingrich and R. B. Hallick, *J. Biol. Chem.* **260,** 16156 (1985).
[g] D. A. Christopher, J. C. Cushman, C. A. Price, and R. B. Hallick, *Curr. Genet.* **14,** 275 (1988).
[h] K. Ohyama, H. Fukuzawa, T. Kohchi, T. Sano, S. Sano, H. Shirai, K. Umesono, Y. Shiki, M. Takeuchi, Z. Chang, S. Aota, H. Inokuchi, and H. Ozeki, *J. Mol. Biol.* **203,**

Footnotes to TABLE II (continued)

281 (1988); K. Umesono, H. Inokuchi, Y. Shiki, M. Takeuchi, Z. Chang, H. Fukuzawa, T. Kohchi, H. Shirai, K. Ohyama, and H. Ozeki, J. Mol. Biol. 203, 299 (1988);
H. Fukuzawa, T. Kohchi, T. Sano, H. Shirai, K. Umesono, H. Inokuchi, H. Ozeki,
and K. Ohyama, J. Mol. Biol. 203, 333 (1988); T. Kohchi, H. Shirai, H. Fukuzawa, T.
Sano, T. Komano, K. Umesono, H. Inokuchi, H. Ozeki, and K. Ohyama, J. Mol. Biol.
203, 353 (1988).

[i] H. Ozeki, K. Ohyama, H. Inokuchi, H. Fukuzawa, T. Kohchi, T. Sano, K. Nakahigashi, and K. Umesono, Cold Spring Harbor Symp. Quant. Biol. 52, 791 (1987).

[j] K. Shinozaki, M. Ohme, M. Tanaka, T. Wakasugi, N. Hayashida, T. Matsubayashi,
N. Zaita, J. Chunwongse, J. Obokata, K. Yamaguchi-Shinozaki, C. Ohto, K. Toro-
zawa, B. Y. Meng, M. Sugita, H. Deno, T. Kamogashira, K. Yamada, J. Kusuda, F.
Takaiwa, A. Kato, N. Tohdoh, H. Shimada, and M. Sugiura, EMBO J. 5, 2043 (1986).

[k] K. Shinozaki, H. Deno, M. Sugita, S. Kuramitsu, and M. Sugiura, Mol. Gen. Genet.
202, 1 (1986).

[l] M. Sugita, K. Shinozaki, and M. Sugiura, Proc. Natl. Acad. Sci. U.S.A. 82, 3557
(1985).

[m] H. D. Ocsiewacz and K. Esser, Curr. Genet. 8, 299 (1984).

[n] E. T. Matsuura, J. M. Domenico, and D. J. Cummings, Curr. Genet. 10, 915 (1986).

[o] H. Neuhaus and G. Link, Curr. Genet. 11, 251 (1987).

[p] F. Michel and A. Jacquier, Cold Spring Harbor Symp. Quant. Biol. 52, 201 (1987).

[q] C. Schmelzer, C. Schmidt, K. May, and R. J. Schweyen, EMBO J. 2, 2047 (1983).

[r] F. Michel and B. Dujon, EMBO J. 2, 33 (1983).

[s] S. G. Bonitz, G. Coruzzi, B. E. Thalenfeld, and A. Tzagoloff, J. Biol. Chem. 255,
11927 (1980).

[t] J. Hennig and R. G. Herrmann, Mol. Gen. Genet. 203, 117 (1986).

[u] B. F. Lang, F. Ahne, and L. Bonen, J. Mol. Biol. 184, 353 (1985).

[v] C. R. Bird, B. Koller, A. D. Auffret, A. K. Huttly, C. J. Howe, T. A. Dyer, and J. C.
Gray, EMBO J. 4, 1381 (1985).

[w] F. Quigley and J. H. Weil, Curr. Genet. 9, 495 (1985).

[x] C. D. Rock, A. Barkan, and W. C. Taylor, Curr. Genet. 12, 69 (1987).

[y] W. Koch, K. Edwards, and H. Kossel, Cell (Cambridge, Mass.) 25, 203 (1981).

[z] E. T. Krebbers, A. Steinmetz, and L. Bogorad, Plant Mol. Biol. 3, 13 (1984).

[aa] T. D. Fox and C. J. Leaver, Cell (Cambridge, Mass.) 26, 315 (1981).

acteristic 14-bp (base-pair) major hairpin, termed helix V, containing a
YG bulge located within 100 nt (nucleotides) of the 3' end of the intron
and separated from the 3' splice site by another helix, helix VI (Fig. 1). All
of the group II introns identified to date (Table II) adhere to these rules.

A number of group II introns are capable of folding into a distinctive
and complex secondary structure containing six helical segments (helices
I–VI), as well as a significant number of conserved nucleotides[11,14,15,24]
(see Refs. 14 and 15 for recent secondary structure diagrams of group II
introns). These "complex" group II introns include the mitochondrial

[24] F. Michel and B. Dujon, EMBO J. 2, 33 (1983).

group II introns and some of the chloroplast introns.[14] The two known examples of self-splicing group II introns, intron 5 of the *Saccharomyces cerevisiae* mitochondrial cytochrome oxidase subunit I gene and intron 1 of the yeast mitochondrial apocytochrome *b* gene,[4–6] are members of this more complex subgroup. Other chloroplast group II introns do not show the complex internal folding structure, but do have the characteristic group II consensus sequences at the splice sites, as well as the 14-bp major hairpin.[23] These introns tend to be shorter than the more complex group II introns described previously, and none have been shown to be self-splicing. It appears that this simpler subclass of group II introns has lost some of the information required for self-splicing and that splicing requires trans-acting RNA, protein, or ribonucleoprotein factors.[23]

Splice sites in group II introns are selected through specific base-pairing interactions. Exon sequences immediately upstream of the 5′ splice site are termed exon-binding sites (EBS) 1 and 2. EBS1 and EBS2 pair with two sequences within the intron, termed intron-binding sites (IBS) 1 and 2 (Table II).[14,15] The EBS1–IBS1 and EBS2–IBS2 interactions are somewhat reminiscent of the IGS–5′ splice site interaction in group I introns. Selection of the 3′ splice site is likely to involve a single base pair, formed between the nucleotide at the extreme 3′ end of the intron and a specific nucleotide located in the single-stranded region joining domains II and III.[14]

A particularly striking result is that trans-splicing of group II introns has been observed *in vivo*.[25] The exons of the *rps12* gene are transcribed from opposite strands of chloroplast DNA; the gene is split within a group II intron. Thus, splicing of group II introns can proceed *in vivo*, when the 5′ and 3′ domains of the intron are on physically separate molecules.

Acknowledgments

The author wishes to thank R. Collins, C. Lemieux, and D. Shub for the communication of unpublished results. This work is supported by grant GM36981 from the National Institutes of Health.

[25] M. Hildebrand, R. B. Hallick, C. W. Passavant, and D. P. Bourque, *Proc. Natl. Acad. Sci. U.S.A.* **85**, 372 (1988).

[37] Compilation of Self-Cleaving Sequences from Plant Virus Satellite RNAs and Other Sources

By GEORGE BRUENING

Introduction

Several RNA-mediated, RNA-cleaving and -joining reactions (see Ref. 1) have 5′-phosphorylated intermediates or products. In contrast, polyribonucleotide chains derived from about 20 sources undergo autolytic cleavage to generate a 5′-hydroxyl group and a 2′,3′-cyclic phosphodiester as the new chain ends. In several instances, the nonenzymatic, self-cleaving phosphotransfer activity has been mapped to sequences of 100 or fewer nucleotide (nt) residues, and the corresponding oligoribonucleotides show the same peculiar instability as the parent molecule, cleaving at the same bond. Self-cleaving sequences that generate a 2′,3′-cyclic phosphodiester are compiled in this chapter, together with information on possible secondary structures of the active molecules. Most of the self-cleaving RNA molecules have their origin in the sequences of satellite RNAs of plant viruses. Two other self-cleaving RNA molecules are derived from another type of plant-infecting agent, a viroid, one self-cleaving RNA molecule is from a RNA of newt, and two self-cleaving RNA molecules are from a virus associated with hepatitis B virus.

A satellite RNA does not replicate when inoculated alone to a plant. A satellite RNA requires infection of the plant host by any one of a group of closely related supporting viruses. The satellite RNA becomes encapsidated in the coat protein of that virus, some as linear molecules, others predominantly as circles. This dependence on a supporting virus and the lack of any extensive nucleotide sequence similarity between the satellite RNA and the genomic RNAs of the plant virus are the defining characteristics of plant virus satellite RNAs. The smallest satellite RNAs, those having a sequence of fewer than 400 nt, do not appear to encode any protein.

In contrast to satellite RNAs, viroids replicate independently of any virus and are not encapsidated. Like the small satellite RNAs, viroid RNAs have fewer than 400 nt and apparently are not messenger RNAs. Viroids occur predominantly in the circular form, and viroids are the simplest, well-defined infectious agents.

---

[1] T. R. Cech and B. L. Bass, *Annu. Rev. Biochem.* **55**, 599 (1986).

The satellite RNAs which are sources of self-cleaving RNAs are associated with viruses that are members, or at least probable members, of three different plant virus groups. The groups and the corresponding satellite RNAs are as follows:

Nepoviruses
    Satellite arabis   sArMV RNA
      mosaic virus RNA
    Satellite tobacco   sTobRV RNA
      ringspot virus RNA
Probable sobemoviruses
    Satellite lucerne   sLTSV RNA
      transient streak
      virus RNA
    Satellite *Solanum nodiflorum*   ssNMV RNA
      mottle virus RNA
    Satellite subterranean   sSCMoV RNA
      clover mottle virus
      RNA
    Satellite velvet   sVTMoV RNA
      tobacco mottle virus
      RNA
Luteovirus
    Satellite barley   sBYDV RNA
      yellow dwarf virus RNA

The nepovirus and luteovirus satellite RNAs are encapsidated in the linear form, whereas the encapsidated forms of the sobemovirus satellite RNAs predominantly are circles. The sobemovirus satellite RNAs have been referred to as "virusoids," because the sobemovirus satellite RNAs share with viroids a circular configuration and compact secondary structure. The one viroid reported to be a source of self-cleaving RNA sequences is avocado sun-blotch viroid (ASBV).

The encapsidated forms of known small satellite RNAs are predominantly of one polarity, which is designated arbitrarily as positive (+). However, tissues in which a small satellite RNA and its associated plant virus are replicating have satellite RNAs of both polarities. Similarly, both polarities of viroid RNA accumulate during infections. The positive polarity of a viroid corresponds to that strand which accumulates in the greatest amount. Both polarities of small satellite RNAs and viroid RNAs occur in extracts of the infected tissue not only as the unit monomeric sequence, but also as tandemly repeated multimeric RNA. Positive polar-

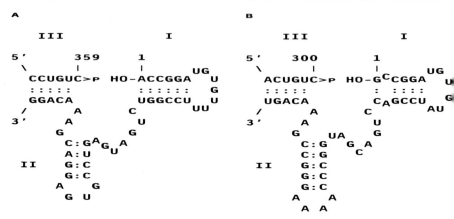

FIG. 1. Terminal nucleotide sequences of (A) positive polarity satellite tobacco ringspot virus RNA [sTobRV(+) RNA] and (B) positive polarity satellite arabis mosaic virus RNA [sArMV(+) RNA]. Sequences are arranged in a three-stem secondary structure, which shows the 3' (nt 359) and 5' (nt 1) ends of the cleaved polyribonucleotide chain juxtaposed. The sequence shown corresponds to nt 354–46 of a circularly permuted sTobRV(+) RNA sequence, which is invariant in four sequenced isolates of sTobRV RNA [J. M. Buzayan, J. S. McNinch, I. R. Schneider, and G. Bruening, *Virology* **160**, 95 (1987)]. The sequence of an isolate of sArMV(+) RNA was determined by Kaper *et al.*[6] Stems I, II, and III are as defined by Forster and Symons.[5]

ity multimers are the presumed precursors of the biologically active monomeric RNA, implying RNA processing.

Fifteen Similar, Self-Cleaving RNA Sequences

The first satellite RNA demonstrated to undergo autolytic processing was the dimeric form of the 359 nt sTobRV(+) RNA purified from encapsidated RNA.[2] The generated monomeric sTobRV(+) RNA was at least as biologically active as the dimeric RNA from which the monomeric RNA was derived. The sequence is presented (Fig. 1A), in the processed form, with the 2'3'-cyclic cytidylate phosphodiester and 5'-adenosine residue as generated by the polyribonucleotide chain-cleaving phosphotransfer reaction. The reaction was promoted by magnesium and spermidine ions, and metal ion chelators reduced the efficiency of processing. Cleavage was not prevented by incubation in detergents or other protein-denaturing conditions. The self-cleavage exhibited by *in vitro,* bacteriophage

[2] G. A. Prody, J. T. Bakos, J. M. Buzayan, I. R. Schneider, and G. Bruening, *Science* **231**, 1577 (1986).

RNA polymerase-synthesized transcripts of sTobRV(+) RNA sequences[3] means that certainly no plant-derived protein is needed for the reaction. Two critical characteristics of genuine autolytic processing are the lack of any protein requirement and the spontaneous generation of the two polyribonucleotide chain products with specific terminal nucleotides and chemical end groups.

The secondary structure form, displayed in Figs. 1–6 and summarized in Fig. 7, has its origin in observations, as cited by Hutchins *et al.*,[4] of sequences within the circular sobemovirus satellite RNA which are similar to those in the terminal regions of sTobRV RNA. These similarities anticipated the cleavage sites in the sobemovirus satellite RNAs and ASBV RNA, as well as a proposed T-shaped secondary structure,[4] termed a hammerhead structure.[5] The three double-helical stems of this structure are designated I, II, and III, according to the convention of Forster and Symons.[5]

The 300-nt sequence of a second nepovirus-associated, small satellite RNA, sArMV RNA, recently was determined.[6] Although evidence for the autolytic processing of sArMV(+) RNA has not been published, the demonstrated 2′,3′-cyclic phosphodiester and the 5′-hydroxyl end groups of the encapsidated RNA and the extensive similarity with sequences of the larger sTobRV(+) RNA imply strongly that sArMV(+) RNA also is self-cleaving. In the sArMV(+) RNA structure of Fig. 1B, stem I has one mismatch, and a uridylate residue occurs just to the 5′ side of stem II. A uridylate is located at this position in only two other isolates, which are sources of self-cleaving RNAs, the sobemovirus-associated satellite RNAs sLTSV-A(+) RNA (the Australian isolate) and sLTSV-N(+) RNA (the New Zealand isolate).

sLTSV RNA-derived, self-cleaving RNAs of both the (+) and (−) polarities are recovered readily in intact form from *in vitro* transcription mixtures. Cleavage occurs when the RNA is heated, snap cooled, and then incubated in a magnesium ion-containing buffer. This step apparently allows the RNA to assume an active conformation.[5,7] Forster and Symons[7] showed that the self-cleavage reaction is not restricted to the full sLTSV(+) RNA sequence. An oligoribonucleotide corresponding to the

[3] J. M. Buzayan, W. L. Gerlach, and G. Bruening, *Proc. Natl. Acad. Sci. U.S.A.* **83,** 8859 (1986).

[4] C. J. Hutchins, P. D. Rathjen, A. C. Forster, and R. H. Symons, *Nucleic Acids Res.* **14,** 3627 (1986).

[5] A. C. Forster and R. H. Symons, *Cell (Cambridge, Mass.)* **49,** 211 (1987).

[6] J. M. Kaper, M. E. Tousignant, and G. Steger, *Biochem. Biophys. Res. Commun.* **154,** 318 (1988).

[7] A. C. Forster and R. H. Symons, *Cell (Cambridge, Mass.)* **50,** 9 (1987).

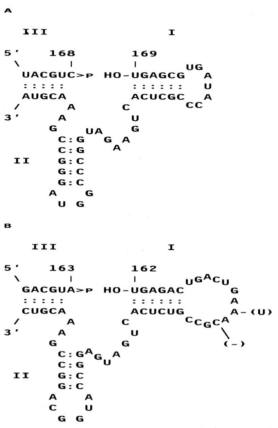

FIG. 2. Self-cleaving (A) positive and (B) negative polarity sequences of the satellite RNA of lucerne transient streak virus. The sequence of the Australian isolate (sLTSV-A RNA) is presented, with two changes in the sequence of the New Zealand isolate [sLTSV-N(−) RNA] in parentheses [P. Keese, G. Bruening, and R. H. Symons, *FEBS Lett.* **159,** 185 (1983)], one being the deletion of a cytidylate residue. Secondary structure models are from Forster and Symons,[5] who reported the cleavage reactions.

51-nt sequence 166 through 216, with an additional cytidylate at the 3′ end, also cleaves after heating and snap cooling. The sequence, presented in Fig. 2A includes the 51-nt sequence and additional nucleotides to give the maximum continuous double helix of stem III.

In contrast to sLTSV RNA, three other sobemovirus-associated satellite RNAs, sSCMoV RNA, sSNMV RNA, and sVTMoV RNA, do not exhibit processing of the negative polarity sequences. For sSNMV(+)

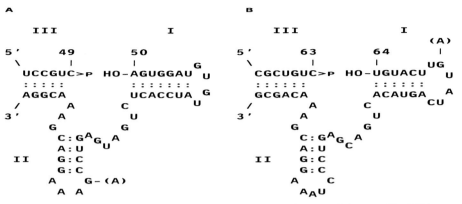

FIG. 3. Self-cleaving, positive polarity sequences of four isolates of three satellite RNAs of presumed sobemoviruses. (A) The secondary structure of sSNMV(+) RNA and sVTMoV(+) RNA. sVTMoV(+) RNA has, in the loop of stem II, a guanylate residue which, in the sSNMV(+) RNA sequence, is an adenylate [J. Haseloff and R. H. Symons, *Nucleic Acids Res.* **10**, 3681 (1982)]. (B) Two sSCMoV(+) RNA isolates have circles of 332 and 388 nt, respectively. These isolates have the same sequence within the region demonstrated to undergo autolytic processing, except that a guanylate in the 322-nt sequence is replaced by an adenylate in the 388-nt sequence as indicated.

RNA and sVTMoV(+) RNA, the site of processing, according to the hammerhead structure (Fig. 3), corresponds to the ends of linear RNA as predicted from the location of a 2′-phosphate group in the circular RNA. This phosphate is considered to be a remnant of the ligation reaction, which resulted in the circularization of these molecules.[8] *In vitro* transcripts, containing sVTMoV(+) RNA[9] or sSCMoV(+) RNA[10] sequences, exhibit self-cleavage which is so efficient that little or no transcript was detected at the end of the transcription reaction.

In vitro transcripts of both polarities of sBYDV RNA exhibit self-cleavage. The sequence for sBYDV(−) RNA conforms well to the three-stemmed consensus of the other satellite RNAs, as indicated in Fig. 4B. However, in sBYDV(+) RNA, the sequence GAAAC, which is invariant in the other RNAs considered in this chapter, is replaced by GAAAU, with a compensating change in the complementary strand of stem III. sBYDV(+) RNA also deviates from the hammerhead structure of other satellite RNAs by a cytidylate, which precedes stem II, and an adenylate, which precedes the GAAAU sequence. These nucleotides are presented

[8] P. A. Kiberstis, J. Haseloff, and D. Zimmern, *EMBO J.* **4**, 817 (1985).
[9] S. P. McNamara and R. H. Symons, personal communication.
[10] C. Davies, J. Haseloff, and R. H. Symons, submitted for publication.

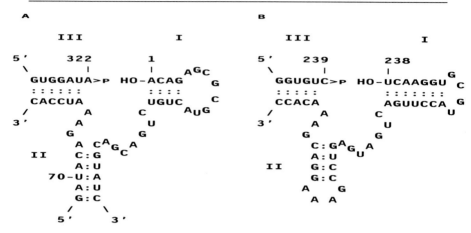

FIG. 4. Sequences and proposed secondary structures for the two polarities of the satellite RNA of barley yellow dwarf virus [(A) sBYDV(+) RNA and (B) sBYDV(−) RNA]. Sequence and secondary structure models are from W. A. Miller, T. Harcus, P. M. Waterhouse, and W. L. Gerlach (submitted for publication).

as a continuation of stem II in Fig. 4A. However, the nucleotides equally could be considered as unpaired inserts, the cytidylate corresponding to the additional uridylate of sArMV(+) RNA and sLTSV(+) RNA. The nucleotide sequences and terminal groups of the sBYDV(+) RNA cleavage products are consistent with the cleavage site of Fig. 4A. As indicated by the sequence numbering, stem II of sBYDV(+) RNA is not terminated by a small loop.

ASBV is unlike other viroids thus far examined both with regard to conserved sequences[11,12] and with the ability of its sequences to undergo autolytic processing. As originally conceived,[4] the secondary structure model for self-cleaving ASBV RNAs has a 3-nt loop at the end of a stem III of only 2 bp (base pairs) in ASBV(+) RNA and 3 bp in ASBV(−) RNA. The likely instability of such structures led Forster et al.[13] to propose and test models for ASBV secondary structure in which the two strands of stem III are derived from two different regions of a multimeric ASBV RNA sequence. A dimeric sequence, for example, may form a so-called double hammerhead.[13] Figure 5 gives the unique portion of such a secondary structure, showing a significantly longer stem III, interrupted by a central pair of adenylates.

[11] P. Keese and R. H. Symons, in "The Viroids" (T. O. Diener, ed.), p. 37. Plenum, New York, 1987.

[12] A. M. Koltunow and M. A. Rezaian, Intervirology, in press (1989).

[13] A. C. Forster, C. Davies, C. C. Sheldon, A. C. Jeffries, and R. H. Symons, Nature (London) 334, 265 (1988).

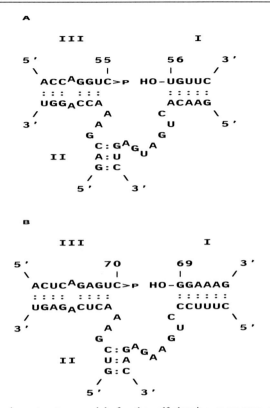

FIG. 5. Secondary structure models for the self-cleaving sequences of (A) ASBV(+) RNA and (B) ASBV(−) RNA. Sequence numbering is according to R. H. Symons [*Nucleic Acids Res*. **9**, 6527 (1981)], and the secondary structure is presented as a portion of a double hammerhead.[13]

A set of related, repetitive DNA sequences of the eastern newt, *Notophthalmus viridescens*, designated as satellite 2 DNA (no direct relationship to satellite RNA as defined previously), is transcribed in several newt tissues. RNA molecules of the repeat size, 330 nt, accumulate.[14] One mechanism by which such RNA could be generated is autolytic cleavage of a transcript of the tandemly repeated DNA sequence. The transcripts have sequences that are consistent with a hammerhead structure, and *in vitro* transcripts, containing this sequence, cleaved at a site which is near or at the expected junction predicted by the hammerhead structure. However, the site of *in vitro* cleavage is more than 45 nt distant

[14] L. M. Epstein, K. A. Mahon, and J. G. Gall, *J. Cell Biol*. **103**, 1137 (1986).

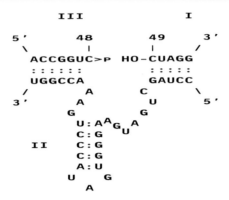

FIG. 6. Secondary structure model for a portion of the transcript of newt satellite 2 DNA. The sequence and sequence numbering are according to Epstein et al.,[14] and the secondary structure is a portion of the double hammerhead, proposed by Forster et al.[13]

from the ends of the RNA molecules recovered from newt tissue.[15] Thus, the biological significance of the *in vitro* reaction is unclear.

As in the case of ASBV, newt satellite 2 transcripts have a sequence that predicts an unstable stem III for a single hammerhead structure. Forster et al.[13] prepared an analog of the newt transcript and demonstrated that its self-cleavage was strongly concentration dependent, implying a double-hammerhead structure as the reactive form of the newt RNA. A portion of the double-hammerhead structure appears in Fig. 6.

Figure 7 presents a consensus, three-stemmed hammerhead structure derived from 15 natural isolates of self-cleaving, or putatively self-cleaving, RNAs, designated in Figs. 1–6. The configuration shown is directly related to the application of these sequences as "ribozymes", i.e., RNA-cleaving RNA molecules.[16] That is, the intact form of the top strand corresponds to the target RNA sequence, and the remaining sequence, with a loop incorporated at the end of stem II, is the trans-acting, endonucleolytic sequence.

The sequence GAAA, present in all these RNAs, is extended to GAAAC in all but sBYDV(+) RNA. In all 15 sequences, the fourth base from the 5' end of the newly cut strand is a purine, which is paired to a pyrimidine. In 14 of 15 sequences, the antipenultimate base from the 3' end is a G of the GUC consensus. In 13 of 15 secondary structures, stem II begins with a C:G base pair and has a G:C base pair in the third

[15] L. M. Epstein and J. G. Gall, *Cell (Cambridge, Mass.)* **48,** 535 (1987).
[16] J. Haseloff and W. L. Gerlach, *Nature (London)* **334,** 585 (1988).

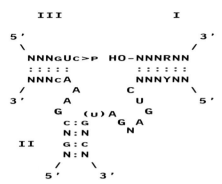

Fig. 7. Consensus, self-cleaving sequence derived from twelve satellite RNAs and three other RNAs. The sequences CUGA, GA, and GAAA, the purine:pyrimidine base pair R:Y, and the three stems are present in the same orientation in all of the structures, although stem I of sArMV(+) RNA and stem II of sBYDV(+) RNA are imperfect. Small capital letters and the parentheses indicate deviations from consensus. In three sequences, a uridylate residue is present between GA and stem II. In three other sequences, the cytidylate of GUC is replaced by an adenylate. In one sequence, the G of GUC is an adenylate. The N between CUGA and GA is a U, A, or C in the fifteen sequences. Other aspects of consensus are considered in the text.

position, the exceptions being sBYDV(+) RNA and newt satellite 2 DNA transcript. Synthetic RNAs with substitutions of some consensus nucleotide residues have retained significant cleavage activity. These substitutions include (1) the G of GUC replaced, with a compensating substitution of its paired residue in stem III; (2) other base pairs replacing the consensus C:G and G:C base pairs of stem II; and (3) pyrimidines at the fourth position of the upper strand of stem I.[13,16–19] Experiments (see Refs. 5, 16, and 20) in which essential sequences were partitioned into separate RNA molecules to produce self-cleaving complexes revealed no consensus for stem length and no requirement for a loop at the end of any stem.

Other Self-Cleaving Sequences

Three well-studied, self-cleaving reactions of RNA generate a 2′,3′-cyclic phosphodiester but do not conform to the hammerhead model. These are the reactions of sTobRV(−) RNA and the two polarities of

[17] J. R. Sampson, F. X. Sullivan, L. S. Behlen, A. B. DiRenzo, and O. C. Uhlenbeck, *Cold Spring Harbor Symp. Quant. Biol.* **52**, 267 (1987).
[18] M. Koizumi, S. Iwai, and E. Ohtsuka, *FEBS Lett.* **228**, 228 (1988).
[19] D. E. Ruffner, S. C. Dahm, and O. C. Uhlenbeck, *Gene* **82**, 33 (1989).
[20] O. C. Uhlenbeck, *Nature (London)* **328**, 596 (1987).

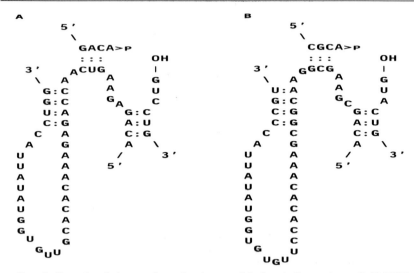

FIG. 8. Two-stranded secondary structure models for similar regions of sTobRV(−) RNA and sArMV(−) RNA. (A) The sequences shown for sTobRV(−) RNA are defined by the truncation experiments of Feldstein *et al.*[23] as the minimum sufficient for functional recognition between the two strands. The 10-nt oligoribonucleotide, which contains the cleaved ApG junction, extends from nt 52 to 43, whereas the second strand corresponds to nt 222 to 177, numbering according to the sequence of J. M. Buzayan, J. S. McNinch, I. R. Schneider, and G. Bruening [*Virology* **160**, 95 (1987)]. The secondary structure of the complex between the two oligoribonucleotides is as suggested by Haseloff and Gerlach,[22] based on insertional mutagenesis experiments and a comparison with the sequence of sArMV(−) RNA. Autolytic processing has not been demonstrated for sArMV(−) RNA (B), and the selection of sArMV(−) RNA sequences shown was guided by the results with sTobRV(−) RNA.

hepatitis δ virus (HDV) RNA. The sTobRV(−) RNA autolytic processing reaction is readily reversed *in vitro*. Linear, monomeric sTobRV(−) RNA spontaneously circularizes in a nonenzymatic ligation reaction in about a 50% yield.[21] Initial attempts to truncate sTobRV(−) RNA sequences, containing the cleavable ApG bond, revealed that either the sequence required for processing corresponds to about half the sTobRV(−) RNA sequence or noncontiguous sequences are active in this process. Insertional mutagenesis experiments of Haseloff and Gerlach[22] support the involvement of noncontiguous sequences in the processing reaction. Feldstein *et al.*[23] demonstrated directly that two sequences participate in

[21] J. M. Buzayan, W. L. Gerlach, and G. Bruening, *Nature* (*London*) **323**, 349 (1986).
[22] J. Haseloff and W. L. Gerlach, *Gene* **82**, 45 (1989).
[23] P. A. Feldstein, J. M. Buzayan, and G. Bruening, *Gene* **82**, 55 (1989).

A

5 ' -GGGUCCGCGUUCCAUCCUUUC

685   686
  | |
UUACCUGAU>P   HO-GGCCGGC

AUGGUCCCAGCCUCCUCGCUGGC

GCCGGCUGGGCAACAUUCCGAGG

GGACCGUCCCUCGGUAAUGG-3 '

B

5 ' -GUCCUUCUUUCCU

901   900
  | |
CUUC>P   HO-GGGUCGGCAUGG

CAUCUCCACCUCCUCGCGGUCCG

ACCUGGGCAUCCGAAGGAGGACG

CAGGUCCACUC-3 '

FIG. 9. Minimum contiguous sequences sufficient for self-cleavage from HDV (A) genomic and (B) antigenomic RNAs.[28] The genomic sequence[24,28] extends from nt 656 to 759, and the antigenomic sequence extends from nt 917 to 832, numbered according to Kuo et al.[24,30]

this reaction. A proposed secondary structure for the self-cleaving sTobRV(−) RNA sequence, shown in Fig. 8, is consistent with the sequence of sArMV(−) RNA. Therefore, sArMV(−) RNA may be expected to undergo autolytic cleavage.

The last two sequences in this compilation are from HDV RNA. HDV replicates in association with hepatitis B virus, a DNA virus. The 1.7-kb (kilobase) HDV genomic RNA is circular and is capable of extensive intramolecular base pairing.[24] Infected tissues have greater than unit length (multimeric) HDV RNAs of both polarities,[25] implying rolling cir-

[24] M. Y.-P. Kuo, J. Goldberg, L. Coates, W. Mason, J. Gerin, and J. Taylor, J. Virol. **62,** 1855 (1988).

[25] P.-J. Chen, G. Kalpana, J. Goldberg, W. Mason, B. Werner, J. Gerin, and J. Taylor, Proc. Natl. Acad. Sci. U.S.A. **83,** 8774 (1986).

cle transcription and a requirement in replication for RNA processing. Certain fragments of cloned HDV sequence cleaved efficiently during the *in vitro* transcription reaction.[26,27] *In vitro* transcripts and the *in vivo* multimeric RNAs, when incubated with magnesium ions *in vitro*,[28] cleaved at the same phosphodiester bond. The cleavage products of the encapsidated RNA polarity, if not dissociated after cleavage, readily ligated to regenerate the original transcript when magnesium ions were removed.[29]

Terminal truncations[28] of both the genomic and antigenomic HDV RNA yielded the autolytically processing sequences presented in Fig. 9. These sequences do not contain the conserved elements of the hammerhead structure such as GAAA. There are no obvious, common secondary structure motifs which can be derived from the genomic and antigenomic sequences and which are similarly located with regard to the ends of the minimum sequences. Nor are there extensive similarities to other autolytically processing sequences that generate a 2′,3′-cyclic phosphodiester. Therefore, the HDV autolytically processing sequences in Fig. 9 are presented without postulated folding. The genomic and antigenomic self-cleaving sequences are from two regions in the HDV circular RNA which very likely are extensively base paired to each other in the folded, circular HDV genomic RNA molecule.[24]

Acknowledgments

I am grateful to P. Keese for valuable suggestions and comments on the manuscript and to W. L. Gerlach, R. H. Symons, O. C. Uhlenbeck, and M. A. Rezaian for receipt of results and manuscripts prior to publication.

[26] L. Sharmeen, M. Y.-P. Kuo, G. Dinter-Gottlieb, and J. Taylor, *J. Virol.* **62**, 2674 (1988).
[27] H.-N. Wu, Y.-J. Lin, F.-P. Lin, S. Makino, M.-F. Chang, and M. M. C. Lai, *Proc. Natl. Acad. Sci. U.S.A.* **86**, 1831 (1989).
[28] M. Y.-P. Kuo, L. Sharmeen, G. Dinter-Gottlieb, and J. Taylor, *J. Virol.* **62**, 4439 (1988).
[29] H.-N. Wu and M. M. C. Lai, *Science* **243**, 652 (1989).
[30] K.-S. Wang, Q.-L. Choo, A. J. Weiner, J.-H. Ou, R. C. Najarian, R. M. Thayer, G. T. Mullenbach, K. J. Denniston, J. L. Gerin, and M. Houghton, *Nature (London)* **323**, 508 (1986).

Author Index

Numbers in parentheses are footnote reference numbers and indicate that an author's work is referred to although the name is not cited in the text.

A

Abelson, J., 50, 63, 67, 69, 110, 113, 194, 443, 512
Abrahams, J. P., 289, 295(4), 300(4), 301, 303(4)
Ackerman, S., 38, 39(95)
Adamietz, P., 13
Adams, D. A., 14, 23(2), 331
Adams, D. C., 531
Adams, D. E., 457, 458(15)
Adams, D. S., 532
Adams, S. A., 410, 413, 417
Adoutte, A., 121
Agabain, N., 164, 171(5), 177, 178(5)
Ahne, F., 544
Akatani, K., 175
Akers, T. G., 285
Alams, J. M., 436
Albergo, D. D., 317, 318(22)
Alberts, B. M., 489
Alexis, M., 311
Allfery, V. G., 410
Aloni, Y., 70
Alonso, A., 531
Altman, S., 29, 231, 419, 424, 425(10, 11), 471
Alzner-Deweerd, B., 154
Anderson, K. W., 511
Anderson, K., 501
Anderson, T., 40
Andrews, N. C., 469
Angerer, L. M., 42, 251
Angerer, R. C., 42
Ansorge, W., 130
Antal, M., 532
Anziano, P. Q., 540
Aota, S., 543
Appel, B., 532
Applequist, J., 314
Ares, M., 531

Arfin, S. M., 31
Armadja, J., 418, 420(2)
Armstrong, J., 42, 48(18)
Arnberg, A. C., 533, 545(5)
Ashburner, M., 492
Assens, C., 532
Astell, R. E., 35
Atger, M., 37
Auffray, C., 122
Auffret, A. D., 544
Aviv, H., 34, 123
Axelrod, V. D., 60, 61(22)

B

Bachellerie, J. P., 121, 163, 221
Baer, M. F., 231
Baer, M., 231
Baguley, B. C., 27
Bakayev, T. G., 442
Bakayev, V. V., 442
Baker, W. C., 230, 272
Bakken, A. H., 481, 483(8), 495(2), 496(8), 497, 509(30)
Bakker, J. P., 107
Bakos, J. T., 548
Ballesta, J. P. G., 103
Banerjee, A. K., 164, 165(1)
Banks, F., 63
Bannwarth, W., 511
Bantle, J. A., 35
Barbosa, E., 176
Bargetzi, J. P., 155
Bark, C., 531
Barkan, A., 544
Baroin, A., 121
Baroudy, B. M., 419, 435(5)
Barrell, B. G., 132, 133, 418, 431, 435(15), 436, 439
Barta, A., 183, 185(18), 189(18), 190(18)

Bartkiewicz, M., 471
Basak, N., 539
Bass, B. L., 546
Baudin, F., 215, 293
Bauerle, R., 495
Bayev, A. A., 103, 104, 105
Beatty, B. R., 481, 483(1)
Beck, E., 531
Beckmann, J. S., 113
Been, M. D., 535
Behlen, L. S., 555
Beinstein, R. M., 469
Bekho, I., 242, 243(21)
Belford, M., 540
Bell, D., 27
Bell, G. I., 105
Bellamare, G., 539
Belort, M., 535
Benended, B. J., 193, 418, 419, 420, 424, 425(11), 431, 432(3), 433, 435(5, 14), 436(15), 440(3, 14), 441(3), 442, 462, 464(20)
Bentley, D. M., 70, 83, 95(10)
Benton, D., 270, 272(32)
Bergdoll, M., 379
Berger, S. L., 3, 4, 6, 7(13), 8, 240, 327, 329, 330, 334, 343, 361, 412
Berget, S. M., 450
Berk, A. J., 334, 338, 339(8, 9), 343, 348
Bernardi, G., 29, 30(25)
Bernstein, L. B., 531
Berquist, P. L., 27
Bevington, P. R., 316, 317(18)
Beyer, A. L., 410, 482, 491, 492(3), 495(16), 498, 499(3, 33), 501, 502(3, 17, 33), 504(3, 17, 33, 34), 505(17), 506, 508, 509
Biernat, J., 214
Biggin, M. D., 130
Billings, P. B., 410
Bilofsky, H. S., 272
Bindereif, A., 454, 455(2), 456, 457(13), 458(13), 462(13)
Bing Kun, J., 297, 298(28)
Bird, C. R., 544
Birkenmeier, C. S., 6, 7(13), 8
Birnboim, H. C., 46
Birnsteil, M., 42
Birnstiel, M. L., 14, 21(6), 361, 511, 532
Bischoff, R., 26, 27(6), 32
Bishop, J. M., 169
Bishop, R., 64

Bitten, R. J., 342
Bjorkroth, B., 501
Black, D. L., 455, 456(6), 457, 458(6, 17), 462(17), 465(6), 466(17), 467(6, 17)
Blakesley, D. E., 31, 32(45)
Blobel, G., 471
Blöcker, H., 418, 420(2)
Blumberg, D. D., 7, 329
Blumenthal, T., 479
Bluthmann, H. B., 241
Bock, C. S., 3
Bock, C.-T., 482, 483(9), 496(9)
Bock, R. M., 26, 27
Boedtker, H., 330
Boffa, L. C., 410
Bogerd, J., 103
Bogorad, L., 540, 544
Boguski, M. S., 155
Bol, J., 298
Bolshoi, G., 264
Bonen, L., 533, 540, 544, 545(5)
Bonitz, S. G., 540, 544
Bonjardin, C. A., 540
Bonnard, G., 540
Bonner, T. I., 242, 243(21)
Bonner, W. M., 331
Boos, K.-S., 38
Boothrayd, M., 177, 178(6, 7), 180(7), 191(7)
Borer, P. N., 264, 307, 319
Bosch, L., 198, 237, 289, 297, 300(4), 302(23), 303(4), 292, 293(15), 295(3, 4), 298
Botstein, D., 364, 510, 516
Boulanger, J., 539
Bourque, D. P., 545
Bouton, A. H., 498, 499(33), 504(33), 509
Boyle, J. A., 469, 479
Bradley, J. E., 13
Branch, A. D., 193, 418, 419, 420, 424, 425(11), 431, 432(3), 433, 435(5, 14), 436(15), 440(3, 14), 441(3), 442, 462, 464(20)
Branch, A., 496
Brand, R. C., 104, 106, 107
Branlant, C., 105, 106(36), 107(36), 108(36), 109(36), 213, 531, 532
Braslauer, K. J., 314, 315(14), 316(14), 318(22)
Breitenberger, C., 539
Brenner, D. J., 242
Breslauer, K. J., 314, 316, 323(20)

Breslow, J. L., 3
Bretner, M., 173
Briand, J. P., 292, 293(16)
Bricknell, P. M., 3
Brimacombe, R., 213, 418, 419, 420(2)
Britten, R. J., 242
Broach, J. R., 514
Brody, E. N., 321, 367
Brody, E., 443
Broker, T. R., 240, 242, 244, 255(25), 257(8, 25), 258(25), 260(25)
Brow, D. A., 532
Brown, A. L., 300, 302(32)
Brown, R. S., 154, 289
Brown, T. A., 228, 264, 289, 295(7), 534, 535(12, 17), 539
Brownlee, G. G., 132, 133, 137(7), 154, 155, 163, 439
Bruce, A. G., 157, 217
Bruck, D., 241
Brucoleri, R. E., 276
Bruening, G., 548, 549, 556
Brugerolle, G., 121
Brunel, C., 443, 532
Bruzik, J. P., 456, 461(11), 464(11), 465(11), 466(11)
Budowski, E. I., 215
Bugianesi, R. L., 36
Buna-stein, M., 121, 126(3)
Burckhardt, J., 14, 21(6)
Burger, G., 539
Burgess, R. R., 356
Burke, J., 534
Burke, J. M., 511, 534, 535, 539, 540
Burks, C., 272
Burn, T. C., 532
Burton, Z. F., 356
Busby, S., 497, 509(30)
Busch, H., 471, 479, 497, 531, 532
Busslinger, M., 532
Butler, E. T., 46
Buzayan, J. M., 548, 549, 556
Bynum, J. W., 3
Byus, C., 5, 123

C

Cabera, C. V., 43
Calbiochem, 307
Calman, A. F., 70

Calvet, J. P., 532
Cameron, G. N., 272
Campbell, C., 31
Campen, T. J., 462
Cantoni, G. L., 431, 435(15), 436(15)
Cantor, C. R., 291, 314
Caput, D., 531
Carbon, P., 532
Carlsson, M., 40
Carpousis, A. J., 57
Carri, M., 531
Carroll, A. R., 300
Carroll, E., III., 418
Carter, A. D., 183
Caruthers, M. H., 186, 266, 307, 320(1), 321(1)
Casesdaban, M. J., 516
Casey, J., 239, 241(4), 242(4), 243(4), 343, 361, 364(14)
Cech, T. R., 228, 265, 272(18), 368, 511, 533, 534(3), 535, 540, 546
Cedergren, R. J., 263, 277
Celantano, J., 192, 193(1), 197(1), 198(1), 206(1), 208(1)
Cesareni, G., 531
Chabot, B., 455, 457, 458(6, 17), 462(17), 465(6), 466(17), 467(6, 17)
Chaconas, G., 158, 175, 356
Chamberlin, M. J., 46, 51, 52, 57
Chambers, J. C., 469
Chang, E. H., 8
Chang, L.-H., 323
Chang, M.-F., 558
Chang, Z., 539, 543, 544
Chanock, R. M., 166
Chen, J.-H., 285
Chen, P.-J., 557
Chen-Kiang, S., 69, 70, 83, 95(9, 11), 96
Cheng, C. C., 163
Cheng, H.-L., 178, 533, 545(4)
Cheng, S.-C., 443, 445(24), 447(24), 450(24)
Cherayil, J. D., 26, 27
Cheroutre, H., 48
Chirgwin, J. M., 3, 5(4), 6(4), 91, 122, 123(14)
Chirikjian, J. G., 211
Chistofori, G., 448, 450(31)
Chitgwin, J. M., 74
Choffat, Y., 532
Choi, Y. D., 410, 413, 417(8, 9, 13)
Chomczynski, P., 13

Choo, Q.-L., 558
Chooi, W. Y., 507
Chow, L. T., 240, 242, 244(8), 251, 255(25), 257(8, 25), 258(25), 260(25)
Christensen, M. E., 410
Christopher, D. A., 543
Chu, F. K., 540
Chung, S. Y., 501
Chung, S., 531
Chunwongse, J., 539, 544
Church, G. M., 42, 451
Ciechanover, 471
Citterich, M. H., 539
Clark, B. F. C., 289
Clarke, B. E., 300, 302(32)
Clarke, S. H., 121
Clayton, D. F., 10, 11(23)
Clayton, W., 127
Clegg, M. T., 543
Coates, L., 557, 558(24)
Cobianchi, F., 504
Coggins, L. W., 483, 485(11), 493(11), 497(11)
Colantuoni, V., 33
Cole, P. E., 315, 322, 325(16)
Coleman, J. E., 53, 57, 59
Colleaux, L., 539
Collins, R. A., 539
Colman, A., 42, 48(18)
Cone, R. D., 531
Conrad, R. C., 479
Contreras, R., 48
Conway, L., 369, 377(1), 379(1)
Cook, R. G., 471
Cool, B., 38, 39(95)
Cornelissen, B., 297
Cortese, R., 33
Coruzzi, G., 540, 544
Cotten, M., 532
Coulsen, A. R., 121, 182, 184(16), 185(16), 368, 437
Couto, J. R., 511, 516(5)
Cox, K. H., 42
Cox, R. A., 3
Cox, R. F., 82
Craft, J., 471
Cremer, N. E., 472
Crkvenjakov, R., 5, 123
Crothers, D. M., 32, 264, 307, 315, 316, 319(2), 322, 325(16), 442

Crowther, J. B., 26, 27(2), 40(2)
Cummings, D. J., 535, 540, 544
Cummings, I. W., 237, 263, 290, 291(13), 301(13)
Currey, K. M., 285, 300, 302(32)
Currey, K., 275, 277(40)
Cushman, J. C., 543

D

D'Aubenton-Carafa, Y., 321, 367
Dahlberg, A. E., 442, 450(8)
Dahlberg, J. E., 14, 531
Dahm, S. C., 555
Damle, V., 314
Dams, E., 230
Daneholt, B., 501, 509
Darnell, J. E. Jr., 69, 70, 82, 83, 96
Darnell, J. E., 9, 10, 11(23), 74, 79
Das, G., 532
Davanloo, P., 51, 52(4)
Davidson, N., 239, 241(4), 242, 243(4), 248, 251, 260(3), 342, 343, 361, 364(14)
Davie, J. R., 495
Davies, C., 551, 552, 554(13), 555(13)
Davies, D. R., 431
Davies, R. W., 228, 239, 253, 260(3), 264, 289, 295(7), 343, 364, 511, 534, 535, 539, 540
de Bruyn, M. H. L., 297
de Haan, M., 540
De Herdt, E., 36
de Jonge, P., 103, 104, 105(21), 106
de Kok, A. J., 106
De Meyer, P., 36
de Peer, Y. V., 230
de Regt, V. C. H. F., 103, 104, 105(16), 106(36), 107(36), 108(36), 109(36)
De Wachter, R., 14, 15, 143, 230
De-Robertis, E. M., 471
DeBorde, D. C., 127
Deckman, I. C., 289, 303(6)
DeGennaro, L. J., 105
DeGrado, W., 42
Degrave, W., 48
Dekker, A. F., 96, 103
DeLeon, D. V., 42
DeLorenzi, M., 532
Dengler, B., 264, 307, 319

Denniston, K. J., 558
Deno, H., 539, 544
Derman, E., 82
DeRoberts, E. M., 531
Desrosiers, R. C., 469
Devereux, J., 271
Diamond, D., 330
Diener, T. O., 552
Dignam, J. D., 184, 377, 445, 446(25), 455, 459(8)
Digweed, M., 297, 298(26)
DiMaio, D., 42
Dingman, C. W., 428, 442, 450(7)
Dinter-Gottlieb, G., 558
DiRenzo, A. B., 555
Doel, M. T., 35
Dognam, J. D., 472
Doly, J., 46
Domenico, J. M., 540, 544
Donis-Keller, H., 154, 155, 158, 175, 201, 206(17), 207(17), 231
Doran, K., 31, 32(45)
Dornburg, R., 26, 27(4), 40(4)
Dorney, D. J., 462
Doty, P., 241
Dougall, D. K., 158, 175
Dove, W. F., 242
Draper, D. E., 54, 55(11), 216, 289, 303(6)
Dreyfuss, G., 410, 413, 417, 443
Drummond, D. R., 42, 48(18)
Dubin, D. T., 164
Dujon, B., 228, 534, 539, 540, 544
Dumas, J.-P., 268
Dumas, P., 294, 298(21), 379
Duncan, R. E., 39
Dunn, J., 51, 52(4)
Durocher, V., 539
Dyer, T. A., 155, 163(6), 544

E

Earley, J., 531
Ebel, J. P., 105, 106(36), 107(36), 108(36), 109(36), 213, 215, 216(4), 292, 293, 379, 531, 532
Ebert, M. H., 476
Edery, I., 165
Edmonds, M., 34, 35(70), 36(70), 177, 178(4), 180, 182(4), 185, 186, 187(1), 188(1), 189(2)

Edstrom, J.-E., 495
Edwards, K., 213, 544
Efstradiatis, A., 158, 175
Egami, F., 433
Egan, Z., 31
Ehresmann, B., 213, 215, 216(4), 293
Ehresmann, C. F., 293
Ehresmann, C., 213, 215, 216(4)
Einerhand, S. W. C., 103
El-Baradi, T. T. A. L., 103
Eldarov, M. A., 104
Engberg, J., 540
England, T. E., 373
Eperon, I. C., 62
Epstein, L. M., 553, 554
Epstein, P., 531, 532
Erdman, V. A., 297, 298(26)
Erickson, J. M., 462
Erikson, B. D., 356
Esser, K., 540, 544
Evangelidis, V., 63
Evans, G. A., 328, 330(4), 331(4)
Expert-Bezançon, A., 213, 216(4)
Eyermann, F., 213, 216(4)

F

Fakan, S., 507
Faller, D. V., 132
Faloona, F. A., 334
Farley, M., 543
Favalaro, J., 339, 347(10)
Favre, A., 418
Feinstein, S. I., 3
Feldmann, R. J., 277
Feldstein, P. A., 556
Felsenfeld, G., 309, 313
Fernandez, J. M., 183
Feunteun, J., 27
Fields, Y., 166
Fiers, W., 14, 15, 48, 143, 338, 462
Filipowicz, W., 173, 531
Financsek, I., 158
Finch, J. T., 289
Fink, T. R., 264
Finn, M., 192, 193(1), 197(1), 198(1), 206(1), 208(1)
Fisher, P. A., 479
Fiske, C. H., 111

Flanagan, J. M., 32
Florentz, C., 292, 293(16), 294, 298(21)
Floyd, T. R., 26, 27(2), 34(2), 40(2)
Foe, V. E., 489, 495
Foldi, P., 26, 27(4), 40(4)
Forbes, D. J., 531
Forbes, J., 106
Forget, B. G., 440
Forman, M. S., 476
Forster, A. C., 549, 550(5), 552, 553(13), 554(13), 555(13)
Fortsch, J., 266
Fox, G. E., 228, 300
Fox, T. D., 544
Fraenkel, D. G., 116
Fraenkel-Conrat, H., 7, 158, 175
Francke, C., 495
Francoeur, A. M., 479
Frank, R., 418, 420(2)
Franklin, R. M., 133, 420, 429(9)
Freier, S. M., 266, 307, 319, 320(1), 321(1), 397
Frelinger, J. A., 13
French, S., 495
Frendewey, C. W., 443
Frendewey, D., 448, 450(31)
Fresco, J. R., 211
Fried, M., 442
Friedemann, T., 270, 272(32)
Friedman, R. M., 8
Fritsch, E. F., 32, 46, 59, 67 113, 195, 351, 372, 376, 450
Fujimura, R. K., 32
Fujita, Y., 32
Fukuzawa, H., 539, 543, 544
Furdon, P., 132
Furth, J. J., 38, 39(95)
Furuichi, Y., 155, 164, 166, 167, 169, 172(14), 173, 175

G

Gaal, A. B., 130
Gabrielsen, O. S., 104
Gait, M. J., 163, 540
Galank, E., 29
Galante, E., 29
Galgo-Dias, L., 532
Gall, J. G., 553, 554

Gall, J., 540
Galli, G., 532
Gallinaro, H., 213, 531, 532
Gamble, H. R., 123, 127(17)
Gardiner, K. J., 231, 233(17)
Gardiner, K., 424, 425(10)
Gardner, M. M., 442
Garrett, R. A., 200, 205(16), 208(16), 212(16)
Garvey, J. S., 472
Gauss, P., 321, 367
Gauthier, A., 539
Gayle, M., 321, 367
Geck, P., 30
Gefter, M. L., 83
Gegenheimer, P., 63
Geiduschek, P., 103
Geliebter, J., 121
Gelinas, R. E., 38, 240, 244(8), 257(8)
Gelpi, C., 476
George, D. G., 230, 272
Georgiev, G. P., 442
Georgiev, O. I., 103, 104, 105
Georgiev, O., 42
Gerard, G. F., 367
Gerhard, W., 121
Gerin, J. L., 419, 435(5), 558
Gerin, J., 557, 558(24)
Gerke, V., 455, 458(5), 466(5), 467(5)
Gerlach, W. L., 549, 552, 554, 555(16), 556, 558
Gershowitz, A., 176
Gesteland, R. F., 531
Ghosh, P. K., 338
Gibson, T. J., 130
Gick, O., 532
Giege, R., 292, 293(16), 294, 298(21), 379
Gilbert, W., 42, 154, 155, 201, 206(17), 207, 214, 292, 297(17), 337, 340(3), 341(3), 342(3), 344(3), 345(3), 360, 361(11), 368(11), 369, 451
Gilham, I., 28
Gilham, P. T., 38, 39(94), 188
Gillam, I. C., 28, 34, 38(18)
Gillum, A. M., 189, 191(29)
Gilmartin, G. M., 511
Gingrich, J. C., 543
Girard, M., 180
Glisin, V., 5, 123
Glotz, C., 419
Goad, W. B., 267

Gold, H. A., 471
Gold, L., 289, 296(6), 303(8), 321, 367
Goldberg, J., 557, 558(24)
Goldman, D., 476
Goldman, E., 31
Gonzalez, I. L., 462
Good, P. J., 368
Goodwin, G. H., 442, 450(8)
Gopinathan, K. P., 122
Gorray, K. C., 41
Gorrieb, E., 477, 478(31)
Gorski, J. L., 462
Gotoh, O., 531
Gott, J. M., 540
Gottschling, D. E., 533
Gouy, M., 265, 268(17), 268
Grabowski, P. J., 42, 43, 50(23), 69, 177, 178(2), 182, 186(17), 188(17), 189(2, 17), 443, 445(23), 533
Graeser, E., 32
Graham, D. A., 13, 242
Gralla, J. D., 57
Gralla, J., 264, 307, 316, 319(2), 322(19)
Gray, D. M., 311
Gray, J. C., 544
Green, M. R., 42, 47(2), 48, 54, 55(12), 69, 164, 171(7), 177, 178, 184, 189(3), 443, 454, 455, 456, 458(7, 13), 462(7, 13)
Greenberg, J. R., 410, 411, 418
Greer, C. L., 63
Griffith, O. M., 3
Griffths, A. D., 62
Griggs, J. R., 269, 276(27)
Gritzmacher, C. A., 479
Grivell, L. A., 533, 540, 545(5)
Groebe, D. R., 62, 321, 367, 441
Groebe, D., 52, 53, 54(9), 55(9), 56(9), 60(9), 62
Grohnamm, K., 360
Gross, H. J., 155, 266
Gross, R. H., 36
Groudine, M., 70, 83, 95(10)
Guarini, L., 33
Gubbins, E. J., 540
Guerrier-Takada, C., 231, 419, 424, 425(10, 11)
Guild, N., 321, 367
Guilley, H., 154
Gunsalus, I. C., 539
Gupta, R. C., 155, 163(7)

Gupta, R., 228
Gurdon, J. B., 506, 509(43)
Gutell, R. R., 105, 238
Gutell, R., 228
Guthrie, C., 228, 468, 511, 512, 513(7), 514(7), 516(5), 531, 532

H

Haber, M., 28
Hadjuolov, A. A., 103, 104, 105
Haeberli, P., 271
Haegeman, G., 338
Haendler, B., 531, 532
Haer, E. J., 495
Hagel, L., 40
Hagemeier, E., 38
Hagenbuchle, O., 93, 367
Hahn, W. E., 35
Hahne, S., 540
Hallick, R. B., 543, 545
Halzorson, H. O., 96
Hames, B. D., 14, 442, 450(8)
Hamkalo, B. A., 482, 483(4, 7, 8), 495, 496, 498(3), 508
Hamlyn, P. H., 163
Hammarstrom, K., 531
Hampel, A. E., 31, 32(43)
Han, J. H., 4
Hanna, M. M., 50
Hanson, D. K., 540
Hanyu, N., 155, 161(8), 163(8)
Harada, F., 531, 532
Harcus, T. 552
Hard, R. E., 33
Hardies, S. C., 31
Hardin, J. A., 469, 471, 476
Hardy, S. F., 69, 177, 178(2), 189(2)
Hart, R., 42
Hartman, G. R., 30
Hartmann, T., 228
Hartmuth, K., 183, 185(18), 189(18), 190(18)
Hartwick, R. A., 26, 27(2), 40(2)
Harvey, R. P., 43
Harward, G. S., 341
Haseloff, J., 551, 554, 555(16), 556
Hashimoto, C., 457, 458(16), 469
Hashumoto, T., 34
Hatfield, G. W., 33

Hay, N., 70
Hayashida, N., 539, 544
Hayat, M. A., 482, 483(8)
Hearst, J. E., 367
Heatherly, D. E., 31
Heckman, J. E., 154, 539
Heinrich, G., 276
Hellman, U., 33
Henderson, R. E. L., 7
Hendrick, J. P., 469
Hendriks, L., 230
Hennighausen, L., 42
Henning, D., 471, 479, 531, 532, 532
Henning, J., 544
Hensgens, L. A. M., 540
Hernandez, R. A., 69
Herrera, R. J., 531
Herrmann, R. G., 544
Heus, H. A., 297
Hieter, P. A., 155
Higashinakagawa, T., 495
Hilbers, C. W., 322
Hildebrand, M., 545
Hilz, H., 13
Hinnebusch, A., 539
Hinterberger, M., 454, 456(4), 457(4), 458(4), 469, 471(2), 475(2), 477, 478(31)
Hirano, H., 29
Hiraoka, N., 51
Hirashima, A., 284
Hirose, T., 128
Hirschman, S. Z., 309
Hirsh, D., 177
Hirth, L., 292, 293(16)
Hitchcock, M. J. M., 8
Hjerten, S., 30, 41
Hodge, L. D., 509
Hoffman, M. L., 532
Hofmann, H., 266
Hogan, J. J., 105
Hogness, D. S., 239
Hohn, T., 41
Holl, J., 511, 540
Hollaender, A., 539
Holland, M. J., 104
Holmes, W. H., 33
Hong, G. F., 130
Honso, M. C., 43
Hood, L. E., 121
Hoopes, B. C., 47

Hopkins, N., 132
Hopper, A. K., 63
Hori, H., 230
Horn, G. T., 31
Horvath, C., 34
Hough, C. J., 31, 32(45)
Houghtan, M., 558
Hovemann, B., 531
Howe, C. J., 544
Howe, J. G., 469
Hsu, M.-T., 59, 83
HsuChen, C.-C., 164
Huang, L. S., 3
Huang, P. H. T., 28
Huang, R. C. C., 82
Hubner, L., 241
Hughes, G. R. V., 469
Hughes, J. M. X., 531
Hunein, F., 127
Hunt, L. T., 230, 272
Huppi, K., 121
HuQu, L., 221
Hurwitz, J., 48
Huse, W. D., 183
Husimi, Y., 322
Hutchins, C. J., 549, 552(4)
Hutchison, N. J., 508
Huttly, A. K., 544
Hutton, J. R., 241, 242, 243(16)
Huysmans, E., 230
Hyman, R. W., 253

I

Iimura, Y., 512
Ikegami, N., 175
Ikemura, T., 14, 15, 19(10), 20, 21, 21(13), 193, 196(6)
Ilan, J., 30
Imai, M., 175
Imamura, T., 40
Imura, N., 155
Inokuchi, H., 539, 543, 544
Inoue, T., 368
Irvine, K. D., 511
Iserentant, D., 338
Ishizaki, Y., 51
Itakura, K., 128
Itoh, M., 532

Itoh, Y., 532
Iwai, S., 555
Iwamura, J., 29
Iyer, V. N., 28

J

Jackle, H., 43
Jacob, M., 213, 531, 532
Jacobson, A. B., 269, 284
Jacobson, K. B., 32
Jacquier, A., 511, 534, 545(14, 15), 539, 540, 544
Jaeger, J. A., 266, 267, 307, 320(1), 321(1)
Jaehning, J. A., 107
Jahnke, P. A., 35
James, B. D., 228, 229(9), 236(9), 238(9), 300, 303
Jamrich, M., 495, 506
Janson, J.-C., 40
Jarell, K. A., 178
Jarrell, K. A., 533, 545(4)
Jarvik, J., 510
Jeanteur, P., 443, 531
Jeffries, A. C., 552, 553(13), 554(13), 555(13)
Jelinek, W., 79, 82
Jemtland, R., 104
Jendrisak, J. J., 531
Jeppesen, P. G. N., 436
Jerome, J. F., 107
Johansson, K., 241
Johnson, B. J. B., 39
Johnson, K. H., 311
Johnson, M. J., 128
Johnson, P. F., 113
Johnson, R. A., 55
Johnson, T. R., 30
Johnston, P., 43
Jones, E. W., 514
Jones, M. H., 228, 531
Jorcano, J. L., 531

K

Kabeck, D. B., 251
Kaesberg, P., 121
Kafatos, F. C., 158, 175, 273
Kahan, B., 531

Kallenbach, N. R., 313, 314
Kallick, D., 324
Kalpana, G., 557
Kambyellis, M. P., 492
Kamen, R., 339, 347(10)
Kamogashira, T., 539
Kan, N. C., 540
Kaneda, S., 531
Kanehisa, M. I., 267
Kaneko, K. J., 511
Kao, S.-Y., 70
Kaper, J. M., 549
Karabin, G. D., 543
Karmas, A., 454, 456(4), 457(4), 458(4), 459(9)
Karn, J., 410
Karpel, R. L., 504
Karsch, T., 540
Kastelein, R. A., 104, 105(21)
Kates, J., 38
Kato, A., 540
Kato, N., 531, 532
Kato, Y., 34
Kaudewitz, F., 511
Kawade, Y., 26, 27(8)
Kawashima, E. H., 128
Kaziro, Y., 158
Kazmaier, M., 532
Kean, J. M., 54, 55(11)
Kecskemethy, N., 38, 39(93), 171
Keene, J. D., 469
Keese, P., 552
Keller, M., 533, 543, 545(23)
Keller, W., 69, 443, 448, 450(31), 455, 459(9)
Kelly, R. B., 28
Kelmers, A. D., 31, 32(40)
Kemper, B., 42
Kempers-Veenstra, A. E., 96, 103
Kermekchiev, M. B., 103, 105
Kessler, S. W., 471
Kester, H., 214
Khym, J. X., 38
Kiberstis, P. A., 551
Kierzek, R., 186, 266, 307, 320(1), 321(1)
Kim, S. H., 262, 289
Kimmel, A. R., 3, 328, 361
Kimura, N., 231
Kinniburg, A. J., 410
Kiper, M., 30
Kirby, K. S., 9

Kireschner, M. W., 531
Kirkegaard, L. H., 7
Kishimoto, T., 83, 95(11)
Kiss, T., 531, 532
Kitamura, T., 34
Klein, B., 31
Klein, I., 99
Kleinschmidt, A. K., 239
Kleitman, D. J., 269, 276(27)
Klemenz, R., 103
Kleppe, K., 357
Kleppe, R. K., 357
Klootwijk, J., 96, 99, 101, 103, 104, 105, 106, 107, 108
Klug, A., 289, 297
Knapp, G., 113, 114, 193, 195(4), 196(4, 8), 197(5), 200(4), 202(4), 204(4), 293
Knipe, D. M., 166
Knipple, D. C., 43
Knonings, D. A. M., 531
Kobori, J. A., 121, 126(3)
Koch, G., 173
Koch, W., 544
Kochel, H. G., 539
Kochetov, N. K., 215
Koehler, K. A., 7
Kohchi, T., 539, 543, 544
Kohne, D. E., 342
Koizumi, M., 555
Kole, R., 132
Koller, B., 544
Koltunow, A. M., 552
Komano, T., 544
Konarskna, M. M., 42, 48, 165, 177, 178(2), 182, 186(17), 188(17), 189(2, 17), 443, 444(20), 445(20, 22, 23), 447(19, 20), 448(19), 449(19), 451(20)
Kondo, M., 36
Konishi, K., 40
Kontani, H., 51
Kopp, D. W., 186
Korf, G. M., 531, 532
Kornberg, R. D., 531
Kossel, H., 38, 213, 544
Kothari, R. M., 31, 32(44)
Krainer, A. R., 48, 69, 177, 178(3), 184, 189(3)
Kraise, M., 177
Kramer, A., 454, 455, 465
Kramer, F. D., 60, 61(22)

Krauter, K., 82
Krayev, A. S., 105
Krayev, V. M., 105
Krebbers, E. T., 544
Kreissman, S. G., 36
Kretznar, L., 531
Krieg, P. A., 42, 47(2), 47(25), 48(3), 54, 55(12)
Kristo, P., 531
Kristofferson, D., 270, 272(32)
Krockmalnic, G., 71, 73(15)
Krol, A., 105, 106(36), 107(36), 108(36), 109(36), 213, 531, 532
Krop, J. M., 101
Krug, M. S., 334
Kruger, K., 533
Krummel, B., 57
Krupp, G., 155
Kruppa, J., 26, 27(4), 40(4)
Kruskal, J. B., 263
Kubli, E., 532
Kuch, U., 540
Kuchino, Y., 155, 161(8), 163(8), 193, 206
Kung, G., 242, 243(21)
Kunitz, M., 155
Kunkel, G. R., 532
Kunkel, T. A., 122
Kuntzel, H., 539
Kuo, M. Y.-P., 557, 558
Kuramitsu, S., 544
Kurilla, M. G., 469
Kusuda, J., 540

L

Labeit, S., 130
Laemmli, U. K., 477
LaFiandra, A. J., 168
LaFiandra, A., 164
Lai, M. M. C., 558
Laird, C. D., 241, 495
Lamb, M. M., 509
Lambowitz, A. M., 539
Lamond, A. I., 443, 445(22, 23)
Lander, J. E., 289
Lane, D. J., 121
Lang, B. F., 540, 544
Langer-Safer, P. R., 508
Lapalme, G., 277

Larionov, V. L., 104
Larson, J. E., 31
Laskey, R. A., 346
Laskowski, M. Sr., 188
Last, J. A., 188
Latchman, D. S., 3
Lawrence, P. A., 43
Lazar, E., 213, 531, 532
Le, S.-Y., 285
Leaver, C. J., 544
Lebovitz, R. M., 184, 377, 445, 446(25), 455, 459(8), 472
Lebowitz, P., 338
Leder, P., 34, 42, 82, 84, 123
Lee, K. M., 40
Lee, M.-C., 114, 193, 195(4), 196(4), 197(5), 200(4), 202(4), 204(4)
Lee, S. I., 469, 504
Leer, R. J., 105
Lehrach, H., 330
Leiby, K. R., 507
Leider, J. M., 122
LeMaster, D. M., 457, 458(17), 462(17), 466(17), 467(17)
Lemieux, C., 539
Leonard, N. J., 7
Leonowicz, A., 32
Lerner, M. R., 469, 471(1), 479
Leschwe, M. A., 471
Leser, G., 507
Leskovac, V., 41
LeStourgeon, W. M., 410
Levin, J. R., 57
Levin, O., 30
Levine, M. D., 264, 307, 319(2)
Levy, C. C., 155
Lewis, J. B., 240, 244, 251, 255(25), 257(8, 25), 258(25), 260(25)
Li, S.-J., 323
Liautard, J. P., 532
Liebold, E. A., 442
Liebschutz, R., 270, 272(32)
Liehard, G. E., 7
Lifson, S., 242
Lilegaug, J. R., 357
Lin, Y.-J., 558
Lindberg, U., 36
Lindquist, R. N., 7
Link, G., 544
Linschiiten, K., 289

Linthorst, H., 298
Lipkin, L. E., 275, 277(40)
Lis, J. T., 46
Little, M. C., 543
Liu, J., 228, 229(9), 236(9), 238(9), 303
Liu, M., 532
Liu, M.-H., 479
Liu, Z., 83, 95(11)
Lizardi, P. M., 531
Lobo, S., 531
Lockard, R. E., 154, 192, 193(1), 197(1), 198(1), 206(1), 208(1)
Lodish, H. F., 471
Loeb, L. A., 122
Loszewski, M. K., 171
Louise, M. L., 127
Lowary, P., 52
Luciw, P. A., 70
Luhrmann, R., 531
Lund, E., 531
Lyon, A., 28

M

Ma, R.-I., 313
Maass, G., 322
McAllister, W. T., 51, 183
McCarthy, B. J., 241
McConaughy, B. L., 241
McCutchan, T. F., 38, 39(94)
McDevitt, M., 42
MacDonald, R. J., 3, 5(4), 6(4), 74, 91, 122, 123(4)
McDonough, K. A., 314
McDowall, A. W., 495
MacGee, J., 154
Machatt, M. A., 213
Macino, G., 535, 539
MacKechnie, C., 96
McKnight, S. L., 482, 489, 492(3), 499(3), 502(3), 504(3), 506, 509
McLaughlin, L. W., 26, 27(6), 32, 34(5)
McLure, W. R., 47
MacManus, J., 214
McNamara, K. J., 532
McNamara, S. P., 551
McPheeters, D. S., 289, 296(6), 303(8)
McPherson, A., 262, 289
Maden, B. E. H., 106. 107

Maderious, A., 70, 83, 95(9)
Maehlum, E., 104
Mager, W. H., 99
Mahler, H. R., 540
Mahon, K. A., 553, 554(14)
Mahowald, A. P., 492
Mainville, S., 263
Maizel, J. V. Jr., 271, 275, 277(40), 285
Makino, S., 558
Maley, F., 540
Maley, G. F., 540
Maly, P., 213
Maniatis, T., 42, 47(2), 46, 48, 54, 55(12), 59, 67, 69, 164, 171(7), 177, 178(3), 184, 189(3), 195, 351, 372, 376, 443, 450
Mankin, A. S., 105
Manley, J. L., 83
Mans, R. J., 367
Manser, T., 531
Marky, L. A., 314, 316, 317, 318(22), 323(20)
Marliere, P., 268
Marmur, J., 241
Marsh, T. L., 231, 233(17)
Marsh, T., 424, 425(10)
Marshall, A. G., 40, 323
Martin, C. T., 53, 57
Martin, F. H., 317, 346
Martin, K., 495
Martin, S. A., 176
Martin, T. E., 410, 507
Martinez, H. M., 269, 301
Martinson, H. G., 29, 30
Maser, R. L., 532
Masiarz, F. R., 105
Mason, W., 557, 558(24)
Mathews, C. Z., 540
Mathews, M. B., 251, 469
Mathis, D., 509
Matsubayashi, T., 539, 544
Matsuura, E., 544
Mattaj, I. W., 479, 531, 532
Matter, L., 471
Matzluff, W. F. Jr., 82
Maulik, S., 270, 272(32)
Maxam, A. M., 154, 155, 337, 340(3), 341(3), 342(3), 344(3), 345(3), 360, 361(11), 368(11)
Maxam, A., 201, 206(17)
Maxwell, I. H., 35
May, K., 544

Mayer, L., 83, 95(11)
Mayrand, S., 410
Mead, D. A., 42
Meckenburg, K. L., 178, 533, 545(4)
Meinkoth, J. L., 328
Meissner, F., 228
Melton, C., 48
Melton, D. A., 42, 43, 47(2, 25), 54, 55(12), 164, 171(7), 194, 328
Melton, D., 455
Melvolt, R. W., 121
Meng, B. Y., 539
Mercer, J. F. B., 35, 37(72)
Merrill, C. R., 476
Mertz, J. E., 368
Messing, J., 341
Meyhack, B., 237
Mezl, V. A., 35, 36(71), 38
Michel, F., 228, 511, 534, 535, 539, 540, 543, 544, 545(14, 15, 23)
Michot, B., 163, 221
Mierendorf, R. C., 54
Migram, E., 37
Miller, G., 469
Miller, O. L. Jr., 481, 482, 483(1, 4, 5), 489, 491, 492(3, 16), 495, 496, 497(5), 498, 499(3, 33), 502(3, 17), 504(3, 17, 33), 505(16, 17), 506, 508, 509
Miller, W. A., 552
Milligan, J. F., 195, 373, 441
Milligan, J., 52, 53, 54(9), 55(9), 56(9), 60(9)
Milstein, C., 163
Milward, S., 28
Mimori, T., 469, 471(2), 475(2), 476, 477, 478(31)
Minns, T. W., 171
Minter, S. J., 511, 535
Mitra, T., 192, 193(1), 197(1), 198(1), 206(1), 208(1)
Miura, K., 173
Miura, K.-I., 166
Miwa, M., 173
Miyake, T., 128
Mizumoto, K., 158
Mizutani, T., 32, 36
Moazed, D., 213, 379
Modelevsky, J. L., 285
Moks, T., 130
Moll, J., 228
Monier, R., 27

Monroy, G., 48
Monstein, H., 531
Monteilhet, C., 539
Montzka, K. A., 469
Moore, C. L., 450
Moore, C., 69, 240, 377
Moras, D., 294, 298(21), 379
Morelli, G., 539
Morgan, G., 497, 509(30)
Morgan, M., 166, 172(14)
Morris, C. E., 183
Morris, C. J. O. R., 33
Morris, T. J., 424, 426(12)
Moss, B., 168, 176
Mougel, M., 213, 215, 216(4), 293
Mount, S. M., 454, 456(4), 458(4), 468, 479, 531
Mount, S., 531
Mous, J., 42
Mowry, K. L., 456, 457, 458(18), 462(12, 13), 469, 532
Muffenegger, T., 471
Mullenbah, G. T., 558
Muller, D. K., 57
Muller, W. E. G., 365
Mullis, K. B., 334
Mundry, K. W., 30
Munro, H. N., 442
Muramatsu, M., 158
Murphy, D., 3
Murphy, E. C. Jr., 82
Murphy, J. T., 531
Murphy, W. J., 177, 178(5)
Murthy, S. C. S., 469
Musakhanov, M. M., 105
Musters, W., 103, 108
Muthukrishman, S., 166, 172(14), 175
Myslinski, E., 532

N

Nadin-Davis, S., 35, 36(71), 38
Naeve, C. W., 127
Nagai, S., 40
Nagawa, F., 231
Najarian, R. C., 558
Nakagawa, T., 417
Nakahigashi, K., 544

Nakamura, M., 476
Nakazato, H., 34, 35(70), 36(70), 180
Naora, H., 35, 37(72)
Narayanswami, S., 508
Narihara, T., 32
Nasseri, M., 240
Nasz, I., 30
Natayan, P., 69
Nathenson, S. G., 121
Naxam, A. M., 206
Neefs, J.-M., 230
Neilson, T., 266
Nelson, H. C. M., 516
Nelson, J. W., 317
Nelson, M. A., 535, 539
Netzker, R., 539
Neuendorf, S. K., 31
Neufeld, B. R., 242
Neuhaus, H., 544
Nevins, J. R., 10, 12(22), 70, 96
Nevins, J., 42
Newton, S. E., 300, 302(32)
Nickerson, J. A., 71, 73(15)
Nicklen, S., 121, 368
Nielsen, H., 540
Niemeter, A., 155
Nigam, S. K., 471
Nikolaev, N., 103, 105
Ninio, J., 265, 268(17, 19), 268, 290
Nishimura, S., 116, 155, 161(8), 163(8), 193, 206
Nishimura, T., 29
Nobrega, F. G., 540
Nohga, K., 471
Nojima, H., 531
Noll, C. H., 40
Noller, H. F., 105, 213, 228, 238, 289, 293, 303(5), 379
Noller, H., 192
Nomiyama, H., 540
Nomoto, A., 155
Nomura, M., 15
Noonan, D., 532
Notario, V., 504
Novelli, G. D., 31, 32(40)
Nowak, G., 32
Nucca, R., 29
Nuleoside, N., 164, 165(1)
Nuss, D. L., 173
Nussinov, R., 269, 276(27)

O

O'Connor, J. P., 111
O'Malley, B. W., 531
Oakley, J. L., 59
Oberg, B., 41
Oberst, R. J., 171
Obokata, J., 539, 544
Ochs, R. L., 471
Ocsiewacz, H. D., 544
Oettgen, A. B., 511
Offenberg, H., 96
Ogden, R. C., 14, 23(2), 114, 193, 197(5), 331
Ohme, M., 539, 544
Ohno, M., 165
Ohshima, Y., 532
Ohto, C., 539, 544
Ohtsuka, E., 555
Ohyama, K., 539, 543, 544
Okada, N., 532
Okamoto, T., 26, 27(8)
Olesen, S. O., 200, 205(16), 208(16), 212(16)
Oliemans, J., 96
Olive, J., 539
Olsen, G. J., 121, 228, 229(9), 236(9), 238, 230, 231, 233(17), 300, 303
Olsson, A., 130
Osawa, S., 230
Osheim, Y. N., 491, 492(16), 495(16), 496(16), 498, 501(34), 502(34), 504(17, 34), 505(16, 17), 506, 508(16, 17), 509(16, 35)
Osterburg, G., 275
Osterman, L. A., 26, 34(1), 37(1), 40(1), 41(1)
Ou, J.-H., 122, 123(15), 558
Oudet, P., 509
Ovauashi, A., 51
Øyen, T. B., 96, 104
Ozeki, H., 14, 20, 539, 543, 544

P

Pace, B., 39, 121, 181, 231, 233(17)
Pace, N. R., 39, 121, 181, 228, 229(9), 230, 231, 233(17), 236(9), 237, 238(9), 300, 303
Pace, N., 231, 424, 425(10)

Padgett, P. A., 42, 48, 69, 165, 177, 178(2), 182, 186(17), 188(17), 189(2, 17), 443
Palese, P., 122
Palukaitis, P., 133
Panayotatos, N., 31
Papanicolaou, C., 265, 268
Parisot, R. F., 471
Parker, K. A., 456, 461(11), 462(11), 464(11), 465(11), 466(11), 475
Parker, R., 468, 511, 512, 513(7), 514(7), 515, 516(5)
Parvin, J. D., 163
Passavant, C. W., 545
Patel, V., 30, 33
Paterson, B. M., 176
Patienl, R. K., 31
Patterson, B., 532
Patterson, T., 495
Paul, C. P., 442
Pavalakis, G., 192, 193(1), 197(1), 198(1), 206(1), 208(1)
Peacock, A. C., 428, 442, 450(7)
Pearson, R. L., 31, 32(40), 154, 163(4), 214
Peattie, D. A., 220, 369, 373(4), 375(4), 377(4)
Peattie, D., 292, 297(17)
Pedersen-Lane, J., 540
Pederson, T., 410, 532
Peebles, C. L., 63, 178, 479, 533, 545(4)
Penman, S., 71, 73(14, 15), 122, 180
Perasso, R., 121
Perlman, P. S., 178, 228, 265, 272(18), 533, 540, 545(4)
Pernemalm, P.-A., 40
Perry, K. K., 164, 171(5)
Persson, H., 42
Persson, T., 36
Peterlin, B. M., 70
Petersheim, M., 316, 317(17), 319(17)
Petes, T.D., 96
Petrillo, M. L., 178, 533, 545(4)
Pettersson, I., 454, 456(4), 457(4), 458(4), 469, 471(2), 475(2), 477, 478(31)
Pettersson, U., 531
Philippsen, P., 322
Philips, R. G., 43
Philipson, L., 41, 241
Piatak, M., 338
Pieczenik, G., 269, 276(27)
Pieler, T., 297, 298(26)

Pikielny, C. W., 443, 447(18), 450(17, 18)
Pilly, D., 155
Piper, P. W., 96
Pittet, A. C., 93
Planta, R. J., 96, 97, 100, 101, 103, 104, 105(16, 21), 106, 107, 108
Platt, T., 132
Pleji, C. W. A., 198, 237, 289, 297, 292, 293(15), 294, 298, 298(15, 21), 300(4, 22), 301, 302(22, 23), 303(4)
Plenum, 539
Poland, D., 313
Poonian, M. S., 36
Popovic, D. A., 33, 41
Potter, B. V. L., 62
Poulson, R., 7
Pousada, C., 532
Preiss, A., 43
Price, C. A., 543
Privalov, P. L., 323
Prody, G. A., 548
Przybyla, A. E., 3, 5(4), 6(4), 74, 91, 122, 123(14)
Puglisi, J. D., 237, 264, 302
Pullman, J. M., 410
Purdom, I. F., 361
Pustell, J., 273
Puvion-Dutilleul, F., 482, 483(10), 487(10), 495(10), 496(10)
Pyeritz, R. E., 28

Q

Qu, L.-H., 121, 163
Quagliariello, E., 540
Quay, W. B., 41
Quigley, F., 544
Quigley, G. J., 262, 289

R

Rahire, M., 539
Rahn, D. R., 469
Rahn, G. M., 237, 263, 290, 291(13), 301(13)
Rait, V. K., 211
RajBhandary, U. L., 154, 189, 191(29), 534, 539
Ralph, R. K., 27

Rand, K. N., 540
Randerath, E., 155, 163(7)
Randles, J. W., 266, 420
Rassi, Z. E., 34
Rathjen, P. D., 549, 552(4)
Rattner, J. B., 495
Raué, H. A., 103
Ray, J. A., 228, 264, 289, 295(7), 534, 535(12), 539
Ray, M., 82
Raymond, B. C., 443, 447(18), 450(18)
Raynal, M.-C., 83, 95(11)
Rebaglaiti, M. R., 42, 47(2), 48(3), 54, 55(12)
Record, M. T., 308
Reddy, R., 479, 531
Reddy, V. B., 338
Reed, R. E., 231
Reed, R., 42, 164, 443
Reeder, R. H., 82, 366, 495, 497, 509(30)
Reedy, R., 471, 532
Reese, R. T., 479
Reeves, W. H., 471
Reich, C., 231, 233(17)
Reichlin, M., 469, 471
Reid, B. R., 33
Reilly, J. D., 177, 178(4), 182(4)
Reinders, R. J., 410
Retél, J., 96, 97
Revzin, A., 442
Reyes, V. M., 67, 69
Reyes, V., 110
Reynolds, C. A., 539
Rezaian, M. A., 552, 558
Rhoads, R. E., 35
Rhodes, D., 289
Ricca, T. L., 323
Rice, C. M., 123
Rich, A., 262, 289, 313
Richards, E. G., 311
Richards, F., 531
Rickwood, D., 14, 442, 450(8)
Riesner, D., 266, 322, 420
Rietveld, K., 198, 237, 289, 292, 293(15), 295, 297(23), 298(15), 300(4), 302(23), 303(4),
Rigby, P. W. J., 3
Rimerman, R. A., 33
Ring, J., 52
Rinke, J., 469, 531
Rio, D. C., 488

Roan, J., 497, 509(30)
Roberts, D. B., 488, 489(13)
Roberts, R. J., 38, 240, 244, 255(25), 257(8, 25), 258(25), 260(25)
Robertson, H. D., 193, 418, 419, 420, 424, 425(11), 431, 432(3), 433, 435(5, 14), 436(15), 440(3, 14), 441(3), 442, 462, 464(20)
Robertus, D. S., 289
Rochiäx, J. D., 539
Rock, C. D., 544
Rodel, G., 511
Roebuck, K., 531
Roeder, R. D., 82, 184, 377, 446(25), 455, 459(8), 472
Rohrer, U., 532
Roiha, H., 532
Roizman, B., 166
Romano, L. J., 51
Romby, P., 213, 215, 216(4), 270, 272(32), 292, 293, 379
Roop, D. R., 531
Rosa, M. D., 469
Rosbash, M., 369, 379(2), 443, 447(18), 450(17, 18)
Rose, M., 516
Rose, S. D., 450
Rosenberg, A., 51, 52(4)
Rosenberg, M., 38, 39, 169, 176, 38, 39
Rosenberge, U. B., 43
Roshash, M., 531
Rossi, J., 512
Roth, J. R., 364
Rothblum, L., 497, 531
Rottman, F., 69
Rougeon, F., 122
Rowlands, D. J., 300, 302(32)
Roychoudhurg, R., 38
Rubin, G. M., 105
Rubin, J. R., 154
Rubtsov, P. M., 105
Rudloff, E., 38
Ruefer, B., 31, 32(43)
Ruezinsky, D., 121
Ruffner, D. E., 555
Ruiz i Altaba, A., 43
Rupp, W. D., 28
Ruskin, B., 48, 69, 177, 178, 184, 189(3), 454, 455, 457(7), 458(7), 462(7, 11)
Rutter, W. J., 4, 64, 74, 91, 105, 122, 123(14)

Ryan, T., 51
Rymond, B. C., 531
Rymond, B., 369, 379(2)
Ryte, V. C., 216
Ryu, W.-S., 368

S

Saba, J., 532
Saffer, L., 495
Sakagi, Y., 540
Sakamoto, H., 165
Sakamoto, K., 231
Salser, W. A., 237, 263, 264, 268(15), 290, 291(13), 301(13), 302(24)
Salser, W., 296
Saluz, H., 532
Salzman, N. P., 189
Salzman, N., 121, 126(3)
Sambrook, J., 195, 351, 372, 376, 450
Sampson, J. R., 555
Sands, J., 533
Sanger, F., 121, 132, 133, 137(7), 182, 184(16), 185(16), 368, 436, 437, 439
Sanger, H. L., 266, 439
Sankoff, D., 263, 277
Sano, S., 543
Sano, T. 543, 544
Santer, M., 367
Sato, K., 433
Sauer, R. T., 516
Saunders, C., 495
Scazzocchio, C., 171, 228, 264, 289, 295(7), 534, 535(12, 17), 539
Schafer, K. P., 171
Schaffner, G., 532
Schaufele, F., 511
Scheer, U., 482, 483, 485(11), 487(10), 493(11), 495(10), 496, 497, 508
Scheraga, H. A., 313
Scheweyen, R. J., 511
Schidkraut, C., 242
Schimmel, P. R., 314
Schmelzer, C., 533, 544, 545(6)
Schmickel, R. D., 462
Schmidt, C., 540, 544
Schmidt, M., 155
Schneider, I. R., 548
Schnolzer, M., 439

Schoffski, A., 241
Schopfer, K., 471
Schrenk, M., 482, 483(9), 496(9)
Schumacher, J., 420
Schweyen, R. J., 533, 545(6), 535, 540, 544
Sebring, E. D., 189
Sedman, S. A., 476
Seeman, N. C., 262, 289, 314
Seeman, N., 313
Seiler, S., 443
Sells, B. H., 361
Seno, T., 531
Setyono, B., 410, 418
Shan, C., 469
Shapiro, B. A., 275, 277(40)
Shapr, P. A., 338, 339(8, 9)
Sharmeen, L., 558
Sharp, P. A., 240, 343, 348, 443, 444(20), 445(20, 22, 23), 447(19, 20), 448(19), 449(19), 450, 451(20), 533
Sharp, P., 177, 178(2), 182, 186(17), 188(17), 189(2, 17)
Shatkin, A. J., 164, 165, 166, 167, 168, 169, 172(14), 173, 175
Sheldon, C. C., 552, 553(13), 554(13), 555(13)
Shenk, T., 418
Sherman, F., 514
Shiki, Y., 539, 543, 544
Shimada, H., 540, 544
Shimotohno, K., 173
Shimura, Y., 165, 231
Shinozaki, K., 539, 544
Shinshi, H., 173
Shirai, H., 544, 543
Shope, R. E., 166
Short, J. M., 183
Shu, M.-D., 469
Shub, D. A., 535, 540
Shub, D., 540
Shulmann, R. G., 322
Sidman, K. E., 230, 272
Siebenlist, U., 369
Sierakowska, H., 173
Silberklang, M., 189, 191(29)
Siliciano, P. G., 228, 511, 513(7), 514(7), 531, 532
Siliciano, P., 468
Simmons, D. T., 122
Simon, M., 239, 260(3)

Simoncsits, A., 154
Sims, H. F., 171
Sinha, N. D., 214
Sinsheimer, R. L., 360
Sirai, H., 539
Siu, J., 121
Sjoberg, B. M., 540, 532
Skoglund, U., 501
Skolnik-David, H., 450
Skuzeski, J. M., 531
Smart, J. E., 251
Smith, A. J. H., 337
Smith, D. E., 479
Smith, E. M., 424, 426(12)
Smith, F. I., 122
Smith, J. D., 511
Smith, K. C., 411, 512(10)
Smith, R. E., 155
Smithies, O., 271
Smits, G., 230
Sninsky, J. J., 188
Sogin, M. L., 121
Sollner-Webb, B., 366, 497, 509(30)
Solnick, D., 504
Solomosy, F., 532, 531
Sommer, R., 275, 540
Sommerville, J., 482, 483, 485(11), 487(10), 493(11), 495(10), 496(9, 10), 497(11), 508
Sonenberg, N., 165
Sorge, J. A., 183
Spector, D., 479, 532
Spiess, E., 531
Spirtz, R. A., 531
Sprinzl, M., 228
Srefano, J. E., 457, 458(15)
Sri-Widada, J., 443, 531, 532
Sriegler, P., 265, 269(16), 270(16)
Srudnicka, G. M., 237, 290, 291(13), 301(13)
Stahl, D. A., 121, 237
Stanely, J., 154, 163(5)
Staudt, L., 121
Stavnezer, E., 169
Steger, G., 266, 549, 307
Steinmetz, A., 540, 544
Steinschneider, A., 158, 175
Steinz, J. A., 367, 454, 455, 456, 457, 458(4, 5, 6, 11, 16, 17, 18), 462(11, 12, 14, 17), 464(11), 465(6, 11), 466(5, 11, 17), 467(5, 6, 17), 468, 469, 471(1, 2), 475, 477, 478(31), 479, 531, 532

Stepherson, P., 379
Stern, S., 213
Stevens, A., 173
Stevens, B. J., 410
Stevens, C., 313
Stiege, W., 419
Stiegler, P., 290, 296(11), 301(11), 302(11)
Stietz, J. A., 192
Stillman, B. W., 251
Stormo, G. D., 289, 296(6), 303(8)
Stormo, G., 321, 367
Strathern, J. N., 514
Strauss, E. C., 121, 122, 123(15)
Strauss, J. H., 122, 123
Stroke, I. L., 531
Strub, K., 532
Studnicka, G. M., 263
Stumph, W. E., 531, 532
Stumph, W., 531
Sturtevant, J. M., 314, 315(14), 316(14)
Stutz, E., 543
Subrahmanyam, C., 531
Suddath, F. L., 262, 289
Suggs, S. V., 128
Sugimoto, N., 266, 307, 320(1), 321(1), 319
Sugimura, T., 173
Sugita, M., 539, 544
Sugiura, H., 540, 544
Suh, D., 531
Sullivan, F. X., 555
Sullivan, N. L., 509
Sussdorf, D. H., 472
Sussman, J. L., 262, 289
Sutton, R. E., 177, 178(6, 7), 180(7), 191(7)
Swanson, M. S., 417
Sylvester, J. E., 462
Symons, R. H., 154, 549, 550(5), 551, 552, 553, 554(13), 555(13), 558
Szalay, A. A., 360
Szczesna, E., 173
Szostak, J. W., 535, 540
Szybalski, W., 239

T

Tabak, H. F., 533, 545(5), 545(5), 540
Tabler, M., 439
Tabor, J. C., 178
Tabor, J. H., 533, 545(4)

Tabor, M. W., 154
Takagi, Y., 540
Takaiwa, F., 540
Takeishi, K., 531, 539, 543, 544
Tamm, J., 511, 512, 516(5)
Tan, E. M., 471, 479
Tanaka, M., 539, 544
Tanaka, Y., 155, 163(6)
Tani, T., 532
Tanner, N. K., 193, 196(7), 228, 265, 272(18)
Tartof, D., 531
Tavak, H. F., 535
Taylor, J., 557, 558
Taylor, W. C., 544
Tebb, G., 532
Thalenfeld, B. E., 540, 544
Thayer, R. M., 558
Thermes, C., 321, 367
Thiebe, R., 322
Thomas, J. D., 479
Thoren, M., 121, 126(3), 239, 343
Tibbets, C., 241
Tierney, W. M., 511
Tinoco, I. Jr., 228, 237, 264, 265, 269, 272(18), 284(29), 302, 307, 313, 314, 315(14), 316(14), 317, 319, 321, 367
Tinoco, I., 346
Tohdoh, N., 540, 544
Tollervey, D., 479, 532
Tomasz, J., 166
Torozawa, K., 539, 544
Toth, M., 532
Tousignant, M. E., 549
Towner, P., 511, 535
Treisman, R., 339, 347(10)
Trendelenburg, M. F., 482, 483(6, 10), 487(10), 495(10), 496(10), 506, 509
Trifonov, E. N., 264
Trimble, J. J., 469
Trinkl, H., 540
Troutt, A., 192, 193(1), 197(1), 198(1), 206(1), 208(1)
Tsai, M. J., 531
Tsanev, R., 495
Tsaneva, I., 495
Tschudi, C., 531
Tuerk, C., 321, 367
Turmel, M., 539
Turner, D. H., 266, 307, 316, 317, 318, 319, 320(1), 321(1)

Tzagoloff, A., 544
Tzagoloff, B. E., 540

U

Uchida, W., 155
Uhlen, M., 130
Uhlenbeck, O. C., 157, 195, 211, 217, 264, 307, 319, 321, 367, 373, 441, 555, 558
Ullu, E., 531
Umesono, K., 539, 543, 544

V

Van Arsdell, S. W., 531
van Batenburg, E., 301
van Belkum, A., 289, 294, 295(4), 297, 298, 300(4), 303(4)
van Boom, J. H., 292, 293(15), 298(15)
van de Berg, M., 192, 193(1), 197(1), 198(1), 206(1), 208(1)
Van De Sande, J. H., 356
van der Horst, G., 533, 540, 545(5)
van der Veer, R., 533, 545(5)
van Eelelen, C. A., 410
van Knippenberg, P. H., 297
van Poelgeest, R., 292, 293(15), 295, 298(15)
Van Santen, P. G., 531
van Venrooji, W. J., 410
Vandenbempt, I., 230
Vanderberghe, A., 230
Vankan, P., 531
Varshavsky, A., 442
Vary, C. P. H., 209, 280
Vasserot, A., 532
Vassilenko, S., 154, 163(5), 211, 216
Venkatesan, S., 176
Verlaan, P., 294, 297, 298(21, 28)
Vidali, G., 410
Vijayraghavan, U., 512
Vlassov, V. V., 379
Vogt, V. M., 335, 338(2), 345(2)
Volckaert, G., 462
Vonurnakis, J. N., 192, 193(1), 197(1), 198(1), 206(1), 208(1)
Vorderwulbecke, T., 228
Vournakis, J. N., 158, 175, 209, 280

W

Wada, A., 322
Wagenmankers, A. J. M., 410
Wahl, G. M., 328, 330, 331(4), 361
Wakasugi, T., 539, 544
Walker, B. W., 410
Walker, T. A., 237
Walker, W. F., 230
Wallace, J. C., 177, 178(4), 180(1), 182(4), 186(1), 187(1), 188, 189(1)
Wallace, R. B., 128
Wang, A. H. J., 262, 289
Wang, K.-S., 558
Wang, L. H., 163
Wang, S., 531
Waranabe, K. K., 356
Ward, D. C., 508
Waring, R. B., 228, 264, 289, 295(7), 511, 534, 535, 539
Warshaw, M. W., 311
Waterhouse, P. M., 552
Watkins, K. P., 164, 171(5), 177, 178(5)
Weaver, R. F., 337
Wei, C. M., 168
Weiben, E. D., 532
Weigert, M., 121
Weil, J. H., 540, 544
Weiner, A. J., 468, 531, 558
Weiner, A., 511
Weiss-Brummer, B., 511
Weissman, C., 337, 338
Welch, R. C., 368
Wells, F. V., 419, 435(5)
Werner, B., 557, 558(24)
Werner, S., 539
West, D. K., 540
Westin, G., 531
Westof, E., 213, 216(4), 379
Wetmur, J. G., 242, 248(19), 342, 343(13)
Wetmur, J., 242
Weymouth, L. A., 122
White, R. L., 239, 343
Whitney, C., 275, 277(40)
Wickens, M., 369, 377, 379
Wicker, R. B., 539
Wiessman, S. M., 440
Wild, M. A., 540
Wilhelm, J. A., 471

Wilinson, L. E., 495
Wilk, H.-E., 171
Williams, A. L. Jr., 269, 284(29)
Williams, K. R., 504
Williamsin, C. L., 511
Wilson, S. H., 504
Wilusz, J., 418, 469
Wise, J. A., 531
Wise, J., 532
Witherell, G. W., 441
Woese, C. R., 228, 230, 231(10), 238, 300
Wolf, K., 540
Wolin, S. L., 469, 471, 479
Wolinsky, S. M., 240
Woodruff, T. K., 417
Wooley, J. C., 531
Wooley, J., 501
Wozney, J. M., 330
Wright, D., 531, 532
Wright, T. R. F., 492
Wu, H.-N., 558
Wurst, R. M., 102, 193(1), 197(1), 198(1), 206(1), 208(1)
Wyatt, J. R., 237, 264, 302

X

Xu, M. Q., 540

Y

Yabuki, S., 322
Yamada, K., 540

Yamaguchi-Shinozaki, K., 539, 544
Yang, S. K., 315, 325(16), 411, 512(10)
Yaniv, M., 418
Yanofsky, C., 132, 280
Yeoman, L. C., 471
Yisraeli, J. K., 194, 455
Youvan, D. C., 367

Z

Zabielski, J., 531
Zachau, H. G., 322
Zahan, R. K., 365
Zaita, N., 539, 544
Zaitlin, M., 133
Zamore, P. D., 455, 457(7), 458(7), 462(7)
Zan-Kowalczewska, M., 173
Zapp, M. L., 450
Zarlenga, D. S., 123, 127(17)
Zaug, A. J., 511, 533
Zeff, R., 121
Zeller, R., 531
Zentgraf, H., 482, 483(9), 496(9), 497
Zhaung, Y., 511
Zhuang, Y., 468
Zillmann, M., 450
Zimmern, D., 121, 551
Zuker, M., 228, 263, 265, 269(16), 270(16), 272(18), 284, 290, 296(11), 301, 302(11)
Zurawski, G., 543, 543
Zwieb, C., 213

Subject Index

A

Absorbance melting curves. *See also*
 Melting curves
 of RNA, 304–325
 buffers, 307–308
 cell path lengths for, 305–306
 data analysis, 309–315
 data collection, 305
 equipment for measuring, 304–305
 experimental methods for acquiring,
 304–309
 extinction coefficient determination,
 309–311
 heating rate for, 305
 helix-to-coil transitions, 313–315
 salt concentration, 308
 sample preparation for, 306–307
 solution conditions, 307
 thermodynamic parameters, 312, 314–
 319
 transitions studied by, 311–313
 UV spectroscopy for, 306–307
 UV wavelengths for, 308–309
Adenovirus, branched oligonucleotides,
 isolation of, from RNA labeled *in vivo*
 with inorganic [^{32}P]phosphate, 178
Adomet, methyl group transfer from, in
 RNA, 165
Adsorption chromatography, of RNA, 29–
 30
Affinity chromatography, of RNA, 34–40
Alphavirus, RNA, isolation of, 122–123
m-Aminophenylboronic acid polyacrylamide column, for chromatography of
 RNA, 40
Antibodies. *See also* Autoantibodies
 for nuclease protection experiments,
 457–458
Antitrimethylguanosine antibodies, immunoprecipitation of RNAs with, in
 nuclease protection experiments, 464–
 467

Arabidopsis thalania. See Crucifer
ASBV. *See* Avocado sun-blotch viroid
Aspergillus nidulans, group I introns, 536
Autoantibodies
 depletion of small RNPs from extracts
 using, 480–481
 immunoprecipitation of ribonucleoproteins with, 468–481
 mammalian small RNPs recognized by,
 469–470
 reactions with small ribonucleoproteins
 from nonmammalian species, 479
 against ribonucleoproteins in mixed sera,
 determination of relative concentrations of, 478–479
Avocado sun-blotch viroid, self-cleaving
 RNA sequence, 547, 549
γ-(4-Azidoanililyl)nucleotide, 390
 synthesis of, 396
8-Azido-ATP, 392
 synthesis of, 401–402
8-Azido-GTP, 392
 synthesis of, 401–402
N-5-Azido-2-nitrobenzoyloxysuccinimide,
 386
3'-Azidonucleoside triphosphates, 392
 synthesis of, 402
8-Azidonucleoside 5'-triphosphate, synthesis of, 401–402
p-Azidophenacyl bromide, 386
α-[(4-Azidophenacyl)thio]nucleotide, 390
 properties of, 394–395
 structure, 390
 synthesis of, 394–396
5'-[(4-Azidophenacyl)thio]phosphoryla-
 denosine, synthesis of, 395
5'-[(4-Azidophenacyl)thio]phosphoryl-
 guanosine, synthesis of, 395
5-[(4-Azidophenacyl)thio]uridine 5'-triphos-
 phate, 391
 synthesis of, 401
β-(4-Azidophenyl)nucleotide, 390
 synthesis of, 395–396

N-(4-Azidophenylthio)phthalimide, 386
5-Azidouridine 5'-triphosphate, 391
  synthesis of, 401

B

Bacillus brevis, RNase P RNA, secondary
  structure, 228–230
Bacillus cereus ribonuclease, as RNA
  structure probe, 197, 201, 203, 207
Bacillus megaterium, RNase P RNA,
  secondary structure, 228–230
Bacillus stearothermophilus, RNase P
  RNA, secondary structure, 228–230
Bacillus subtilis, RNase P RNA, second-
  ary structure, 228–230, 231–235
Bacteriophage, RNA polymerase, 51. *See
  also* SP6/T7 transcription system
Bacteriophage SPO1, group I introns, 538
Bacteriophage SP6. *See also* SP6 RNA
  polymerase
Bacteriophage T3. *See also* T3 RNA
  polymerase
Bacteriophage T4, group I introns, 538
Bacteriophage T7. *See also* T7 RNA
  polymerase
  promoter/pre-tRNA^{Phe} gene system,
    construction of, 64–67
  sequences of synthetic oligonu-
    cleotides used, 64–66
Barley, group II introns, 541
Barley yellow dwarf virus, satellite RNA,
  551–552, 554–555
Base pairing. *See also* RNA, structure
  demonstration of, through oligomuta-
    genesis to synthesize compensatory
    changes in predicted RNA helices,
    511, 513–515
BD-cellulose, 28
Bean. *See also Phaseolus vulgaris*
  U2 RNA, sequences of, 524, 525
Bentonite, as RNase adsorbent, 7
Bifunctional cross-linking reagents, 384–
  387
Bifurcation loop, 290
B myeloma cells, nuclei, isolation of, 86
BND-cellulose, 28
Boronyl columns, for chromatography of
  RNA, 38–40

Branched oligonucleotides. *See* Branch
  point
Branch point
  base composition of, determination of,
    188
  branching nucleotide, determination of,
    188
  characterization of, 185–189
    enzymatic and chemical reactions for,
      185–186
  composition, determination of, 185–186
  high-voltage paper electrophoresis, 187,
    189–191
  identification of, 185–186
  processing of, for secondary analysis,
    186–187
  quantitation of, 185–186, 189
  from RNA labeled *in vivo* with inorganic
    [^{32}P]phosphate, isolation of, 178–
    182
  from RNA lariats synthesized *in vitro*,
    isolation of, 182–185
  separation of, 189
  structure, determination of, 185–186
Broad bean, U6 RNA, sequences of, 529
5-Bromo-UTP, 391
Bulge loop, 263, 290, 320
  free energy parameters for, 321

C

Calorimetry, melting studies, 323–324
Campylobacter pylori, tRNAs, 3' end
  labeling, 196
Candida, tRNAs, 3' end labeling, 196
Cap-binding protein, 165
Capped structures, 158, 164–165
  characterization of, 165–176
  isolation of, 168–171
    methods, 168–174
  DBAE-cellulose column chromatogra-
    phy, 169–171
  DEAE-cellulose column chromatogra-
    phy, 168–170
  distribution of, 164
  enzymatic cleavage, 171–174
  at 5' terminus, 164–165
    removal of, 158
  nomenclature, 164

obtained from viral transcription reactions, 166–168
paper chromatography, 171–172
paper electrophoresis, 171–172
PEI-cellulose thin-layer chromatography, 171
radiolabeling. *See* Messenger RNA, radiolabeling
selective radiolabeling of specific sites, 166–167
stabilization of mRNA, 164
cDNA
cloning, 25
oligo(dT)-cellulose chromatography, 36
Cell-free translation systems, RNA–protein complexes in, study of, using UV-induced RNA–protein crosslinking, 418
Chicken
liver
U4 RNA, sequences of, 527
U5 RNA, sequences of, 528
oviduct, U4 RNA, sequences of, 527
U1 RNA, sequences of, 522
U2 RNA, sequences of, 524
Chlamydomonas eugametos, group I introns, 536
Chlamydomonas moewussi, group I introns, 536
Chlamydomonas reinhardtii, group I introns, 536
Chromatin, nicking of, with DNase I, 91
Chromatin spreading method, 481–509
adaptation to different cell types, 495–497
applications of, 481–482
contrast enhancement, 492–493
rotary metal shadowing, 493
staining material on grids, 492–493
for *Drosophila*, 487–492
analysis of data, 497–506
conversion of length measurements to molecular estimates, 506
identification of polymerase I, II, and III genes, 497–499
identification of specific genes in, 508–509
RNP fibril mapping, 498–501
solutions for spreading cells, 486
structures related to splicing on nascent transcripts, 504–506
types of ribonucleoprotein structure seen, 501–504
electron microscopy protocol, 493–495
identification of specific genes in, 508–509
and immunoelectron microscopy, 507–508
materials, 487
overview of, 482–483
preparation of carbon support film on grids, 483–486
with *in situ* hybridization, 508
specific applications to improve information yield of, 506–509
for *Xenopus*, identification of specific genes in, 509
Chromatographic techniques, 26. *See also specific technique*
Class I introns. *See* Group I introns
Class II introns. *See* Group II introns
Cloverleaf structure
computer prediction of, 302
of tRNA, 290
CMCT. *See* 1-Cyclohexyl-3-(2-morpholinoethyl)carbodiimide metho-*p*-toluene sulfonate
Column chromatography, of RNA, 25–41
Compensatory base changes, 300. *See also* Covariations
Computer-assisted covariance analyses, 238
Computer prediction of RNA structure, 262–288, 290, 295–296, 301–303
Computer programs
BIOFLD, 270, 271
CRAYFOLD, 270–273
CRUSOE, 268–269
confidence and reliability, 284, 285
experimental data and constrained folding, 279, 281
DRAW, 275, 277
FOLDVAX, 271
GCGFOLD, 271, 272, 279
confidence and reliability, 285
experimental data and constrained folding, 279–284
global-folding prediction with, 238

MOLECULE, 276–277
MONTECARLO, 269–270
confidence and reliability, 284–285
PCFOLD, 270–273
experimental data and constrained folding, 279–284
output and display, 278
PCFOLD2, experimental data and constrained folding, 281
recursive, 268–269
REGION table, 277
experimental data and constrained folding, 282
RNAFOLD, 270–273
confidence and reliability, 284–286
experimental data and constrained folding, 279–284
output and display, 277–278
prediction of common folding, 288
refolding procedure, 273–274
Salser energy input file for, 265
RNA folding, 267–270
confidence and reliability, 284–286
display, 273–279
experimental data and constrained folding, 279–284
output, 273
prediction of common folding, 288
with pseudoknotting considered, 301–303
and searching for pseudoknots in RNA, 295–296
SEQL, 267–268
adjustable parameters of, 267
experimental data and constrained folding, 282
prediction of common folding, 288
SQUIGGLES, 275
Covariations, demonstration of, 300
Cross-linked RNA–protein complexes
analysis of, 406–409
cleavage of, 408–409
isolation of, 406–409
transfer to nitrocellulose and nylon membranes, 407
Cross-linked viroid RNA, novel oligonucleotide of, 431–432
Cross-linking. See Photoaffinity cross-linking; Ultraviolet- induced cross-linking

Crucifer, U2 RNA, sequences of, 524, 525
1-Cyclohexyl-3-(2-morpholinoethyl)carbodiimide metho-p-toluene sulfonate
as modifying reagent, 379
as RNA structure probe, 215, 222, 223
Cytoplasm, preparation of, using ribonucleoside–vanadyl complexes, 8–9
Cytoplasmic polyhedrosis virus, in vitro transcription of capped mRNA, 166–167

D

Dangling ends, 320
DBAE-cellulose columns, for chromatography of RNA, 39–40
DBAE chromatography
of branched oligonucleotides from RNA of HeLa cells, 179, 181–182
of capped structures, 169–171
DEAE-cellulose, 26
benzoylated, 28
naphthoylated, 28
DEAE-cellulose chromatography
of capped structures, 168–170
of products of RNase digestion of poly(A)⁺ nuclear and cytoplasmic RNA of HeLa cells, 179, 182
DEAE-Fractogel, for purification of small RNAs, 59
DEAE-Sephadex, 26
Debranching enzyme, 177–178
Deoxynucleotides, as substrates for T7 RNA polymerase, 61–62
DEPC. See Diethyl pyrocarbonate
4,4'-Diazidobiphenyl, 386
Diazidonaphthalene, 386
Dictyostelium discoideum
U2 RNA, sequences of, 524, 525
U3 RNA, sequences of, 526
Diethylaminoethane, 26–27. See also DEAE
Diethyl pyrocarbonate
as modifying reagent, 369–370
precautions with, 350–351
as RNase inhibitor, 350
as RNA structure probe, 215, 216, 223, 292, 296–300
Differential scanning calorimetry, melting studies, 323–324

Dihydroxyboronyl columns, for chromatography of RNA, 39
N-[*N'*-(*m*-Dihydroxyborylphenyl)succina-
myl]aminoethyl cellulose. *See* DBAE-
cellulose
Dimethyl sulfate
as modifying reagent, 379
as RNA structure probe, 215, 216, 222,
223, 292, 296–300
Dinoflagellates, U5 RNA, sequences of,
528
4,4'-Dithiobisphenylazide, 386
DMS. *See* Dimethyl sulfate
DNA
cloned, preparation and isolation of
primers from, 354–361
double-stranded
melting temperature (T_m), 240
strand-separation temperature (T_{ss}),
240
thermal stability of, 240–242
effect of duplex length, 242
effect of partial homology, 242
effects of base composition and
formamide concentration, 240–
241
effects of electrolytes, 241–242
labeling
at 5' end, 337–338, 340, 342
at 3' end, 337
melting temperature, 343
effective, under nonstandard conditions (T_m^*), 241–243
empirical determination of, in
R-loop buffer–formamide, 249–
250
single-stranded, probe, preparation of,
340–342
S1-protected, autoradiography, 345–
346
uniformly labeled, 337, 342
Drosophila melanogaster
chromatin spreading method for, 487–
492
embryo chromatin, procedure for
spreading, 490–491
embryo collection, 488–489
as experimental system, 487–488
identification of specific genes in, and
chromatin spreading method, 509

oocyte follicle cell chromatin, procedure
for spreading, 491–492
transcriptionally active chromatin, electron microscopic study using Miller
chromatin spreading method, 481–
509
U1 RNA, sequences of, 522
U2 RNA, sequences of, 524
U4 RNA, sequences of, 527
U6 RNA, sequences of, 529
Duplex formation, free energy of, 319–321

E

Eastern newt, satellite 2 DNA, 553–555
Electron microscopy. *See also* Chromatin
spreading method
of ribonucleoprotein complexes, on
nascent RNA, 481–509
RNA mapping by, 239–261
ENU. *See* Ethylnitrosourea
Escherichia coli
α mRNA, pseudoknot in, 296
message-specific 16 S RNA, 517
mutant defective in tRNA biosynthesis,
purification of tRNA precursors
from, 14–15
M1 RNA, UV-cross-linking study of,
424–426
RNase P, secondary structure, 231–235
RNase P RNA, secondary structure,
228–230
tRNA precursor, hydroxylapatite chromatography, 29
tRNAs, 3' end labeling, 196
Ethyl 4-azidophenyl-1,4-dithiobutyrimi-
date-HCl, 386
Ethylnitrosourea
as modifying reagent, 379
as RNA structure probe, 215, 216, 223
Euglena gracilis, group II introns, 541
Exon-binding sites, 545
Exons
characterization, 325
nuclease protection assay, 325

F

Fast-protein liquid chromatography, 26
4-Fluoro-3-nitrophenylazide, 386

Fly. *See Drosophila melanogaster*
Foot and mouth disease virus, RNA,
  secondary structure, 300
Formamide, in RNA–DNA hybridization,
  242–243
FPLC. *See* Fast-protein liquid chromatog-
  raphy
Freeze-thawing, of splicing reactions, 448–
  449
Friend erythroleukemia cells, U1 RNA,
  sequences of, 522
Frog. *See Xenopus laevis*

G

Genetic analyses, of RNA interactions,
  510–517
  applications of, 517
  genetic system for, 510–513
  starting mutations, 512–513
  selection of suppressors *in vivo*, 515–
    516
  strategy for, 510–511
  synthesis of compensatory changes *in*
    vitro, 513–515
Group I introns
  classification of, 536–538
  families of, 535
  internal guide sequence, 535
  occurrence of, 533
  sequences characteristic of, 534–535
  sequences of, 536–538
  structure, 533–535
    phylogenetic comparisons of, 228
  subclasses of, 535
Group II introns
  classification of, 541–543
  occurrence of, 533
  self-splicing of, 544–545
  sequences, 541–543
  sequences characteristic of, 534–544
  splicing of, 545
  structure, 533–545
    phylogenetic comparisons of, 228
Guanidinium thiocyanate
  isolation of RNA with, 3–6, 74–78,
    123
  solution, preparation of, 3–4

H

Hairpin
  computer program for predicting, 267–
    268
  definition of, 263
  in phylogenetic comparisons of second-
    ary structure, 236–238
  variability in structure of, 236–237
Hairpin loops, 263, 290, 320
  screening for pseudoknots, 295–296
Hammerhead structures, 62, 554–555
HeLa cells
  analysis of hnRNA labeled in isolated
    nuclei from, 90. *See also* Pulse
    labeling
  branched oligonucleotides, isolation of,
    from RNA labeled *in vivo* with
    inorganic [^{32}P]phosphate, 178, 180–
    182
  culture of, for UV-induced RNA–protein
    cross-linking study, 413
  5 S RNA, RNA secondary analysis,
    439–441
  growth, 180
  labeling
    for UV-induced RNA–protein cross-
      linking study, 413
    with [^{32}P]phosphate, 180
  monolayers, UV irradiation of, 413–414
  nuclei, isolation of, 86, 180
  RNA
    fractionation, 180
    isolation of, 180
  tRNAs, 3' end labeling, 196
  U1 RNA, sequences of, 522
  U2 RNA, sequences of, 524
  U3 RNA, sequences of, 526
  U4 RNA, sequences of, 527
  U5 RNA, sequences of, 528
  U6 RNA, sequences of, 529
  U7 RNA, sequences of, 530
Helix, definition of, 263
Heparin, added to splicing reactions, 448–
  449
Hepatitis δ agent
  self-cleaving RNA, 556–558
  UV-treated RNA transcripts of
    cross-link-specific oligonucleotide
      detected in, 433, 434

RNase T1 digestion at elevated temperature, 433–435
Heterogeneous nuclear ribonucleoprotein isolation of, 410
in spliceosome formation, 443
Heterogeneous nuclear RNA, pulse labeling of, in isolated nuclei, 82–96
Heterogeneous nuclear RNA-binding proteins, identification of, in vivo, using UV-induced RNA–protein cross-linking, 410–417
High-performance liquid chromatography, 26
for purification of small RNAs, 59
Homochromatography, 155
ascending, for RNA fingerprinting, 132–133, 147–150, 153
HPLC. See High-performance liquid chromatography
H pseudoknot, 291–294
characteristics of, 297
connecting loops, 297–299
properties of, 294–295
Hydrazine, as modifying reagent, 369–370
Hydrophobic interaction chromatography, of RNA, 33–34
Hydroxylapatite chromatography, of RNA, 29–30
N-Hydroxysuccinimidyl-4-azidobenzoate, 386
Hyperchromicity, 304

I

Immunoelectron microscopy, chromatin spreading method with, 507–508
Immunoprecipitation
of protected fragments, in ribonuclease protection experiments, 454–468
of ribonucleoproteins, with autoantibodies, 468–481
analysis of immunoprecipitated proteins, 477–478
analysis of immunoprecipitated RNAs, 475–477
antibodies for, 471–472
cell extracts for, 472–473
interpretation of immunoprecipitation data, 479–480
protocol for, 473–475

Inosine triphosphate, as substrate for T7 RNA polymerase, 61
In situ hybridization, chromatin spreading method with, 508
Interior loop, 263, 290, 320
Internal loop. See Interior loop
Intron-binding sites, 545
Introns. See also Group I introns; Group II introns
mapping, nuclease digestion for, 325–333
Ionic chromatography, 26–28

J

Jurkat cells, analysis of hnRNA labeled in isolated nuclei from, 90, 94–95. See also Pulse labeling

K

Kethoxal, as modifying reagent, 379
Kluyveromyces fragilis, group I introns, 536
Kluyveromyces thermotolerans, group I introns, 536
K1 cells, U1 RNA, sequences of, 522

L

Lariats, 25, 132, 177, 443, 444, 533
release of branch points from, 185
synthesis of, in nuclear extracts, 184–185
synthesized in vitro, isolation of branch points from, 182–185
Long-range pairings, in phylogenetic comparisons of secondary structure, 236–238
Loops. See also Bifurcation loop; Bulge loop; Hairpin loops; Interior loop; Multiloop; R loops; Stem-loop structure
destabilizing energies, 265–266
free-energy parameters for, 321
nomenclature of, 263
in pseudoknotting, 291
vs. orthodox loops, 297–299
Lucerne transient streak virus, satellite RNA, 549–550, 552
Luteovirus, satellite RNA, 547

M

Macaloid, as RNase adsorbent, 7
Magnesium, effects on RNA secondary
    structure, 205–206, 212
Marchantia polymorpha
    group I introns, 536
    group II introns, 541–542
Maxham–Gilbert sequencing reactions,
    337
MB nuclease, as RNA structure probe,
    202, 203, 205, 209
Melting
    calorimetry studies, 323–324
    of complex RNA molecules, 324–325
    NMR studies, 322–323
Melting curves. See also Absorbance
    melting curves
    definition of, 304
    temperature-jump methods for, 322
5-Mercaptouridine 5'-triphosphate, 391
    synthesis of, 399–401
Messenger ribonucleoprotein
    free, study of poly(A)-binding proteins
        in, using UV-induced RNA–protein
        cross-linking, 418
    isolation of, 410
Messenger RNA
    capped, in vitro synthesis, by viral
        transcription, 166–168
    eukaryotic, sequence analysis, 163
    5'-terminal cap structure, removal of,
        178–180
    nuclear precursor
        introns, 533
        splicing of, 443
    oligo(dT)-cellulose chromatography, 36
    precursors
        32P-labeled transcripts
            preparation of, 183–184
            purification of, 184
        preparation of
            from mammalian cell nuclei, 69–82
            procedure for, 72–82
            solutions for, 70–71
        splicing of, 443
    rabbit β-globin, computer prediction of
        structure, 296
    radiolabeling
        with 3H by 2'-O-methylation, 176

with 3H by NaB3H4 after periodate
        oxidation, 174–175
    with 32P and/or 3H by guanylylation,
        175–176
    with 32P by phosphorylation after
        decapping, 175
    reversed-phase chromatography, 31–32
    splicing, steps in, 443–444
    structure, 262
    3' end formation, analysis of, 377–379
    transport, 70
    trans-splicing reaction, 177
    two-dimensional gel electrophoresis of,
        21–23
Messenger RNA-binding proteins, identifi-
    cation of, in vivo, using UV-induced
    RNA–protein cross-linking, 410–417
Messenger RNA (mRNA)-decapping en-
    zyme, yeast, cleavage of capped
    structures, 174
Micrococcal nuclease, pretreatment of
    extract or fraction, in nuclease protec-
    tion experiments, 464–466
Miller chromatin spreading method, 481–
    509
Modification interference
    applications of, 369
    buffers for, 370–372
    chemical modification of RNA, 373–375
        amount of RNA required, 373
        purine modification (A + G), 374
        pyrimidine modification, 374–375
    cleavage of modified RNA with aniline
        and gel electrophoresis, 375–376
    enzymes for, 370
    materials for, 370–372
    messenger RNA 3' end formation, analy-
        sis of, 377–379
    method for, 369–370
    purification of RNA, 376–377
    reagents, 369–370, 372, 379
    strategy, 369–371
Mouse
    Friend leukemia, U8 RNA, sequences
        of, 530
    kidney, U5 RNA, sequences of, 528
    kidney lymphoma, U4 RNA, sequences
        of, 527
    liver
        U2 RNA, sequences of, 524

U6 RNA, sequences of, 529
sperm, U1 RNA, sequences of, 522
U7 RNA, sequences of, 530
Multibranched loop, 263
Multiloop, 263
Mung bean nuclease, as RNA structure
probe, 197–198

N

NACS, 32
Native gel electrophoresis
analysis of splicing complexes with,
442–453
applications of, 442–443
Native polyacrylamide gels
modification of, 449–450
preparation of, standard protocol for,
448–449
Nepovirus, satellite RNA, 547
Neurospora crassa
group I introns, 536, 537
mitochondrial NADH dehydrogenase
subunit 4L intron, 535
Nicotiana tabacum
group I introns, 537
group II introns, 542
Ninio's rules, for RNA folding energy,
265–266
p-Nitrophenyl-2-diazo-3,3,3-trifluoropro-
prionate, 386
Northern blot analysis, of polyadenylated
RNA, 81–82
Northern hybridization, 25
Notophthalmus viridescens. See Eastern
newt
Nuclear magnetic resonance, melting
studies, 322–323
Nuclear RNA. *See also* Messenger RNA,
nuclear precursor; RNA, small nuclear
from HeLa cells
ribonuclease digestion of, 179–181
isolation of, 10–13, 73–79
labeled *in vivo*
fractionation and analysis of, 75
poly(A) selection of, 74, 76
methylation, kinetics and regulation of,
69–70
newly synthesized, labeling of, 72–73

polyadenylation, kinetics and regulation
of, 69–70
processing, kinetics and regulation of,
69–70
splicing, kinetics and regulation of, 69–
70
Nuclease. *See also* Ribonuclease
double-stranded structure specificity,
197–198
with no sequence specificity, 197–198
for RNA structure probing, 196–199
sequence specificity, 197–199
single-stranded structure specificity,
197–199
Nuclease digestion, for mapping introns,
325–333
analysis of results, 326–327, 330–334
artifacts, 333
cDNA for, 327–328
experiments with labeled RNA, 330–331
experiments with unlabeled RNA, 331–
333
genomic RNA for, 328
materials for, 327–329
methods, 329–330
nucleic acid requirements, 327–328
and polymerase chain reaction, 334
principle of method, 325–326
reagents, 329
starting materials for, 327–328
troubleshooting, 333
Nuclease M1, in RNA sequencing, 162
by postlabeling method, 156–157
Nuclease protection assay, 325
analysis of protected fragments, by
double-label fingerprinting, 456, 464
determination of nature of protecting
factor, 464
for determination of RNA–protein and
RNA–ribonucleoprotein interactions, 454–468
analysis of protected fragments, 462–
464
antibodies for immunoprecipitation,
457–458
digestion with RNase and immunoprecipitation, 454, 456–457
extracts or fractions for, 454–455
incubation of radiolabeled preRNA
and extract, 456

micrococcal nuclease pretreatment, 465

order of addition of RNase and antibody, 458

precursor RNA for, 455–456

protocol for, 458–462

RNase H pretreatment, 465

using antitrimethylguanosine antibodies, 465–467

with mutant transcripts, 467–468

of polyadenylated RNA, 81–82

protected–immunoprecipitated fragments, interpreting significance of, 467–468

Nuclease protection mapping, 348

Nuclease P1, branch points. *See* Branch point

Nuclei
alterations in, during pulse labeling of heterogeneous nuclear RNA, 95–96

extraction of RNA from, 12–13, 91–92

isolated, pulse labeling of heterogeneous nuclear RNA in, 82–96

isolation of, 86–87, 90

lysis of, 12–13

preparation of
with detergents, 11–12

by hypotonic lysis, 11

Nucleic acid chromatography system, 32

Nucleic acid denaturation, and UV absorbance, 304

Nucleoside 5'-triphosphate γ-azidoanilidates, synthesis of, 396

Nucleotide pyrophosphatase
cleavage of capped structures, 171–173

HeLa, m⁷G-specific, cleavage of capped structures, 172–173

Nucleotidyltransferase, in 3' end labeling of transfer RNA, 195–196

O

Oligo(dA)-cellulose, 36

Oligo(dC)-cellulose, 36

Oligo(dG)-cellulose, 36

Oligo(dI)-cellulose, 36

Oligo(dT), attached to glycerated, porous glass, 36

Oligo(dT)-cellulose, 36

affinity chromatography of RNA with, 34–36

sources, 34

storage, 414–415

usage, 414–415

Oligo(dT) chromatography
of hnRNPS or mRNPs, 410

of polyadenylated RNA, 79

poly(A)-specific fractionation of UV-cross-linked RNPs on, 414–416

of RNA, 123

of UV-cross-linked RNPs, 412

Oligonucleotides, for primer extension analysis of RNA, 352–354

advantages of, 352

purification of, 353–354

P

Pancreatic ribonuclease
A, digestion of UV-cross-linked RNAs, 432–436

digestion of internally labeled RNA, 132, 134–135

in nuclease protection experiments, 457

Pea. *See Pisum sativum*

PEI-cellulose thin-layer chromatography, of capped structures, 171

Phaseolus vulgaris, U1 RNA, sequences of, 522

Photoaffinity cross-linking. *See also* Ultraviolet-induced cross-linking
approaches, 383–393

of RNA–protein, 383–409

with bifunctional cross-linking reagents, 384–387

nucleotide analogs for, 389–393

selection and synthesis of, 394–402

with photoreactive nucleotide analogs
isolation and analysis of cross-linked complexes, 406–409

isolation of analog-containing RNA, 404–405

method, 405–406

synthesis of analog-containing RNA, 402–404

by placement of photoreactive cross-linking group on purified protein and addition of protein to RNA, 387–89

by placement of photoreactive cross-linking group within RNA molecule and addition of RNA to protein, 389–93
selection of light source, 393–394
Photoreactive nucleotide analogs
for cross-linking study of RNA–protein, 389–393. *See also specific nucleotide analog*
initiating, 394–396
internal, 396–401
RNA–protein cross-linking with, 402–406
synthesis of, 394–402
terminating, 401–402
structures, 390–392
Physarum microplasmodia, U6 RNA, sequences of, 529
Physarum polycephalum, group I introns, 537
Pisum sativum, U5 RNA, sequences of, 528
Placenta, human
U1 RNA, sequences of, 522
U2 RNA, sequences of, 524
U3 RNA, sequences of, 526
Placental ribonuclease inhibitor, 330
Plaskon bead, alternatives to, 32
Plasmid pBSAd1, 445, 452, 453
Plasmid pBSAL4, 445
Plasmid pBSU1, 445
Plasmid pBSU2b, 445
Plasmid pBSU4, 445
Plasmid pBSU6a, 445
Plasmid pORCS, 108
Plasmid pSPU5b, 445
Plasmids, transcription vectors containing SP6 and T7 promoters, 43–46
Podospora anserina
group I introns, 537
group II introns, 542
mitochondrial NADH dehydrogenase subunit I gene, first intron of, 535
Poly(A)-agarose, 40
Polyacrylamide gel electrophoresis, of yeast RNA, 114–116
Polyadenylation signal AAUAAA, identification of, using UV- induced RNA–protein cross-linking, 418
Poly(C)-agarose, 40

Poly(G)-agarose, 40
Poly(I)-agarose, 40
Poly(I)·poly(C)-agarose, 40
Poly(U)-cellulose, affinity chromatography of RNA with, 36–40
Poly(U)-Sepharose
affinity chromatography of RNA with, 36–40
source, 36
Poly(U)-Sepharose chromatography
column preparation, 79–81
column regeneration, 81
elution of poly(A)^f RNAs, 81
loading of RNA sample, 81
of polyadenylated RNA, 79–82
Precursor RNA. *See also* Messenger RNA, nuclear precursor; Transfer RNA, precursor; Yeast, ribosomal RNA precursors; Yeast, transfer RNA precursors
for nuclease protection experiments, 455–456
incubation with extract, 456
Pre-tRNA. *See* Transfer RNA, precursor
Primer extension
analysis of RNA
characterization of multiple primer extension products, 368–369
oligonucleotides for, 352–354
restriction fragment primers for, 354–361
for mapping UV-induced RNA–RNA cross-links, 438–439
reverse transcription for, 364–367
Protein, purified, placement of photoreactive cross-linking group on, 387–389
Protein sequencing, of cross-linked RNA–protein complexes, 409
Pseudoknots, 237–238, 289–303. *See also* H pseudoknot
biophysical studies of, need for, 303
characteristics of, 291–294
definition of, 264, 289, 291
detection of, 296–300
chemical probes, 296–298
chemical probing in presence and absence of Mg^{2+}, 296–297
and phylogenetic sequence comparisons, 300
NMR of, need for, 303

occurrence of, 289, 303
prediction of, 300–303
representative example of, 291–294
searching for, in RNA, 295–296
X-ray diffraction data, need for, 303
Pulse labeling, of heterogeneous nuclear
RNA, 82–96
analysis of data, 93–95
equipment, 85–86
extraction of RNAs from nuclei, 91–92
isolation of nuclei, 86
procedure, 85–93
RNA–DNA hybridization, 92–93
solutions for, 83–85
transcription *in vivo*, 86–91
P1 nuclease, digestion of RNA, for release
of branch points from lariats, 185

R

Radioactive ink, preparation of, 422
Rat
brain
U1 RNA, sequences of, 522
U2 RNA, sequences of, 524
U4 RNA, sequences of, 527
U5 RNA, sequences of, 528
liver, U3 RNA, sequences of, 526
Novikoff hepatoma
U1 RNA, sequences of, 522
U2 RNA, sequences of, 524
U3 RNA, sequences of, 526
U4 RNA, sequences of, 527
U5 RNA, sequences of, 528
U6 RNA, sequences of, 529
U8 RNA, sequences of, 530
R–D heteroduplexes, 240
electron microscopic analyses of, 240
Reovirus, *in vitro* transcription of capped
mRNA, 166–168
Restriction enzyme, for cleavage of plas-
mid DNA templates, 54–55
Restriction fragment primers, for primer
extension analysis of RNA, 354–361
5' end labeling, 354
with kinase, 357–358
with phosphatase, 356–357
5' overhangs, 355–356
isolation of, 354–355
labeled
isolation of, 358–359

removal of second labeled end, 358–359
preparation of, 354–355
strand separation, 359–361
by denaturing gel electrophoresis, 359–360
by nondenaturing gel electrophoresis, 360–361
3' end labeling, 354, 358
Reversed-phase chromatography, of RNA, 31–32
Reverse transcriptase
avian, 123–124
dideoxy sequencing of RNA using, 121–130
advantages of, 121–122
analysis of reaction products on sequencing gels, 129–130
buffers for, 123–126
development of, 121
disadvantages of, 122
enzyme reaction conditions, 124–126
primers
annealing to template, 128–129
GC content and position effects, 126
labeled, chain-termination reactions
using, 129
labeling with [γ^{32}P]ATP and T4
polynucleotide kinase, 126–127
length of, 126
polyacrylamide gel electrophoresis, 127–128
prelabeling before chain-termination
reaction, 126, 129
purification of, 127–128
selection of, 126
unlabeled, chain-termination reactions using, and [α^{-32}P]dATP or
[α^{-35}S]dATP, 126, 129
reaction protocols, 126–129
reagents for, 123–126
RNA template preparation, 122–123
stock solutions for, preparation of, 124
forms of, 123–124
murine, 123–124
in RNA structure probing experiments, 224–227
limitations of, 227
Reverse transcription
modification interference with, 379

premature termination of, 367
for primer extension, 364–367
Ribonuclease. *See also* Nuclease; Pancreatic ribonuclease; Ribonuclease digestion; *specific enzyme*
inhibitors, 3, 6–7, 74, 329
nitrocellulose filtration for, 329
for probing RNA secondary and tertiary structure, 192
in RNA fingerprinting, 130–132
for RNA sequencing by postlabeling method, 155–156
Ribonuclease A
digestion of terminally labeled RNA, 160
for RNA sequencing by postlabeling method, 155–157
as RNA structure probe, 197–199, 201–203, 207–209
Ribonuclease *Bc*
digestion of terminally labeled RNA, 160, 162
for RNA sequencing by postlabeling method, 156–157
Ribonuclease CL3
digestion of terminally labeled RNA, 160, 162
for RNA sequencing by postlabeling method, 156
as RNA structure probe, 197, 201
Ribonuclease digestion
of cytoplasmic and nuclear RNA from HeLa cells, 179–181
in nuclease protection experiments, 456–457
order of, in protocol, 458
in RNA fingerprinting, 132–137
of ultraviolet-cross-linked RNA–protein complexes, 415–416
Ribonuclease H. *See also* Nuclease digestion, for mapping introns
selective destruction of RNAs with, in nuclease protection experiments, 464–465
Ribonuclease III, cleavage site, 130–131
Ribonuclease P, RNA, secondary structure, phylogenetic comparisons of, 228–238
Ribonuclease *Phy*I
digestion of terminally labeled RNA, 160, 162

for RNA sequencing by postlabeling method, 156–157
Ribonuclease *Phy*M
digestion of terminally labeled RNA, 160
for RNA sequencing by postlabeling method, 156–157
as RNA structure probe, 197, 201
Ribonuclease-resistant oligonucleotides, from cross-linked RNAs, generation of, 431–437
Ribonuclease S7
in RNA sequencing, 162
by postlabeling method, 156–157
Ribonuclease T1
digestion of internally labeled RNA, 132, 134
digestion of terminally labeled RNA, 152, 159
digestion of UV-cross-linked RNAs, 431–437
in nuclease protection experiments, 455–457
protection of mouse histone pre-mRNA, and immunoprecipitation with anti-Sm antibodies, 462–464, 467
protection of α_1-sarcin region of 28 S RNA, and immunoprecipitation by anti-(U3)RNP antibodies, 459–466
for RNA sequencing by postlabeling method, 155–156
as RNA structure probe, 197–199, 201–205, 208–209, 215, 216, 218, 298–299
Ribonuclease T2
branch points. *See* Branch point
digestion of RNA, for release of branch points from lariats, 185
in RNA sequencing, 162
by postlabeling method, 156
as RNA structure probe, 197–198, 205
Ribonuclease U2
digestion of terminally labeled RNA, 159–160, 162
for RNA sequencing by postlabeling method, 156
as RNA structure probe, 197–199, 201, 202, 207–209
Ribonuclease V1, as RNA structure probe, 197–198, 204–205, 208–209, 215, 216, 218

Ribonucleoprotein
  complexes
    containing unmodified RNA and
      protein, protein–RNA cross-
      linking study, with bifunctional
      cross-linking reagent, 384
    large, generation of, during pre-mRNA
      splicing, 443
    on nascent RNA, electron microscopy
      of, 481–509
    protein–RNA associations, photoaffin-
      ity cross-linking experiments for
      studying, 383
    small, in prerNA splicing, 443–444
  UV-cross-linked, 410–417
  fibril mapping, 498–501
  immunoprecipitation of, with autoanti-
    bodies, 468–481
  particles, poly(A)-containing, affinity
    chromatography, 36
  small, characteristics of, 468–469
  small cytoplasmic, 469
  small nuclear, 469
    electrophoretic separation from splice-
      osomes and Northern hybridiza-
      tion, 442, 444–445, 450–453
    structure probing, 213–227
    U1, protection–immunoprecipitation
      experiments with, 466–467
  small nucleolar, 469
    U3, interaction with pre-rRNA, nuclease
      protection experiment, 458–468
Ribonucleoside–vanadyl complex
  preparation of cytoplasm using, 8–9
  as RNase inhibitor for RNA preparation,
    6–7
  source, 7
Ribosomal RNA
  computer-assisted covariance analyses,
    238
  gel electrophoresis, 21–23
  homologous sequences, 229–231
  precursors. See Yeast, ribosomal RNA
    precursors
  reversed-phase chromatography, 31–32
  structure, phylogenetic comparisons of,
    228
  UV-induced RNA–RNA cross-links
    introduced into, mapping of, 420
  yeast, 96

R loops, 239
  electron microscopic analyses of, 240
  electron microscopic structure, interpre-
    tation of, 254–259
  formation, 250–252
  in RNA mapping, 244–246
RNA. See also Messenger RNA; Precursor RNA; Ribosomal RNA; Self-
  cleaving RNA molecules; Transfer
  RNA
  absorbance melting curves of, 304–325
  acidic phenol extraction, 74–75, 78–79,
    91
  alternatively spliced, quantitation of
    relative abundance of, 339–340
  base pairing. See RNA, structure
  biotinylated, synthesis of, 49–50
  branched
    detection of, 177–178
    discovery of, 177
  capped transcripts, synthesis of, 42, 48–
    50
  capping, 194–196
  chain-termination sequencing of, 121–
    130
  characterization of, 25
    with S1 nuclease analysis, 334–347
  chemical modification of, 373–375
  column chromatography of, 25–41
  comparative sequence analysis, 122
  complementary sequence pairs in, 235–
    238
  complex
    mapping of, 339–340
    melting curve, 315
    melting studies of, 324–325
  containing bromo-UTP or thio-UTP
    nucleotides, synthesis of, 49–50
  containing photoreactive nucleotide
    analogs, isolation of, 404–405
  cyclic phosphate groups, in RNA
    fingerprinting, 154
  cytoplasmic
    biogenesis of, 69
    of HeLa cells
      DEAE-cellulose chromatography of
        products of RNase digestion of,
        179, 182
    isolation of, 9–10, 73–79
    solutions for, 7–8

labeled *in vivo*
  fractionation and analysis of, 75
  poly(A) selection of, 74, 76
  ribonuclease digestion of, 179–181
dideoxy sequencing of, using reverse
  transcriptase, 121–130
electrophoretic transfer to nylon membrane for Northern hybridization,
  450–451
extraction of, from virions, 122–123
5' end
  labeling, 158–159, 193–195, 222, 372–
    373
  with polynucleotide kinase and
    [γ-³²P]ATP, 193–195
  using guanylyltransferase and
    [α-³²P]ATP, 194–195
  mapping of, 347, 349
  with S1 nuclease and primer exten-
    sion, 338–339
5'-³²P-labeled nucleotide, analysis of, 161
folded structure
  evaluation of, 321
  prediction of, 321
folding, 262. *See also* RNA, secondary
  structure
  energy rules, 264–267
fractionation, 73–79
gel electrophoresis, 14–25
  acrylamide and urea concentration
    shift combination method, 14–25
  equipment, 16, 24–25
  fractionation of intermediate RNAs,
    20–21
  fractionation of large RNAs, 21–23
  fractionation of small RNAs, 17–20
  reagents for, 16–17
  recovery of RNA from gels, 23–24
homologous sequences, 228, 229–230
  computer algorithm for, and structure
    prediction, 288
hydroxide ladders, 206–208
incomplete transcripts of, 50
intermediate size, fractionation of, 20–21
isolation of, 3–13, 193
  with chaotropic agents, 3–6
  with guanidinium salts, 13
  with guanidinium thiocyanate, 3–6,
    74–78, 91–92, 123
  and phenol, combination, 13

with phenol, 13
from subcellular compartments, 3, 6–
  13
by subcellular fractionation, 3
  techniques, 3, 13
from yeast cells, 113–116
labeled fragments, in enzymatic sequencing, analysis of, 162
labeled *in vivo* with inorganic [³²P]phos-
  phate, isolation of branch points
  from, 178–182
labeling, 419
  with [¹⁴C]uridine, 72
  with ortho[³²P]phosphate, 72–73
large
  fractionation of, 21–23, 25
  RNase digests of, fingerprinting of,
    14
  sequence analysis, 163
limited alkaline hydrolysis, 92
melting behavior of
  determination of, 304–319. *See also*
    Absorbance melting curves
  prediction of, 319–321
methylation, and cap structures, 164
minimum energy folding of, 262–264
modification interference analysis using,
  369–370
modified transcripts, synthesis of, 49–
  50
nuclear. *See* Nuclear RNA
nuclease contamination, 194
overlapping segments, computer prediction of, 284–285
pancreatic RNase-resistant oligonucleotides, fingerprints of, 150–154
photoreactive sites, 418
placement of photoreactive cross-linking
  group within, 389–393
poly(A), affinity chromatography, 34–
  40
poly(A)⁺
  isolation of, 13
  two-dimensional gel separation of, 22–
    23
polyacrylamide gel electrophoresis, 377
poly(A) selection of, 74, 76, 79–82
postlabeling, 419–420
postnuclear (Magic Wash) fraction, 73–
  79

primer extension analysis, 338–339, 347–
    369
  applications of, 347
  buffers for, 350–351
  characterization of multiple primer
      extension products, 368–369
  extension, 364–367
    artifacts, 367
    increased sensitivity with continu-
        ous labeling, 367–368
    interpretation, 366
  gels for, 351–352
  hybridization of primer to transcripts
      of interest, 361–364
    aqueous hybridization, 363
    controls, 362–363
    with formamide, 363–364
    temperature for, 361–363
    theory, 361
  materials, 348–352
  oligonucleotide primers, 348, 352–354
  primer selection and preparation, 352–
      361
  reagents for, 350
  restriction fragment primers, 354–361
  sequence determination of extension
      products, 368
  technique for, 347–349
  variations, 367–369
purification of, 193, 376–377
  by column chromatography, 25–41
  by gel techniques, 14–25
rapid column chromatography, 376–377
RNase T1-resistant oligonucleotides,
    fingerprints of, 150–154
secondary and tertiary structure
  chemical approaches to, 192
  enzymatic approach to, 192–212
    enzymes for, 196–199
    experimental design, 211–212
    protocol, 201–206
    requirements for, 193–201
    solutions and reagents for, 199–201
    standard reaction conditions, 206–
        207
    structure probing reactions, 208–211
    troubleshooting, 207–208
  experiments for studying, 192–193
  factors affecting, 205–206, 211–212
  and function, experimental approaches
      to studying, 212–213

rationale for studying, 192–193
secondary structure, 262. See also
    Pseudoknots
  elementary elements of, 290
  elements of, 319–320
  free energy for, 319–321
  phylogenetic comparisons of, 227–239,
      300
    alignment of sequences, 229, 233–
        234
    selection of organisms for, 229–
        233
  prediction of, 319
separation into nuclear, postnuclear
    (Magic Wash) and cytoplasmic
    fractions, 73–79
small
  elution of, from polyacrylamide gels,
      429
  fractionation of, 17–20, 132–133
  purification of, 59
  sequence analysis, 163
  synthesis of, using T7 RNA poly-
      merase, 51–62
  UV-cross-linking studies, 419
small nuclear
  detection of modifications and/or
      cleavages using primer extension,
      221–226
    chemical probes, 222–226
    enzymatic cleavages, 223–224
    5′ end labeling and gel fractionation
        of primer, 222
    gel fractionation of extended
        primers, 225–226
    primer annealing and elongation
        with reverse transcriptase,
        224–225
    probing N-1-G and N-3-U with
        CMCT, 223
    probing of snRNAs–snRNPs, 223
  sequences, compilation of, 521–530
  structure, phylogenetic comparisons
      of, 228
  structure probing, 213–227
    buffers for chemical probes, 214
    chemicals and enzymes, 213–214
    methods, 214–226
    3′-end-labeled
      chemical probes for detection of
          modification, 215–216

direct detection of chemical modifications or enzymatic cleavage on, 214–221

enzymatic cleavages and base-specific modifications, 216, 218–220

enzymatic probes for detection of modification, 216–221

isolation of, on polyacrylamide gel, 216, 217–218

size fractionation on polyacrylamide gels, 216, 220–221

3′-end-labeling, with [5′-^{32}P]pCp, 216, 217, 220

spliced, mapping of, 338–339

structure

analysis of, 25

computer prediction of, 262–288, 290, 295–296, 301–303

effects of ionic conditions on, 205–206

model building using computer graphics, 294

nomenclature, 263–264

phylogenetic comparisons, 212, 294

three-dimensional, 289

prediction of, 286–288

synthesis in vitro, 42–44

basic transcription reaction, 43–47

DNA template, preparation of, 43–47

plasmid DNA templates for, 445

SP6 RNA polymerase system for, 445

SP6/T7 transcription system for, 46–47

T3 RNA polymerase system for, 445

T7 RNA polymerase system for, 51–62, 445

terminally labeled

alkaline hydrolysis, 160

partial digestion of, 159–161

terminal phosphate groups, in RNA fingerprinting, 153–154

tertiary structure, analysis of, by mapping of UV-sensitive sites, 418–419

thermodynamic stability of, determination of, 312, 314–319. See also Absorbance melting curves

3′ end labeling, 157–158, 195–196, 373

basic CCA addition procedure, 195–196

with T4 RNA ligase, 195, 196

3′-terminal nucleotide, analysis of, 161

total, purification of, guanidinium thiocyanate method for, 4–5

transcriptional analysis, 82–83, 95–96

transcripts

priming of, 60–61

purification of, 59

two-dimensional gel electrophoresis, types of, 14–16

uniquely end-labeled, 193–196

unlabeled, ribonuclease digestion and kinase labeling, 135–137

UV-sensitive sites, analysis of, 441–442

U1, sequences of, 522–523

U2

sequences of, 524–525

in spliceosome formation, 443, 444

U3, sequences of, 526

U4

sequences of, 527

in spliceosome formation, 443

U5

sequences of, 528

in spliceosome formation, 443, 444

U6

sequences of, 529

in spliceosome formation, 443, 444

U7, sequences of, 530

U8, sequences of, 530

viral, nuclear dwell time in human cells, 95–96

Y branch, 177

RNA–DNA heteroduplexes, forms of, 239

RNA–DNA hybridization, 92–93, 260–261

buffer for, 243, 247

combination procedures, 251–252, 261

DNA for, 247–248

formamide in, 242–243, 247

glassware, utensils, and water baths for, 246

glyoxal in, 247

materials for, 246

nucleic acid concentrations for, 243–244

protocol for, 250–252

RNA for, 248–249

for S1 nuclease mapping, 335–336

temperature for, 243

unstable duplex structure formed by, 346

using double-stranded DNA probes, 343–344

temperature, 343–344

using single-stranded DNA probes, 335, 338, 340–343
water in, 247
RNA fingerprinting, 130–155, 431, 436
  cyclic phosphate groups in, 154
  digestion and kinase labeling step, 132, 135–137
  first-dimensional separation in, 132, 137–147
    acid–urea gels used for, 132, 143–147
    ascending RNA homochromatography, 132–133, 147–150
    fractionation on cellulose acetate strips, 132, 136–143
    electrophoresis of, 132, 142
    transfer to DEAE thin-layer plates, 132, 142–143, 145–147
  of HeLa 5 S RNA, 439–441
  for identification of fragments generated in protection–immunoprecipitation assays, 456, 460–464
  interpretation of, 150–154
  preparation of RNase digests for first-dimensional separation, 132, 135
  procedure, 132–133
  ribonuclease digestion step, 132–137
    materials and methods for, 133–134
  role of, in RNA processing, 130–132
  second-dimensional separation in, 132–133, 147–150
  terminal phosphate groups of oligonucleotides in, 153–154
  of uniquely end-labeled RNA, 193
RNA helix, breathing of, 226
RNA mapping, 292–293
  by electron microscopy, 239–261
    data analysis, 260–261
    evaluation of specimens, 253–254
    interpretation of structures, 254–260
    mounting samples on grids, 252–253
    photography, 254
    procedures, 252–254
    reagents for, 249
    tracing, 254
    R-loop method, 244–246
    by RNA–single-stranded DNA heteroduplex method, 245–246
RNA probes, advantages of, 42
RNA–protein complexes, in cell-free translation systems, study of, using

UV-induced RNA–protein cross-linking, 418
RNA–protein interactions
  chemical and enzymatic approaches to, 192
  genetic methods for study of, 510–517
  photoaffinity cross-linking methods for studying, 383–409
RNA–RNA cross-links
  extraction of, from polyacrylamide gels
    elution of small RNAs, 429–430
    with RNAs longer than 70 bases, 429
  purification of, by two-dimensional gel electrophoresis, 419, 420, 426–429
    nondenaturing gels for first-dimensional separation, 426–427
    urea gels for second-dimensional separation, 427–429
  UV treatment of RNA in gel strip, 427
UV-induced
  generation of ribonuclease-resistant oligonucleotides from, 431–437
    conditions for, 432–433
    digestion and kinase labeling of nonradioactive RNAs, 436–437
    digestion of radiolabeled RNAs, 433–436
    digestion under standard conditions, 432–433
    pancreatic RNase A digestion, 434–436
    RNase T1 digestion at elevated temperature, 433–434
    RNase T1 digestion under standard conditions, 433
    special conditions for RNase T1 partial digestion, 436
  mapping, 419, 431–441
    by primer extension, 438–439
    one-dimensional oligonucleotide analysis, 431
    using polyacrylamide gels, 437–438
    and RNA secondary analysis, 431, 439–441
    two-dimensional oligonucleotide analysis, 431
Whatman CF11 cellulose chromatography, 430–431
RNA–RNA interactions, genetic methods for study of, 510–517
RNase. See Ribonuclease

RNA secondary analysis, for mapping UV-induced RNA–RNA cross-links, 435, 439–441
RNA sequences. *See also* RNA, homologous sequences
comparison of, 321
phylogenetic comparisons, 229, 233–234
in pseudoknot detection, 300
U1 to U8, 522–530
RNA sequencing, 192–193, 201, 206–208, 262. *See also* Reverse transcriptase, dideoxy sequencing of RNA using; RNA secondary analysis
assignment of modified nucleotides in, 163
chemical modification–degradation method, 163
covariation analysis, 238
of cross-linked RNA–protein complexes, 409
enzymatic method, 154–163
principle of, 155
reading, 162
ribonuclease for, 155–156
terminal nucleotide analysis, 163
identification of capped nucleotides by, 161
by postlabeling method, 155–157
postlabeling techniques for, 154–155
prediction of structure from, 238–239
RNA sequencing kits, 161, 201
RNA–single-stranded DNA heteroduplex
electron microscopic structure, interpretation of, 255, 258–260
electron microscopy of, 261
formation, 251
RNA splicing, 25, 42, 164, 240, 533
intermediates, 132, 177
steps in, 443–444
RPC-5, 31
RPC-5 ANALOG, 32

S

Saccharomyces carlsbergensis, 104
Saccharomyces cerevisiae. See Yeast
Salmonella typhimurium, RNase P RNA, secondary structure, 228–230
Salser's rules, for RNA folding energy, 265–267

sArMV. *See* Satellite arabis mosaic virus
Satellite arabis mosaic virus RNA, 548–549, 552, 555–557
Satellite RNA, 546
plant viruses associated with, 547
polarity, 547–548
Satellite tobacco ring-spot virus RNA, 548–549, 555–557
sBYDV. *See* Barley yellow dwarf virus, satellite RNA
Schizosaccharomyces, tRNAs, 3' end labeling, 196
Schizosaccharomyces pombe
group I introns, 537–538
group II introns, 543
Sea urchin, U7 RNA, sequences of, 530
Self-cleaving RNA, 548–558
consensus sequence (hammerhead structure), 554–555
sources of, 546
Silk worm. *See Bombyx mori*
Sinapis alba, group II introns, 542
Single-copy genes, relative transcription of, analysis of, 83
Size-exclusion chromatography, of RNA, 40–41
sLTSV. *See* Lucerne transient streak virus, satellite RNA
Sobemovirus, satellite RNA, 547, 549, 550–551
Sodium bisulfite, as RNA structure probe, 297
Splicing complexes
analysis of, by native gel electrophoresis, 442–453
monitoring of, 445
native gel electrophoresis
electrophoretic transfer to nylon membrane, 450–451
heparin treatment with, 448–449
preparation of samples, 446–448
Northern hybridization, 451–453
preRNA substrates for, 444–445
SP6 RNA polymerase, 51
SP6/T7 transcription system, 42–50
sSCMoV RNA, 550–551
sSNMV RNA, 550–551
Stack, definition of, 263
Stem, definition of, 263
Stem-loop structure, 289
Stem region, 290

sTobRV. *See* Satellite tobacco ring-spot virus
N-Succinimidyl 6-(4′-azido-2′-nitrophenylamino)hexanoate, 386
N-Succinimidyl (4-azidophenyl)-1,3′-dithiopropionate, 386
Sulfosuccinimidyl 2-(*m*-azido-*o*-nitrobenzamido)ethyl-1,3′-dithiopropionate, 386
Sulfosuccinimidyl 6-(4′-azido-2′-nitrophenylamino)hexanoate, 386
Sulfosuccinimidyl(4-azidophenyldithio)propionate, 386
Sulfosuccinimidyl 2-(*p*-azidosalicylamido)ethyl-1,2′-dithiopropionate, 386
sVTMoV RNA, 550–551
S1 nuclease
    characterization of RNA molecules with, 334–347
    control for, 346–347
    pitfalls of, 346–347
    for quantitation of RNA, 347
    for quantitation of specific RNA in mixture, 336–337
    recovery of products, by ethanol precipitation, 345
    strategy for, 335–336
    S1 nuclease digestion step, 344–347
    using end-labeled probes, 337–338
    as RNA structure probe, 197–198, 215, 216, 218, 298–299

T

T cells, nuclei, isolation of, 86
Temperature-jump methods, for melting curves, 322
Tetrahymena, U5 RNA, sequences of, 528
Tetrahymena cosmopolitanis, group I introns, 538
Tetrahymena hyperangularis, group I introns, 538
Tetrahymena malaccensis, group I introns, 538
Tetrahymena pigmentosa, group I introns, 538
Tetrahymena sonneborni, group I introns, 538
Tetrahymena thermophila
    group I introns, 538
    nuclear large rRNA introns, 535

Thin-layer chromatography, of uniquely end-labeled RNA, 193
4-Thiouridine 5′-triphosphate, 391
    synthesis of, 396–398
Tobacco mosaic virus, RNA, pseudoknot in, 296
Transcription pausing–termination, analysis of, 83
Transcription vectors, 43–45
Transfer RNA
    computer-assisted covariance analyses, 238
    end-mature, intron-containing, synthesis of, 63–69
    gel electrophoresis, 17–20
    hydrophobic interaction chromatography, 33–34
    hydroxylapatite chromatography, 29–30
    precursor. *See also* Yeast, transfer RNA precursors
        from *E. coli*, hydroxylapatite chromatography, 29
        from *E. coli* mutant defective in tRNA biosynthesis, purification of, 14–15
    gene
        linked to bacteriophage T7 promoter, construction of, 64–67
        mutations, introduction of, 63
    synthesis
        in vitro, 63–69
        methods, 63
        by T7 RNA polymerase, 67–69
        in vivo preparation of, in yeast, 110–117
    purification of, 14–15
    reversed-phase chromatography, 31–32
    sequence analysis, 163
    structure, phylogenetic comparisons of, 228
Triticum aestivum, group II introns, 543
Trypanosoma brucei
    branched oligonucleotides, isolation of, from RNA labeled *in vivo* with inorganic [^{32}P]phosphate, 178
    poly(A)$^+$ RNA, isolation of capped structures from, 171
Trypanosoma brucei gambiense, U2 RNA, sequences of, 524, 525
TSK DEAE-PW, 27

TSK DEAE-SW, 27
Turner's rules, for RNA folding energy, 266–267
Turnip yellow mosaic virus, RNA
pseudoknots, 297–299
computer prediction of, 301–302
secondary structure, 292–294
T3 RNA polymerase, 51
T7 RNA polymerase. *See also* SP6/T7
transcription system
in vitro synthesis of RNA using, 51–62
accuracy of transcription, 62
buffers, 55–56
materials, 51–56
methods, 56–59
modified nucleoside triphosphates as
substrates, 61–62
nucleoside triphosphates, 55
optimal reaction conditions, determining, 57–59
plasmid DNA templates, 54–55
priming of transcripts, 60–61
purification of RNA transcripts, 59
sequence variation in +1 to +6 region, effect on reaction yield, 60
synthetic DNA templates, 51–54
transcription reactions, 56–57
nucleoside triphosphate substrate specificity of, 61–62
pre-tRNA synthesis by, 67–69
purification of, from *E. coli* containing
pAR219, 51–52

U

Ultraviolet-cross-linked RNA–protein
complexes, RNase digestion of, 415–
416
Ultraviolet-induced cross-linking, 383
overview of, 418–421
of RNA–protein
in vitro, 417–418
in vivo, 410–418
adaptation for cells grown in suspension culture, 416–417
applications of, 410
buffers for, 412
cell culture and labeling, 413
for detection of effects of transcriptional modulators, 417

for detection of effects of viral
infection, 417
extraction of cross-linked ribonucleoproteins, 414
limitations of, 411–412
poly(A)-specific fractionation by
chromatography on oligo(dT)-
cellulose, 410–412, 414–416
principle of, 411
procedure, 412–418
reagents, 412
UV-induced protein–protein cross-
linking with, 417
UV irradiation of cell monolayers,
413–414
of RNA–RNA, 418–442
applications of, 418–419
autoradiography, 422–423
exposure of gels, 422–423
radioactive ink for, 422
exposure time, 420
extraction of RNA from polyacrylamide gels, 429–430
for function studies, 419, 420
introduction of covalent cross-link,
423–426
irradiation in a drop, 424–426
irradiation in nondenaturing gel, 425–
426
mapping of cross-links, 431–441
materials, 421–423
methods, 421–423
precautions, 423–424
preparation of carrier RNA, 422
preparation of glassware, 421–422
RNA purification by Whatman CF11
cellulose chromatography, 430–
431
for structure-function studies, 421
suitable RNA for, 418–420
technical considerations, 420–421
two-dimensional gel electrophoresis
in, 419, 420, 426–429
within tertiary structure of RNA, 420–
421

V

Vaccinia virus
in vitro transcription of capped mRNA,
166–168

2'-O-methyltransferase, 176
mRNA (guanine-7-)methyltransferase,
175–176
mRNA guanylyltransferase, 175–176
Vanadyladenosine, 412, 414
preparation of, 416
Vicia faba, group I introns, 538
Viroid RNA, 546
Virusoids, 547

W

Whatman CF11 cellulose chromatography,
RNA purification by, 430–431

X

Xenopus laevis
identification of specific genes in, and
chromatin spreading method, 509
oocytes, RNA interactions in, genetic
analysis of, 511
tRNAs, 3' end labeling, 196
U1 RNA, sequences of, 522
U2 RNA, sequences of, 524
U5 RNA, sequences of, 528
U6 RNA, sequences of, 529

Y

Yeast
group I introns, 537
group II introns, 542–543, 545
mRNA-decapping enzyme, cleavage of
capped structures, 174
nuclear pre-mRNA splicing, genetic
analysis of, 511–517
nuclei, isolation of, 107
preribosomes
identification of, 100
isolation of, 100–101
ribosomal RNA, 96
maturation, timing of modifications in,
106–107
mutational analysis in vivo, 108
pseudouridylations, 106–107
ribosomal RNA precursors
characterization of, 103–108
extraction from cells, 98–99
extraction from spheroplasts, 99
fractionation of, 99
identification of, 98–100

labeling, for structural analysis, 101–
103
mapping of processing sites, 103–104
mapping of transcription initiation site,
103
methylation, 106–107
in isolated nuclei, 107–108
modification of, 106–108
nomenclature, 97
nontranscribed spacers including 5 S
rRNA gene, 104
Northern blot hybridization, 98
and preribosomes
characterization of, 103–108
isolation of, 96–103
primary structure of, 104–105
pulse labeling in vivo, 98–100
species of, 96–97
synthesis in vitro, 103
transcribed spacer sequences, 104–105
ribosome biogenesis, study of, new
approaches for, 108–109
RNA
gel electrophoresis of, 22–23
isolation of, 113–116
polyacrylamide gel electrophoresis,
114–116
radioactive labeling of, 113, 116–117
SNR3 RNA, sequences of, 530
strain M304, growth of, 112–113
transfer RNA^{Phe}, 297
transfer RNA precursors
in vitro splicing assay system for, 63
in vitro synthesis, 110
in vivo preparation of, 110–117
media for, 110
reagents for, 110–112
solutions for, 110–112
polyacrylamide gel electrophoresis,
114–116
specific activity determination, 116
3' end labeling, 195–196
U1 RNA, sequences of, 523
U2 RNA, sequences of, 525
U3 RNA, sequences of, 526
U5 RNA, sequences of, 528

Z

Zea mays
group I introns, 538
group II introns, 543